PRINCIPLES OF DISTRIBUTED PROCESSING	INTRODUCTION TO TELEPROCESSING	TELEMATIC SOCIETY
OTEX			
GN OF MPUTER GUES	COMPUTER NETWORKS AND DISTRIBUTED PROCESSING	INTRODUCTION TO COMPUTER NETWORKS	TELE-COMMUNICATIONS AND THE COMPUTER (second edition)
AMMING -TIME R SYSTEMS	DESIGN AND STRATEGY FOR DISTRIBUTED PROCESSING	TELEPROCESSING NETWORK ORGANIZATION	COMMUNICATIONS SATELLITE SYSTEMS
GN OF L-TIME R SYSTEMS	DISTRIBUTED FILE AND DATA-BASE DESIGN	SYSTEMS ANALYSIS FOR DATA TRANSMISSION	FUTURE DEVELOPMENTS IN TELE-COMMUNICATIONS (second edition)

TELECOMMUNICATIONS
AND
THE
COMPUTER

A *James Martin* BOOK

Prentice-Hall
Series in Automatic Computation

AHO, ed., *Currents in the Theory of Computing*
AHO and ULLMAN, *The Theory of Parsing, Translation, and Compiling,*
   Volume I: *Parsing;* Volume II: *Compiling*
ANDREE, *Computer Programming: Techniques, Analysis, and Mathematics*
ANSELONE, *Collectively Compact Operator Approximation Theory*
   and Applications to Integral Equations
AVRIEL, *Nonlinear Programming: Analysis and Methods*
BENNETT, JR., *Scientific and Engineering Problem-Solving with the Computer*
BLAAUW, *Digital System Implementation*
BLUMENTHAL, *Management Information Systems*
BRENT, *Algorithms for Minimization without Derivatives*
BRINCH HANSEN, *Operating System Principles*
BRZOZOWSKI and YOELL, *Digital Networks*
COFFMAN and DENNING, *Operating Systems Theory*
CRESS, et al., *FORTRAN IV with WATFOR and WATFIV*
DAHLQUIST, BJÖRCK, and ANDERSON, *Numerical Methods*
DANIEL, *The Approximate Minimization of Functionals*
DEO, *Graph Theory with Applications to Engineering and Computer Science*
DESMONDE, *Computers and Their Uses,* 2nd ed.
DIJKSTRA, *A Discipline of Programming*
DRUMMOND, *Evaluation and Measurement Techniques for Digital Computer Systems*
ECKHOUSE, *Minicomputer Systems: Organization and Programming (PDP-11)*
FIKE, *Computer Evaluation of Mathematical Functions*
FIKE, *PL/1 for Scientific Programmers*
FORSYTHE and MOLER, *Computer Solution of Linear Algebraic Systems*
GEAR, *Numerical Initial Value Problems in Ordinary Differential Equations*
GILL, *Applied Algebra for the Computer Sciences*
GORDON, *System Simulation*
GRISWOLD, *String and List Processing in SNOBOL4: Techniques and Applications*
HANSEN, *A Table of Series and Products*
HARTMANIS and STEARNS, *Algebraic Structure Theory of Sequential Machines*
HILBURN and JULICH, *Microcomputers/Microprocessor: Hardware, Software, and Applications*
JACOBY, et al., *Iterative Methods for Nonlinear Optimization Problems*
JOHNSON, *System Structure in Data, Programs, and Computers*
KIVIAT, et al., *The SIMSCRIPT II Programming Language*
LAWSON and HANSON, *Solving Least Squares Problems*
LORIN, *Parallelism in Hardware and Software: Real and Apparent Concurrency*
LOUDEN and LEDIN, *Programming the IBM 1130,* 2nd ed.
MARTIN, *Computer Data-Base Organization*
MARTIN, *Design of Man-Computer Dialogues*
MARTIN, *Design of Real-Time Computer Systems*
MARTIN, *Future Developments in Telecommunications*
MARTIN, *Introduction to Teleprocessing*

MARTIN, *Principles of Data-Base Management*
MARTIN, *Programming Real-Time Computing Systems*
MARTIN, *Security, Accuracy, and Privacy in Computer Systems*
MARTIN, *Systems Analysis for Data Transmission*
MARTIN, *Telecommunications and the Computer*, 2nd ed.
MARTIN, *Teleprocessing Network Organization*
MARTIN and NORMAN, *The Computerized Society*
MCKEEMAN, et al., *A Compiler Generator*
MEYERS, *Time-Sharing Computation in the Social Sciences*
MINSKY, *Computation: Finite and Infinite Machines*
NIEVERGELT, et al., *Computer Approaches to Mathematical Problems*
PLANE and MCMILLAN, *Discrete Optimization*
POLIVKA and PAKIN, *APL: The Language and Its Usage*
PRITSKER and KIVIAT, *Simulation with GASP II: A FORTRAN-based Simulation Language*
PYLYSHYN, ed., *Perspectives on the Computer Revolution*
RICH, *Internal Sorting Methods Illustrated with PL/1 Programs*
RUDD, *Assembly Language Programming and the IBM 360 and 370 Computers*
SACKMAN and CITRENBAUM, eds., *On-Line Planning: Towards Creative Problem-Solving*
SALTON, ed., *The SMART Retrieval System: Experiments in Automatic Document Processing*
SAMMET, *Programming Languages: History and Fundamentals*
SCHAEFER, *A Mathematical Theory of Global Program Optimization*
SCHULTZ, *Spline Analysis*
SCHWARZ, et al., *Numerical Analysis of Symmetric Matrices*
SHAH, *Engineering Simulation Using Small Scientific Computers*
SHAW, *The Logical Design of Operating Systems*
SHERMAN, *Techniques in Computer Programming*
SIMON and SIKLOSSY, eds., *Representation and Meaning:*
    Experiments with Information Processing Systems
STERBENZ, *Floating-Point Computation*
STOUTEMYER, *PL/1 Programming for Engineering and Science*
STRANG and FIX, *An Analysis of the Finite Element Method*
STROUD, *Approximate Calculation of Multiple Integrals*
TANENBAUM, *Structured Computer Organization*
TAVISS, ed., *The Computer Impact*
UHR, *Pattern Recognition, Learning, and Thought:*
    Computer-Programmed Models of Higher Mental Processes
VAN TASSEL, *Computer Security Management*
VARGA, *Matrix Iterative Analysis*
WAITE, *Implementing Software for Non-Numeric Application*
WILKINSON, *Rounding Errors in Algebraic Processes*
WIRTH, *Algorithms + Data Structures = Programs*
WIRTH, *Systematic Programming: An Introduction*
YEH, ed., *Applied Computation Theory: Analysis, Design, Modeling*

TELECOM

MUNICATIONS AND THE COMPUTER

Second Edition

JAMES MARTIN

IBM Systems Research Institute

PRENTICE-HALL, INC., Englewood Cliffs, New Jersey

Library of Congress Cataloging in Publication Data

MARTIN, JAMES, (date)
  Telecommunications and the computer.

  Bibliography
  Includes index.
    1. Telecommunication.   2. Data transmission systems.
I. Title.
TK5101.M326   1976   621.38   75-37800
ISBN 0-13-902494-8

Telecommunications and the Computer, 2nd edition
James Martin

Jacket photograph courtesy of Western Union.

© 1976 by Prentice-Hall, Inc.
Englewood Cliffs, N. J.

19   18   17   16   15   14   13   12

Printed in the United States of America.

PRENTICE-HALL INTERNATIONAL, INC. *London*
PRENTICE-HALL OF AUSTRALIA PTY. LIMITED, *Sydney*
PRENTICE-HALL OF CANADA, LTD., *Toronto*
PRENTICE-HALL OF INDIA PRIVATE LIMITED, *New Delhi*
PRENTICE-HALL OF JAPAN, INC., *Tokyo*
PRENTICE-HALL OF SOUTHEAST ASIA PTE. LTD., *Singapore*

TO CORINTHIA

CONTENTS

PART **III**   SWITCHING

PREFACE

The marriage of telecommunications and computer technology is one of the most exciting developments of this century and will eventually change the entire fabric of society. The new types of networks are beginning to grow at a rapid rate.

This means that almost all persons in the computer field need to understand telecommunications, and many people in the telecommunications field need to understand how their facilities are being used by computers. This book attempts to meet the needs of both groups and is the basis of a course on the subject given at IBM. It is highly desirable that the subject matter should be extensively taught elsewhere.

Many computer personnel have insufficient knowledge of the technology of telecommunications. This lack of knowledge is sometimes keenly felt and has often impeded the creation of valuable systems approaches.

It is easy to obtain literature on the computer manufacturers' hardware, but surprisingly difficult to acquire knowledge of the common carriers' equipment. To satisfy this need, this book concentrates on common carrier technology. Detailed discussion of the computer manufacturers' devices and the means for organizing the flow of data in computer networks is the subject of other books shown in the front end papers.

The early part of the book can be read by the interested layman and management. It is intended that some of the more detailed sections of the book should be used for repeated reference, and hence should be owned by computer personnel.

ACKNOWLEDGEMENTS   Any book about the state of the art in a complex technology draws material from a vast number of sources. While many of these are referenced in the text, it is impossible to include all of the pioneering projects that have contributed to the new uses of telecommunications. To the many systems engineers who contributed to this body of knowledge, the author is indebted.

The author is very grateful for the time spent reviewing and criticizing the manuscript by Mr. K. L. Smith in London, Mr. J. W. Greenwood and Mr. H. Newton in New York, Mr. "Bill" Accosta and Mr. F. A. Hatfield in the Western Electric plant at Columbus, Ohio, and several unknown members of the Bell Telephone Laboratories.

The author is particularly grateful to Miss Cora Tangney for her help in manuscript preparation, and to his wife for editing and indexing.

James Martin

INDEX OF BASIC CONCEPTS

The basic concepts, principles and terms that are explained in this book are listed here along with the page on which an introductory explanation or definition of them is given. There is a complete index at the end of the book.

PART **I** **INTRODUCTION**

1 THE FUTURE

This is a book about the technology of the present day. However, rather than begin it, as is often the case with such books, with a survey of the history of the subject, it is probably more interesting to talk about the future.

There are forces for change of shattering magnitude inherent in today's technology. The public perceives them only dimly with a vague feeling of uneasiness.

We have built a society, especially in the Western world, which is highly dependent on technologies, some of which have a limited future. The world supply of petroleum is running out and it will continue to rise in cost as long as we seek it in increasingly inaccessible places. Certain of the vital minerals needed to produce the plethora of goods that characterize Western consumption are running out. Pollution is growing and will become increasingly expensive to control. The world population is growing inexorably and will soon exceed the earth's capacity to sustain it in the manner of the past. The aspirations of the earth's billions are growing such that it is physically impossible to satisfy them by spreading traditional technology.

And yet we live in an age of immense technological riches. Never before in man's long journey has such a panorama opened up as that which can be seen in today's research laboratories. We have the capability to work what would have looked like miracles in an earlier age. However, as in the parables of childhood, we ought to be careful about just which miracles we select.

One of the most exciting technological developments of this exciting century is the marriage of the engineering of telecommunications to that of the computer industry. Both of these fields are developing at a fast and furious rate. Calculated speculation about what either of them is likely to lead to brings awe-inspiring conclusions. Either the computer industry or the telecommunications industry, alone, is capable of bringing about changes in our society, in the working habits and in the government of people, that will change ways of life throughout the world. But the two techniques complement each other. In combination they add power to each other. Telecommunication links will bring the capabilities of the computers and the information in data banks to the millions of locations where they can be used, and computers in return will control the immense switching centers and help divide the enormous capacity of the new linkages into usable channels.

In England, in the eighteenth century, the spinning jenny and a variety of weaving inventions portended a revolution in clothmaking. Around the same time the steam engine was developed. These two inventions also complemented each other. Either, by itself, would have caused changes, but in combination they brought about an upheaval that was to alter drastically the lives of all the people involved. The attractive villages with their cottage industry gave way to the dark satanic mills of the early industrial revolution. The pounding new machines dominated men's lives. Today, as then, the new technologies sweep across our society too fast for the sociologists and politicians to plan the type of world they want to build or to comprehend the new potentials.

Small computers are now being mass-produced in microscopic circuitry. A microprocessor, small enough to carry in your pocket, can have as much power as a typical "second generation" processor like the IBM 1401, and four times as much memory. Unlike a second generation processor it needs no air-conditioning, no large power supply, and is unlikely to fail. Like printing newspapers, if a large enough quantity is made, the cost of each becomes very low. Before long the semiconductor industry will be mass-producing such machines at less than $10 each.

We are moving into an era when it will cost less to buy a microprocessor than to fill your car with gasoline.

At the other end of the scale are computers growing rapidly in size and power, attached to even larger data banks. There are several storage units on the market capable of holding more than a trillion (10^{12}) bits of data. The cost of making data directly accessible to a computer is dropping. Before long the cost of a paperback will be greater than the cost of storing the same information in a computer storage.

In telecommunications we have long been familiar with the data rates of telegraphy. Wires have transmitted bits fast enough to operate typewriterlike terminals. Now very high speed digital transmission links are coming into operation. A new Bell System radio link operating into White Plains, New York, carries 1.9 billion bits per second—enough bits to transmit the contents of the Bible in one hundredth of a second. A new cable system installed in New York transmits 2.7 billion bits per second. A new type of link, a waveguide pipe, also being installed by Bell, transmits 16 billion bits per second.

Particularly impressive in its potential (and much misunderstood) is the communications satellite. Satellites, slightly different in their design from today's commercial satellites, can carry signals of very high capacity directly to corporate premises. We now know how to build satellites with enough capacity to provide every man, woman, and child with a computer terminal if that were a desirable objective.

The industrial revolution of the history books depended primarily on man's building machines to harness physical power, such as the steam engine, and applying this to a range of different applications. The computer era we are now embarking upon, often referred to as the second industrial revolution, depends on harnessing electronic logic and stored information and applying it to a wide variety of applications. Man first built an extension of his muscles which gave him power enormously greater than his own frail limbs. He then built an extension of his brain which is giving him mechanized memory of unimaginable magnitude and logic capabilities which can process the data in this memory with absolute accuracy and at speeds millions of times greater than his own head-scratching thought. When this second revolution has run its course it will have wrought a change in mankind's environment even greater than the first.

An essential ingredient of economic growth has always been the means of communication or transportation. Those nations which have had the best facilities for distribution have tended to be those which rose fastest to economic power. The first industrial revolution was able to sustain its mushroom-like growth because transportation facilities were built. Canals were dug, and weary horses spent their lives lugging barges of coal from the new mines to the factories. Many factories clustered round the coal fields, and whole segments of the population moved there into gloomy, jerry-built rows of houses. It took many years for the means of transportation to become adequate for the needs of the new industry. Later the railways were built, and then electricity was invented so that overhead wires carried the necessary power across hill and dale.

Britain built the largest empire in history because it excelled in the means of transportation, its navy. America became the world's richest nation after it built great railroads and, later, the finest highway network and a trucking industry more efficient than other country's.

Communications links are a vital part of society's infrastructure. In a computerized society, we look towards electronic communications. Manufacturing will be increasingly run by process control computers and production robots. The paperwork will be handled increasingly by data processing. Management will have an insatiable desire for information and those with the best information will succeed over their competition. Vast data bases will be built up and will serve many locations. Computer networks will lace together the corporate and government facilities. Small computers may become as common as adding machines but many personnel will need to use the *million-dollar* machines and these may be far away, a shared resource.

Many uses of the machines will need *speed*. A managing director wants a quick answer to his question so that he can make a decision. A doctor wants a quick analysis of electrocardiac or other data. A factory shop floor is to be controlled in "real time" as its events are taking place. There is not time to write a letter to the computer center.

In many ways telecommunications will act as a substitute for the increasingly expensive *physical* transportation. There will be no need to send corporate or government mail physically; it will be cheaper to send it electronically. There will be no need to mail invoices; one computer can invoice another electronically. Banks will not mail checks to one another creating physical sorting problems; electronic fund transfer networks are coming into operation. A manager need not travel across the country to present his views or seek facts; he will do this electronically with his corporate video links. A researcher will use computers to search distant libraries, and data networks to explore diverse data bases.

As in the past, the nations which have the best communication facilities may have the highest economic growth. America may be better off than Europe in this respect. The corporations which have the best electronic communications may surpass their competition. He who runs the information runs the show.

In the industrial revolution the means for distribution lagged seriously behind the need for it. Today there is such a lag with computing. In the first decade of widely accepted commercial usage of computers there was hardly any data transmission, except in a handful of pioneering experimental installations. Now, data transmission is fast becoming accepted, but there are problems. The telecommunication links and switching were mainly designed for handling telephone conversations, not computer data. With appropriate attachments they can be made to transmit data, but the electronic links needed for running industry and government are fundamentally different from the public telephone network. A modern society, therefore, needs new facilities including fast switched data networks, and communications satellites which operate with small earth antennas on corporate premises.

The future developments of data transmission are likely to come from two sources. First the telecommunication facilities themselves will improve. Second, our ability to use them will develop greatly. Let us look at the former first.

THE INCREASE IN TELECOMMUNICATION CAPACITY   As the need for more transmission capacity grows, many more links will be built. In addition to this, new inventions and engineering developments will greatly increase the capacity of the links. The major telecommunication highways will be able to carry much larger volumes of data than today. The voice-carrying capacity of the larger intercity trunk systems has grown almost exactly exponentially since 1920, or so, as shown in Fig. 1.1.

The *data* rate that could be sent over one telephone circuit has also grown. In the 1930s, 600 bits per second (bps) could be sent over one voice channel. In the 1950s, 1200 bps could be sent. In the 1960s, 4800 bps could be sent. In the 1970s, PCM voice channels (discussed in Chapter 14) transmitted 56,000 bps of speech. Using these numbers, Fig. 1.2 shows how the bit rates of channels have increased. Figure 1.2 plots the major milestones in transmission technology since the genesis of telecommunications 1½ centuries ago.

In 1819 Oersted discovered the relation between magnetism and electricity, and Ampere, Faraday, and others continued this work in 1820. Shortly after this

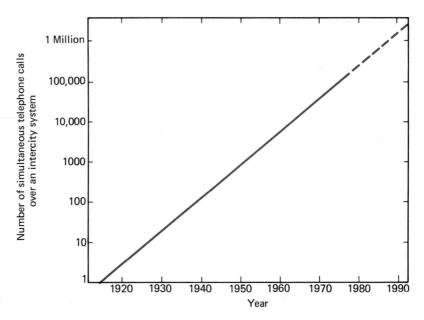

Figure 1.1 The number of telephone calls that can be carried by the highest capacity intercity system has grown exponentially.

the first systems for transmitting information were prepared using suspended magnetized needles. In 1834 Gauss and Weber strung wire over the roofs of Göttingen to make a telegraph system in which information was conveyed by the slow swinging of magnetic needles. A 40-mile telegraph line was set up between Baltimore and Washington by Morse in 1844. The information was coded in dots and dashes, and a steel pen made indentations in a paper strip. In 1849 the first slow telegraph printer link was set up, and by the early 1860s speeds of about 15 bits per second were being achieved. In 1874 Baudot invented a "multiplexor" system (Chapter 12) which enables up to six signals from telegraph machines to be transmitted together over the same line.

Telephone lines were first constructed in the 1890s. In 1918 the first "carrier" technique (Chapter 8) was used, sending several voice channels over one wire pair. In the 1940s coaxial cables were laid down to carry large numbers of voice channels, and in the 1950s microwave links (Chapter 9) were built. Today in the United States a single long-distance telephone cable or microwave route carries many thousands of voice channels and is growing rapidly.

The channel capacity is almost certainly going to go on increasing very rapidly for the rest of the century and possibly far beyond. The waveguides and optical fibers now under development ensure this. By the year 2000 the largest telecommunications cables will probably transmit several hundred billion bits per second, using many hundreds of fine glass fibers packed into one cable with laser transmission.

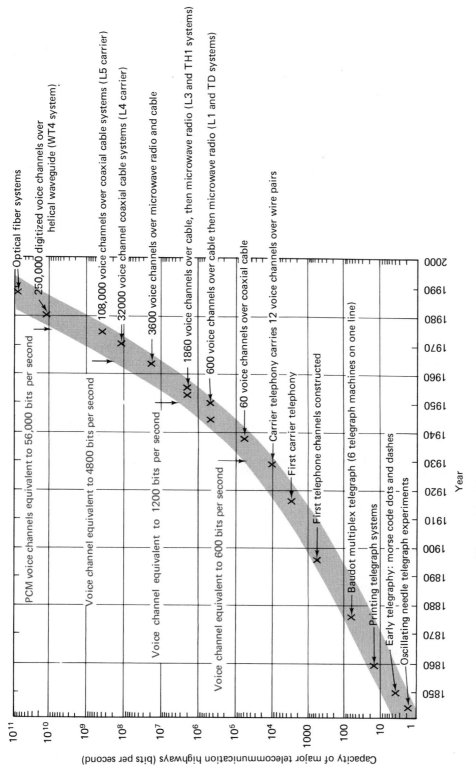

Figure 1.2 The sequence of inventions in telecommunications.

Where might this increase come from in terms of physics? The speed of motor cars will not increase indefinitely because of physical limitations. Could there be physical limitations that will prevent us climbing the curve in Fig. 1.2?

In general, the increase in channel capacity of the past has been achieved by increasing the *range of frequencies* with which signals are transmitted. Indeed, as is discussed in Chapter 16, the quantity of information we can send down a channel is approximately proportional to its *bandwidth* or range of frequencies usable. Figure 1.3 shows the electromagnetic spectrum. Just as we may have a rainbow-colored spectrum of light by separating the visible frequencies, so all the frequencies of physics may be laid out in a spectrum; this is drawn in Fig. 1.3. Visible light is only a small part of this wide spread of frequencies which travel through electrical conductors, or through the air. Higher frequencies can be transmitted through space in the form of radio waves. As the frequencies become higher the waves suffer less diffraction and tend to travel in straight lines. Long radio waves are reflected by the ionosphere. Microwave signals are not, and therefore they must be sent between antennas that are normally within sight of each other. Because microwave signals start at a higher frequency, they can carry a proportionately higher information content. If we increase the frequency, the waves become infrared, or heat radiation. If we increase it further, it becomes visible light. The increase of the frequency of light beyond the visible spectrum turns it into ultraviolet radiation, then x-rays, and at very high frequencies, γ-rays.

How much of this spectrum have we used for transmitting information? Only a small portion of it as indicated in Fig. 1.3. How much will we use in the future? No one can be sure, but it seems likely that new inventions of which we have no inkling today will utilize many parts of this spectrum higher than at present. Already, for example, laser beams have been used experimentally. These are very much higher in the spectrum than anything used for public telecommunications. An x-ray laser has already been made to work. We have no idea how we would amplify or modulate (Chapter 11) the frequencies high on this scale. But then some of the systems in our laboratories now were entirely beyond the technologies of scientists 20 years ago.

Note that the scale of frequencies in Fig. 1.3 is logarithmic. If we drew a diagram with a *linear* scale so that that part of the spectrum used today for commercial communication links occupied the width of this page, then the part of the spectrum as yet unused would require a sheet of paper stretching from earth to a point a hundred times the distance of the sun! Theoretically the quantity of information the medium can carry is proportional to the distance on this scale.

The theoretical capacities suggested by examining the electromagnetic spectrum enormously outstrip the requirements of the dotted line in Fig. 1.1. Probably we are only beginning to utilize telecommunications, just as we are only beginning to utilize computer potentialities. These are the toddler's first halting steps.

In addition to the channel capacity increasing, the number of channels built can be increased to fulfill any future we can dream of. This would not be true if we were talking about radio links. The radio wavelengths are already becoming notoriously overloaded in parts. Microwave systems are a little better than earlier

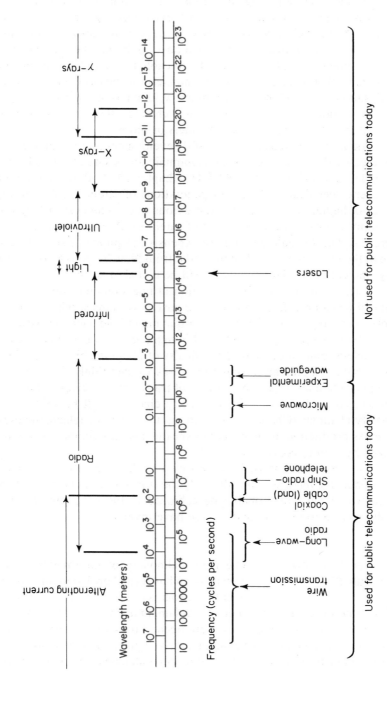

Figure 1.3 The electromagnetic spectrum.

9

forms of radio. Here the signal is transmitted in a narrow beam with little scattering so that separate transmissions can coexist fairly near each other without interference. Nevertheless major cities have severe microwave conjestion. There is virtually no limit, however, other than cost, to the number of coaxial cable channels that could be built into a city. Each cable in use today is capable of carrying more than 10,000 voice (telephone) channels, or the equivalent. The largest cable carries more than 100,000 . For tomorrow's technology, there is again virtually no limit to the number of helical waveguide channels that could be built, each carrying a quarter of a million voice channels or more.

Another spectacularly successful development that has arrived quickly, precipitated by the space race, is the communications satellite. The Early Bird satellite and its successor, INTELSAT II (Chapter 20), both worked better than their designers' specifications and new generations of satellites followed rapidly, each design of higher capacity than the previous ones. The cost of long-distance telecommunications will drop because of satellites. A novel feature in satellite economics is that it should be no more expensive to phone or send data to the other side of the world than it will be to send it from New York to Boston. Where necessary, the computer data highways will be global in scale.

There is no foreseeable limitation in telecommunication capacity. We can build all we can pay for. However, the cost of systems serving the majority of the public is enormous, and the one factor that might hold back the growth of telecommunications is scarcity of capital. Although today systems analysts can be heard to complain sometimes about the difficulty of obtaining high-speed lines, channels *can* be built fast enough for most purposes we can think of. As the capacities of individual channels increase, the cost of transmitting a given quantity of information will come down. The high-capacity channels will be split into many low-capacity channels (Chapter 12). Also, as the usage of telecommunications goes up, mass-production methods will lower the cost. Telecommunications is characterized by major economies of scale.

The next question is: How will we use this capacity? This question is exciting enough even if we consider only the transmission channels of today. Compared with what will probably be achieved in the next few decades, we have hardly started to use the facilities we have invented.

THE INFORMATION EXPLOSION   The new technology of information processing and transmission has turned up just in time for the needs of scientific progress. Looking at the history of technology one observes a number of fundamental inventions which arrived in one field of research just when they were needed to permit progress elsewhere. The thermionic vacuum tube, for example, arrived just in time to allow development in telecommunications. The moon landing depended on a variety of newly developed technologies, and if any one of them had been developed a few decades later, the landing would probably have been impossible. We have reached a state now in man's learning when the quantity of information being generated in industry, in governments, and in the academic world is reaching alarming proportions. The press euphemistically calls it the "information explosion," but that is not a good term because explosions quickly end their violent growth. The growth of man's information has no end in prospect, only greater growth.

The sum total of human knowledge changed very slowly prior to the relatively recent beginnings of scientific thought. But it has been estimated that by 1800 it was doubling every 50 years; by 1950, doubling every 10 years; and by 1970, doubling every 5 years.

Certainly, the number of documents an employee in the computer industry would like to read is going up by leaps and bounds. The computers of the 1950s had one manual of operation which contained all that had to be known about the machine. . . . Then came the "second generation" of computers, and before long there were ten manuals that described parts of the software for the machine. The "third generation" had well over 100.

Future generations may not have manuals at all but may well contain all the information, ready for display in their own data base.

It is happening in other fields too. The weight of the drawings of a jet plane has been stated to be greater than the weight of the plane. In many fields of research, even one as old as medicine, more papers have been published since World War II than in all prior human history.

Automated means of filing and indexing research papers, engineering drawings, and other types of documents is becoming essential and must be coupled with means of searching for, and retrieving, the required information. Telecommunication links will enable people needing information to search with computer assistance through what is available. In searching for reports on a given topic the user carries on a two-way dialogue at a terminal with the distant computer. The computer might suggest more precise categorizations of what he was seeking. He might browse through many data indices, titles, and abstracts before he finds what he wanted. Library systems will be made "public" so that a large number of people can use them, and a large variety of people can contribute to the building up of the data in the system.

Banks of data on many subjects will grow in the years to come. The biochemist will be enabled to check what reports have been written on a particular topic, the lawyer will have the mass of literature he needs electronically available to him, and the patent agent will be enabled to carry out a search in "real time." The New York Times morgue has been computerized so that writers may explore what articles on a given subject have appeared in that and other news media. Eventually all manner of information sources will be connected to data networks, making them available to researchers and other users throughout the world.

BOX 1.1 Some of the data kept in government computer systems. More than 2000 government systems in the U.S. now contain personal information about individuals.

1. *Personal identity data.* Social security number, recognizable physical characteristics, address, race, relatives, etc.

2. *Personal employment data.* Employer, previous employers, occupation, earning, etc.

3. *Personal education data.* Schools, courses, degrees, achievements, etc.

4. *Welfare data.* Periods of welfare, aid received, basis for aid, etc.

5. *Health data.* Physical deficiencies, reportable diseases, immunization, x-ray data, etc.

6. *Tax data.* Details of income tax returns, dependents, investments, life insurance, etc.

7. *Voters' registration data.*

8. *Licenses and permits.* Type, number, date issued, issuing agency, expiration date, etc.

9. *Law enforcement data.* Records of offenses, outstanding warrants, ex-convict registration, suspects, missing persons, etc.

10. *Court actions.* Plaintiff and defendant, court, date, type of action, result.

11. *Probation/parole data.* Probation number, agency, court, offense, terms, conditions.

12. *Confinement data.* Type of confinement, period, place, reason, treatment, escapes, etc.

13. *Registered personal property.* Cars, motorcycles, boats, firearms, radios (in the United Kingdom), dogs, ambulances, elevators, etc.

14. *Vehicle registration files.* Owner, license number, make, engine, body, axles, wheel base, whether registered in other states, etc.

15. *Land details.* Zoning, uses, assessed value, taxable value, deeds, water and mineral rights, drainage, productivity data, soil type, details of buildings, owners, sales, etc.

16. *Owner/occupant files.* Names, address, licenses, occupation group, race, place of work, children in school, income group, vehicles owned, police records, etc.

17. *Street-section records.* Name, class of street, length, limits, intersections, surface, drainage, traffic data, parking data, street lights, curbs, sewers, signals, accidents, etc.

GOVERNMENT
INFORMATION

Government and police records are also growing alarmingly fast. If one accepted the civil servants' statements of what records are necessary to run our increasingly complex society, then, without automation, we would soon all be employed governing and policing ourselves and would have no time to do anything else.

Box 1.1 lists the data in computers in local and state government [1].

Many other items are stored by government bodies in addition to these. Now there is emphasis on obtaining access to them more readily and minimizing the work of keeping the files up to date. This will be done mainly by storing the data on direct-access files and providing telecommunication links to all their likely users. There will probably be many different locations where data are stored. Data which are used largely within one county will be stored in that county to minimize transmission costs. And so, as in other applications, a network will grow up for transferring data from one place to another as and when they are needed. The locations of the files, and the question of what data are stored where, will be determined largely by the balance of cost between data storage and data transmission.

Spectacular use is already being made of data transmission in some police systems. Police, stopping a suspect, radio his name and other data to an operator at a computer terminal. The computer gives information about whether the suspect is wanted, whether his car is a stolen one, or whether he has failed to pay his parking tickets. In this way a policeman can interrogate a data bank from his patrol car. Most police data banks contain information for a limited geographical zone. Soon they will become nationwide, and worldwide.

Much of this sounds 1984ish. It is difficult to escape the conclusion that in the era of data transmission much of the preceding personal documentation will be generally accessible to government officials. We can pass laws to stop Big Brother from watching us too closely if we act quickly and some such laws have been passed. Our data files should not be accessible to just anybody. Computers, like other inventions that change civilization, are a force for great good or great evil depending on how countries use them. There is no doubt about how the systems we envisage would have been used in Germany in the 1930s. Sometimes great civilizations fall under despotic rule, and when the computer techniques we describe are perfected, the totalitarian state will have a new means of maintaining control.

INDUSTRY

As the mechanical processes of industry become increasingly handled by automated machines, so the running of industry will be increasingly concerned with information.

A massive amount of data is collected and digested by computers in the running of a modern chemical plant or a fully automated semiconductor factory. In a less automated plant with items of work moving between machine tools, workers enter data about the jobs into computer terminals on the shop floor. The computer tracks the jobs, the machine tool loading, the parts inventory, and other parameters which enable computer and management together to run the factory as efficiently as possible. When exceptional conditions arise, as they do constantly in most factories, these can be dealt with immediately, the computers notifying the appropriate persons. As far as possible the computers will try to detect problems such as stock shortages or machine overloads before they happen.

In commercial situations computers similarly attempt to collect the information and maintain the data bases needed for efficient management. A computer may schedule operations within a geographically dispersed organization, and reschedule them constantly as new requirements occur. A computer can control the selling of airline seats all over the world. Immediate insurance quotations can be given, and claims and payments handled more quickly. Salesmen's orders from all over the country can be processed in time for the following day's production. The computer can control a steel mill and optimize its efficiency. It can monitor a commercial process in the same way it monitors a manned space flight. Traffic in a turbulent city has been speeded up by computers which maneuver it into groups of vehicles and change traffic lights at the best moment; in a similar way, the events in an industrial process can be regulated in an optimum manner. The benefits of such control are not necessarily the traditional saving of manpower. Often they accrue from increased efficiency, higher yield, better customer service, or more profitable utilization of facilities.

Many companies are widely dispersed geographically, and so data links connect separate plants, link warehouses, connect large numbers of sales offices or service centers to a central information system, and so on. As in other fields, paperwork is being replaced by electronic storage and data transmission. The airline office does away with its filing cabinets of cards for each passenger. Industry branch offices enter and retrieve data on screen units, and a computer has centralized information on all the new developments.

A board room or *management control room* in industry in the future may be designed with computer display screens and individual consoles for communicating with the computer.

Management control rooms are being built in which a group of men assist the computer in controlling the flow of work through the organization. Such rooms have a number of functions. First, they are necessary in many systems to control errors: often errors made in human input to the system, or errors found in the filed data. Second, they handle exceptional conditions that arise and require human intervention. Third, they represent a recognition that it is not necessarily the best policy to make the computer do *all* the processing that is needed. There are some transactions or situations that need human judgment, and so the computer requests help and transmits details to the man with the relevant experience or judgment ability. Fourth, they enable management to obtain information it needs or to test the effect of possible actions. Many members of management will not have the ability to communicate with the computer fluently, and so they do this via the staff in the control room.

Many large organizations now send more than a million transactions a day over their data networks. Most such data networks have been built using telephone lines. Now, however, it is recognized that specialized data networks could be more efficient in some situations, possibly using new facilities such as communication satellites.

FINANCE　　　　　　　　Some industries do not deal in physical goods but in information. These include banking, insurance, travel agencies, stock brokerage companies, education, consulting firms, police departments, intelligence, medical, and data base services, and the entertainment industry. Such industries will be dramatically changed by the new information technologies including data networks, cable television, satellites, mass data storage systems, and microcomputers. The banking industry is an example.

The nature of the payments mechanism in some countries will swing from being predominantly paper-oriented to predominantly electronic, with vast quantities of financial transactions traveling over data networks. Some of the world's largest data networks will be involved in this application. Some large banks are now planning private networks with tens of thousands of terminals. There are more than 1400 banks in the U.S. and eventually they must be interlinked into nationwide networks for transferring money. Many institutions other than banks handle money, hold deposits, and offer credit. The financial data networks affect all such institutions and present sudden new opportunities that will generate fierce competition in the money-handling business. Before long in the U.S. many billions of messages per year will be passing over the financial networks.

The replacement of gold by paper money and of paper money by checks were each revolutionary in their day. Now we must become used to financial transfers occurring in the form of electronic pulses on data links. The paperwork associated with the transaction will now merely inform us about the transaction, rather than represent the transaction itself. Nor will it have to be punched into cards and fed into a receiving computer. The eventual consequences of the simple idea of automatic credit transfer will be enormous. Vast random-access computer files in banks will hold full details of all accounts. As a transaction is entered into the system, transmitted data will cause the appropriate amount to be deducted from an account in one computer and added to an account in another. Eventually, the financial community will become one vast network of electronic files with data links carrying information between them.

FOUR TYPES　　　　　There are four main types of electronic fund transfer
OF EFT　　　　　　　(EFT) representing successive steps towards an EFT society. The *first* involves transfers of money between banks, to carry out clearing operations. *Second*, there are transfers between the computers of other organizations and the bank computers. A corporation may pay its salaries, for example, by giving a tape or transmitting salary information to a bank clearing center which distributes the money to the appropriate accounts.

Third, the general public use terminals to obtain banking services. These terminals include cash-dispensing machines in the streets. There are a variety of such terminals with different functions, and bankers refer to them as CBCTs (Customer Bank Communication Terminals). They threaten to play havoc with the traditional structure of banking, at least in the U.S.

To operate the CBCTs, customers are equipped with machine-readable bank cards. These cards, which look like credit cards, make possible the *fourth* and ultimate phase of electronic fund transfer, in which consumers pay for goods and services in restaurants and stores by using their bank cards or similar cards provided by American Express, large retail chains, petroleum companies and other organizations. Today's credit card devices (which create paperwork) are replaced by terminals which accept the new machine-readable cards. Thousands of such machines are already in use.

EFTS (Electronic Fund Transfer Systems) thus describes a wide variety of different computer systems, but in general has become synonymous with advanced new technical directions in banking.

The present payments mechanism is highly labor intensive. Credit cards have increased, not decreased, the quantity of paperwork and manual operations. Labor costs are rising and it is becoming more difficult to obtain workers for dull, boring, but high-accuracy tasks. It has been estimated that the overall cost of using credit cards exceeds 50 cents per transaction in the U.S., and that the cost of equivalent EFT transactions could be dropped to 7 cents.

Furthermore EFT can make cash available to bankers faster, and time is money, especially with today's high interest rates. The volume of checks alone in the U.S. is about $20 trillion per year. If the money from these could be available to banks one day earlier on average, because of faster processing and clearing, that represents a float of $54 billion per year. It is worth installing some expensive automation schemes to capture a portion of this float.

CUSTOMER ACTIVATED   Using a bank card at an appropriate terminal, a bank
TERMINALS   customer can inquire about the status of his accounts.
                                       He can deposit or withdraw cash, borrow money if it is
not in his account, or transfer money between different types of accounts. In fact he can do virtually everything that he would previously have done by going to a branch of the bank, standing in a queue, and talking to a teller. If all of a bank's customers used bank cards and terminals, would the bank need tellers? Perhaps it would only need officers, who deal with situations needing human interaction and decisions. Banks will probably close some of their branches in expensive city streets and still give their customers more convenient service because the automated teller terminals are becoming located in stores, shopping plazas, airports, factory cafeterias, and office buildings. Furthermore, the customers could obtain cash or other banking services when the bank was closed. Some customers have an initial hostility to banking by machine, but once

used to its convenience few want to go back to queuing in marble-pillared branches.

The prospect of doing away with the bank teller is revolutionary enough, but another implication of automated teller terminals threatens to play havoc with the entire structure of banking. Banking in the U.S. has traditionally been regulated by state and federal laws saying where a bank may have its branches. The McFadden Act of 1927 prohibits interstate branching and makes national banks conform to restrictive state laws. No bank can have branches in more than one state; some can operate only within a city; some can have no branches other than at the head office location. Persons living near state boundaries, for example near New York City, are sometimes flooded with advertising from banks they cannot use. In 1974 the Comptroller of the Currency, who regulates banking activity, made the ruling that a remote terminal which customers use does not constitute a bank "branch." Following this ruling, banks rapidly started to spread their tenacles into geographical areas from which they had earlier been excluded. The controversial ruling was then challenged in the courts and partially reversed, but nevertheless it seems certain that the structure of American banking will change fundamentally.

New York's First National City Bank has almost 300 branches in New York State and cannot have branches elsewhere. However, it has several thousand bank-card terminals in the New York area with some across the state boundary in New Jersey.

Banking terminals can be operated by organizations other than banks. Several savings and loan associations operate them. Consumer finance companies, large chain stores, gasoline companies, credit unions, and other organizations, collectively extend more consumer credit than the banks. Many of these operate credit checking terminals and some have applied for permission to hold customer balances in which case their terminals could have most of the functions of a banking terminal. The banks, in other words, could face electronic competition from a variety of other organizations.

Of particular importance, terminals connected by telephone lines to bank computers are being placed in stores, restaurants, supermarkets, hotels — in fact, anywhere where bank customers make payments. The payments can then be completed without the use of currency or checks, funds being transferred without paper processing from the customer's account to the payee's account. Bank of America already plans a network with 30,000 such terminals in California on-line to two computer centers.

A nation's financial operations will thus become laced together with vast data networks. Some of the fund transfer networks are growing worldwide. Many localized systems of banking terminals are growing up. It will become necessary to interconnect them and this is not unlike the problem of hooking together many localized telephone companies. Some of the terminals are expensive and so will need to be shared by many banks. There must be national standards for the terminals that are used. A nationwide network may develop that serves many different banks, as AT & T Long Lines serves many different telephone companies. A small-town bank, like a small-town telephone company, will be able to say to its customers "We can hook you into the world."

MULTIPLE   Not only will financial networks lace the world, but
NETWORKS   also networks for many other types of purposes. For
                    travel and hotel reservations, for insurance, for crime control, for ordering goods, for obtaining facts, for sending mail electronically, networks will link together diverse geographically scattered terminals and systems. Firms should no longer send paper orders or invoices to other firms. The firms' computers should communicate directly.

Some of these networks will be constructed from *leased* telephone lines, data lines, or satellite channels. Some will use the *public* telephone network or data networks designed to be shared by large numbers of simultaneous users.

THE SYMBIOTIC AGE   We are moving into an age when intelligent men in all
                            walks of life will need, and constantly use, their computer terminals — this will be a *symbiotic age* when the limited brain of man is supplemented by the vast data banks and logic power of distant machines. All the professions will have their own data banks and sometimes their own languages. The nonprofessional man will use the terminals for doing calculations, for working out his tax returns, for computer dating, for planning vacations, or just for sheer entertainment. The systems will grow, multiply, and interlink.

The user, then, is likely to have a catalogue of remote systems which he can dial for different operations. He will, we hope, use his same input/output machine, or possibly a group of machines, for all the various computers he communicates with.

The dialogue the user employs in "conversing" with the machine needs very careful design. Some of the systems can by communicated with only by programming. However, if they are to be used by the mass public, or by management, clerks, salesmen, and so on, in industry, it must be possible to communicate with them in a much easier form than a programming language. The computer must speak or display the language of its user and must have a simple means of interpreting what the user wants to tell it. When computer dialogues are designed which the mass public enjoy using, the computer manufacturer revenues are going to soar.

There are certain thought processes which will always remain in the domain of human rather than machine activity. Others, however, are better done by machine. When the two can be efficiently combined, the result is more powerful than either process on its own.

The computer can store vast quantities of data and retrieve individual items quickly. It can carry out well-defined calculations or logic processes at enormous speed. It can keep files up to date and take controlling actions the moment they are required. The repertoire of procedures which can be built in a computer does not become distorted or forgotten as in the human brain. The human brain, on the other hand, can select goals and criteria, consider approaches, detect relevance, and formulate questions and hypotheses. It can handle the unforeseen.

With the logic power of the computer and mass-information retrieval available to it, the capability of the human brain for tackling certain types of problems can become immensely greater.

Much research is needed, and is being done, on the combined use of man and computers. Using time-sharing techniques, a man may use a part share in a powerful computer and its files, at a fraction of the cost of the whole system. The term *man-machine symbiosis* is used to describe this new type of thinking—part machine, part human.

PUBLIC USE OF COMPUTERS   We could list countless applications of computing-plus-data-transmission. They reach into almost all walks of life, almost all industries and professions.

It now becomes apparent that we are going to need a *public* use of computers. People everywhere must be able to dial up a computing facility appropriate to their needs. They may have a portion of a file reserved for them personally in a distant machine. The grocer on the street corner will transmit from his cash register and have his accounts done. Designers, statisticians, and engineers will use terminals and store programs or information that will be of use to them later.

There will be "on-line" files of innumerable kinds of data, with automated means for searching them. Commercial and economic data files will be interrogated for business planning. A private investor can check his hypotheses about how to make money on the stock exchange against files of data on past stock movements. One visualizes the clergyman of the future preparing his Sunday sermon at a computer terminal with ability to retrieve appropriate literature and perhaps past sermons. The result will probably be better than some of the sermons we get today. The architects, the electronic designers, the market researchers, and many other professional people are already having application-oriented programs written for them.

Design processes done by hand or with a batch-processing computer, take a long time, so long that the designer would be restricted in the full use of his imagination or inventive ability. Most design processes are 1% inspiration or art and 99% calculation and the laborious working out of detail. The object of using a real-time computer for design is to take away as much as possible of the tedious work and enable the designer to observe as quickly as possible the effects of his ideas. In a trial-and-error process, if there is a long time lag between the trial and the error, the designer can lose much of his original ideas. Where it is possible to explore the effect of these ideas in real time, much more fruitful and exciting thinking can be stimulated.

The surprising effectiveness of computers used for *teaching* has already been demonstrated. Many pupils can be handled at once and different subjects taught. The computer gives individual treatment to students and modifies its behavior according to their progress in a much more adaptable way than simpler machines.

The computer used in this way becomes a type of storehouse for human learning and thinking. An effective teaching program can be used throughout the world in many types of computer and is likely to be constantly improved as student reactions are observed.

Languages for writing teaching programs are now being developed so that a professional educator with little knowledge of computers can write them. As the potentialities of this become understood and developed, it is likely that many educators throughout the world will set to work writing and improving computerized instruction courses.

MEDICINE   *Doctors* will one day make use of a distant computer as a help in diagnosis. A patient's symptoms will be transmitted to the distant machine. The machine will make suggestions to the doctor and perhaps ask for additional information. Electrocardiograph results have been transmitted from patients' bedsides over the telephone for immediate analysis by a distant computer. The computer would not in any way replace the human qualities of the doctor, but it would add to his limited store of information. The computer can also carry on a type of information-retrieval work, but with more logical or analytical programming than straightforward information retrieval.

Computers are used to store and automatically retrieve patient case histories. The case history accessible to the computer contains details of the patient's past diseases and symptoms, diagnostic tests performed, inoculations, drugs given, the effectiveness of treatments attempted, side effects, and so on. This might be retrieved by the doctor to help him in his diagnosis, and might even in the future be processed by the computer so that the accumulated experience of the medical profession might be of help in the case.

The case histories accumulated in this way provide an excellent base of information from which medical statistics would be gathered. Medical research will use this body of data widely. The effects of new drugs will be monitored. Undesirable side effects may be detected as quickly as possible and communicated to the doctors who need to know them. As with other personal information files, laws need to be passed restricting access to this data; otherwise the individual is going to lose much of the privacy he cherishes today. If all the organizations who want to record things about an individual have their way, much of his past life is going to be accessible to the machines.

As in other fields many opportunities open up when the public rather than a doctor use the transmission terminals. In some working systems the public carry on a dialogue with a computer terminal to obtain preliminary medical help and in some cases a preliminary diagnosis. Other "telemedicine" systems are designed to give patients access to medical help which is not available in their vicinity. By means of a video link a remote doctor or specialist can examine the patient. Equipment in a local clinic can be used by a nurse or can transmit medical data about the patient to the distant doctor.

TERMINALS IN THE HOME
Today only a small number of persons have a computer terminal in their home, but sooner or later this is going to become a mass market. Perhaps, in the beginning, they will be paid for by the firms the users work for. Perhaps advertising, catalogue scanning, and computer-assisted purchasing will help to pay for them. Perhaps the cable television operators. Perhaps people will buy them because they are the latest and greatest status symbol. They need not be expensive. A touchtone telephone will suffice for some applications, with voice responses being generated by the computer. A television and simple keyboard would suffice for others.

The airline industry, the automobile industry, telecommunications, and other complex technical industries all spent two decades or so of limited growth but then expanded rapidly when there was general acceptance and usage of their product by the public. Probably data transmission is going to help this happen to the computer industry also.

Certainly when the layman, the dedicated amateur, the experimenting engineer, and the inventive academician can all tinker with terminals, the quantity of computer time that will be *wasted* will be phenomenal. Computing is like a narcotic. Once their programs begin to work, people cannot let it alone. If Colman's made a fortune out of the mustard left on the side of the plate, how much more will the computer manufacturer make out of time wasted at on-line terminals?

A television-like terminal in the home could be used for many purposes. It could give access to a wide variety of data banks and computers. The user might store details for his tax return, learn French, scan the local lending library files, or play chess with a computer. It is likely that "newspapers" will be presented in this way in the future. The user will quickly flip through the pages, or indices, for what he wants to read. With satellite communications he might read the London "papers" for no more cost than the New York ones.

The television screen of the future is likely to become much bigger than today. Screens several feet in width are already available. By increasing the television transmission bandwidth, large screens could be filled with high-quality pictures. The user could relax in his armchair and watch such a screen, using a detached keyboard in his lap.

Whereas a man, perhaps, uses his home terminal for writing programs, his wife might go shopping with it. In America all manner of highly colored catalogues may become available the moment the new medium exists, and there will be varied enticements for exploring them. Very elaborate presentations of products will become possible. Perhaps critical consumer guides will also become automated to aid the exploration. Having scanned the relevant catalogues and inspected pictures of the goods in detail, the shopper can then use the same terminal to order items. The money will be automatically deducted from her bank record and added to that of the seller. Again with satellite transmission, dialing a shopping catalogue in a machine thousands of miles away will be no more expensive than dialing one in the next town, and one visualizes New Yorkers "shopping" in Paris and Tokyo.

The cable television systems of today, like other localized transmission systems, will eventually be linked together into nationwide and multinational networks. Two-way transmission will be available, opening up endless possibilities other than mere passive television.

The world news coverage could be scanned selectively like a magazine, some news being in film clips, some in photographs, and most in text form. The Sears-Roebuck catalogue could have film sequences in it. The user of a screen information medium would be free to "turn the pages," to use the index, to select and reject.

We are beginning to see devices in which the contents of magnetic tape libraries are stored on-line. The equivalent of a tape of data can be automatically loaded when requested. It would be possible to put music or film libraries on-line, with their contents accessible on demand by a computer, and their users miles away on cable TV on satellite links.

Dialable video channels are in use in some schools. Business video calls have many possible uses. Eventually many office buildings will have their own screen units or teleconference studio.

The range of users will extend into many fields. We will find enthusiastic users of data communications adding programs and data to the files that are for public use. Dedicated amateurs will probably begin to write and sell programs to the computer companies. We must achieve mass participation in the enormous quantity of programming work that lies ahead if such schemes are to flower fully. In computer-assisted teaching alone the programming work ahead is of unimaginable magnitude. The technique can achieve maturity only after thousands of talented and dedicated teachers have been harnessed into the work of preparing the programs. This is equally true in other fields.

This technology is bringing the computer to the masses. People everywhere will be able to participate in using and building up an enormous quantity of computerized information and logic.

We are at the beginning of a chain reaction. The ingredients already exist. The fuse has been lit. It is clear that we now have a tool so powerful that it will take many decades for us to use it to its full potential. But its full potential is far beyond our cleverest imagining today.

GROWTH
As we commented at the start of the chapter, there are limits to growth in many of our existing consumer patterns. New consumer patterns will emerge out of new technology. There are no foreseeable limits to growth in telecommunications or electronic technology. The mass-produced semiconductor wafers of the 1980s will contain more than a million transistors. New satellites and optical fibers could transmit all the information man could use. Such facilities consume little energy compared with the automobile, and use little of our limited minerals. A satellite generates its own energy from sunlight. There are no limits near in the consumption of information, the growth of culture, or the development of the mind of man. The new information channels which can access the world's data banks, film libraries, computer-assisted instruction programs, and digitized encyclopedias with built-in film clips, will become available to mankind everywhere.

Societies growing towards new forms of greatness have often had an image of what might become possible – a vision which inspires the young and draws out the best in men. It is vitally important to understand that the immense riches of today's technology permit us to create visions of a better world. It is possible now to build a world without pollution, without massive destruction of nature's beauty, without human drudgery, in which destructive consumer patterns are avoided, and in which the mind of man can be nourished as never before in history so that it can soar to new forms of greatness.

REFERENCE

1. Edward F. R. Mason and Raymond J. Mason, *A Data Processing System for State and Local Governments,* Prentice-Hall, Englewood Cliffs, N.J.; 1963.

2 ORGANIZATIONS INVOLVED IN TELECOMMUNICATIONS

Before describing the technicalities of the telecommunications links, we will discuss briefly some of the organizations and regulating authorities involved with this industry.

THE UNITED STATES COMMON CARRIERS The companies in America which furnish communication services to the public are referred to as "common carriers." Legally, the name applies to all companies who undertake to carry goods for all persons indiscriminately. Reference to the communication companies by this name dates back to days when messages were carried by a stagecoach clattering through the sagebrush.

The telecommunication common carriers offer facilities for the transmission of voice, data, facsimile, television, telemetry, and telephoto pictures. Some of them are now offering on-line computer services also, and it is possible that their business could branch out widely in this direction. Many computer manufacturers and independent companies are also offering on-line computer services to the general public and so run the risk of being classed as public utilities for this part of their business in the future and being subject to the various government controls on public utilities.

It is surprising to note that there are nearly 2000 telecommunication common carriers in the United States. Many other countries have only one such organization, which is run by the government. Most of these common carriers are very small. Only about 250 of them have more than 5000 subscribers. The largest common carrier is the American Telephone and Telegraph Company (AT&T) whose subsidiaries and associated companies operate 83% of the telephones installed in the United States.

Table 2.1 lists the eleven largest U.S. telephone companies and shows their proportions of the installed telephones.

Table 2.1   Principal telephone holding companies in the
United States.†

Rank	Holding Company	Headquarters	Percentage of U.S. Telephones
1	American Telephone & Telegraph	New York City	83.07
2	General Telephone & Electronics	New York City	7.81
3	United Utilities	Kansas City	1.93
4	Continental Telephone	St. Louis	1.13
5	Central Telephone & Utilities	Lincoln, Nebraska	0.78
6	Mid-Continent Telephone Corp.	Hudson, Ohio	0.43
7	Rochester Telephone Corp.	Rochester, New York	0.43
8	Puerto Rico Telephone Co.	San Juan	0.24
9	Lincoln Telephone & Telegraph Co.	Lincoln, Nebraska	0.18
10	Commonwealth Telephone Co.	Dallas, Pennsylvania	0.11
11	Florida Telephone Corp.	Ocala, Florida	0.09
Next 10 companies			0.50
Remaining 1800 or so companies			3.30

†AT&T has 83% of U.S. telephones but serves only about one third of the geographical area.

The Bell System (U.S.A.)

AT&T, affectionately known as "Ma Bell," is the world's largest corporation. It employs over a million people. Its assets are over 3 times greater than General Motors, America's largest industrial corporation. Its revenues are almost 20 times those of Bank of America, America's largest financial institution.

"The Bell System" refers to the vast network of telephone and data circuits with many switching offices and to the television and other links which are operated across the United States by AT&T and its subsidiaries and associated companies. AT&T owns all or the majority of the stock in most of the 23 Bell operating companies, and its Long Lines Department provides much of the U.S. interstate long-distance service. AT&T also owns the Western Electric Company. AT&T and Western Electric own the Bell Telephone Laboratories. Western Electric is the main manufacturing company for the Bell System. It manufactures and installs most of the equipment in use. The Bell Telephone Laboratories, which is claimed to be the world's biggest research organization, has done much of the development work that has made today's telecommunications possible. It was there that the transistor was invented, Shannon's work on information theory was done, the solar battery was invented, and the first communication satellite, Telstar, was designed and built. Figure 2.1 shows the organization of the Bell System.

Many of the line types made available by the Bell System are summarized in Chapter 18. AT&T leases private lines of a wide range of capacities. Data can be transmitted either over these leased lines or over the public telephone network.

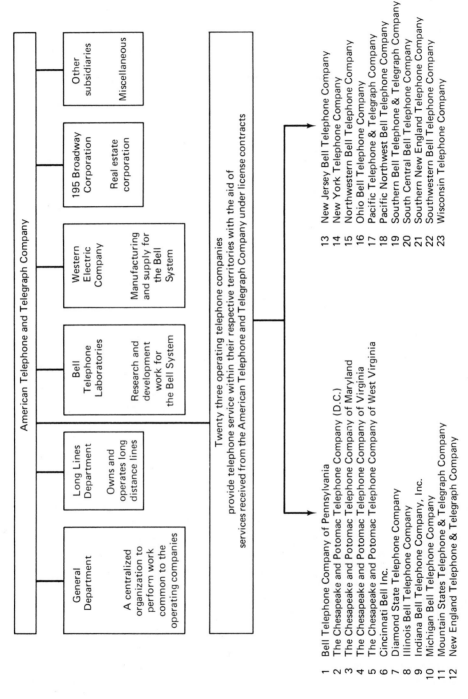

American Telephone and Telegraph Company

General Department	Long Lines Department	Bell Telephone Laboratories	Western Electric Company	195 Broadway Corporation	Other subsidiaries
A centralized organization to perform work common to the operating companies	Owns and operates long distance lines	Research and development work for the Bell System	Manufacturing and supply for the Bell System	Real estate corporation	Miscellaneous

Twenty three operating telephone companies provide telephone service within their respective territories with the aid of services received from the American Telephone and Telegraph Company under license contracts

1 Bell Telephone Company of Pennsylvania
2 The Chesapeake and Potomac Telephone Company (D.C.)
3 The Chesapeake and Potomac Telephone Company of Maryland
4 The Chesapeake and Potomac Telephone Company of Virginia
5 The Chesapeake and Potomac Telephone Company of West Virginia
6 Cincinnati Bell Inc.
7 Diamond State Telephone Company
8 Illinois Bell Telephone Company
9 Indiana Bell Telephone Company, Inc.
10 Michigan Bell Telephone Company
11 Mountain States Telephone & Telegraph Company
12 New England Telephone & Telegraph Company

13 New Jersey Bell Telephone Company
14 New York Telephone Company
15 Northwestern Bell Telephone Company
16 Ohio Bell Telephone Company
17 Pacific Telephone & Telegraph Company
18 Pacific Northwest Bell Telephone Company
19 Southern Bell Telephone & Telegraph Company
20 South Central Bell Telephone Company
21 Southern New England Telephone Company
22 Southwestern Bell Telephone Company
23 Wisconsin Telephone Company

Geographical areas in which the 23 operating telephone companies operate:

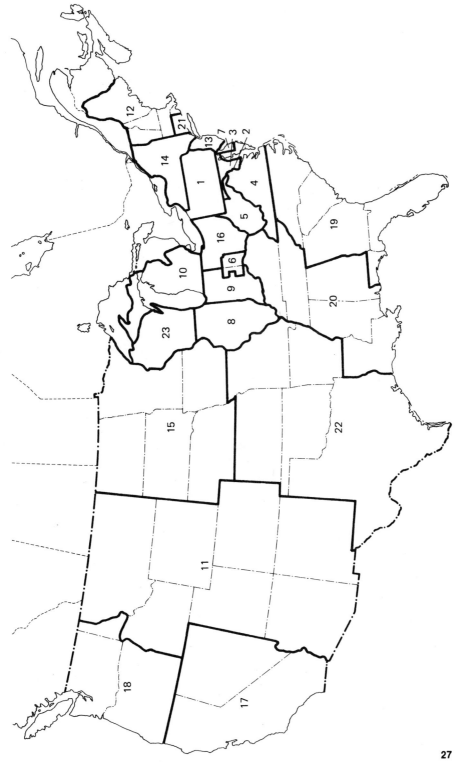

Figure 2.1 The organization of the Bell System.

General Telephone and
Electronics Corporation (U.S.A.)

General Telephone and Electronics' (GTE) network of telecommunication facilities is known as the General System. This has about 7.8% of American telephones, the second biggest slice of the telephone business after AT&T. General System equipment is compatible with Bell System equipment in most areas, to allow direct interconnection. The system offers a wide range of data services, as well as telephones.

Its organizational structure is similar to the Bell System. It has quite complete vertical integration and manufactures all the major components in a telephone system except cable. It buys cable and some other equipment from Western Electric. Its two manufacturing subsidiaries are the Automated Electric Company and the Lenkurt Electric Company, who also manufacture equipment for the rest of the telephone industry. Unlike AT&T, GTE is free to have a nontelephone subsidiary – Sylvania.

The "Independents"

The non-AT&T telephone companies are referred to as "independent" telephone companies. These serve more than half of the U.S. geographical area and have about 17% of the telephones. The number of independents has been decreasing ever since 1920 when there were 9211 of them, but their combined revenue has been growing more rapidly than that of AT&T. The independents today are bigger than AT&T was in 1940. AT&T is barred from acquiring independent telephone companies by the *Kingsbury Agreement* of 1913.

Virtually all the independents interconnect with the Bell System and transmit signals compatible with it. Bell sets the standards for the entire U.S. telephone industry and has a large staff keeping the independents informed. Revenues for long-distance calls are divided between the independents involved and the Bell Long Lines Department. The operations of a telephone company are spelled out in *Bell System Practices* (BSP), and the independents adopt this with a few of their own modifications.

The United States Independent
Telephone Association (USITA)

Most of the independents are members of the United States Independent Telephone Association, which coordinates their practices. The tariffs are established by committees, and literature is distributed to the member companies. Most states also have independent telephone associations. These and USITA are highly influential and have played an important part in the development of the independent telephone industry.

Western Union (U.S.A.)

The Western Union Telegraph Company has provided America with telegraph links since the days of the Wild West. It operates a national telegraph message service to all parts of the United States. Western Union also leases private communication links, and it operates two public dial-up telegraph networks, a *telex* network compatible with the worldwide telex network, and the *TWX (Teletypewriter Exchange)* network, which is bought from AT&T. The number of telegrams in the United States is falling rapidly, but as this business declines the telex and TWX business is growing. The Western Union leased-line facilities now include voice, and data and facsimile services of a wide range of speeds. Many of these are summarized in Chapter 18.

Recently Western Union has done much experimenting with on-line computer systems and now offers a wide range of computer services which might be extended widely in the future. It also acquired the PS Newswire Association. To do this it had to form a holding company, Western Union Corporation, which is separate from the common carrier, the Western Union Telegraph Company. Western Union International Inc. is also a separate company which handles international cablegrams and data traffic.

Western Union distinguished itself in 1974 by having the United States' first two domestic satellites launched — the WESTAR Satellites. Western Union leases satellite channels as well as using them to carry its subscriber traffic.

**GOVERNMENT
AGENCIES**

The Federal Communications Commission

With so many common carriers, many of which monopolize the services they offer, it is necessary to have some regulating authority. There is at least one such authority for each American state, as well as a national authority for controlling interstate lines and foreign facilities originating in the United States. The latter is the Federal Communications Commission (FCC).

The FCC is an independent federal agency which regulates radio, television, telephone, telegraph, and other transmissions by wire or radio. The powers of the FCC are defined in the Communications Act (1934). It was created to

> "regulate interstate and foreign commerce in communication by wire and radio
> so as to make available, so far as possible, to all the people of the United States
> a rapid, efficient, nationwide and worldwide wire and radio communications
> service with adequate facilities at reasonable charges."

The FCC has jurisdiction over *interstate* and *foreign* telecommunications

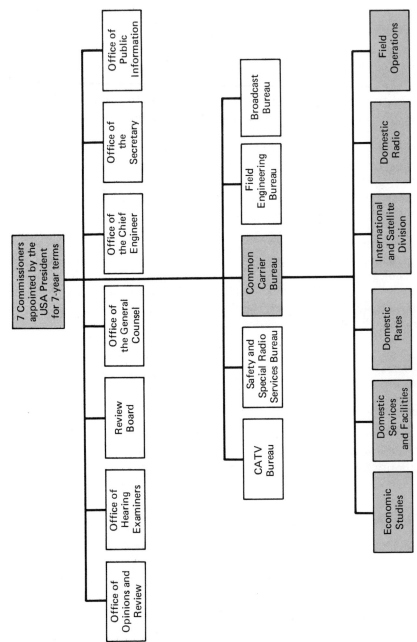

Figure 2.2 The organization of the United States Federal Communications Commission (FCC).

but not telecommunications within a state. The latter are regulated by the State Public Utility Commissions.

Figure 2.2 shows the organization of the FCC.

Every subject common carrier must have its plans for facilities offered to the public approved by the FCC before they come into effect. To achieve this, schedules must be filed with the FCC giving details of the intended service, the charges, classifications, regulations, and so on. These documents are called *tariffs,* and they form the basis of the contract between the common carrier and the user. It is intended that the FCC should regulate these for the benefit of the public and that all common carriers provide service at "reasonable charges on reasonable request." Some of the tariffs become standards which other companies use. There is no requirement for a company to file a new telephone or telegraph tariff if it can use an existing one from another carrier. For this reason some of the more important tariffs are common to many suppliers.

The State Public Utility Commissions

What the FCC does for interstate links and foreign links originating in the United States, the State Public Utility Commissions do for links within one state. Different states have different tariffs for the same grade of service, and there can be a wide difference in the prices of facilities from one state to another. Interstate tariffs, however, are uniform across America.

The Office of Telecommunications Policy

The Office of Telecommunications Policy (OTP) was set up by President Nixon in 1970 because telecommunications technology was advancing so rapidly that it was outpacing the ability of government to formulate policy. The OTP was created to formulate plans, policies, and programs designed to maximize the value of telecommunications to the public interest, the U.S. economy, and national security.

The OTP has no executive authority. It can issue recommendations, not directives. Its recommendations appear to have influenced the FCC to stimulate more competition in an industry which monopolistic practices might make slothful. It appears to have stimulated the launching of domestic communication satellites, the growth of the specialized common carriers, the rise of the value-added common carriers, and the regulations encouraging the growth of cable television operators.

THE SPECIALIZED COMMON CARRIERS   In the early 1970s a new breed of common carriers developed which do not install telephones or telegraph machines but provide specialized transmission facilities. The would-be "specialized" carriers seized upon the rigidity of the U.S. telephone system, claiming that separate microwave and other transmission links were needed for specialized purposes, including data transmission. In 1970, the FCC, summarizing the conclusion of a lengthy inquiry into computers and telecommunications stated that there was

> "dissatisfaction on the part of the computer industry and by many data users who had been attempting to adapt their requirements to existing (communications) services."

In 1971, after intense opposition from AT&T, the FCC commissioners voted favorably on the concept of specialized carriers, saying:

> "The entry of new carriers would have the effect of dispersing somewhat the burdens, risks and initiatives involved in supplying the rapidly growing market for new and specialized (communications) services."

This sudden injection of competition into an industry which for decades had little competition has been dramatic in its effect. There are many aspiring new carriers, and they have triggered new tariffs and service offerings (such as Bell's Dataphone Digital Service, DDS) from the old carriers. The telecommunications industry is in more turmoil than for decades. In the United States, a land of entrepreneurs, the way to become a millionaire in the next ten years may be in telecommunications!

Microwave Communications Inc. (MCI)

The pioneer and pacesetter of the specialized carriers was MCI. In 1969, after six years of legal battling (the telecommunications lawyers are making a fortune), the FCC gave MCI permission to build a microwave system between St. Louis and Chicago. This historic decision triggered a flood of 1900 new microwave station applications.

MCI was soon a $100 million group of corporations building a nationwide microwave network and selling a wide variety of bandwidths to any customer who could use them. (See Fig. 2.3). MCI's prices, like those of the other specialized carriers, were attractively lower than those of the established carriers.

AT&T bitterly accused MCI of "cream skimming," that is, providing service only to those parts of the country where there would be maximum profit, whereas Bell had to provide a similar service to its entire geographical area. The FCC ruled that in order to enable new corporations to come into business they were indeed allowed to skim the cream of the market.

MCI originally described themselves as a "nontelephone" common carrier. It soon became clear, however, that their main business revenue was to come from corporate telephone service, and their marketing drive became oriented largely toward this.

MCI "War Room" in Washington, D.C. From here all circuits in the MCI system are monitored.

Figure 2.3 Facilities of MCI, the corporation which initiated the U.S. specialized common carrier industry. Most specialized common carrier rates are somewhat lower than those of the main telephone companies. (Photos by Harry Newton.)

MCI microwave towers in New York City.

Other Specialized Common Carriers

There are now many other specialized carriers competing to form the core of this new segment of the industry. As is always the case when a new industry grows there will probably be many mergers and takeovers. Figure 2.4 shows the networks of two of the major companies.

A particularly interesting breed of specialized carrier is that concerned with satellite communications. Several new carriers in addition to existing carriers have been given permission to operate a satellite network in the United States. SBS, Satellite Business Systems, a subsidiary of IBM, Comsat, and Eatna, has been given permission to operate a system with small earth stations on user premises, carrying data, voice, and video.

THE VALUE-ADDED COMMON CARRIERS In 1973 the industry took another new turn, further enhancing the marriage between telecommunications and computing. The first value-added common carriers (VACCs) came into existence. A value-added carrier does not construct any telecommunication links. Instead he leases links from other carriers and creates a network with sophisticated computer control to provide new types of telecommunication services.

The first value-added common carrier to receive FCC approval was Packet Communications Inc. (PCI) which intended to build a network providing fast packet-switched data communications (explained in Chapter 24) between its customers' computers or computer terminals. At the time of writing PCI has not constructed its network and has ceased operating.

The second FCC approval went to Graphnet Systems Inc., who became the first operational value-added common carrier. Graphnet operates a store-and-forward network to interconnect facsimile machines and to transmit from computers, telegraph machines, and data terminals to facsimile machines. A Graphnet subscriber can deliver mail, documents, or computer data to other subscribers with facsimile equipment.

Figure 2.5 shows the proposed value-added network (VAN) of the Telenet Communication Corporation. This, like PCI's proposal, is a network interconnecting computers and terminals. It is based on the ARPA network (Fig. 4.9) and uses similar software, giving a very fast network response time.

The value-added networks take advantage of economies of scale. When many users share the same wideband communication channels they can communicate more cheaply. Also the communication path can be made more reliable, because failures can be bypassed. Users who could not otherwise communicate, because they have incompatible machines, can be connected. Graphnet will interconnect incompatible facsimile machines, and Telenet will interconnect incompatible computers.

The network of Microwave Communications Inc., offering leased analog channels of a wide range of bandwidths.

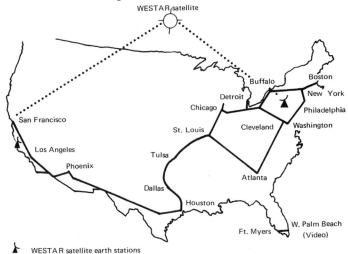

The Southern Pacific Communications Co. operates a microwave network, like MCI, offering a wide range of leased analog bandwidths.

Datran operated a switched and point-to-point data network subdividing a 44 megabit microwave signal into channels of speeds usable in data processing. This service was taken over by Southern Pacific Communications in 1977, and the networks merged.

Figure 2.4   The networks of two of the largest specialized common carriers. The plans of this new industry are changing rapidly.

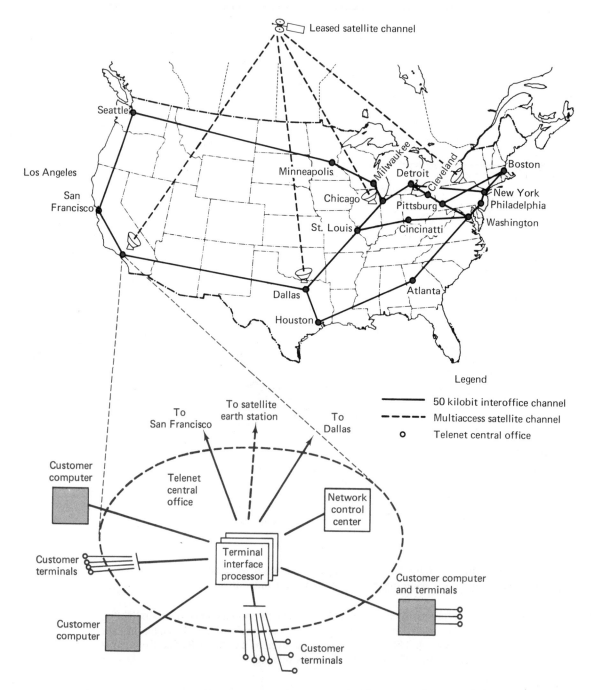

Figure 2.5 Telenet's proposed packet-switching network — one of the first "value-added common carriers." (Reproduced from the Telenet FCC filing, 1973.)

The value-added common carriers do not *process* the messages they transmit and store. If processing takes place, then the service is a computing or time-sharing network, not merely a common carrier.

THE INTERCONNECT   Prior to 1968 it was illegal for subscribers to connect
INDUSTRY   their own devices to telephone company lines. After a
long legal battle the "Carterphone ruling" permitted
the Carter Electronics Corporation to connect its mobile radio system direct to
the Bell System. The FCC concluded that

> "a customer desiring to use an interconnecting device. . .should be able to do
> so, so long as the interconnection does not adversely affect the telephone com-
> pany's operations or the telephone system's utility for others. . . .The appro-
> priate remedy is to. . .permit the carriers if they so desire to propose new tariffs
> which will protect the telephone system against harmful devices and they may
> specify technical standards if they wish."

It was then permitted to connect devices to the Bell System via a small white box on the wall called a DAA (direct access adaptor), which is essentially an isolation transformer. Later the need to use a DAA was eliminated in most situations.

A large new industry began to grow after the Carterphone ruling. Everything from computers to decorator telephones can now be connected to the telephone network. Terminal manufacturers are putting modems under the covers. Computerized telephone exchanges, monitoring equipment, and network control equipment are being manufactured by organizations other than the common carriers. The entire telecommunication facilities of a corporation can now be independently designed to serve the overall communication needs of the corporation.

The interconnect decision gave would-be telecommunication users freedom to invent.

TELECOMMUNICATION
ORGANIZATIONS IN
OTHER COUNTRIES

Canada

The structure of the industry in Canada is somewhat similar to the United States except that some of Canada's 1600 common carriers are government-owned. Canada has had no equivalent of the Carterphone ruling and has no independent value-added common carriers. A value-added or "intelligent" data network, using packet switching, is being operated by the Trans-Canada Telephone System (TCTS), a company owned by Canada's eight largest regional telephone companies, and provides users with the DATAPAC service for sending data. TCTS plans and coordinates joint system efforts and allocates toll revenues among members, rather like AT&T's Long Lines Department in the United States. Another nationwide telecommunications carrier is Canadian National/ Canadian Pacific Telecommunications (CN/CPT). CN/CPT offers private services and public message telex and telegraph service.

Sixty-nine percent of Canada's telephones are operated by Bell Canada, which is entirely separate from the U.S.A. Bell. Twelve percent of the telephones are operated by GTE (United States). Four percent are operated by small independent common carriers and almost 16% by provincial or municipal governments.

Telesat Canada, a mixed private and government-owned corporation, established the free world's first domestic communications satellite system in 1972.

The British Post Office Corporation (BPOC)

In many countries the job of delivering mail and providing telecommunications is undertaken by the same organization. This came about because the first use of telecommunications was in telegraph message delivery, which was an extension of the mail. It may make sense again, before long, as it may become cheaper to transmit letters in facsimile form and reproduce them than to deliver them physically. The British Post Office Corporation consists of one organization for handling mail and a separate organization for handling telecommunications.

In 1969 it changed its status from a government department to a nationalized commercial company. It has its own research laboratories, though most of its equipment is manufactured by private industry.

For data transmission, the BPOC offers a variety of links under the heading of datal services. It operates a telex of dial-up teleprinters and paper tape machines, it provides facilities for sending data over the dial-up public telephone network, and it offers private leased lines of varying speeds. It provides private wideband circuits with a capacity 12 to 60 times voice circuits between about 180 selected towns. The delay in obtaining these facilities, however, is sometimes much greater than in the United States. The lines are not always compatible with the American ones, as will be discussed in later chapters, and sometimes equipment that works well in the United States cannot be installed in Britain.

Telecommunication Departments in Other Countries

Most of the countries of the world have a government-controlled monopoly providing their telecommunications, like Britain's post office, and offering or planning to offer facilities for data transmission which are broadly similar. In some countries the organization is a civil service department. In others it is a nationalized corporation. In Germany there is the Deutschen Bundespost and in France the Postes Téléphonique et Télégraphique. In most countries, the finance comes from government, not public, sources. For the Swedish Telecommunications Administration, for example, the state, through the Riksdag (Parliament), decides the amount of investments and makes the necessary grants. Some countries manufacture most of their own telecommunication equipment. Others import it. India, because of her currency problems, tries to manufacture as much as possible internally. The Indian Posts and Telegraphs Department has an active research department which works closely with private firms, such as Indian Telephone Industries Ltd. and Hindustan Cables Ltd., to produce modern equipment at low cost.

There is much duplication among the research of different countries. Certainly in Europe, close technical cooperation seems to be needed between the different countries' telecommunications organizations, as in other fields, if Europe is to compete with America. As will be mentioned in later chapters, there are certain minor incompatibilities between national networks that are causing increasing expense now that international telecommunication is growing so rapidly.

INTERNATIONAL STANDARDIZATION

The International Telecommunications Union (ITU)

Although incompatibilities exist, the degree of *compatibility* is remarkable. This is largely due to the International Telecommunications Union. This organization, centered in Switzerland, has 124 member countries throughout the world. Its consultative committees carry out very detailed studies of world telecommunications and make recommendations for standardization. The recommendations are put into practice widely throughout the world, with some notable dissensions.

There are three main organizations within the ITU: *The International Frequency Registration Board,* which attempts to register and standardize radio-frequency assignments and to assist in the elimination of harmful radio-frequency interference on the world's radio communications circuits, *The Consultative Committee on International Radio* (CCIR), which deals with other standards for radio, especially long-distance radio telecommunications; and *The Consultative Committee on International Telegraphy and Telephony* (CCITT).

Figure 2.6 shows the organization and functions of the ITU.

Comité Consultatif International Télégraphique et Téléphonique (CCITT)

The Consultative Committee on International Telegraphy and Telephony, based in Geneva, is divided into a number of study groups which make recommendations on various different aspects of telephony and telegraphy. There are

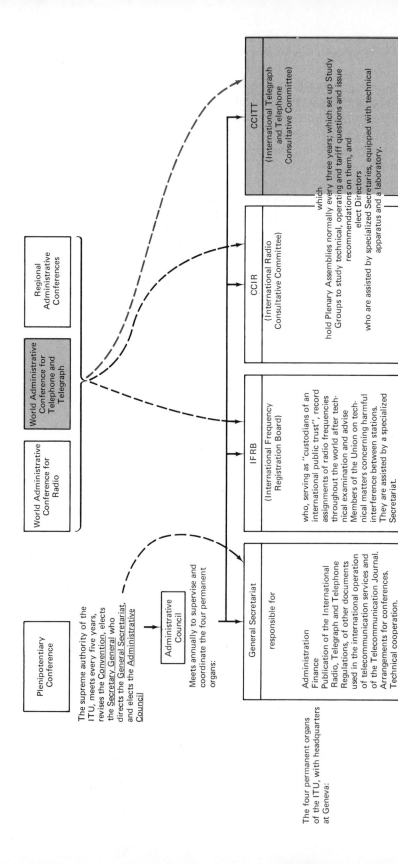

Figure 2.6 The organization and functions of the International Telecommunications Union, ITU.

study groups, for example, on telegraphy transmission, performance, telegraph switching, alphabetic telegraph apparatus, telephone channels, telephone switching and signaling, noise, and several others. The study group which is perhaps the most concerned with the subject of this book is the special committee on data transmission. It has produced reports of great thoroughness, giving recommendations for standards for data transmission. They are widely accepted. Some examples of CCITT standards will be given later in the book.

INTERNATIONAL ORGANIZATIONS

International Telephone and Telegraph Corporation

ITT is the world's largest company engaged internationally in operating telecommunication services and manufacturing telecommunication equipment. It has been diversified in the last decade to include electronics and other products and services such as hotels and Avis Rent-a-Car. It embraces more than 150 associated companies and has activities in 57 countries, manufacturing telecommunications equipment in many of them. It operates a worldwide communication network of thousands of cable, radio, and satellite circuits.

ITT connects telegraph users in the United States and other countries by means of radio and cable facilities. It also provides communications between certain foreign countries. The United States has three "gateway" cities for international traffic — New York, Washington, and San Francisco. At these cities the Bell System or Western Union lines, or those of other carriers, connect to the ITT international lines. Subscribers in the gateway cities can connect directly to the ITT lines; otherwise they are routed in on lines of other carriers.

ITT operates an international telex system. An ITT telex subscriber in America can dial numbers overseas, in London or Rome, for example, without manual intervention. It also operates a radio-telephone service, leases private international lines, and has over 110 traffic offices in countries outside the United States.

RCA Communications

The Radio Corporation of America Communications, a wholly owned subsidiary of RCA, also provides international circuits. It provides worldwide telegraph service, telex service to 105 countries, and radio-photo service to 53 foreign cities. Fully automatic telex dialing is available between several countries, and RCA Communications also provide private leased international telegraphy lines, radiotelephone circuits in the Pacific area, and ship-to-shore communications.

The Communication Satellite Corporation, COMSAT (U.S.A.)

The COMmunications SATellite Corporation (COMSAT) is a private company responsible for the launching and operation in America of commercial communication satellites with worldwide coverage. It works in conjunction and cooperation with other countries to achieve this and to share in the ownership and operation of a worldwide system of satellites and earth stations. COMSAT files applications with the FCC for the launching of its satellites. The first satellite it launched was the Early Bird satellite in 1965. COMSAT is a company chartered by the American Congress in the Communications Satellite Act of 1962. It is not a government agency but a shareholder-owned company, with widely distributed stock. The role of COMSAT in the global satellite system will steadily decline as functions are assumed by INTELSAT. A subsidiary of COMSAT, Comsat General is involved in COMSAT domestic satellites in the United States.

The International Telecommunications Satellite Consortium, INTELSAT

The INternational TELecommunications SATellite Consortium (INTELSAT) is, in effect, a partnership of owners in the Early Bird system and its successors, the INTELSAT II, III, and IV satellites. Many countries are members, and agreements are open for signing by any of the 124 nations which are members of the International Telecommunications Union.

The international agreements were formulated in 1964 whereby it became possible for all nations to use, and share in the development of, one satellite system. The participating organization from each country shares in the financing and owning of the satellites with their tracking and control equipment, but not in the earth stations. The latter are owned by the individual countries or corporations who use them. The participating countries receive revenues derived from share in the ownership.

As new countries join, the percentage interest of each member drops, but the 1964 agreements state that COMSAT, the American partner, cannot have its interest drop below 50.6%

Some countries, for example, Italy and Japan, have set up a company similar to COMSAT to participate in INTELSAT. In most countries, the telecommunications organizations, such as the Post Office in Britain or the Deutschen Bundespost in Germany, are the INTELSAT participants.

Countries who are not members of INTELSAT, and who have not signed the agreements, can still use the satellites by paying a lease cost for the circuits.

COMSAT, in addition to being the U.S. participant in INTELSAT and a member of its governing body, has the role of manager in the design, development, and operation of the satellites. COMSAT is engaged in research and development work on satellite systems. It submits the satellite programs to be implemented for approval by an INTELSAT committee.

3 TERMINALS AND CODES

A vast ever-growing array of machines can be attached to telecommunication lines for transmitting and receiving data. In this chapter we will give examples of these devices. The reader who is familiar with computing hardware may skip this chapter.

A device for feeding data to or receiving data from a distant computer is called a *terminal* (perhaps an unfortunate choice of a word because line termination equipment in general has been called a terminal; "microwave terminal," for example, refers to the electronic equipment at the end of a microwave link).

Data transmission can be

1. Between computer and computer.

2. Between terminal and computer.

3. Between terminal and terminal.

In any of these links there can be intermediary network devices such as concentrators and switches.

TERMINALS   Terminals can be devices into which data are entered by human operators or devices that collect data automatically from instruments. Terminals designed for human use may permit a fast two-way "conversation" with the computer or may be a remote equivalent of the computer-room input/output devices. The people in Fig. 3.1 are using terminals to communicate with a computer. Paper-tape readers and punched-card readers may provide input over communication lines. Printers may provide the output. Figure 3.2 shows non-real-time terminals.

Most computer peripherals can be taken out of the computer room and attached to a communication line. They can have a typewriter added or a keyboard or screen display, and then they are called conversational terminals.

Figure 3.1 Interactive terminals in all walks of life: vast data networks will be needed.

Banks, savings-and-loan associations and credit unions are extending their services directly to consumers by installing terminals in retail stores. The terminals are used to validate credit cards or personal checks, or the bank computer can accept an electronic transfer of funds.

A Bunker Ramo terminal in a stock broker's office connected by a voice line to a remote computer. The terminal displays facts and figures, and charts composed of dots, all up-to-the-minute.

44

Terminals used for teaching connected to a remote computer using a teaching program.

Terminals in a police department helping to keep track of the emergencies taking place in a city. They are connected by six voice lines to a distant computer.

Figure 3.2   Terminals for batch transmission.

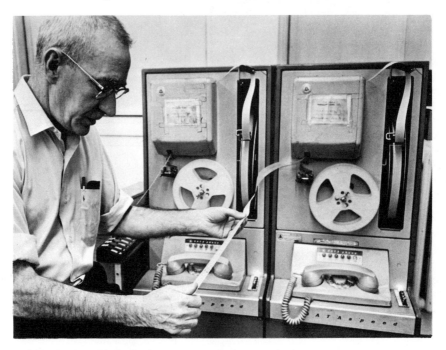

A Bell System Dataspeed tape transmission terminal. A remote location can be dialed and data transmitted at voice line speed.

A Univac printer which can be connected to a communication line. It can receive and print at the maximum speed of a voice grade line.

A minicomputer which can both process data and transmit it to a larger computer for filing. Data may be stored by the minicomputer for night-time transmission. (Photo NCR.)

An NCR terminal which can collect data on a small tape cartridge for later batch transmission.

With a control console, they can be called an off-line data-preparation terminal, or with a cluster of manual-input devices, they can be called a data-collection system. Tape cartridges or small disks can be added. Logic circuitry and memory can be added in varying quantities. Microprogramming may be used or a stored-program minicomputer. The endless variations on these terminal possibilities are further proliferated by the fact that the components can be supplied from many hundreds of peripheral equipment manufacturers. The units are often built in a modular fashion so that, as with a hi-fi system, a variety of different devices may be added. The user's choice can be complex.

The information, whether from automatic devices or from manually operated keyboards, may be transmitted immediately to the computer or may be stored in some medium for transmission at a later moment. The preparation of data, in other words, may be *on-line* or *off-line*. Readings of instruments, for example, may be punched into paper tape, which is later transmitted to the computer. Similarly, data collected from manually operated devices may be punched into paper tape. In this case, the terminal of the computer is an on-line paper-tape reader. The output may also make use of an interim medium, such as paper tape or punched cards, or it may directly control the environment in question. Very often it is necessary to make a printed copy of the computer output for later analysis. In this case, part of the terminal equipment may be a typewriter or printer.

Table 3.1 lists some of the common types of terminal devices. Many machines use combinations of these facilities.

Table 3.1

Document Transmission Terminals	Human-Input Terminals	Answerback Devices and Displays
Paper-tape readers/punches	Typewriterlike keyboard	Typewriter
Card readers/punches	Special keyboard	Printer
Magnetic card readers	Matrix keyboard	Teleprinter
Badge readers	Lever set	Passbook printer
	Rotary switches	Display screen
Optical document readers	Push buttons	Display tubes
Magnetic-ink character readers	Teleprinter	Light panel
	Telephone dial	Microfilm or film-strip projector
Mark-sensing devices	Touchtone telephone keyboard	
Microfilm		Big display boards
		Graph plotter
Plate readers	Light pen with display tube	Strip recorder
Facsimile machines		Dials
	Coupled stylus	Telephone voice answerback
Magnetic-tape units	Facsimile machine	
Tape cassettes	Plate reader	Facsimile machine
Magnetic disks	Badge reader	

BOX 3.1 Types of teletype machines

Seven types of on-line teletype devices can be obtained. In American terminology these are referred to with abbreviated initials as follows:

1. *TD, transmitter distributor.* Reads and transmits punched paper tape.

2. *ROTR, receive-only typing reperforator.* Receives the signal and punches paper tape, printing characters on the edge of the tape, or over the perforations for chadless tape (chadless tape does not have the paper completely removed when a hole is punched but instead remains attached, like a flap, so that the entire tape can be printed over).

3. *RT, reperforator-transmitter* (in effect a TD + ROTR). Combination of a transmitter distributor and a typing reperforator. Sometimes used as a store-and-forward device in manual message switching.

4. *RO, receive-only page printer.* Receives the signal and prints continuous stationery. It has no keyboard but does have page-positioning keys.

5. *KSR, keyboard sending and receiving unit.* Receives the signal and prints as with an RO. There is a keyboard on the same device which transmits data.

6. *KTR, keyboard typing reperforator.* Receives the signal and punches and prints paper tape as with the ROTR. In addition, the keyboard can transmit a signal, punching and printing paper tape at the same time. No tape reader or page printer.

7. *ASR, automatic sending and receiving unit.* This combines all the preceding devices into one machine containing a keyboard, page printer, reperforator, and tape transmitter. Paper tape can be prepared off-line. This can take place while printing is occurring from the transmission line.

Another machine used in conjunction with these is a device for punching paper tape off-line, ready for transmission. This is called a perforator.

British General Post Office teleprinter No. 15 for operation over 50-baud telegraph or telex lines (66 words per minute), using Baudot code (Fig. 3.6). *(Courtesy of H. M. Postmaster General, England.)*

Note: Upper case H may be stop or #

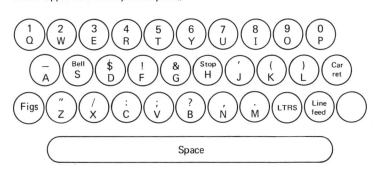

Teleprinter keyboard using 5-bit CCITT Alphabet No. 2
(Baudot code) (American version)

Figure 3.3 Telegraph machines.

AT&T No. 33 ASR teletypewriter station, operating at 100 words per minute. It can transmit directly from keyboard or paper tape. It receives either by page printing or punching paper tape. Tape can be punched from the keyboard for transmission later. It uses the U.S. ASCII code. *(Courtesy of AT&T)*

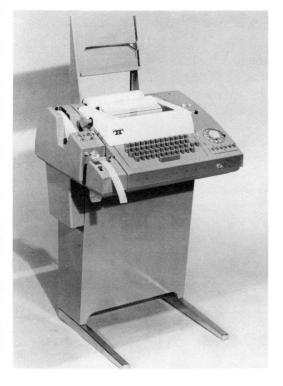

AT&T DATASPEED RO Receive-only printer. It prints at speeds up to 1200 words per minute, using Data-phone or voice grade lines, and employs the U.S. ASCII code.

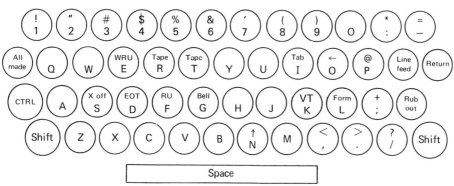

Teletypewriter keyboard using 7-bit CCITT Alphabet No. 5 (U.S. ASCII code).

There can be a major difference in design between terminals designed for conversational operation and terminals not designed for such "dialogue." Very careful attention needs to be paid to the design of the man-machine dialogue [1], and terminal design is part of that. The terminal may have specially labeled keys; a mark on a screen, called a cursor, which the operator can move around; a light pen for pointing at the screen; or other facilities for making man-machine interaction more fluent. An associated local microcomputer may assist in making the man-machine dialogue more efficacious.

TELEGRAPH EQUIPMENT

From its earlier days the telecommunications industry has produced a range of machines for telegraphy. These are called *teletype* machines in North America. The word teleprinter is common elsewhere. Box 3.1 lists the categories of teletype machines.

Figure 3.3 shows some telegraph machines.

Telegraph machines transmit at a speed much lower than the capacity of a voice line. Common teletype speeds in North America are 75 and 150 bits per second. Common speeds elsewhere in the world are 100 and 200 bits per second. The international telex network and many European lines operate at 50 bits per second.

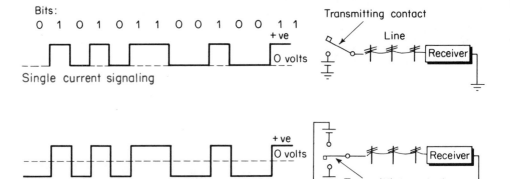

Figure 3.4

Telegraph signals are formed simply by switching an electrical current on and off or by reversing its direction of flow. The upper part of Fig. 3.4 shows a *single-current* telegraph signal in which the information is coded by switching the current on and off at particular times. The lower part shows the same signal on a *double-current* telegraph system in which positive and negative potentials are applied at one end of the line, thus reversing the direction of current flow. The means for producing these current changes are some form of make-and-break contact. The instruments for sending signals, such as teleprinters or paper-tape readers, make and break the circuit at appropriate times. The on and off pulses form a code which is appropriately interpreted by the receiving device. Single-current telegraph signaling is also known as "neutral" or "unipolar" signaling. Double-current signaling is called "polar" or "bipolar." In the United States, neutral signaling is more common than bipolar signaling. In Europe, bipolar (there usually called "double-current") signaling is most common. The adaptors for connecting telegraph equipment to computers may be planned to use either neutral or bipolar signals, whichever is appropriate.

TELEGRAPH   At one time most telegraph signals were carried on
CIRCUITS   overhead wires like those seen in old Western movies.
                       In Wall Street the sky used to be filled with overhead
wires. When telephony grew up it soon outgrew telegraphy, and when there were many more telephone circuits than telegraph it became economical to send the telegraph signals over telephone links. Today this trend has gone much farther, and many telegraph signals can travel together over one path, being manipulated by the complex electronics that are part of today's telephone plant. The simple electromagnetic repeaters that were once used for telegraphy have been largely replaced by the amplifiers and modulation and multiplexing equipment described in the following chapters.

A circuit designed for telegraph (not telephone) often had only a single wire spanning the distance. This is called an earth-return circuit. As in the diagrams in Fig. 3.4, each end of the line was connected to earth, and this completed the circuit. One pole of the power supply at each station was connected to earth. This was done by making a connection to the protective lead sheaths of underground signal cables, or to a metal main water pipe, or by burying plates, lead strips, or driving spikes into the earth. The resistance of such an earth-return path is very small.

Other telegraph circuits were obtained as a by-product of telephone circuits. Telegraph signals may be sent down any two telephone channels by a process called "superposing." Each telephone channel ends on a transformer winding. The midpoints of these windings were taken and used as an extra pair of wires down which telegraph signals could be sent without interfering with the telephone signals or without the telephone signals interfering with the telegraph. The extra circuit obtained this way was called a "phantom circuit."

Over a single wire it was common to use equipment which could not transmit in both directions at the same time. The signals would travel either east to west or west to east. With double-current signaling, relays could be used to separate the east-to-west and west-to-east signals. This was referred to as "simplex" transmission in Europe and as "half-duplex" in America. A more ingenious arrangement of relays would permit "full-duplex" operation, i.e., transmission in both directions at the same time.

START-STOP
INSTRUMENTS

Figure 3.3 shows the keyboards of some typical instruments using telegraph circuits.

Such devices normally use start-stop transmission. The character generated by one key depression begins with a START bit and ends with a STOP bit or bits. The STOP condition is a "mark" or positive voltage on the line. The START bit is a "zero" bit or "space," i.e., no voltage on the line or negative voltage with double-current signaling.

The positive voltage of the STOP condition will remain on the line until the next character starts, as indicated in Fig. 3.5, which shows 5-bit and 7-bit telegraphy characters. In other words, the line's idle condition is a "mark" rather than a "space." As soon as a space is detected the receiving device starts and will then be in synchronization with the transmitting device. The STOP condition thus has a minimum duration, which is 1.42 (or 1.5 or 2) times the length of the other bits but which could go on indefinately. If the machine is printing or transmitting continuously, there will be no gap between characters, and the STOP condition will last only long enough to clearly separate the characters.

With start-stop transmission, a new character can begin at any time. There is an indeterminate period between one character and the next. When on character ends, the receiving device will wait idly for the start of the next character. The transmitter and the receiver are then exactly in phase and will remain in phase while the character is sent. The receiver will thus be able to attach the correct meaning to each bit it receives. Box 3.2 illustrates start-stop operation.

Figure 3.5   Typical character structures for start-stop (asynchronous) transmission.

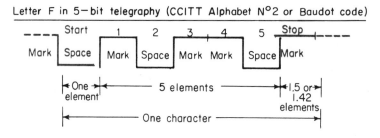

Letter F in 5−bit telegraphy (CCITT Alphabet N°2 or Baudot code)

Figure 5 for 8 bit telegraph machines (CCITT Alphabet N° 5 or ASC II code)

BUFFERED   Terminals using modern electronics can be equipped
TERMINALS   with a buffer. A buffer is a small memory which holds a
                       block of data while it is being transmitted, received,
keyed in, displayed, or otherwise manipulated.

When data are transmitted to and from buffers they can be transmitted *a
block at a time*. There is no need, then, to send a special signal to indicate the start
of each character. The data bits can be packed tightly together inside the block
with regular, synchronized timing. This is referred to as *synchronous transmission*. A higher data rate can be transmitted than with start-stop transmission because there is no need to stop and start between characters.

INTELLIGENT   An emerging technology of the 1970s is micro-
TERMINALS   miniature electronic logic circuitry and memory. It is
                       highly reliable and can be mass-produced at low cost.
The public has seen the effects of this circuitry in products such as electronic calculators. It has brought a revolution in terminal design.

First it became economical to put buffers in terminals and build synchronous
rather than start-stop devices. It became economical to use more efficient line-control techniques and permit more terminals to share the same line.

Advanced terminals steadily became more "intelligent", containing increasing quantities of logic and memory. The terminals could carry out certain types of
computations or operations which assisted in the man-machine dialogue. The terminal could be made easier to use, and sometimes faster in its speed of response.
Increasingly, functions were handled in the terminals or their local controllers,
which previously would have required transmission to the remote computer. Consequently the number of messages transmitted to and from the computer could be
less, and sometimes the number of bits in a message could be much less.

The ultimate consequence of this trend was for the terminals or their controller to contain a small stored-program computer. This was programmed for a
variety of functions related to the application of the system. Often the terminal
user had no need to know that there was a small computer under the cover of the
terminal. The use of such peripheral computers is referred to as *distributed
intelligence*. It will be one of the main trends in teleprocessing for years hence.

The more powerful the computing facilities at the terminal location, the less
is the necessity to transmit to a remote control computer. There remain two basic
reasons for transmitting:

1. Greater computing power is needed than that at the terminal location.

2. A central file facility or data base is needed.

As the peripheral computers become more powerful so, increasingly, the
reason for interactive teleprocessing will be to use a remote *data base* rather
than to use remote computing power.

BOX 3.2   Electromechanical start-stop telegraph machines

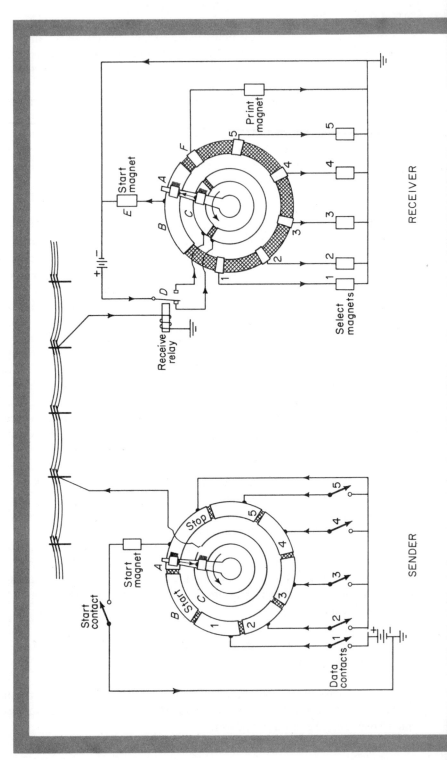

The above diagram illustrates the principle of operation of electromechanical *start-stop* telegraph machines.

The sending device and the receiving device each have an armature, *A*, which can rotate at a constant speed when a clutch connects it to an electric motor in these machines. Brush contacts on the armatures connect an outer ring of contacts, *B*, to an inner ring, *C*. In the above diagram, the armature is in its **STOP** position. Consequently, current from the battery in the sender flows via the outer contact labeled STOP through the armature and onto the line. This current causes the contact of a receiver relay, *D*, to be in the position shown, and thus no current from the battery in the receiver flows to any of the operating magnets shown.

Now suppose that a girl operating the sending device presses the **H** key on the keyboard. In accordance with the code in Fig. 2.6 the **3** and **5** data contacts now close. Also the **START** contact closes and the armature of the sender begins to rotate counterclockwise. The armature connects the contacts on the outer ring to the line in sequence: **START, 1, 2, 3, 4, 5, STOP.** The **3, 5,** and **STOP** contacts convey current; the others do not. As soon as the armature of the sender travels to its **START** contact, the positive currents on the line ceases. The contact of the

relay, *D*, flips across, and positive current from the battery of the receiver flows through the inner ring, *C*, armature, *A*, and outer ring, *B*, to the start magnet, *E*. The start magnet causes the armature of the receiver to rotate counterclockwise at the same speed as that of the receiver.

When the receiver armature is over contact **3**, current will again flow on the line. Receiver relay, *D*, will again move to the position shown in the figure, and so as the armature of the receiver passes over the **3** contact on the outer ring, current will flow to the select magnet **3**. Similarly, select magnet **5** will be operated. These will cause the letter **H** to be selected on the type mechanism of the receiver. When the sender armature reaches its **STOP** contact, the receiver armature will pass over the contact, **F**, and current will flow to the print magnet. The letter **H** will be printed. Both armatures will then come to rest in the position shown in the figure, unless the sender is ready to transmit another character immediately.

There are several variations of this basic operation principle, but the speeds and timing have become standardized. This means that a variety of different telegraphy machines can communicate with one another. A computer can send data to a telegraph machine by sending pulses with the same timing as the sender in the diagram.

Table 3.2 Commonly used transmission speeds.

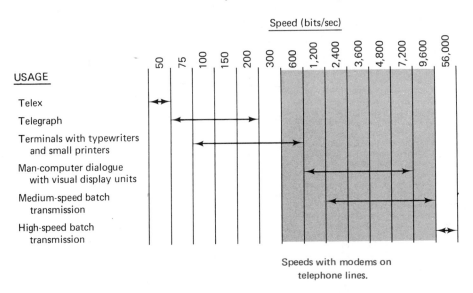

Speeds with modems on
telephone lines.

Further discussion of distributed intelligence is beyond the scope of this book and is discussed in Reference 2. A high proportion of the terminals in use are still "unintelligent" machines.

SPEEDS OF TRANSMISSION

Different types of terminals and uses have different requirements for transmission speeds. Table 3.2 lists commonly used speeds. The bit rates listed at the top of the table are the most common operational speeds of today's data transmission links.

DATA CODES

It is desirable to have internationally agreed-upon code alphabets for transmission of data so that different computers and terminals can intercommunicate.

During the 1960s an alarming proliferation of data transmission codes developed [3]. Some of these codes have now dropped into disuse, and two codes predominate, each with minor deviations for local needs. The first is standardized by the CCITT [4] as their *International Alphabet No. 2:* This is a 5-bit code based on the Murray code used decades ago for telegraphy and now used for telex transmission around the world. It is commonly (and incorrectly †) known as the Baudot code in the United States. The second is standardized by the CCITT [5] and also by the ISO (International Organization for Standardization) [6] as the *Inter-*

---

†Donald Murray's code is related to CCITT Alphabet No. 2; Baudot's work produced a different code structure which resulted in CCITT Alphabet No. 1. There is little resemblance between the two codes except that they both used 5 bits per character.

Letters shift	Figures shift	Number of symbol	5-unit international code No. 2 (used by telex machines)
A	—	1	Z Z A A A
B	?	2	Z A A Z Z
C	:	3	A Z Z Z A
D	Who are you?	4	Z A A Z A
E	3	5	Z A A A A
F	*	6	Z A Z Z A
G	*	7	A Z A Z Z
H	*	8	A A Z A Z
I	8	9	A Z Z A A
J	Bell	10	Z Z A Z A
K	(	11	Z Z Z Z A
L	)	12	A Z A A Z
M	.	13	A A Z Z Z
N	,	14	A A Z Z A
O	9	15	A A A Z Z
P	0	16	A Z Z A Z
Q	1	17	Z Z Z A Z
R	4	18	A Z A Z A
S	'	19	Z A Z A A
T	5	20	A A A A Z
U	7	21	Z Z Z A A
V	=	22	A Z Z Z Z
W	2	23	Z Z A A Z
X	/	24	Z A Z Z Z
Y	6	25	Z A Z A Z
Z	+	26	Z A A A Z
< Carriage return		27	A A A Z A
≡ Line feed		28	A Z A A A
↓ Letters		29	Z Z Z Z Z
↑ Figures		30	Z Z A Z Z
■ Space		31	A A Z A A
(Not used)		32	A A A A A
Signal repetition			—
Signal α			(Permanent A polarity)
Signal β			(Permanent Z polarity)

*: Not used internationally; reserved for national allocation.

EQUIVALENT MEANINGS OF THE ABOVE A AND Z CONDITIONS

	A Condition	Z Condition
Bits:	0	1
Start-stop code:	Space (start condition)	Mark (stop condition)
Holes (perforations) in paper tape:	No hole	Hole
Single-current signaling:	No voltage	+ ve voltage
Double-current signaling:	− ve voltage	+ ve voltage
Amplitude modulation*	Tone-off	Tone-on
Frequency modulation*	High frequency	Low frequency
Phase modulation with* reference phase	Opposite phase to reference phase	Reference phase
Differential phase* modulation	Inversion of the phase	No inversion

Figure 3.6 International telegraph alphabets, CCITT recommendations R44 and S12. The 5-bit alphabet is the code used by Telex machines.

*: Explained in Chapter 11.

national Alphabet No. 5. This is a 7-bit code of which the U.S. ASCII code (American Standard Code for Information Interchange — pronounced "askey") is a version.

BITS PER   If we transmit n bits such as the ON and OFF pulses
CHARACTER   of telegraphy, we can, in theory, code 2^n different com-
                          binations into these pulses. The "characters" that are
sent by data transmission often contain 5, 6, 7, or 8 bits. Five bits can give 32 different characters, 6 bits, 64; 7 bits, 128; and 8 bits, 256. Some of the combinations, however, are reserved for special-purpose *control characters,* which have such functions as indicating the end of a record, making a teleprinter carriage return, and a variety of other operations. An n bit code may therefore transmit substantially less than 2^n characters.

ESCAPE   Telegraphy has commonly used a 5-bit code, as shown
CHARACTERS   in Fig. 3.6. Such a code gives only 32 bit combina-
                          tions; however, letters-shift and figures-shift charac-
ters are used to extend the range of characters. When a *figures-shift* character is sent, the characters that follow it are uppercase characters until a *letters-shift* character is sent. Similarly, the characters following a *letters shift* are letters until a *figures shift* is sent. The *letters-shift* and *figures-shift* characters must be recognized in either case. Figure 3.7 shows the use of *figures shift* and *letters shift.*

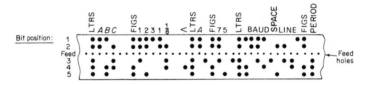

Figure 3.7 Five-channel paper tape illustrating the use of figures and letters shift in telex code (CCITT alphabet number 2).

This is an example of what is known as an *escape* mechanism in a code. The *letters-shift* and *figures-shift* characters are referred to as *escape* characters. By using escape characters or combinations of characters, the total number of possible characters in a code can be greatly increased. Sometimes the escape character changes the meaning of all the other characters following it until an additional escape character is received, as in the above example. In other codes, the escape character changes the meaning of only the one character that follows it. Thus special characters can be inserted at the wish of the user.

An important factor in the transmission of computer data is that the computer may have to send *every possible combination* of bits. This situation occurs in the transmission of both data and programs. It is not possible with the code shown in Fig. 3.6; it is possible using the character DLE (Data Link Escape) or TC (Transmission Control) in the codes related to the CCITT Alphabet No. 5, shown in Figs. 3.9 to 3.11.

EQUIVALENT REPRESENTATIONS

The *A* and *Z* conditions shown in Fig. 3.6 could be interpreted as meaning *zero* and *one* bit, respectively, and represented by voltage changes as shown in Fig. 3.4. Alternatively, the *A* condition could be a hole in paper tape and the *Z* condition *no hole,* as shown in Fig. 3.7. There are several other binary conditions which

Figure 3.8 North American uses of the 5-bit alphabet which differ in the *figures* shift.

● Denotes positive current or the Z condition of Fig 3–6

Code signals:

Start	1	2	3	4	5	Stop	Lower case	Upper case: CCITT standard international telegraph alphabet No.2 **used for telex**	Upper case: **North American** teletype commercial keyboard	Upper case: A T & T fractions keyboard	Upper case: Weather keyboard
	●	●				●	A	–	–	–	↑
	●			●	●	●	B	?	?	$\frac{5}{8}$	⊕
		●	●	●		●	C	:	:	$\frac{1}{8}$	○
	●			●		●	D	Who are you?	$	$	↗
	●					●	E	3	3	3	3
	●		●	●		●	F	Note 1	!	$\frac{1}{4}$	→
		●		●	●	●	G	Note 1	&	&	↘
			●		●	●	H	Note 1	#		↓
		●	●			●	I	8	8	8	8
	●	●		●		●	J	Bell	Bell	'	↙
	●	●	●	●		●	K	(	(	$\frac{1}{2}$	←
		●			●	●	L	)	)	$\frac{3}{4}$	↖
			●	●	●	●	M	.	.	.	.
			●	●		●	N	,	,	$\frac{7}{8}$	⦶
				●	●	●	O	9	9	9	9
		●	●		●	●	P	0	0	0	ø
	●	●	●		●	●	Q	1	1	1	1
		●		●		●	R	4	4	4	4
	●		●			●	S	,	,	Bell	Bell
					●	●	T	5	5	5	5
	●	●	●			●	U	7	7	7	7
		●	●	●	●	●	V	=	;	$\frac{3}{8}$	◯
	●	●			●	●	W	2	2	2	2
	●		●	●	●	●	X	/	/	/	/
	●		●		●	●	Y	6	6	6	6
	●				●	●	Z	+	"	"	+
						●	Blank				–
	●	●	●	●	●	●	Letters shift ↓				
	●	●		●	●	●	Figures shift ↑				
			●			●	Space ■				
				●		●	Carriage return <				
		●				●	Line feed ≡				

Note 1. Not allocated internationally; available to each country for internal use.

the CCITT recommends should be regarded as equivalent, and these are shown at the bottom of Fig. 3.6.

The codes in Fig. 3.6 are normally transmitted with *start-stop* transmission, the start bit being an *A* condition and the stop bits being a *Z*, as shown in Fig. 3.5.

North American practice in telecommunications frequently differs slightly from the rest of the world, often for the reason that the North American carriers led the world and then had too much invested to change when the world formulated standards. There are some minor differences between the telegraph codes used in North America and elsewhere. Figure 3.8 shows the American varients of CCITT Alphabet No. 2. The Western Union teleprinter code differs from the international code in the following uppercase characters:

	Uppercase Character	
Letter	*Western Union*	*CCITT*
D	$	Who are you?
F	!	Private allocation
G	&	Private allocation
H	#	Private allocation
V	;	=
Z	"	+

Some telegraph services use sequences of characters for certain functions, for example, to indicate the start or end of a message. Table 3.3 lists the sequences in common use. These are standardized sequences, again following a CCITT recommendation.

Table 3.3   The use of sequences of telegraph characters for special purposes.

Purpose of Sequence	Sequence	Notes
Start of message	ZCZC	Used in retransmission systems and store-and-
End of telegram	ZZZZ or ++++	forward devices
End of message	NNNN	Used to prevent the retransmission of delay signals
Suppression of delay signals	HHHH	used for error-corrected radio-telegraph channels
Connection of reperforator or equivalent device	CCCC	Switching into the circuit by remote control a reperforator or equivalent device
Disconnection of reperforator or equivalent device	FFFF	Switching out of the circuit by remote control a reperforator or equivalent device.
Connection of data equipment	SSSS	Switching into the circuit by remote control data transmission equipment
Ready-for-test signal	KKKK	For automatic tests of transmission quality

Bit positions 5, 6, 7:

Bit positions 1, 2, 3, 4:		000	100	010	110	001	101	011	111
		0	1	2	3	4	5	6	7
0000	0	NUL	TC₇ (DLE)	SP	0	③	P	' ④	p
1000	1	TC₁ (SOH)	DC₁	!	1	A	Q	a	q
0100	2	TC₂ (STX)	DC₂	" ⑥	2	B	R	b	r
1100	3	TC₃ (ETX)	DC₃	£(#) ②	3	C	S	c	s
0010	4	TC₄ (EOT)	DC₄	$ ¤ ②	4	D	T	d	t
1010	5	TC₅ (ENQ)	TC₈ (NAK)	%	5	E	U	e	u
0110	6	TC₆ (ACK)	TC₉ (SYN)	&	6	F	V	f	v
1110	7	BEL	TC₁₀ (ETB)	' ⑥	7	G	W	g	w
0001	8	FE₀ (BS)	CAN	(	8	H	X	h	x
1001	9	FE₁ (HT)	EM	)	9	I	Y	i	y
0101	10	FE₂ (LF) ①	SUB	*	:	J	Z	j	z
1101	11	FE₃ (VT) ①	ESC	+	;	K	③	k	③
0011	12	FE₄ (FF) ①	IS₄ (FS)	, ⑥	<	L	③	l	③
1011	13	FE₅ (CR) ①	IS₃ (GS)	–	=	M	③	m	③
0111	14	SO	IS₂ (RS)	.	>	N	^ ④⑥	n	④⑤
1111	15	SI	IS₁ (US)	/	?	O	–	o	DEL

Note: The 10 shaded positions can be allocated to different characters in different <u>versions</u> of the code

Positions 2/3 and 2/4 have alternate meanings of £ or #, and $ or ¤.
See notes on following page.

Figure 3.9 CCITT alphabet number 5. (Reproduced from CCITT Recommendation V.3.)

1. The Format Effectors are intended for equipment in which horizontal and vertical movements are effected separately. If equipment requires the action of carriage return to be combined with a vertical movement, the Format Effector for that vertical movement may be used to effect the combined movement. For example, if new line (symbol NL, equivalent to CR + LF) is required, FE_2 shall be used to represent it. This substitution requires agreement between the sender and the recipient of the data.

The use of these combined functions may be restricted for international transmission on general switched telecommunication networks (telegraph and telephone networks).

2. The symbol £ is assigned to position 2/3 and the symbol $ is assigned to position 2/4. In a situation where there is no requirement for the symbol £ the symbol # (number sign) may be used in position 2/3. Where there is no requirement for the symbol $ the symbol ¤ (currency sign) may be used in position 2/4. The chosen allocations of symbols to these positions for international information interchange should be agreed between the interested parties. It should be noted that, unless otherwise agreed between sender and recipient, the symbols £, $ or ¤ do not designate the currency of a specific country.

3. National use positions. The allocations of characters to these positions lies within the responsibility of national standardization bodies. These positions are primarily intended for alphabet extensions. If they are not required for that purpose, they may be used for symbols.

4. Positions 5/14, 6/0 and 7/14 are provided for the symbols upward arrow head, grave accent and overline. However, these positions may be used for other graphical characters when it is necessary to have 8, 9 or 10 positions for national use.

5. Position 7/14 is used for the graphic character ‾ (overline), the graphical representation of which may vary according to national use to represent ~ (tilde) or another diacritical sign provided that there is no risk of confusion with another graphic character included in the table.

6. The graphic characters in positions 2/2, 2/7, 2/12 and 5/14 have respectively the significance of quotation mark, apostrophe, comma and upward arrow head; however, these characters take on the significance of the diacritical signs diaeresis, acute accent, cedilla and circumflex accent when they are preceded or followed by the backspace character (0/8).

Figure 3.9   (Continued)

SEVEN-BIT CHARACTERS

Much data transmission now uses 7 bits (of data) per character. The CCITT Alphabet No. 5 is a widely accepted 7-bit code which permits minor national variations. It is shown in Fig. 3.9.

Ten characters can be varied, as indicated in Fig. 3.9, to produce *versions* or dialects of the CCITT Alphabet No. 5. Two important versions are shown in Figs. 3.10 and 3.11. Figure 3.10 shows a version for international transmission. Figure 3.11 shows the U.S. ASCII code, which is the most widely used code for data transmission and telegraphy in North America. ASCII is an ANSI (American National Standard Institution) standard.

Bit positions 5, 6, 7:

Bit positions 1, 2, 3, 4:		000	100	010	110	001	101	011	111
		0	1	2	3	4	5	6	7
0000	0	NUL	TC_7 (DLE)	SP	0	@	P	`	p
1000	1	TC_1 (SOH)	DC_1	!	1	A	Q	a	q
0100	2	TC_2 (STX)	DC_2	''	2	B	R	b	r
1100	3	TC_3 (ETX)	DC_3	#	3	C	S	c	s
0010	4	TC_4 (EOT)	DC_4	¤	4	D	T	d	t
1010	5	TC_5 (ENQ)	TC_8 (NAK)	%	5	E	U	e	u
0110	6	TC_6 (ACK)	TC_9 (SYN)	&	6	F	V	f	v
1110	7	BEL	TC_{10} (ETB)	'	7	G	W	g	w
0001	8	FE_0 (BS)	CAN	(	8	H	X	h	x
1001	9	FE_1 (HT)	EM	)	9	I	Y	i	y
0101	10	FE_2 (LF)	SUB	*	:	J	Z	j	z
1101	11	FE_3 (VT)	ESC	+	;	K	[	k	{
0011	12	FE_4 (FF)	IS_4 (FS)	,	<	L	\	l	|
1011	13	FE_5 (CR)	IS_3 (GS)	−	=	M	]	m	}
0111	14	SO	IS_2 (RS)	.	>	N	^	n	−
1111	15	SI	IS_1 (US)	/	?	O	_	o	DEL

Figure 3.10 A version of the CCITT alphabet number 5. The *International Reference* version for international interchange.

Bit positions 5, 6, 7:

Bit positions 1, 2, 3, 4:		000	100	010	110	001	101	011	111
		0	1	2	3	4	5	6	7
0000	0	NUL	DLE	SP	0	@	P	`	p
1000	1	SOH	DC1	!	1	A	Q	a	q
0100	2	STX	DC2	"	2	B	R	b	r
1100	3	ETX	DC3	#	3	C	S	c	s
0010	4	EOT	DC4	$	4	D	T	d	t
1010	5	ENQ	NAK	%	5	E	U	e	u
0110	6	ACK	SYN	&	6	F	V	f	v
1110	7	BEL	ETB	'	7	G	W	g	w
0001	8	BS	CAN	(	8	H	X	h	x
1001	9	HT	EM	)	9	I	Y	i	y
0101	10	LF	SUB	*	:	J	Z	j	z
1101	11	VT	ESC	+	;	K	[	k	{
0011	12	FF	FS	,	<	L	\	l	¦
1011	13	CR	GS	—	=	M	]	m	}
0111	14	SO	RS	.	>	N	^	n	~
1111	15	SI	US	/	?	O	_	o	DEL

Figure 3.11 The U. S. ASCII code. This is the U. S. national version of CCITT alphabet number 5. The control characters are explained opposite.

NUL (Null): No character. Used for filling in time or filling space on tape when there is no data.

SOH (Start of Heading): Used to indicate the start of a heading which may contain address or routing information.

STX (Start of Text): Used to indicate the start of the text and so also indicates the end of the heading.

ETX (End of Text): Used to terminate the text which was started with STX.

EOT (End of Transmission): Indicates the end of a transmission, which may have included one or more "texts" with their headings.

ENQ (Enquiry): A request for a response from a remote station. It may be used as a "WHO ARE YOU?" request for a station to identify itself.

ACK (Acknowledge): A character transmitted by a receiving device as an affirmation response to a sender. It is used as a positive response to polling messages.

BEL (Bell): Used when there is need to call human attention. It may control alarm or attention devices.

BS (Backspace): Indicates movement of the printing mechanism or display cursor backwards in one position.

HT (Horizontal Tab): Indicates movement of the printing mechanism or display cursor forward to the next preassigned "tab" or stopping position.

LF (Line Feed): Indicates movement of the printing mechanism or display cursor to the start of the next line.

VT (Vertical Tab): Indicates movement of the printing mechanism or display cursor to the next of a series of preassigned printing lines.

FF (Form Feed): Indicates movement of the printing mechanism or display cursor to the starting position of the next page, form, or screen.

CR (Carriage Return): Indicates movement of the printing mechanism or display cursor to the starting position of the same line.

SO (Shift Out): Indicates that the code combinations which follow shall be interpreted as *outside* of the standard character set until a SHIFT IN character is reached.

SI (Shift In): Indicates that the code combinations which follow shall be interpreted according to the standard character set.

DLE (Data Link Escape): A character which shall change the meaning of one or more contiguously following characters. It can provide supplementary controls, or permits the sending of data characters having any bit combination.

DC1, DC2, DC3 and DC4 (Device Controls): Characters for the control of ancillary devices or special terminal features.

NAK (Negative Acknowledgment): A character transmitted by a receiving device as a negative response to a sender. It is used as a negative response to polling messages.

SYN (Synchronous/Idle): Used as a synchronous transmission system to achieve synchronization. When no data is being sent a synchronous transmission system may send SYN characters continuously.

ETB (End of Transmission Block): Indicates the end of a block of data for communication purposes. It is used for blocking data where the block structure is not necessarily related to the processing format.

CAN (Cancel): Indicates that the data which precedes it in a message or block should be disregarded (usually because an error has been detected).

EM (End of Medium): Indicates the physical end of a card, tape or other medium, or the end of the required or used portion of the medium.

SUB (Substitute): Substituted for a character that is found to be erroneous or invalid.

ESC (Escape): A character intended to provide code extension in that it gives a specified number of contiguously following characters an alternate meaning.

FS (File Separator):	Information separators to be used
GS (Group Separator):	in an optional
RS (Record Separator):	manner except
US (United Separator):	that their hierarchy shall be FS (the most inclusive) to US (the least inclusive).

SP (Space): A nonprinting character used to separate words, or to move the the printing mechanism or display cursor forward by one position.

DEL (Delete): Used to obliterate unwanted characters (for example, on paper tape by punching a hole in *every* bit position).

REFERENCES

1. James Martin, *Design of Man-Computer Dialogues,* Prentice-Hall, Engle-wood Cliffs, N.J., 1973.

2. James Martin, *Systems Analysis for Data Transmission*, Prentice-Hall, Engle-wood Cliffs, N.J., 1972.

3. James Martin, *Introduction to Teleprocessing,* Prentice-Hall, Englewood Cliffs, N.J., 1972, Chapter 6.

4. *CCITT Recommendations C.7, C.8,* and *C.12,* International Telecommunications Union, Geneva, 1964.

5. *CCITT Recommendation V.3, Green Book,* Vol. VIII: Data Transmission, International Telecommunications Union, Geneva, 1973.

6. *International Standards Organization Standard ISO/R646.*

4 SYSTEMS WHICH USE DATA TRANSMISSION

Data transmission systems are built for a wide variety of purposes and differ accordingly in the way they function.

Probably the most common form of system in the future will be one in which people at terminals communicate with a distant computer. The computer will usually respond to them quickly. Often a dialogue takes place between the terminal user and the remote computer.

ON-LINE AND OFF-LINE SYSTEMS

Many systems are not interactive in nature but are required merely to move a quantity of data from one point to another. The communication links in this case may be on-line to a computer or off-line. On-line means that they go directly into the computer, with the computer controlling the transmission. Off-line means that telecommunication data do not go directly into the computer but are written onto magnetic tape or disk or are punched into paper tape or cards for later processing.

An on-line system may be defined as one in which the input data enter the computer directly from the point of origination and/or output data are transmitted directly to where they are used. The intermediate stages of punching data onto cards or paper tape or of writing magnetic tape or off-line printing are avoided.

INTERACTIVE AND NONINTERACTIVE SYSTEMS

Off-line systems are not "interactive." Because no computer is directly connected at the location the data are sent to, no data response will be received from that location, although simple control signals may be received to control the mechanical functioning of the devices and to indicate whether the transmission has been found free of errors.

Some on-line systems are also noninteractive. The computer may merely receive a batch transmission and may have no need to respond to it. Sometimes it may take a hash total at the end of a batch transmission to ensure that no data have been garbled, and so the only interactive response is confirmation of correct receipts of the transmission.

Most transmissions from human operators at terminals are interactive; in fact, it is bad design not to give a response to an operator and to leave him wondering whether his input has reached the computer or not. In inexpensive terminals, however, there may be no mechanism for responding. Such is the case with some factory data-collection terminals into which a worker may insert a machine-readable badge and set some keys or dials. The systems may respond with a light or mechanical action indicating simply that it has received the message.

For noninteractive systems or systems that give a very rudimentary response, the data will flow in one direction only. The transmission system is not normally designed to be entirely one-way because a small trickle of control signals

Figure 4.1 Common categories of data transmission systems.

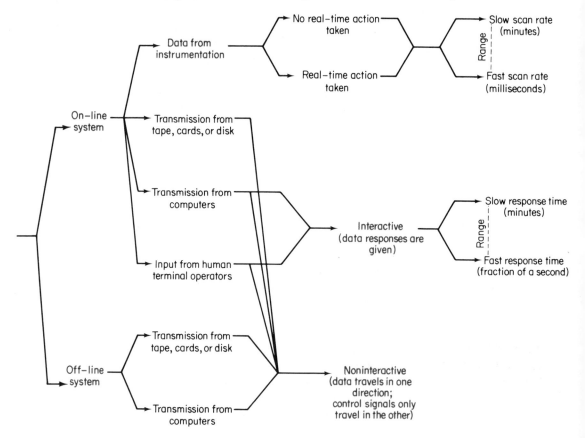

going in the other direction is needed. Occasionally, in telemetry, the use of radio makes two-way transmission difficult, and a purely one-way link is used. An interactive system, on the other hand, can have a high flow of data in both directions. Factors such as this will sometimes affect the data transmission techniques used.

Figure 4.1 lists some common categories of data transmission systems.

QUANTITY OF DATA TRANSMITTED

The quantity of data required to be transmitted varies enormously from one system to another. At one extreme we find whole files being transmitted, and at the other all that is needed is to indicate a YES/NO condition and one bit would suffice. Sometimes a single transaction is transmitted, and sometimes a batch of many transactions. When no immediate action is required on data that are gathered, the information may be collected and transmitted in a batch, which is less expensive than sending it a transaction at a time. Sometimes the contents of entire magnetic tapes or disks are sent. Payroll data or machine-shop schedules, for example, may be transmitted from one location to another. When a "dialogue" takes place between a terminal user and a computer, the message sizes then depend on the design structure of the dialogue (see Chapter 5). The input to the computer is one statement by the user, and the output is the computer's response, which may range from being a one-character confirmation to a screenful of data or a printed listing.

TIME FOR TRANSMISSION

Sometimes it is necessary to transmit the data quickly. The speed required depends on the system. A system for relaying one-way messages such as telegrams may be required to deliver the message in an hour or so. It would be convenient to have it done faster, but no major economic need to do so exists. When batches of data are sent for batch processing on a distant computer, a delivery time longer than 1 hour is sometimes acceptable. However, where a man-computer dialogue is taking place, the responses must be returned to the man sufficiently quickly so as not to impede his train of thought. Response times between 1 and 5 seconds are typical. In real-time systems in which a machine or process is being controlled, response times can vary from a few milliseconds to many minutes.

Response time for a terminal operator can be defined as *that time interval from the operator's pressing the last key of the input to the terminal's typing or displaying the first character of the response.* For different types of situations, response time can be defined similarly: *the interval between an event and the system's response to the event.*

Systems differ widely in their response-time requirements, and the response time needed can, in turn, have a major effect on the design of the data transmission networks.

Where systems are not interactive, we might specify a "delivery time" rather than a response time. Delivery time refers to situations in which the data are flowing in one direction and can be defined as *that time interval from the start of transmission at the transmitting terminal to the completion of reception at the receiving terminal.*

The question of how fast the response time must be can have a major effect on the cost of the communication subsystem. For a dialogue in which the operator must maintain a continuing chain of thought, responses within 2 seconds are usually needed. Where the input is a simple inquiry, or a request for a listing, or for a program to be run, the response time may be longer, sometimes much longer. Where machinery is being controlled, a faster response is sometimes necessary, and where one computer is requesting data from another computer, it may be needed very quickly so that the requesting computer itself can achieve a specified response time.

Figure 4.2 shows some of the common requirements for delivery time or response time and for quantities of data transmitted.

The block labeled "terminal dialogue systems," for example, indicates a response-time requirement of from 1 to 10 seconds and a message size ranging from 1 character (usually 7 or 8 bits) to about 4000 characters (around 30,000 bits). A few exceptional cases will extend beyond the block shown (as with all the blocks on the diagram).

The block labeled "terminal inquiry systems" extends on the time scale to 100 seconds. A wait of this duration would be quite unacceptable in a dialogue situation, but where a storekeeper or foreman (for example) is making a simple request for a listing, a wait of a minute may not cause much inconvenience. In all such cases, the cost of providing different response times must be related to the requirements of the user.

Systems for delivering messages or cables, commonly called *message-switching systems,* are referred to with the block labeled "message delivery systems." With the apparent deterioration of the mail services, it seems likely that facsimile mail transmission will grow in popularity. Certainly it will be the fastest way of delivering mail. Delivery times up to 1 day will often be acceptable for the bulk transmission of data. Bulk transmission of data extends to the extreme right of the diagram because sometimes very large quantities of data are sent.

There is not much point in using data transmission over a line so slow that the delivery time is more than 24 hours. In these cases, mail, delivery van, or air transportation is usually cheaper.

In the years ahead it is probable that data transmission will be used in areas outside the blocks shown in Fig. 4.2. To some extent the areas in which it has been used to date have been determined by the speeds of available transmission lines. The transmission speed equals the delivery time of one-way messages in Fig. 4.2 (vertical axis) divided by the number of bits transmitted (horizontal axis). For most of the applications shown in the figure, the speed of teletype and tele-

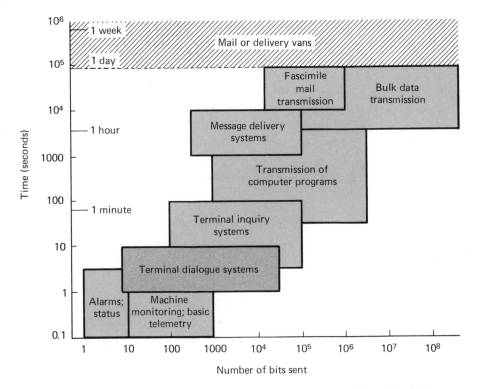

Figure 4.2　Desirable delivery times and typical quantities of data for common uses of data transmission. Terminal dialogue systems are the most common.

phone lines – speeds up to, say, 4800 bits per second – is sufficient. Higher-speed lines are required today mainly when a very large bulk of data must be sent.

The computer industry changes fast, and it seems probable that those areas in which a computer responds to a machine or another computer, rather than to a person, will require different transmission facilities. One computer may request from another distant one a program or portion of a file for immediate processing. One computer will pass jobs on to another because it does not have the capability to handle them itself, or for load-sharing reasons. The requirement for access time in obtaining data from disks – the time taken to "seek" and read a record – may in the future become the time requirement for obtaining data from a distant computer over telecommunication links. Small local computers will sometimes pass their compiling work on to bigger distant machines with better facilities. The initial program loading of small local computers will often be done by a large remote computer.

The computers and file units that are connected by cabling in the machine room today may be connected by telecommunication links tomorrow if channels of suitable speed are available.

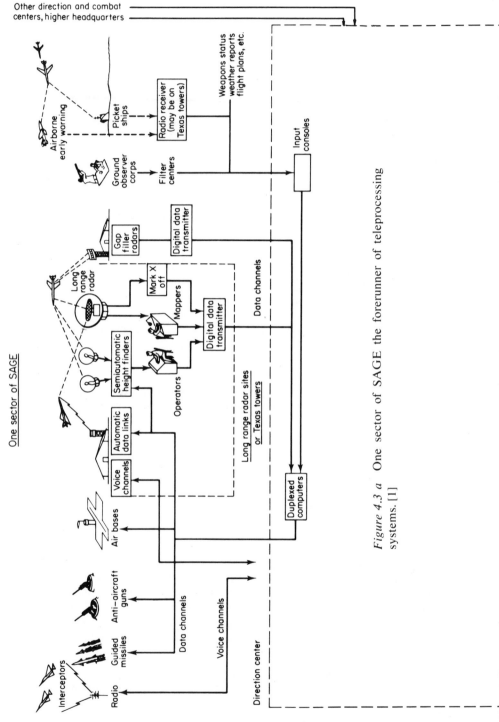

Figure 4.3 a One sector of SAGE the forerunner of teleprocessing systems. [1]

Situation display New England coastline.

Fourth floor of direction center.

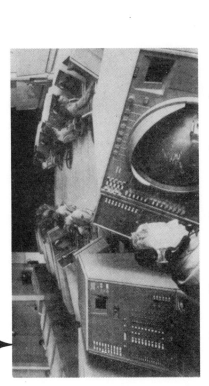

Air surveillance room. Operators using consoles attached to the computers.

Figure 4.3 b One floor of the direction center contained operational rooms for surveillance, identification, status input, weapons assignment and control, and command functions. In one room up to fifty operators used consoles such as those shown.

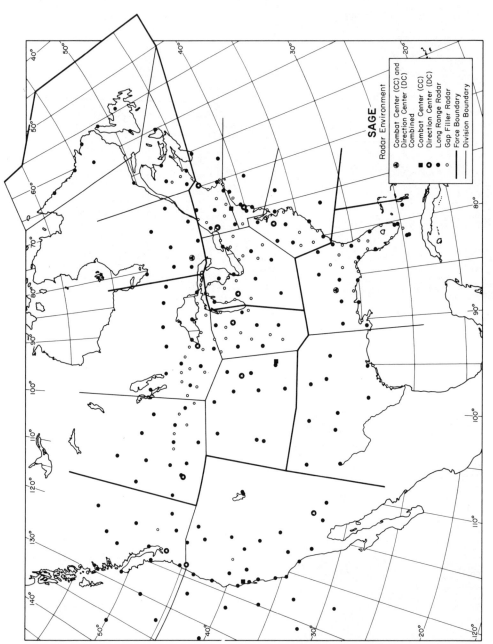

Figure 4.4 (Courtesy *SAGE*.)

MILITARY
SYSTEMS

To a large extent, the pioneering work for systems combining telecommunications terminal users and computers was done in the military. One of the first, and still one of the most spectacular, was SAGE, the U.S. Air Force Semi-Automatic Ground Environment, designed in the early 1950s to protect the United States from surprise air attack.

SAGE was designed to maintain a constant watch on the air space over North America, provide early warning of airborne attack, and give its Air Force operators the information needed for conducting an air battle. Its input came over data transmission links from a variety of radars which unceasingly swept the skies of the continent. These included the CADIN line (CAnaDian Integration North) and many other land-based radars. Input also came from observation aircraft, ground observer corps stations, and picket ships. The computers digested this constant stream of data and prepared displays for their output screens. In addition, the system contained aircraft flight plans, weapons and base status reports, weather data, and other information. Figure 4.3 gives a highly simplified diagram of one sector of SAGE, and Figure 4.4 gives the division of the United States into SAGE sectors with the approximate locations of the radar sites and computer centers. These locations were all connected to the computer, and the computers were interlinked by telegraph and telephone lines carrying digital data.

Air Force personnel sat in front of screens on console units designed for their different responsibilities. They communicated with the computer by using a light pen—a forerunner of the light pen on today's commercial systems. They could request displays of particular situations and could command the machines to compile special displays for their use. The computer, on its own initiative, would flash on the screen displays of particular urgency. If an unidentified aircraft approached, the system notified command personnel, who would then use the screen to investigate and, if necessary, dispatch interceptor aircraft or activate weapons.

Since SAGE, many command and control systems have been installed and planned. The U.S. Marine Corps built the Marine Tactical Data System (MTDS), which has in effect a mobile SAGE, transportable to an operational theater by helicopters. The U.S. Navy built the Naval Tactical Data System (NTDS), which had computers and screens on board ships linked together on a worldwide basis to form a vast mobile command and control system. The Pentagon could obtain on its screens positions and the status of its ships anywhere, and could display data that were being used by the Navy on the other side of the world.

The U.S. Army, meanwhile, planned a command and control system of immense complexity to be used by every level of command and for handling intelligence and logistics information. Information on tactical operations, fire support, personnel, and other administrative matters was kept up to date in the system, and the output was available either on individual consoles or large group displays.

The BMEWS system (Ballistic Missile Early Warning System) uses giant radar units for the detection of potentially hostile missiles aimed at the West. Its duplexed computers in the United States and England scan the radar signals they receive ceaselessly and notify appropriate military personnel if they conclude that an attack might have been launched. It is these computers that would give us 20 minutes of grace prior to nuclear devastation, in which time we could launch our counterattack and do whatever else needed to be done in that interval.

More spectacular in its technology is the ABM (Anti Ballistic Missile) System in which computers attempt to shoot down the missiles as they scatter multiple warheads toward their targets at speeds close to 20,000 miles per hour. The ABM is deployed only in one or two locations. The missiles attempt to deceive the ABM by sending out many "decoys" to draw attention from the real warheads, and ABM tries to compute which are real and which are decoys.

The Pentagon wished to be on-line to these and the many other military systems, so that the President and Chiefs of Staff can display details of all action. The task of linking all the systems to a central information system is one of Herculean magnitude because of the incompatibilities among their different techniques, record formats, and means of display. Telecommunications networks, gigantic in scale, are used by the military with many more communications satellites than used by civilian authorities.

In addition to the communication links connecting the various military systems, the U.S. Defense Communications Agency has two vast telecommunications networks: AUTOVON for voice, and AUTODIN for data. These were first installed internally in the United States but then became part of a rapidly developing global system. The networks must give precedence to priority traffic. When a general telephones, his call must, if necessary, preempt lower-priority calls. The system must be able to handle the unusually high traffic volumes that may occur in times of emergency. It must present no unusually attractive target to potential attack, and so its facilities are as well distributed as possible. Above all, it must have sufficiently widespread alternative routings to survive massive nuclear attack. A configuration of trunk groups in the form of multiple overlapping hexagons and diagonals links the many switching centers with many alternative paths.

No doubt the other side also makes spectacular use of computer systems, and the prospect of them grappling in some future life-and-death struggle with satellite jamming, decoys, software bugs, and attacks on computer centers may make the history books of the future complicated.

INDUSTRIAL SYSTEMS

As SAGE and its successors become operational, the vision of command and control systems in industry began to grow. Systems were perceived which would give information to all levels of management and help in the decision-making processes of running a factory on a far-flung corporate operation. The costs had to drop by a large factor before such systems were economically viable.

Piece by piece, however, interactive and off-line systems were built in industry, and then interlinked by telecommunications, as in Figure 4.5.

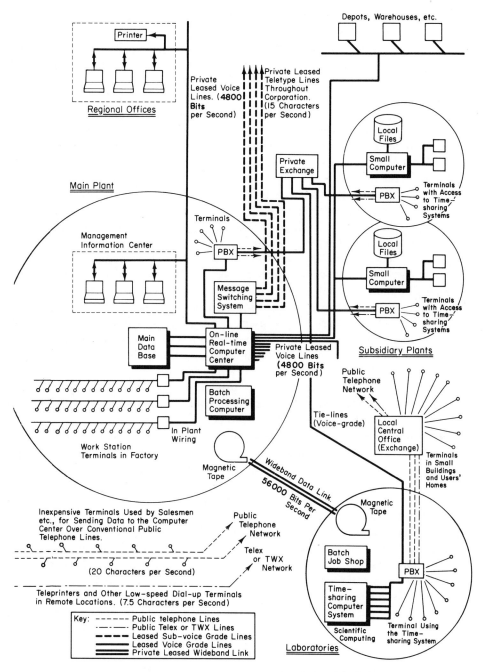

Regional Offices

Private Leased Voice Lines. (**4800 Bits** per Second)

Depots, Warehouses, etc.

Private Leased Teletype Lines Throughout Corporation. (15 Characters per Second)

Main Plant

Private Exchange

Local Files

Small Computer

PBX

Terminals with Access to Time-sharing Systems

Management Information Center

Terminals

PBX

Message Switching System

Local Files

Small Computer

PBX

Terminals with Access to Time-sharing Systems

Subsidiary Plants

Main Data Base

On-line Real-time Computer Center

Private Leased Voice Lines (4800 Bits per Second)

Batch Processing Computer

In Plant Wiring

Work Station Terminals in Factory

Tie-lines (Voice-grade)

Public Telephone Network

Local Central Office (Exchange)

Terminals in Small Buildings and Users' Homes

Magnetic Tape

Wideband Data Link 56000 Bits Per Second

Magnetic Tape

Inexpensive Terminals Used by Salesmen etc., for Sending Data to the Computer Center Over Conventional Public Telephone Lines.

Public Telephone Network

Telex or TWX Network

(20 Characters per Second)

Batch Job Shop

Time-sharing Computer System

Scientific Computing

PBX

Terminal Using the Time-sharing System

Teleprinters and Other Low-speed Dial-up Terminals in Remote Locations. (7.5 Characters per Second)

Key: ----- Public telephone Lines
      -·-·- Public Telex or TWX Lines
      ----- Leased Sub-voice Grade Lines
      ━━━━ Leased Voice Grade Lines
      ══════ Private Leased Wideband Link

Laboratories

Figure 4.5 Large computer center in main plant with links to computers in subsidiaries and laboratory. An in-plant data collection system and on-line terminals in sales offices and depots.

Whereas some computer systems are global in scale, others use communication links within one building only. Some extend to distant locations: other factories, warehouses, and head offices; terminals are placed in branches and sales offices. The functions of these systems vary enormously from one system to another. Corporations differ widely in their methods, their needs, and their attitude toward automation.

The events in a production cycle are never exactly predictable, and in order to maintain control over the various stages in the production process, data about goods received, operations completed, items that transfer location, and so on, are fed back into the computer. The computer constructs and updates records of the status of the events that must be controlled. These may be records giving the levels of stock and details relating to the reordering of stock, records giving the status of each customer order, records giving details of the loading of each machine, and records relating to the work given to each employee. In general, each type of facility that is employed in the production process may have a record giving its current status or loading, and the jobs to be done at each work center may be recorded with indications of priority. These records are kept up to date. They may be used either by the computer or by management and foremen in planning, scheduling, progress chasing, reordering, and so on. The computer uses them for such functions as producing work tickets giving instructions to operators, giving loading instructions, handling new orders, and notification of exceptional conditions or situations that have become critical and need expediting. Management uses them for displaying the facts they need.

This type of mechanism may be used in the operation not only of a manufacturing process but of any environment in which a large number of discrete events must be controlled in order to achieve certain results.

In a railroad, for example, the movement and use of the cars must be constantly rescheduled to convey the maximum quantity of goods in an acceptable time with the minimum cost. Details of each car, its contents, its actual movements, and its future intended movements are stored in the computer files, and these records used by the computer or by management, or more probably by both, to optimize the operation. In an airline, the scheduling of aircraft operations, crew movements, and maintenance is a severe problem aggravated by circumstances that cannot be planned for in detail, such as the need to ground aircraft for unscheduled maintenance, crew changes due to sickness, fog, and other holdups.

The transmission and processing techniques required depend on the time scale on which events take place. The time for job completion, for example, varies widely from one industry to another. In some firms the main criterion is using the machines fully. In others it is desirable to rush orders through the production shops as quickly as possible.

In many production shops, the events that must be controlled occur in rapid succession even though individual jobs take several hours. The control system that governs them must be at least as dynamic as the events it is controlling. A typical system needs to respond in seconds to human inquiries and data input.

Rescheduling, new job allocation, or other response to events may be fast enough if it happens in half an hour, or in some factories, half a day.

Basically, the following elements are needed in controlling a factory shop floor:

1. A means of deciding what parts are to be made and by when. This is obtained by breaking down the orders or forecast and comparing this with what is in stock.
2. A means of scheduling: building a schedule of work for each tool and worker. These schedules will have to be constantly modified as events take place.
3. A means of issuing the orders to the shop floor.
4. A means of collecting data about the status of events on the shop floor.
5. A set of records giving the current status of orders, schedules, jobs, machines, and workers.
6. A means of communicating data to management when they are needed for planning purposes.
7. A means of dealing with exceptional conditions and sorting out things that go wrong. This must never be deemphasized. It is folly to imagine that when a computer is in control events will run like clockwork.

One could enlarge this list considerably by extending it into other areas beyond the shop-floor operation.

One of the first steps into telecommunications is often to put terminals for data collection near each work station on the shop floor. Using these terminals the worker reports when he starts and completes a job. If he leaves his work, he reports the percentage completion of the job. He identifies himself by inserting his badge into the machine. The jobs and the machine tools may also have machine-readable means of identification. He may also enter details such as scrap values, stock shortages, and so on.

Many exceptional conditions arise such as material running out, tools breaking, and the workers entering invalid data. These are reported immediately to staff in a control room, which has one or more representatives on the shop floor who can walk over to the work stations to sort out any problems. The control-room staff alerts the shop-floor representative by radio, and he goes to the relevant work station. From the work station he can talk by telephone to his control-room partner, who sits at a computer terminal and can interrogate the system.

The dispatching of work to the shop floor and the giving of instructions to workers might also take place via terminals. When one job is completed the computer issues the next job for that work station, always keeping at least two jobs ahead in order to give the worker or foreman time for preparation. Again the control room is an essential buffer needed to take care of errors and other problems.

The control-room staff also carries on functions such as answering management inquiries, maintaining file integrity by correcting errors if they are discovered, modifying schedules, and perhaps in an advanced system testing the effect of

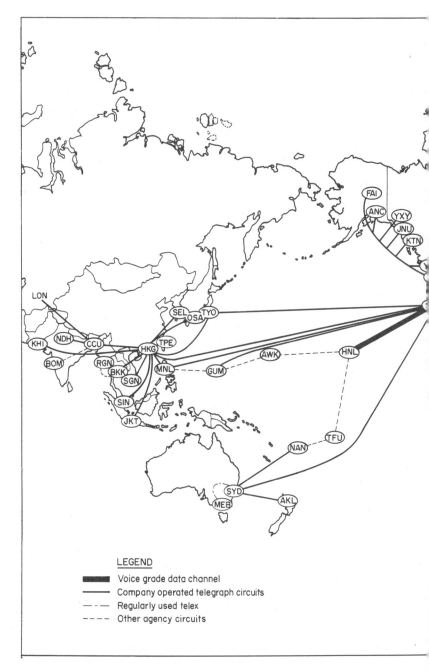

Figure 4.6 Pan American's world-wide communications network as used with their Panamac computer system before the advent of satellite communications. (Courtesy Pan American.)

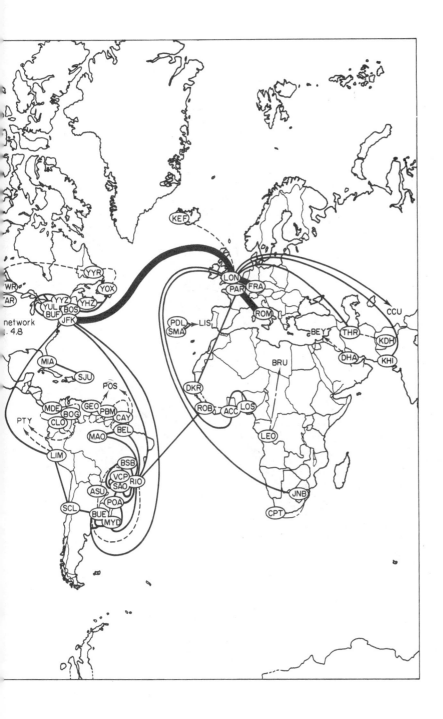

network
4.8

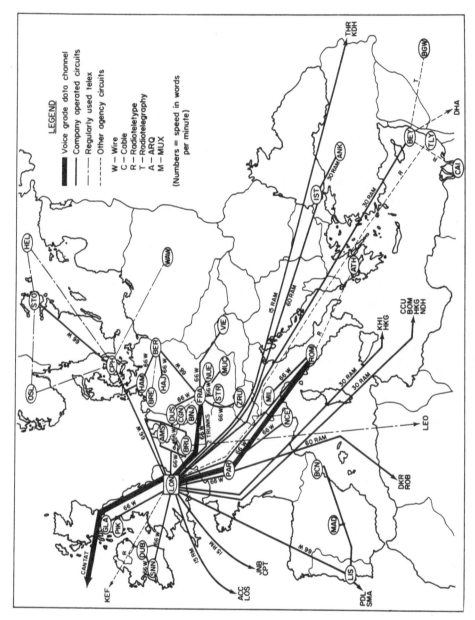

Figure 4.7 Pan American's world-wide communications network in Europe. (Courtesy Pan American.)

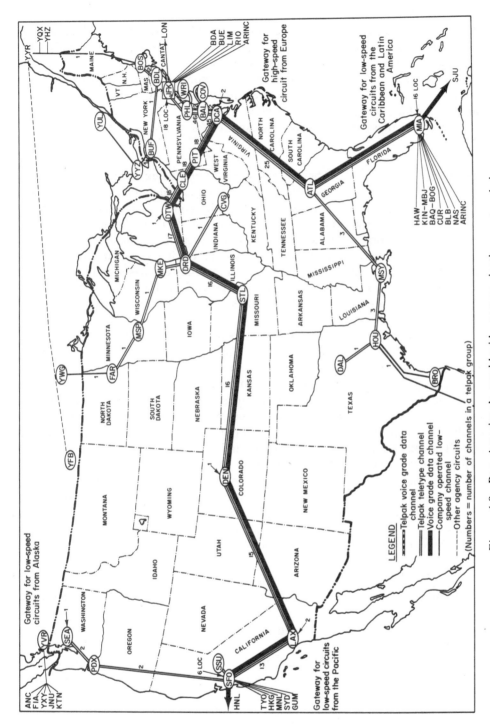

Figure 4.8 Pan American's world-wide communications network in America. (Courtesy Pan American.)

85

possible management decisions using techniques such as simulation. Management might want to know, for example, what the effect on other orders will be if a certain large order is accepted or brought forward.

A typical control room might have a small number of printers, perhaps teleprinters, on which the exception notices are printed, and some terminals with cathode-ray tube screens for fast display of records, schedules, and so on. The controllers use these terminals for changing data in the system.

AIRLINE
RESERVATIONS

There are many systems using computers and telecommunications for special functions such as banking, transferring funds electronically, and booking hotels.

Spectacular use of telecommunications is found in worldwide airline reservation systems. Some of these have a thousand or more terminals, all obtaining very quick responses on communication links that circle much of the world. They employ most of the different communication facilities discussed in this book.

Figure 4.6 shows the worldwide data network of Pan American in a simplified form. Figures 4.7 and 4.8 show details of the American and European parts of that network.

The real-time terminals of Pan American are on leased voice-grade lines with many terminals connected via concentrators to one line. The voice-grade lines in Fig. 4.6 cover America and Europe and extend from Honolulu to San Juan. To take them farther than this would have been uneconomical. The rest of the world is reached, therefore, by leased telegraph lines and telex. These lower-speed circuits often travel their long distances by radio and where this is so have a much higher proportion of data errors than the voice-grade lines or land telegraph lines. The telegraph circuits are used for administrative messages of all types, as well as for reservations. They go through message-switching centers some of which are manually operated or semiautomatic, and this means that there is a delay in some reservation signals reaching the computer. Special procedures therefore have to be programmed for the offices that do not have conversational terminals.

There are three main objectives in designing a reservation system:

1. To improve the service given to passengers and potential passengers.
2. To save staff in sales offices and in control offices where reservations are processed and space on aircraft controlled.
3. To improve the load factor on flights.

When a system handles cargo reservations, the objectives here are to save staff and to optimize the loading of cargo. Many airline systems carry out functions other than reservations, including crew scheduling, passenger check-in at airports, load and trim calculations prior to a plane's takeoff, scheduling of maintenance and breakdown services, and so on. Box 4.1 lists a typical set of functions.

Most reservation systems use terminals with keyboards and screens. When a passenger wishing to make a booking telephones the airline or walks into its office, he talks to an agent operating such a terminal. The agent does not use a timetable but obtains information about possible flights on his screen. The passenger might say, for example, that he wants to travel to Los Angeles on March 25 at about 5 P.M. The agent keys in this fact in a coded form, and on the screen appears the four best flights with seats to fill the requirement. If for some reason none of these is suitable, it can be instructed to scan outwards from this time and produce the next four, and so on. The screen tells the agent fares and all the other details he needs to know.

When it has been agreed on which flight the passenger will travel, the agent then enters details about the booking into the system. The passenger's name, home and business telephone numbers, ticketing arrangements, and other pertinent information are sent to the computer and stored on its disk files.

The computer checks that all the pertinent details have been entered. It makes various checks for validity. It may compare the name with other names booked to ensure that it is not a duplicate booking. When all the details are entered correctly and checked, the computer stores this information in the relevant files.

Passengers may call later to confirm, change, or cancel a booking. This happens with the majority of seats sold. While a passenger is on the telephone, his details will be retrieved from the system, and the agent will make the desired changes. The appropriate seat inventory will be reduced when cancelations or changes are made. The system will check that this is a valid reduction: that the seat being canceled has in fact been booked. Tight checks and controls ensure that most of the errors which occur on a noncomputerized airline reservation cannot happen here.

The response time of the system to each agent's action must be low enough to permit efficient "conversation" with the computer while talking to a passenger on the telephone. The contract on a typical system says that 90% of all messages must have a response of less than 3 seconds. In fact, on most working systems, the majority of messages receive a reply in a second or two.

INTERLINKED SYSTEMS

When many on-line systems exist for functions such as banking and ticket reservations, the next logical step is for separate systems to be interlinked. An airline agent should be able to interrogate the reservation systems of other airlines so that multi-airline journeys can be planned and booked. A travel agent should have access to systems which book hotels, cars, theater tickets, boats, trains, and airline seats. An electronic fund transfer system should interlink the computers of many banks. Computers doing corporate payrolls should be able to send the money electronically to the computers of employees' banks, via a clearing system. Computerized cash registers in supermarkets and stores have been connected by telecommunications to bank computers. A society is evolving which will employ vast networks of machines connected by telecommunications.

BOX 4.1   Functions of an airline reservation system

Function 1. Giving flight information to distant sales points, and especially answering requests about what seats are available.

Function 2. Centralized inventory control of seats booked and canceled at distant locations. This has been done without passengers' names being used, but to be done efficiently these are needed to eliminate duplicate bookings and invalid cancellations.

Function 3. Control of space allocation to offices not on-line to the computer or offices of other airlines. Small offices or offices in remote parts of the world, to which leased voice-grade lines are still too expensive, are likely to be off-line. They send telegraph signals to the computer, and these may take many hours to reach it. From a knowledge of the booking patterns on given routes, allotments of seats to be sold by various offices must be set, and as the bookings build up "status" limits must be set.

Function 4. Mechanization of passenger files. These have previously been maintained manually in the sales offices. To keep them in the files of the distant computer and maintain them in a real-time manner, using the facilities needed for the preceding functions, will give major labor cost savings and improve the service given to passengers.

Function 5. Waitlisting, reconfirmation, checking ticket time limits, and other operations concerned with manipulation of the passenger files. If, for example, a passenger does not reconfirm or collect his ticket when he should, the computer notifies the appropriate sales office.

Function 6. Provision of special facilities for the passenger such as renting a car or booking hotels in a distant location, providing a baby carrier, wheelchair, facilities for pets, and so on.

Function 7. Message switching. Airlines need large message-switching centers for routing off-line teletype bookings and other messages to the control points. Many of the teletype booking messages will be from other airlines. A computer can be used for this as part of the reservation system.

Function 8. Passenger check-in at airports. Manifests giving details of passengers are sent to the airport check-in desks. The check-in clerks have the facility to make inquiries on the distant computer and use this in allocating seats.

Function 9. Load and trim calculations done in a real-time fashion just before the airplane takes off can be combined with passenger check-in. Weights of passengers' baggage and estimated weights of passengers

BOX 4.1 (Continued)

are used for the calculations. The results determine last-minute acceptance of cargo and passengers.

Function 10. Cargo reservations may also be controlled by the system, and the weight and approximate volume of cargo used in the load and trim calculations.

Not all these functions are mechanized on all airline reservation systems. Today, however, it is probably economical to put all of them on the same system in large airlines.

TIME-SHARING SYSTEMS

Most systems with manually operated terminals are time-shared, meaning that more than one user is using them at the same time. When the machine pauses in the processing of one user's item, it switches its attention to another user.

The term "time sharing," however, is commonly used to refer to a system in which the users are *independent,* each using the terminal as though it were the console of a computer and composing, testing, and executing programs of his own at the terminal. Each user feels as though he were the only person using the system. The programs of one terminal user are quite unrelated to those of other users.

In many systems the users do not program the system at the terminal, and neither are the users independent. They are each *using* the programs in the computer in a related manner. They may, for example, be insurance or railroad clerks possibly using the same programs and possibly the same file areas. In many time-shared systems, however, the computing facility is being divided between separate users who can program *whatever they wish* on it, independently of one another. Four categories of on-line systems, differing by the degree of independence of the users, are in common use:

1. Systems that carry out a carefully specified and limited function, for example, banking systems or airline reservation systems in which all terminal users can update the same files.

2. Systems for a specific limited function in which the user has personal independent files or in which shared files can be read but not updated by general terminal users.

3. Systems in which programmers can program anything they wish at the terminals, providing they all use the same language, an interpreter or compiler for this being in the machine. Normally each user has the same type of terminal.

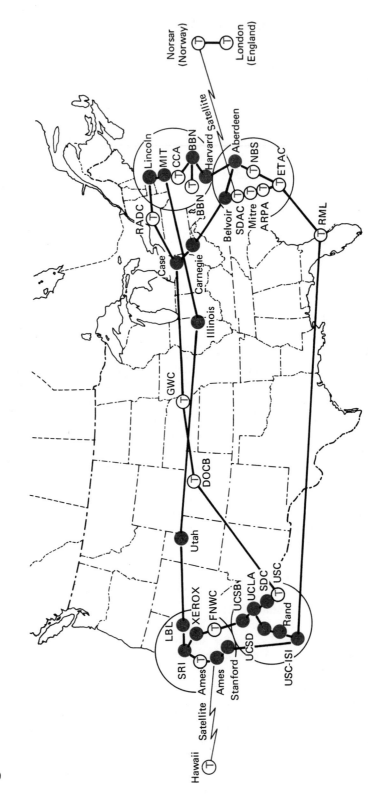

Figure 4.9 a Geographical layout of the ARPA network, 1974.

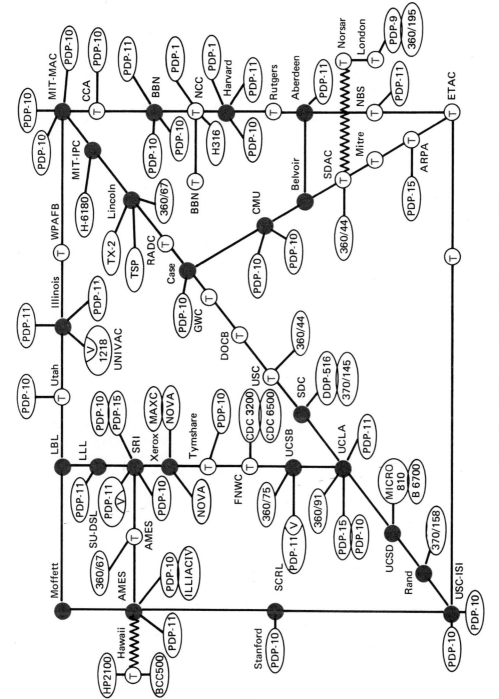

Figure 4.9 b Logical layout of the ARPA network, 1974.

91

4. Systems in which programmers can program anything by using a variety of different languages. Often a wide variety of terminals is permitted also.

The last two types of systems are, in effect, dividing up the computing facility "timewise" and giving the pieces to different programmers to do what they want with the pieces. The files are also divided according to use. It cannot be known beforehand how much space in the files the various users will occupy. This is quite different from category 1, in which the file size and organization are planned in detail.

TIME-SHARING COMMUNICATION LINES   The reason time sharing is so important in computers is that human keying rate and reading rate are very much slower than computer speeds. Furthermore, the human being requires lengthy pauses to think between transactions. The key to using the computer efficiently for real-time dialogue operations is to make it divide its time between many users. *The same is true with telecommunications.* A voice line could transmit many more bits than a terminal operator generally employs in an alphanumeric dialogue with a computer. The key to using it efficiently is to time-share the line between many users.

The need to share communication lines will become greater in the future, as facilities capable of transmitting higher bit rates will be employed. Today's telephone lines are commonly made to carry data rates of 2400 to 9600 bits per second. However, on the Bell System T1 carrier, of which millions of miles are already installed, the telephone voice channel is equivalent to 56,000 bits per second (as discussed in Chapter 14).

The degree of independence of users on a shared communication line varies. On some shared lines, all users must employ the same terminal type with the same line-control procedure. On others, they can be different but must use the same character code. With some systems they can be entirely independent, using different codes. The users may all be of one type, in one organization—reservation agents linked to an airline system, for example. On the other hand, they may be different types of users but may share communication facilities installed within one organization. Finally they might be entirely independent users sharing public data transmission facilities. It is in the latter case that the largest savings may result, and eventually public data networks will play a very important part in a nation's data processing.

RESOURCE-SHARING NETWORKS   "Time sharing" often refers to the sharing of a single computer. The user's terminal may, however, be attached not to one computer but to a data transmission network which links many diverse resources. The network may provide the user with rapid access to distant computers, programs, libraries, information-retrieval systems, data bases, message-delivery facilities, mail-printing facilities, and so on. The cost of small computers is dropping sufficiently that the reason for using telecommunications is often not to gain access to a computer but to gain access to these other resources.

One of the best known resource-sharing networks is ARPANET, a network using packet switching (discussed in Chapter 24) to give its users fast access to a wide variety of resources. The users time-share both the telecommunications and the distant resources they employ. ARPANET is illustrated in Fig. 4.9, and its growth is shown in Fig. 24.6. It was constructed with Department of Defense funds to interlink major university and other computer centers with ARPA (Advanced Research Projects Agency) grants. A user sitting at a terminal anywhere on the ARPANET can gain access to any of the diverse collection of computers shown in Fig. 4.9, part b, and their far more diverse libraries of programs. The transmission time on the network is sufficiently fast that for most interactive uses it does not degrade the response time significantly.

The university professors who live with the ARPANET become very accustomed to employing its remote facilities and sending messages to one another. They complain of feeling isolated when they are at locations where they use terminals not linked to their network.

Possibly the way computing will evolve in the future is that expensive or important data-processing resources such as large computers, data bases, and program libraries will be designed primarily for interconnection to networks. They will be locked away in fireproof, theftproof, sabotageproof, operatorproof vaults and except for maintenance will be accessible only by the telecommunication cables that pass through the walls of the vaults.

But the users will have access to the libraries of the world and the world's most powerful data-processing facilities.

REFERENCE

1. R. R. Everett, C. A. Zroket, H. B. Benington, *SAGE—A Data Processing System for Air Defense*, Eastern Joint Computer Conference, 1957.

5 MAN-COMPUTER DIALOGUE

A system in which a man communicates with a distant computer in a conversational fashion is called an *interactive* system. While many applications of data transmission are for the one-way delivery of data, the fastest growing use of telecommunications is for interactive systems.

Interactive systems make different types of demands on the transmission and switching facilities to the traditional uses of telecommunications such as telephony, telegraphy, broadcasting, and batch data transmission. So different are these demands that new types of public networks are coming into existence in some countries to meet them. A whole new body of technology and theory is necessary, and is emerging.

Prior to the existence of new types of networks, interactive systems have to be built from today's telecommunications facilities, which are largely oriented to what the common carriers refer to as POTS—plain old telephone service. The ingrained wisdom of decades in the telecommunications administrations relates to POTS, and most administrations consider that their bread-and-butter revenue for the foreseeable future will come from POTS. It is often difficult for experienced telephone engineers to accept that man-computer conversations are very different from telephone conversations.

In this chapter we will give a collection of illustrations of man-computer conversations. Box 6.1 in the next chapter summarizes the difference between the requirements for telephone conversations and for computers and their users.

MAN-COMPUTER DIALOGUE   To bring the power of computers and the information in their data banks to the maximum number of people, careful attention must be paid to the man-machine interface. Increasingly, during the next decade, man must become the prime focus of computer systems design. The computer is there to serve him, to obtain information for him, and to help him do his job. The ease with which he communicates with it will determine the extent to which he uses it. Whether or not he uses it powerfully will depend on the man-machine language available to him and how well he is able to understand it. For the ordinary manager and for many other types of computer users, remarkably little has been done to provide an efficient man-machine interface as yet.

The interface is going to differ greatly from one man to another and from one machine to another. Different applications will need fundamentally different types of dialogue structures [1]. Some are very complex and need a high level of intelligence; others are simple. In some applications today, one can observe the "man-in-the-street" who has never touched a terminal before sitting down at one and carrying on a successful, if simple, dialogue with a computer. On the other hand, one also finds terminals being thrown out a few months after installation because the intended user never learned to communicate successfully with the system.

The man-computer dialogue should be a starting point in the design of the data communication facilities in many real-time systems. A systems analyst must design the structure of the dialogue — in other words, determine what the man says to the machine and what the machine says back to the man. In some cases a formal programming language is used, such as FORTRAN or BASIC. The majority of terminal users, however, are nonprogrammers; they are clerks, travel agents, brokers, factory workers, managers, and persons in all walks of life. Some are intelligent and highly trained; some are not. The dialogue they employ for communicating with a terminal must be more straightforward than a programming language and must be tailored to their psychology and ability.

"Dialogue," then is a generic word for a preplanned man-machine interaction; it encompasses formal programming languages, languages for interrogating a data base, and innumerable nonformal conversational interchanges many of which are designed for one specific application.

SIX EXAMPLES　　　　　Figures 5.1 to 5.6 give six typical examples of dia-
OF DIALOGUES　　　　　logues in which a man is using a distant computer.
　　　　　　　　　　　　They are chosen to illustrate types of dialogues in common use. The last of the six is less likely to take place over communication links than the other five, lergely because the communication channels in existence are not yet organized, in most cases, in a suitable fashion for it.

Dialogue Specimen 1 (Fig. 5.1)

The first example takes place at a teletype machine at the frustratingly slow speed of 10 characters per second. The operator is using a programming language, BASIC, to do a simple calculation in order to plot a set of curves of the function

$$P(T) = e^{-[(1-p)T/S]}$$

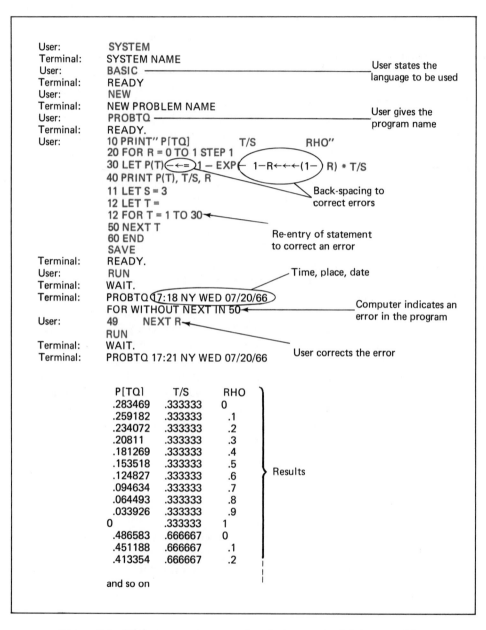

User:	SYSTEM		
Terminal:	SYSTEM NAME		
User:	BASIC ————————————————————		User states the language to be used
Terminal:	READY		
User:	NEW		
Terminal:	NEW PROBLEM NAME		
User:	PROBTQ ————————————————		User gives the program name
Terminal:	READY.		
User:	10 PRINT" P[TQ] T/S RHO"		
	20 FOR R = 0 TO 1 STEP 1		
	30 LET P(T)⟨←=⟩1 − EXP⟨− 1−R←←←(1−⟩ R) * T/S		
	40 PRINT P(T), T/S, R		
	11 LET S = 3		Back-spacing to correct errors
	12 LET T =		
	12 FOR T = 1 TO 30 ←		Re-entry of statement to correct an error
	50 NEXT T		
	60 END		
	SAVE		
Terminal:	READY.		
User:	RUN		Time, place, date
Terminal:	WAIT.		
Terminal:	PROBTQ ⟨7:18 NY WED 07/20/66⟩		Computer indicates an error in the program
	FOR WITHOUT NEXT IN 50 ←		
User:	49 NEXT R ←		
	RUN		
Terminal:	WAIT.		User corrects the error
Terminal:	PROBTQ 17:21 NY WED 07/20/66		

P[TQ]	T/S	RHO	
.283469	.333333	0	
.259182	.333333	.1	
.234072	.333333	.2	
.20811	.333333	.3	
.181269	.333333	.4	
.153518	.333333	.5	Results
.124827	.333333	.6	
.094634	.333333	.7	
.064493	.333333	.8	
.033926	.333333	.9	
0	.333333	1	
.486583	.666667	0	
.451188	.666667	.1	
.413354	.666667	.2	

and so on

Figure 5.1   Dialogue specimen number 1. A user at a teletype machine doing a calculation using the programming language BASIC.

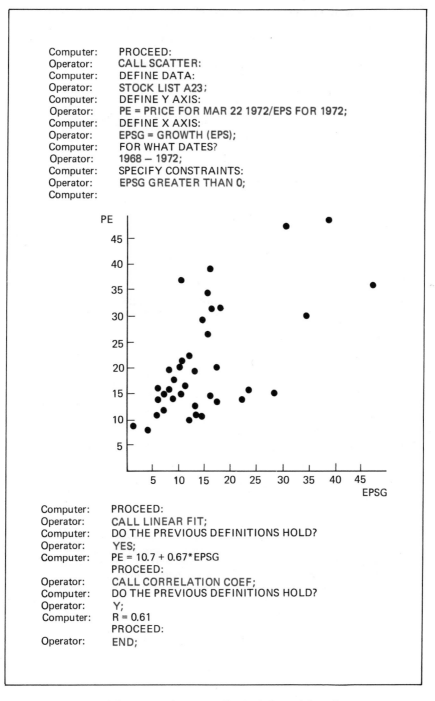

```
Computer:    PROCEED:
Operator:    CALL SCATTER:
Computer:    DEFINE DATA:
Operator:    STOCK LIST A23;
Computer:    DEFINE Y AXIS:
Operator:    PE = PRICE FOR MAR 22 1972/EPS FOR 1972;
Computer:    DEFINE X AXIS:
Operator:    EPSG = GROWTH (EPS);
Computer:    FOR WHAT DATES?
Operator:    1968 — 1972;
Computer:    SPECIFY CONSTRAINTS:
Operator:    EPSG GREATER THAN 0;
Computer:
```

```
Computer:    PROCEED:
Operator:    CALL LINEAR FIT;
Computer:    DO THE PREVIOUS DEFINITIONS HOLD?
Operator:    YES;
Computer:    PE = 10.7 + 0.67*EPSG
             PROCEED:
Operator:    CALL CORRELATION COEF;
Computer:    DO THE PREVIOUS DEFINITIONS HOLD?
Operator:    Y;
Computer:    R = 0.61
             PROCEED:
Operator:    END;
```

Figure 5.2 Dialogue specimen number 2. A financial analyst at a type-writer-like terminal using a program designed for him to carry out stock analyses [2].

A Mr. Goldsmith telephones the airline to say that he wants to modify his reservation. The agent he speaks to instructs her terminal to display his record—Flight 21 on the 25th of March. She types:

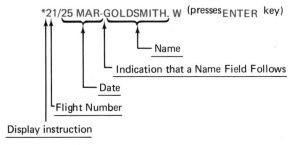

*21/25 MAR-GOLDSMITH. W (presses ENTER key)

— Name
— Indication that a Name Field Follows
— Date
— Flight Number
Display instruction

The computer types details of his record:

GOLDSMITH. W

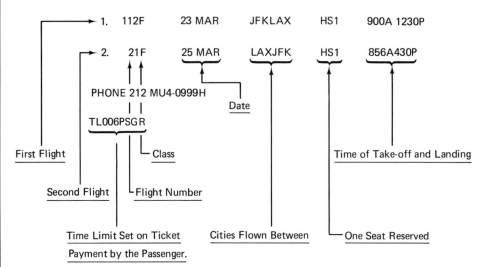

1.	112F	23 MAR	JFKLAX	HS1	900A 1230P
2.	21F	25 MAR	LAXJFK	HS1	856A430P

PHONE 212 MU4-0999H
TL006PSGR
Date

First Flight
Second Flight
Class
Flight Number
Time of Take-off and Landing

Time Limit Set on Ticket
Payment by the Passenger.
Cities Flown Between
One Seat Reserved

The operator checks that this is indeed the required record, and the passenger indicates that he wishes to change the second flight of his journey. He wants to fly back from Los Anglees (LAX) to New York's Kennedy Airport (JFK) on the 26th March rather than the 25th.

The agent informs the computer that the second flight of the journey is to be changed. She types:

X2   (presses ENTER key)

Cancel   — Line 2

Figure 5.3   Dialogue specimen number 3. An airline reservation agent using a screen terminal designed for that application.

The computer responds:

NEXT SEG ENTRY REPLACES 2

She types:

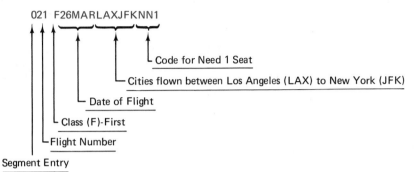

021 F26MARLAXJFKNN1

Code for Need 1 Seat

Cities flown between Los Angeles (LAX) to New York (JFK)

Date of Flight

Class (F)-First

Flight Number

Segment Entry

This statement is a request for a seat on the 26th of March rather than the 25th. "NN1" is a code meaning that the agent needs one seat. The computer replies with a replacement for line 2 above:

2. 21F   26MAR   LAXJFK   HS1   845A430P

The operator attempts to book this seat by pressing the "E" key meaning "END TRANSACTION":

E

The computer asks who requested this modification:

WHO MADE CHANGE

The operator replies that the passenger himself requested it. The "6" key means "change made by" . . . :

6   PSGR

Change   Passenger
made by

The computer indicates that this message has been received by placing an asterisk at the end of it.

6PSGR*

The operator again presses the "E" key, and this time the computer carries out the "END TRANSACTION" operation and updates the appropriate files. It responds:

OK   1232A

Booking   Time
Made

The first action is to inform the computer that BASIC is the language to be used. The operator makes several mistakes as he keys in statements. Some he notices and corrects himself. Some the computer catches, and tells him. He obtains his results much quicker than with a desk calculator or with a noninteractive computer to which he has to submit a job and receive the results hours later.

(It is interesting to note that computations such as this can be done with ease on pocket computers such as the HP55 or HP65. For much computation not requiring remote files of data, telecommunications will be unnecessary. Teleprocessing will increasingly be used to gain access to data, not access to computing power.)

Dialogue Specimen 2 (Fig. 5.2)

The second example shows a financial analyst using a typewriterlike terminal [2]. He is not a programmer but is using a dialogue programmed for him which permits him to do various kinds of stock analyses. The terminal responds at 15 characters per second, which is still frustratingly slow and which places constraints on the designers to limit the lengths of responses.

Dialogue Specimen 3 (Fig. 5.3)

Figure 5.3 shows an airline agent's dialogue. He is talking to potential passengers on the telephone at the same time he is conversing with the terminal to handle the passenger's requirements. He needs to be able to act promptly, obtaining quick responses from the distant computer. The computer may be thousands of miles away on a network such as that in Fig. 4.6.

In this case the terminal is a visual display unit designed for the application. Some of the keys have been specially labeled to assist in the dialogue. The terminal can display responses at a rate of several hundred characters per second, so the operator is not frustrated by slow responses as in the previous two specimens.

Dialogue Specimen 4 (Fig. 5.4)

Figure 5.4 shows a dialogue in which an operator is changing customer information as in the previous case, but here the operator has learned no special mnemonics or formats. Each screen tells the operator exactly how to respond. This form of dialogue is suitable for a casual operator who uses the terminal only occasionally and who is not highly trained in the dialogue usage, as in the previous case.

The increased ease-of-use of the dialogue is paid for by the increased number of characters it requires. If all the characters of Fig. 5.4 are transmitted from the central computer, this constitutes a much heavier communications load than with the brief dialogue structure of Fig. 5.3. However if the terminal is "intelligent" and has a local microprocessor for generating the responses, a much smaller number of characters would be transmitted.

Screen:	ENTER CUSTOMER NUMBER IF KNOWN. IF NOT TYPE "N"
User:	N
Screen:	TYPE FIRST TWO LETTERS OF CUSTOMER'S NAME
User:	GO
Screen:	WHICH CUSTOMER? ENTER LINE NUMBER
	1. GOLDBERG SIDNEY & CO.
	2. GOLDEN KAUFMAN SALES CO.
	3. GOLDSMITH WALTER & CO.
	4. GOODMAN ABE
	5. NONE OF THE ABOVE
User:	3
Screen:	CUSTOMER NUMBER: 17-8341
	GOLDSMITH WALTER & CO.
	745, OLD STONE HIGHWAY, SPRINGS,
	EAST HAMPTON, N.Y., 11937
	REPRESENTATIVE E. S. ROPLEY #543271
	PRESS ENTER IF CORRECT
	PRESS 1 IF INCORRECT
User:	Presses ENTER key.
Screen:	WHICH INFORMATION IS TO BE CHANGED?
	ENTER LINE NUMBER

	1. NEW ORDER		9. INVOICE TO ADDRESS
	2. CANCEL ORDER		10. SHIP TO ADDRESS
	3. ORDER QUANTITY		11. CUSTOMER ADDRESS
	4. DELIVERY DATE		12. CUSTOMER NAME
	5. PRIORITY CODE		13. MISC. INSTRUCTIONS
	6. ORDER DETAILS		14. REPRESENTATIVE DETAILS
	7. DISCOUNT		15. BRANCH OFFICE
	8. TAX DETAILS		16. OTHER

User:	4
Screen:	WHICH ORDER IS TO BE CHANGED?
	ENTER LINE NUMBER

	ORDER #	ITEMS			ORDER DATE	DEL. DATE
1	573824	87289	00241	00029	4/1/77	4/1/79
		00310	77139	04126		
		00900				
2	575931	00391	00700	00421	10/1/77	4/1/78
3	575828	00750	00900		12/1/77	12/1/78
4	579003	02107			6/1/77	8/1/78

User:	2
Screen:	ORDER NUMBER 575931

ITEM #	DESCRIPTION	QTY.
00391	MAGNETIC D EXTRACTOR 7 BR	5
00700	LINK ADAPTOR ADDL	5
00421	CTC DIRECT CONTROL	5

	ORDER DATE 10/1/77 DELIV DATE 4/1/79
	ENTER NEW REQUIRED DELIV DATE (XX/XX/XX)
User:	10/01/78
Screen:	EARLIEST DELIV DATE POSSIBLE AT PRIORITY 3 IS
	2/1/78
	IF THIS IS REQUESTED ENTER "Y"
User:	Y

Figure 5.4 Dialogue specimen number 4. A visual display unit user carrying out a clerical operation concerned with sales orders.

The terminal presents the operator with a "form" to fill in:

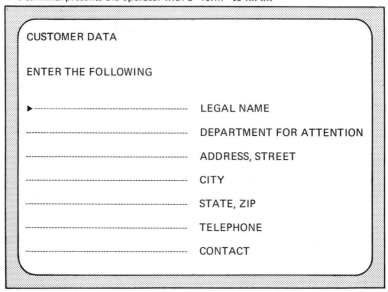

The operator types data into the form:

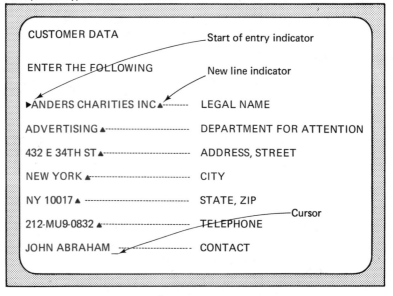

Figure 5.5 Dialogue specimen number 5. A data entry dialogue using a visual display unit. It could be on-line or off-line.

The terminal must respond with a data rate of several hundred characters per second; otherwise this form of dialogue would be very slow and frustrating for the operator. The responses of the operator are very brief, so if the response time and data rate are fast enough, a dialogue such as this can be speedy and effective.

Dialogue Specimen 5 (Fig. 5.5)

Much data have to be entered into computer systems, and this dialogue shows an operator "filling in a form" on a screen to enter data.

It could take place either *off-line* or *on-line* to a distant computer. If it is on-line, the computer can be programmed to detect operator errors, where possible, when the operator makes them, If it is off-line, interactive error control cannot take place, and errors much be corrected later. The on-line dialogue needs the data to be sent to the computer when they are entered. With the off-line dialogue it can be stored for transmission in a batch. The "form" which the operator fills in will not be transmitted if it can be stored at terminal location (for example, in a terminal controller).

Dialogue Specimen 6 (Fig. 5.6)

Figure 5.6 shows a graphics terminal using a computer programmed for designing and testing logic circuits [3].

The user employs a *light pen,* which he can point to the screen to select items or to indicate where items should be positioned. He selects circuits from a circuit library and positions them on the screen. He moves the light pen to wire them together. He can then test the operations of the circuit he has designed with the computer simulating its behavior. He modifies it at the screen until it behaves correctly. Then, using the last of the panels shown, he documents what he has designed.

A graphics user of this type may flip rapidly from one type of panel display to another, so he needs a fast data transmission rate if he is remote from the computer he is using. Most such graphics applications have avoided using interactive data transmission, so far, because the high-speed bursts of data required by graphics are too expensive to transmit. This is unfortunate because many interesting computer uses would result if a user could have occasional access to remote computers with graphics programs.

**TRANSMISSION
SPEEDS**

The reader should relate the transmissions taking place in the above illustrations to the speeds of communication lines that might be used. Table 3.2 listed typical transmission speeds. Typical speeds for terminals used in the preceding illustrations would be as follows:

2

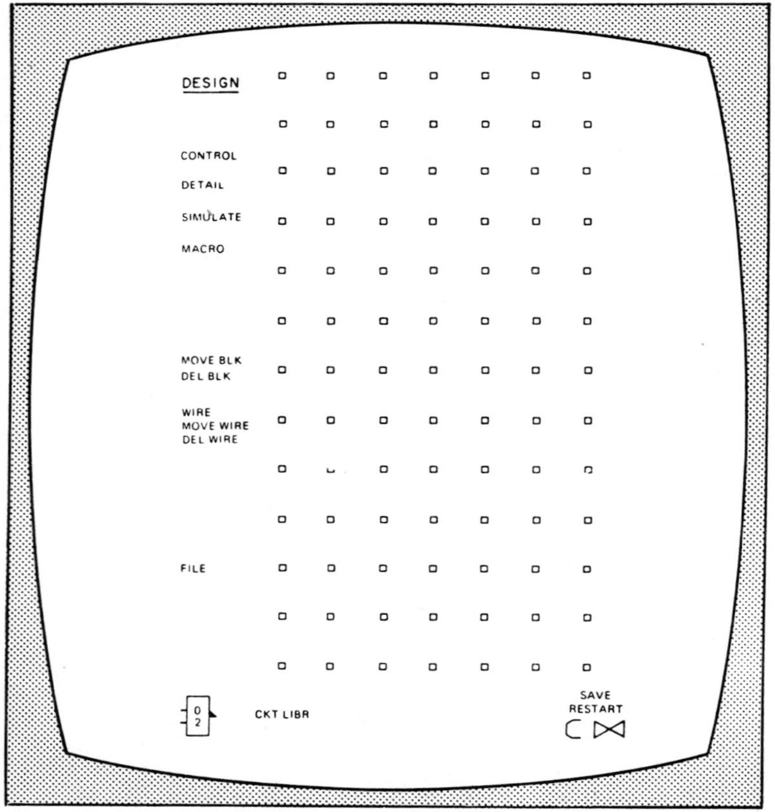

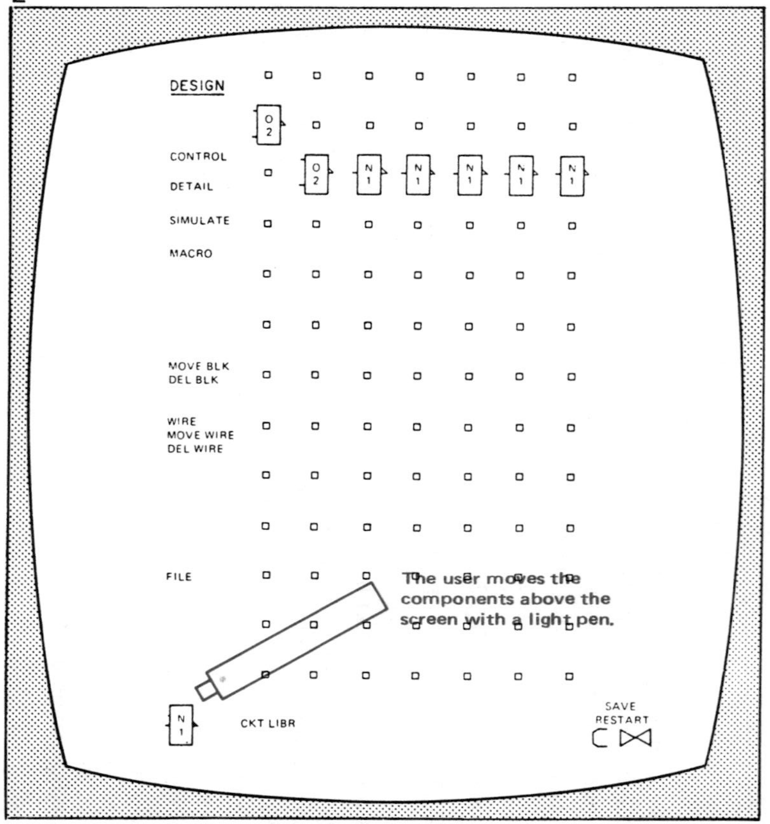

DESIGN

CONTROL
DETAIL
SIMULATE
MACRO

MOVE BLK
DEL BLK

WIRE
MOVE WIRE
DEL WIRE

FILE

CKT LIBR

SAVE
RESTART

The user moves the components above the screen with a light pen.

1

DESIGN

CONTROL
DETAIL
SIMULATE
MACRO

MOVE BLK
DEL BLK

WIRE
MOVE WIRE
DEL WIRE

FILE

CKT LIBR

SAVE
RESTART

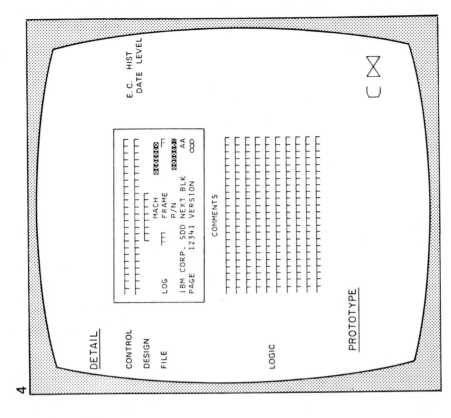

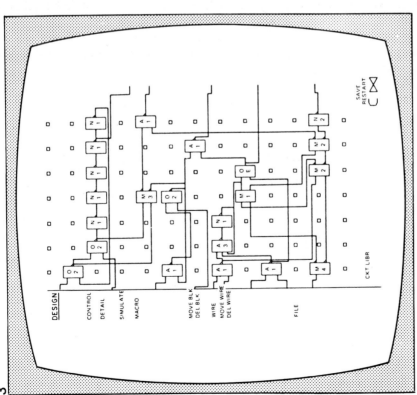

Figure 5.6 Dialogue specimen number 6. A logic circuit being designed with a graphics program. The user interacts with the screen with a keyboard and lightpen [3].

Specimen	Application	Terminal Type	Typical Speed
1	Calculation in BASIC	Teletype	10 characters/sec
2	Stock analysis	Typewriterlike device	15 characters/sec
3	Airline reservations	Application-built display	2400 or 4800 bits/sec
4	Sales order	Visual display unit	4800 bits/sec
5	Data entry	Data entry unit	9600 bits/sec
6	Circuit design	Graphics terminal	56,000 bits/sec

SPORADIC TRANSMISSION In dialogues transmission takes place in a sporadic fashion. The user will often pause to read a response or to think. When he types, he may do so slowly.

Figure 5.7 gives a time scale showing the timing of the airlines agent's dialogue. The terminal has a buffer. Because of this, the line is not occupied for the time the agent is keying data in but only for the time it takes to transmit those data. It will be seen from Fig. 5.7 that the line is occupied for only a small proportion of the total time. Furthermore, there is no transmission from this terminal for the next 2 minutes.

In fact, a leased voice line to a terminal such as this is likely to transmit at 4800 bits per second. The occupancy of the line shown in Fig. 5.7 assumes a speed of only 480 bits per second. This was done in order that the transmission bursts could be drawn at all. The reader should imagine that, in fact, they are one tenth of the thickness of those in the figure.

A similar diagram could be drawn for their dialogues. If only one dialogue takes place on a line, then that line is idle most of the time. The following is a calculation of the transmission efficiencies with one user transmitting over typical lines:

Specimen	Application	Average Quantity of Bits Transmitted in 1000 Seconds	Typical Line Speed (*bits/sec*)	Efficiency
1	Calculation in BASIC	8,000	100	0.08
			4,800	0.0017
2	Stock analysis	10,000	150	0.067
			4,800	0.0021
3	Airline reservations	4,000	4,800	0.0021
4	Sales order			
	a. With simple terminal	40,000	4,800	0.0094
	b. With programmable terminal	2,000	4,800	0.0004
5	Data entry	3,000	4,800	0.0021
6	Circuit design	200,000 (only new panels are transmitted)	56,000	0.0036

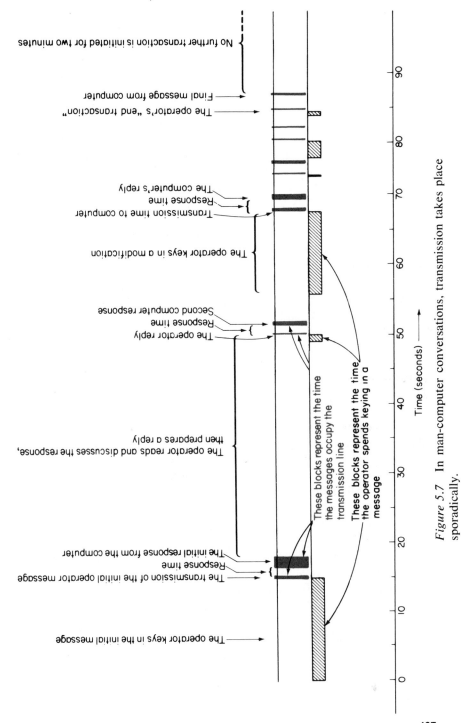

The operator keys in the initial message

The transmission of the initial operator message
Response time
The initial response from the computer

These blocks represent the time the messages occupy the transmission line

These blocks represent the time the operator spends keying in a message

The operator reads and discusses the response, then prepares a reply

The operator reply
Response time
Second computer response

The operator keys in a modification
Transmission time to computer
Response time
The computer's reply

The operator's "end transaction"
Final message from computer

} No further transaction is initiated for two minutes

Time (seconds)

0   10   20   30   40   50   60   70   80   90

Figure 5.7   In man-computer conversations, transmission takes place sporadically.

The middle column above gives the *average* number of bits transmitted in a typical 1000 seconds of each dialogue. This number is much lower than the number of bits that *could* have been transmitted because of the pauses in the transmission; 4800 bits per second has been used as the typical speed of a voice line.

SHARING   To improve the efficiency with which lines are utilized, many terminals must share the same line. There are two ways in which sharing can be accomplished. First, a variety of devices is available from computer and terminal manufacturers. These are discussed elsewhere [4]. Second, the common carriers can provide a network structured to facilitate sharing and tariff such that the user is charged only for his share. Common carrier networks such as this did not exist prior to the mid-1970s but are now beginning to emerge.

The simplest way to share a line is to subdivide its bit stream into channels which operate at smaller bit streams. A 4800-bit-per-second channel could be split into ten 480-bit-per-second (or slightly less) channels, for example. This simple subdivision (called *multiplexing* and discussed in Chapter 12) is not an adequate answer to the problem, however. A typical dialogue with a screen needs *zero* bits per second at one time and several thousand bits per second at another time in order to fill the screen up sufficiently quickly.

PEAK/AVERAGE   The need for sporadic transmission in a dialogue can
RATIOS   be expressed as the ratio of the *peak* transmission rate that is needed to the *average* transmission rate. If we forget for a moment about the physical speeds of lines, the peak transmission rate is determined by the psychological needs of the dialogue user. The dialogue in Fig. 5.4, for example, would be slow and frustrating if the screens from the computer took much more than 2 seconds to be transmitted. In other words, a transmission speed of several thousand bits per second is desirable. This peak transmission rate may be divided by the average transmission rate to obtain a peak/average ratio for the dialogue. Table 5.1 gives approximate estimates of peak/average ratios for the six dialogues illustrated.

It will be observed that the peak/average ratio is a characteristic of the dialogue design. It varies substantially for different dialogues. Often it is as high as 1000. The trend away from dialogues on teletype machines to more powerful dialogues on display units, graphics terminals, and distributed-intelligence systems often results in a higher peak/average ratio. This characteristic is very different from human telephone conversations which have a peak/average ratio of 1.

The higher the peak/average ratio, the greater the inefficiency of using a transmission channel *which transmits at a constant fixed rate*. Instead a channel is needed which transmits bursts of data when they are required with a suitable short response time. Such a channel can be derived either by using suitable equipment

Table 5.1

Specimen	Application	Average Rate (*bits/sec*)	Desirable Peak Rate (*bits/sec*)	Peak/Average Ratio
1	Calculation in BASIC	8	100	12.5
2	Stock analysis	10	150	15
3	Airline reservations	10	2,000	200
4	Sales order			
	a. With simple terminal	40	4,000	100
	b. With programmable terminal	2	4,000	2000
5	Data entry	15	1500	100
6	Circuit design	200	200,000	1000

to share a fixed-rate channel or by using a common carrier network architecture which allocates capacity on a rapidly time-varying basis, as discussed in Chapters 24 and 25.

REFERENCES

1. Dialogues used with data transmission are discussed in detail in James Martin, *Design of Man-Computer Dialogues,* Prentice-Hall, Englewood Cliffs, N.J., 1973.

2. J. J. Gal, "Man-Machine Interactive Systems and Their Application to Financial Analysis," *Financial Analysts Journal* (May–June 1966). The dialogue illustration from this article is used in this chapter with minor modifications.

3. William H. Sass and Stephen P. Krosner, *An 1130-Based Logic Layout and Evaluation System,* IBM Internal Paper, IBM Corp., Kingston, N.Y., 1970. More details of this dialogue are given in Reference 1 above.

4. The alternative methods of line organization are discussed in the author's books *Teleprocessing Network Organization;* and James Martin, *Systems Analysis for Data Transmission,* Prentice-Hall, Englewood Cliffs, N.J., 1970 and 1972.

PART **II**   **TRANSMISSION**

6 THE STRUCTURE OF TELECOMMUNICATION NETWORKS

Basically, there are two ways in which information of any type can be transmitted over telecommunication media: analog or digital.

Analog means that the amplitude of the transmitted signal varies over a continuous range. Oscillating signals are normally transmitted and the frequency of the oscillation can also vary over a continuous range. Both the sound you hear and the light you see are analog signals spread over a range of frequencies. Sound, as any hi-fi enthusiast knows, consists of a spread of frequencies from about 30 to 15,000 hertz (cycles per second†), or, for people with very good ears, 20,000 Hz. Sound cannot be heard below 30, and it cannot be heard above 20,000. If we wanted to transmit high-fidelity music along the telephone wires into your home (which is technically possible), we would send a continuous range of frequencies from 30 to 20,000 Hz. The current on the wire would vary continuously in the same way as air pressure does in the sound you hear.

The telephone company, in the interests of economy, transmits a range of frequencies that may vary from about 300 to 3400 Hz only. This is enough to make a person's voice recognizable and intelligible. When telephone signals travel over lengthy channels, they are packed together, or multiplexed, so that one channel can carry as many such signals as possible. To do this your voice might have been raised in frequency from 300–3100 to 60,300–63,100 Hz. Your neighbor's voice might have been raised from 64,300 to 67,100. In this way, they can travel together without interfering with one another, but both are still transmitted in an analog form — that is, as a continuous signal in a continuous range of frequencies.

Digital transmission means that a stream of on/off pulses are sent like data

---

†Today the term "cycles per second" is no longer used. One hertz (Hz) means the same as 1 cycle per second. One kilohertz (kHz) is 1000 cycles per second (kilocycles); 1 megahertz (MHz) is 1 million cycles per second (megacycles); and 1 gigahertz (GHz) is 1 billion cycles per second.

travelling in computer circuits. The pulses are referred to as bits. It is possible today to transmit an extremely high bit rate.

Figure 6.1 shows an analog and a digital signal. *A transmission path can be*

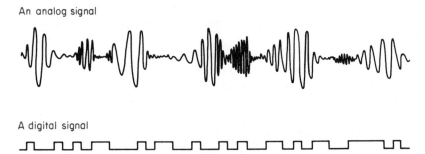

An analog signal

A digital signal

Figure 6.1   Any information can be transmitted in either an analog or a digital form.

designed to carry either one or the other. As we shall see in the chapters that follow, this fact applies to all types of transmission paths: wire pairs, high-capacity coaxial cables, microwave radio links, satellites, and other transmission media.

It is important to understand that any type of information can be transmitted in either an analog or digital form.

Most telephone channels today are analog channels, capable of transmitting a certain range of frequencies. If we send computer data over them, we have to convert that digital bit stream into an analog signal using a special device known as a modem. This converts the data into a continuous range of frequencies – the same range as the telephone voice. In this way, we can use any of the world's analog channels for sending digital data.

On the other hand, where digital channels have been constructed, it is possible to transmit the human voice over them by converting it into a digital form. Similarly, any analog signal can be digitized for transmission in this manner. We can convert hi-fi music, television pictures, temperature readings, the output of a copying machine, or any other analog signal into a bit stream. High-fidelity music would need a larger number of bits per second than telephone sound. Television would need a much higher bit rate than sound transmission. The bit rate needed is dependent on the bandwidth, or range of frequencies, of the analog signal as well as on the number of different amplitude levels we want to be able to reproduce.

Almost all the world's telephone plant grew up using analog transmission. Much of the transmission will remain so for years to come because of the multibillions of dollars tied up in such equipment. However, many millions of channel miles of digital channels are now operating, mostly designed to carry digitized speech. Digital technology is rapidly evolving, and major advantages in digital transmission are beginning to emerge, as we will discuss in Chapter 14.

THE TELEPHONE NETWORK

The Bell System has been described in AT&T publicity as "the world's most complicated machine." When a telephone call is made it may travel over many different types of channels, and complex switching facilities are needed to set up its path.

Telecommunication systems can be divided into four main parts:

1. *Instruments.* The term "instrument" is used for the device the subscriber employs to originate and receive signals. The vast majority of instruments are telephone handsets. Today, however, an endless array of other devices are being attached to telephone lines, including the terminals used for data transmission.

2. *Local loops.* These are the cables which enter the subscriber's premises. On a telephone network they connect the telephone handsets or other devices to the local switching office (central office). Telephone and telegraph loops today consist of wire-pair cables. Every subscriber has his own pair of wires to the local switching office, and nobody else uses these unless he is on a party line. The United States has several hundred million miles of telephone subscriber loops. Coaxial cables are also laid into homes by cable television organizations, and these cables have many potential uses other than television.

3. *Switching facilities.* An elaborate network of switching offices enables any telephone to be connected to almost every other telephone. The switching and control functions will eventually be carried out entirely by computers, but modernization proceeds slowly, and electromechanical switching facilities will continue to exist in telephone systems for many years.

4. *Trunk circuits.* Trunk circuits are transmission links which interconnect the switching offices. Such links normally carry more than one telephone call. On high-traffic-routes they carry many thousands of calls simultaneously. A variety of different transmission media is employed, including wire pairs, coaxial cables, microwave radio, and satellites. (In Chapter 9 we will discuss transmission media.)

In the Bell System the cost breakdown between these areas is approximately

switching:	45%
instruments:	23%
trunk circuits:	17%
local loops:	15%

The technology of transmission has traditionally been managed separately from the technology of switching. In this part of the book we will discuss transmission, and in Part III we will discuss switching. With new technologies now emerging, including satellites and time-division switching (chapters 15, 25, and 26), the technologies of switching and transmission become fascinatingly interrelated.

NONVOICE TRAFFIC   For half a century telephone technology has dominated telecommunications. There have been a thou-
sand times as many telephone subscribers as other types of telecommunication
users. Consequent economics of scale have dictated that telegraph and data traffic
should be converted to a form in which they can travel over the telephone system.

In the 1970s new types of common carriers are beginning to emerge, some
with a desire to build their own nontelephone networks. Separate data transmission networks for computer users are being built or discussed in most industrial
countries. Some of these will operate by attaching new types of equipment to the

BOX 6.1   Telephone traffic and computer traffic
have characteristics so different that
different network architectures are
needed

Telephone Users	Computers and Their Terminals
• Require a fixed capacity channel.	• Require a very wide spread of channel capacities ranging from a few bits per second to (ideally) millions of bits per second.
• Always carry out a two-way conversation.	• One-way or two-way transmission.
• Tolerant of noise on the channel.	• Data must be delivered without errors.
• Transmit or listen continuously until the call is disconnected.	• In a man-computer dialogue, transmission is in bursts.
• Require immediate delivery of the signal.	• In non-real-time data transmission the data can be delivered later, when convenient.
• The transmission rate is constant.	• In a man-computer dialogue the mean number of bits per second is usually low, but the peak requirement is often high. The peak-to-average ratio is often as high as 1000.
• The time to set up the connection can range from a few seconds to 1 minute.	• Sometimes it is desirable that the connection be set up in a second or so.

telephone networks. Others will employ new transmission networks, physically separate from the telephone networks.

It cannot be stressed too strongly that computer users have fundamentally different characteristics and requirements to plain old telephone service. Box 6.1 summarizes the differences. A particularly important difference is the peak-to-average transmission rate requirement, which, as we discussed in the previous chapter, is 1 in telephone conversations but can be 1000 or higher in man-computer dialogues.

Economies can arise from designing networks to carry both voice and data. However to force computer traffic to fit into channels designed for single telephone calls both severely limits the potential of the computer and is likely to waste the channel capacity. New network architectures are being employed for data transmission, and a whole new body of theory is emerging.

TELEPHONE CONNECTIONS   Figure 6.2 illustrates an intercity telephone connection. When the subscriber at the left of the diagram picks up his telephone, hears a dial tone, and is in contact with his local central office; he will be connected directly without any trunk being involved. If he dials a subscriber in a different city, the central office will switch the subscriber loop to a *toll-connecting trunk,* thus connecting the call to a toll office. The toll offices are interconnected with *intertoll trunks,* and this toll network establishes the connections between towns. There may be several intermediate switching offices in the path which is set up between the two toll offices.

There are about 2000 toll and intermediate switching offices in the toll network of the United States and about 25,000 central offices. About 100 million local loops are connected to the central offices.

The long-distance trunks are all *four-wire.* The term *four-wire* implies that there are two wires carrying the signals in one direction and two wires carrying

Figure 6.2 An intercity telephone connection showing 2-wire and 4-wire paths.

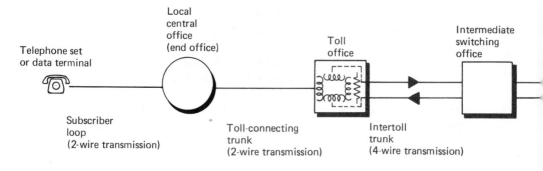

them in the other direction. There are many repeaters (amplifiers) on these circuits, amplifying the entire group of calls in both directions. In reality the circuit may not consist of wire pairs but of higher capacity transmission media, however, the historical term *four-wire* is still used and implies separate circuits in each direction.

In normal telephone service the local loops are *two-wire* circuits, on which a single telephone call can be transmitted in both directions over the same pair of wires. The toll-connecting trunk can be either a two-wire or a four-wire circuit, as illustrated in Fig. 6.2. When a two-wire line is connected to a four-wire line a special circuit is required to join them. For special purposes a four-wire local loop can be used. Some data transmission machines require a four-wire rather than a two-wire connection.

SIMPLEX, DUPLEX, AND HALF-DUPLEX LINES Transmission lines are classed as simplex, half-duplex, and full-duplex. In North America, these terms have the following meanings:

Simplex lines transmit in one direction only.

Half-duplex lines can transmit in either direction but only in one direction at once.

Full-duplex lines transmit in both directions at the same time. One full-duplex line is thus equivalent to two simplex or half-duplex lines used in opposite directions.

The above are meanings in current usage throughout most of the world's computer industry. Unfortunately, however, the International Telecommunications Union defines the first two terms differently—as follows:

Simplex (circuit). A circuit permitting the transmission of signals in either direction but not simultaneously.

Half-duplex (circuit). A circuit designed for duplex operation but which because of the nature of the terminal equipment can be operated alternately only.

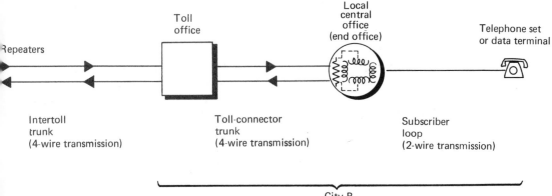

"Simplex" and "half-duplex" are thus used differently by European tele-communications engineers and computer manufacturers (especially American ones) using the same facilities. The ITU and CCITT are not consistent and sometimes use the North American wording.

Throughout this book, the words will be used with the former meanings exclusively.

Simplex lines (American meaning) are not generally used in data transmission because even if the data are being sent in only one direction, control signals are normally sent back to the transmitting machine to tell it that the receiving machine is ready or is receiving the data correctly. Commonly, error signals (positive or negative acknowledgment) are sent back so that there can be retransmission of messages damaged by communication line errors. Many data transmission links use half-duplex lines. This allows control signals to be sent and two-way "conversational" transmission to occur. On some systems full-duplex lines can give more efficient use of the lines at little extra line costs. A full-duplex line often costs little more than a half-duplex line. Data transmission machines which can take full advantage of full-duplex lines have been more expensive than those which use half-duplex lines. Half-duplex transmission is therefore more common at present, though this might well change.

A four-wire circuit can always be used in a full-duplex fashion (if full-duplex terminals are connected to it). A two-wire circuit is often used in a half-duplex fashion. A two-wire circuit can be used in a full-duplex fashion if its amplifiers are designed to permit simultaneous transmission in both directions. Two-wire local loops can be used for full-duplex transmission. To do so, one needs a special modem which splits the band of frequencies used into two parts, each for transmitting in one direction. The maximum speed available would be less than with a similar four-wire line. Sometimes the frequency band on a two-wire line is split to give a high speed in one direction and a low-speed return path, and the directions may be reversed. The specifications of data sets which connect a data-handling machine to the line state whether they operate in a half-duplex or full-duplex manner and whether they need a two-wire or four-wire circuit.

To achieve full-duplex transmission, then, it is necessary to have a suitable line which could in some cases be a two-wire line and also to have data sets suitably designed, taking into account whether the circuit is two-wire or four-wire.

LOCAL DISTRIBUTION   The reader should distinguish in his mind between local distribution facilities and trunk networks.

Trunk networks carry signals between the common carrier offices. Local distribution facilities carry signals from the subscriber's instrument or terminal to the local common carrier office and back. The outermost links of Fig. 6.2 are the local distribution facilities. The new common carriers now emerging also need local distribution facilities, and, as on telephone systems, these function differently from the trunk facilities.

Because the trunks interconnect the common carrier offices, sophisticated equipment can be used for organizing them, such as switching equipment or equipment for combining many signals. The new value-added common carriers, for example, use computers to interleave the data packets that are sent and select the route over which to transmit them. The transmission links between common carrier offices can have a high capacity because they carry large traffic volumes. High capacity facilities such as coaxial cable, microwave radio and satellites link the offices. The specialized common carriers built cost-effective trunking systems which carry signals thousands of miles but then have the problem: How can the signal be delivered the last few miles between the carrier office and the subscribers premises?

It is a little like the mail delivery system. Large sacks of letters can be moved efficiently across the nation between sorting offices, but the local delivery has to be done separately and less efficiently by a mailman.

The telephone system has a pair of wires (local loop) going from its local telephone offices to each subscriber as illustrated in Fig. 6.3. In most cases only one subscriber is connected to this pair or wires. The utilization of the local loop is very low. It stands idle most of the day, and when it does transmit it is used at a fraction of its capacity. However, the wire pairs exist, going into almost every home and office building in North America and having a total book value in North America of between $10 and $20 billion.

The telephone companies claim that local loop operation is a "natural monopoly." It would not make sense for competing organizations to run separate pairs of wires into subscribers' homes. Other ways of transmitting speech and data to/from the end users are possible however. These include using CATV television cables or various types of radio. A single CATV cable interconnects many hundreds of subscribers and hence is installed at a lower cost per subscriber than the telephone loops each serving one subscriber. More commercial buildings in the future will have their own satellite, microwave, or millimeterwave radio antennas, bypassing local loops.

Today, local distribution is a problem for the new common carriers. They can build trunking systems economically but often cannot build the last leg to and from the subscriber economically. They are often forced to rely on telephone company facilities for local distribution. The networks shown in Figs. 2.3 and 2.6 show trunking systems but omit the local loops. When new carriers want to deliver signals other than speech, the telephone company might not have the type of local distribution they need. They may, for example, want to deliver high-speed data of facsimile signals, but the telephone loop can carry these only if the electronics on it is modified.

Local distribution is the Achilles' heel of many new uses of telecommunications.

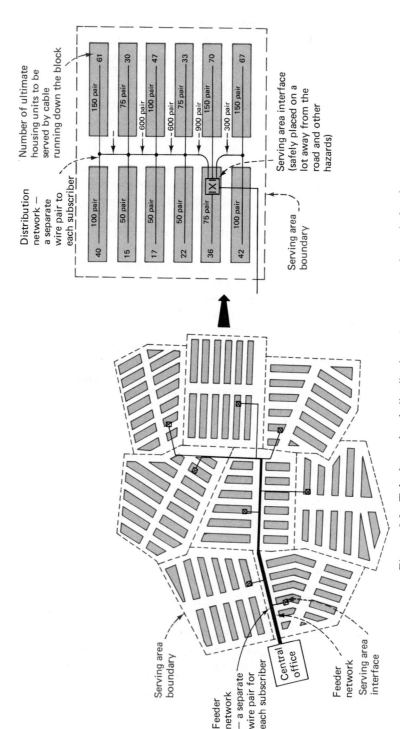

Figure 6.3 Telephone local distribution. A separate wire pair local loop is normally connected from the central office to each subscriber. The local loops can be made to carry a data rate of 56,000 bits per second or higher. (Courtesy AT&T.)

CORPORATE Local loops to a company building often terminate at a
TELEPHONE switching facility in the building which has extension
FACILITIES lines to telephones. The switching may be done manually by an operator using cords with jack plugs, or it
may be automatic, in which case the user can dial his own calls. A manual switching facility is referred to as a PMBX (private manual branch exchange), and an
automatic switching facility is referred to as a PABX (private automatic branch
exchange). PBX (private branch exchange) implies either. Automatic PBXs are
sufficiently common that the term PBX is often used to imply a PABX, especially
in North American usage.

A *tie line,* or tie-trunk, is a private, leased, communication line between two
or more private branch exchanges. Many companies have a leased system of telecommunication lines with switching facilities. To telephone a person in a distant
company location, an employee must first obtain the appropriate tie line to that
person's private branch exchange. On an automatic system this is done by dialing
a tie-line code before the extension number, thus:

<table>
<tr><td>8</td><td>444</td><td>7215</td></tr>
<tr><td>number to obtain
access to tie-line
network</td><td>location
number</td><td>extension
number</td></tr>
</table>

The tie line (or lines) to that location may be busy, in which case a "busy
tone" ("engaged signal" in British parlance) can be heard before the extension
number is dialed.

All the lines above can be used for data transmission as well as voice. Where
one voice line is used the transmission speed is limited to a few thousand bits per
second depending, as will be discussed later, on the equipment used at each end of
the line. This is the data capacity of a telephone line. Often, however, the tie-line
system has more than one voice line connecting two locations. Some tariffs make
a broadband channel, giving many voice lines, considerably less expensive than
leasing the same number of individual voice channels.

One typical large corporation uses its tie-line groups in this manner as a
broadband data network during its second and third shifts. The lines are transferred to data automatically, as a group, and so telephone calls in progress on
those lines may possibly be interrupted. The corporation tie-line directory contains the following note:

> "When a transfer is about to occur, callers will hear a special interrupt
> tone that is introduced during a conversation. Callers hearing this tone hang up
> and redial the call after a 30-second interval. Normally, these transfers will
> only occur after 6:00 P.M. Eastern Time."

7 DC SIGNALING

There are two ways in which signals are sent down a communication line. They may be sent as they are, without modification, or they may be superimposed upon a higher frequency waveform which "carries" them. This chapter is concerned with the former, in which the current on the line contains relatively low-frequency components.

BASEBAND SIGNALING

The transmission of signals at their originating frequencies is called "baseband" signaling to differentiate it from the carrier frequency transmission discussed in the following chapters. Subscriber loops (see Fig. 6.2) carry baseband signals. Longer links such as the lines between toll offices do not carry baseband signals. They use higher frequencies to carry many signals simultaneously. Privately laid wires in an office or factory may also carry baseband signals. In this chapter we will discuss dc (direct-current) signals such as those in Fig. 7.1 carried in a baseband fashion.

As the squared-edged pulses travel along the communication line they are distorted by the line. The pulses received at the other end are far from square-edged, and if the line is too long, the signals too weak, or the transmitting speed too great, the received signal may be unrecognizable and wrongly interpreted by the machine at the other end. The factors that cause this distortion in direct-current signaling will still be at work with alternating-current (ac) signaling, dealt with later.

The factors mainly responsible for the distortion are the *capacitance, inductance,* and *leakage* associated with the line used for transmission. There are other factors such as the noise impulses that interfere with signaling, but these will be discussed in later chapters. There is one advantage to sending signals in a dc form as in Fig. 7.1: It is simple and inexpensive. If the signals from a data-processing machine are converted to an ac form in which they are "carried" by higher frequencies, this needs a device called a modem at each end of the line to convert the signals to ac form and then convert them back again after transmission. Such a device for low-speed transmission typically costs about $20 to $40 monthly rental. A time-sharing system with 500 low-speed terminals (only a few of which are in use at any one time) would be likely to pay, then, about $20,000 to $40,000 per month for modems. Figure 7.1 shows a rack of such modems for a typical time-sharing system. If dc signaling were used, this cost could be avoided.

DC signals can be sent over wire pairs of a few miles in length at speeds of up to 300 bits per second without harmful distortion. The speed can be increased enormously by using small coaxial cables rather than wire pairs or by having a regenerative repeater every few thousand feet on a wire pair. In many systems a large number of typewriter-speed terminals within a localized area, say 3 miles across, could be connected to a time-sharing system or to a concentrator without modems. While today most common carrier lines require modems, it is possible that the wire pairs which connect to all locations with a telephone could be used over a limited area for dc signaling as in the earlier days of telegraphy. A low-cost exchange for data signals used in this way has been developed.

Baseband dc signals cannot be sent over lines with amplifiers.

Figure 7.1   Some time-sharing computer systems have large numbers of data sets (modems on lines to the users' terminals). It is possible to avoid this expense in some cases where the terminals are in the same locality by designing the line control with DC signaling. (Photo by author.)

LINE CAPACITANCE   Let us first discuss the capacitance of the line. This is
the main cause of distortion. As we raise the voltage at
one end of the line there is some delay before the voltage at the other end rises by
an equivalent amount. The cable acts rather like a water hose. Because of its
"capacity," when electricity is applied at one end, a certain amount is needed to
fill it up before the result is detectable at the other end. Let us suppose that a short
section of line may be represented by the resistance, R, and electrical capacity, C,
in Fig. 7.2.

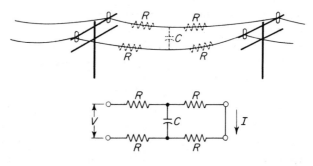

Figure 7.2

As is shown in the elementary electricity books, when a battery of voltage
V is connected across a capacitor of capacity C and resistance R the current, I,
flows until the condensor is fully charged. It starts at a high value, momentarily
($I = V/R$), and falls until it approaches zero, at which time the capacitor is fully
charged.

The equation for the current at a time, t, after the voltage was applied is

$$I = \frac{V}{R} e^{-t/RC}$$

Similarly, the current, I, received in Fig. 7.2, is

$$I = \frac{V}{R} (1 - e^{-t/RC})$$

If the voltage at the sending end is now suddenly replaced by a short circuit, the current at the receiving end does not cease instantly but dies slowly
because of the discharge of the capacitance. The equation for the falling current is

$$I = \frac{V}{R} e^{-t/RC}$$

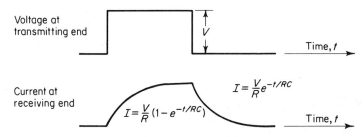

Figure 7.3   Distortion of a dc pulse due to line capacitance.

Time Constant

This buildup and decay of current is shown in Fig. 7.3. The square-edged pulse transmitted is nothing like square-edged when it is received at the other end.

The constant RC in the preceding equations is referred to as the time constant of the circuit and is a measure of the time taken for the current to build up to $1 - 1/e = 0.6321$ of its final value, or to decay to $1/e = 0.3679$ of its original value. The time of buildup or decay on these levels is thus proportional to both the resistance and capacitance of the line.

Where the transmission line consists of an open pair of wires hanging from telegraph poles, the capacitance in question is between these two conductors separated by several inches of air. As in the case of an ordinary electrical capacitor, the capacitance is larger if the size of the conductors is larger and also when the distance between the conductors is smaller. If the wires are in a cable, as is now usually the case, the capacitance will be much greater because the wires are closer together and the insulating material has a higher dielectric constant than air. When the first submarine cables were laid in the sea they had a much poorer performance than was anticipated because the presence of seawater increased the effective capacitance of the cable.

The preceding illustration is oversimplified. However, it serves to illustrate the point that if the pulses representing data are too short in duration, or if there are too many pulses sent per second, then the pulses become indistinguishable when they are received, as illustrated in Fig. 7.4.

Figure 7.4   The effect of a capacitance in the communication line. The faster the pulse rate, the more difficult it becomes to interpret the received signal.

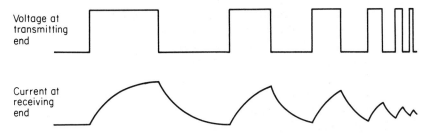

In practice, there will be resistance across the receiving end, such as the resistance of the relays or electromagnets of the receiving equipment. The transmitting end will also have some resistance. Both of these increase the time constant.

LINE INDUCTANCE   In addition to capacitance and resistance, there is also inductance associated with the line, although this is generally of less importance than the capacitance.

Inductance in a circuit resists the sudden buildup of current. Current in an inductive circuit produces a magnetic field. A coil of wire in a circuit is an inductance; it is a much stronger inductance if it surrounds magnetic material. Large inductances which form part of electronic circuits are constructed by wrapping many thousands of turns of fine wire around a heavy, laminated, highly magnetic core.

When a battery of voltage V is applied across a circuit with resistance R and inductance L the current grows as shown in the equation

$$I = \frac{V}{R}\left(1 - e^{-(R/L)t}\right)$$

When the battery is replaced by a short circuit the current does not cease immediately but dies away slowly as the magnetic field built up by the current collapses. The energy stored in the magnetic field is dissipated in the circuit. The current falls as shown in the equation

$$I = \frac{V}{R}\,e^{-(R/L)t}$$

The time constant here is L/R.

The reader might think of the communication line as being something like a fireman's hose. Suppose that the fireman wishes to send data down his hose. (He is not really a fireman but an espionage agent who has an accomplice inside the building which the hose goes into.) He attempts to send the data by means of a piston. As he pushes and pulls the piston at one end of the hose, the pulses are transmitted to a receiving piston at the other end. If the hose were absolutely rigid and the water in it absolutely incompressible, the movement of the receiving piston would follow the movement of the transmitting piston exactly. Again, if the water had no viscosity and moved completely without friction, the piston would be able to transmit at a very high speed. However, the hose is not rigid. It is slightly elastic and the water with air bubbles in the hose is slightly compressible, so the receiving piston does not follow the movement of the transmitting piston exactly. Furthermore, there is viscosity and friction; therefore the piston cannot move and transmit at limitless speed. These properties are very loosely analogous to the capacitance and inductance of a communication line.

If our James Bond fireman were to attempt to transmit at a fairly slow speed, say 1 pulse every 5 seconds, the receiving piston would follow the movement of the transmitting piston faithfully enough to recognize each pulse. As he increased his transmission speed, however, the signal distortion would become greater. If he were to transmit at 2 pulses per second his accomplice might be able to receive this correctly. At 10 pulses per second it would need sensitive and sophisticated receiving equipment to detect the pulses without error. On a communication line, the natural capacitance and inductance of the line would have a more serious effect, the higher the frequency of transmission. The longer the line, the worse the effect. In addition, there is the noise of the line. Suppose that the fire hose is vibrating due to the motion of the pump nearby. At high transmission speeds the strength of the received pulses becomes comparable in magnitude with this "noise," and errors in the interpretation of the data will occur.

ARRIVAL CURVE   The properties of a given dc telegraph line are somewhat more complex than would be suggested by the preceding simple equations, and the transient effects of resistance, capacitance, inductance, and also leakage from the line due to not quite perfect insulation may be summarized in an "arrival curve." Such a curve is illustrated in Fig. 7.5.

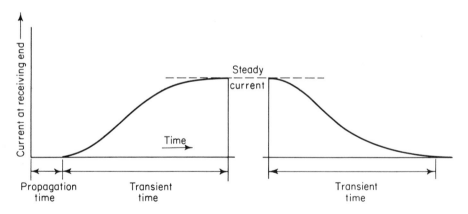

Figure 7.5   Arrival curve for a typical telegraph line (unloaded).

This shows the growth of current at the receiving end of a line when a steady voltage is suddenly applied across the transmitting end. The right-hand side of the diagram shows the decay of current when the voltage is then suddenly removed. There is an interval between the switching on of the voltage and the start of the current growth at the other end. This is referred to as the propagation time and is in effect the time the signal takes to pass down the line. The periods during which the current is building up, or decaying, are called the "transient times." From the shape of the arrival curve for a particular line, its response to any voltage change may be evaluated. The exact shape of the arrival curve is difficult to calculate the-

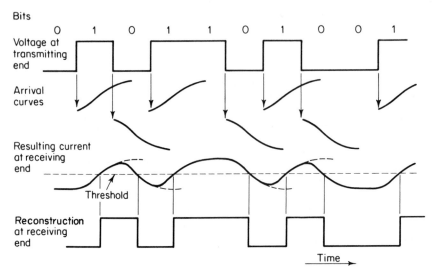

Bits

Voltage at transmitting end

Arrival curves

Resulting current at receiving end

Threshold

Reconstruction at receiving end

Time

Figure 7.6 Baseband transmission of data. The received current permits the original bit pattern to be reconstructed, for example, by means of a relay. If the transmission rate over the same line were doubled, the relay at the receiving end would not correctly reconstruct the bit pattern.

Figure 7.7 Bias distortion. Because of the shape of the arrival curve, an ill-adjusted receiver threshold can cause "bias distortion." Either the marks (bits) become elongated and the spaces (0 bits) shortened or vice versa.

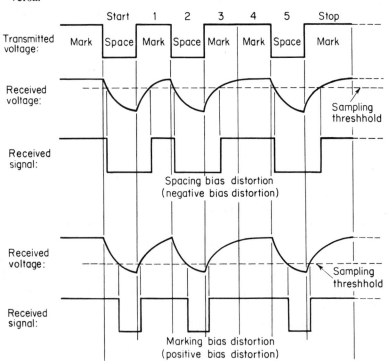

Transmitted voltage:

Received voltage:

Sampling threshhold

Received signal:

Spacing bias distortion
(negative bias distortion)

Received voltage:

Sampling threshhold

Received signal:

Marking bias distortion
(positive bias distortion)

128

oretically for a particular line but may be measured experimentally. It can be recorded with an oscillograph.

Knowing the shape of the arrival curve for a given line, the distortion that a bit pattern will suffer may be plotted by adding and subtracting the arrival curve for the various input voltage changes. This is done in Fig. 7.6. The resulting current at the receiving end may be used to operate a relay. The threshold current for relay opening and closing is the dashed line through the curve of received current in Fig. 7.6. When the current is greater than this, the relay closes, thus reconstructing a sharp voltage change equivalent to the input.

If the bit rate transmitted were twice as great as that in Fig. 7.6, it would not be possible to set a threshold level for relay operation that would give the resulting correct bit pattern. The input pulse duration would be too small relative to the transient time of the line.

This was a condition which prevented the correct operation of some of the early submarine cables. The first cable from England to France was laid in 1850 after some difficulty. A receiving device was connected to it by an excited group on the French coast, and a message was sent from the founder of the cable-laying company to Prince Louis Napoleon Bonaparte. The transient time of the cable in water was greater than had been anticipated, however, and to the group's bewilderment the receiver produced gibberish. Before the reason for the trouble was discovered moreover, a fisherman's anchor pulled on the line, and he hauled it aboard his ship in astonishment and cut out a section to show his friends!

KELVIN'S LAW   The transient time is affected mainly by the capacitance, resistance, and length of the line. In 1855 Kelvin produced a famous law of telegraphy saying that for a line of negligible inductance and leakage and negligible terminal impedances, the maximum operating speed is inversely proportional to CRl^2, where C is the capacitance per unit length, R is the resistance per unit length, and l is the length.

To transmit at a given rate, therefore, a given line must not exceed a certain length. If it is desired to transmit over a distance greater than this, a repeater must be inserted in the line.

Any device which reconstructs and retransmits a teletype or other data signal is called a *regenerative repeater*. This is true of modern electronic circuits as well as of the older relay circuits. A repeater which is not regenerative simply amplifies the signal and corrects certain types of distortion that have occurred but does not reconstruct a new, sharp-edged pulse train.

BIAS DISTORTION   Because of the shape of the arrival curve, distortion can occur in the length of the bits received. This is illustrated in Fig. 7.7. As a "mark" changes to a "space" there will be a slight delay in the falling off of the received voltage, and so the receiving machine will begin its "space" slightly late. Similarly the buildup of voltage when a "space" changes to a "mark" will not be instantaneous, and so the receiving machine will begin its "mark" slightly late. Let us suppose that the sampling threshold of the receiving circuit is set slightly too high. As shown in Fig. 7.7 this will cause the space-to-mark transition to be slightly later than the mark-to-space transition. Hence the received spaces will be longer than the marks. Similarly, if the threshold is set too low, the marks will be elongated at the expense of the spaces.

This effect will be greater the higher the transmission speed and the higher the value of CRl^2. It may also become bad when there are several telegraph links in tandem. If the bias distortion becomes too bad, this will result in the incorrect reception of data.

LOADING   The losses of the type described are much higher in twisted wire-pair circuits than in open-wire lines because the capacitance is greater. There are a number of ways of reducing CR, and hence reducing the loss. The resistance can be decreased by using larger cable conductors. The capacitance can be decreased by increasing the separation of the conductors. Both of these approaches are used on wires strung between telegraph poles, but there is a practical limit as to how far apart the wires can be and how heavy or expensive they can be. In cables, enlarging the conductor or the spacing between conductors decreases the number of separate circuits that can be carried in one cable, and so again there is a limit to what can be done.

Heaviside in 1887 proved that the distortion can be minimized by causing the relationship between the inductance and capacitance of the line to satisfy the equation

$$RC = LG$$

where L is the inductance per unit length,

R is the resistance per unit length,

C is the capacitance per unit length, and

G is the shunt conductance per unit length, sometimes known as the leakance, a measure of the leakage between conductors on the line.

RC has already been made as small as possible, but it is still large compared with LG on a conventional telegraph or telephone line. The inductance and leakage are both very small. It is undesirable to increase the leakage of the line as this would diminish the signal; therefore, to satisfy the equation $RC = LG$, the inductance is increased.

The first important application of Heaviside's work was to submarine cables. These had no repeaters, and so distortion was a severe problem. To increase their inductance, iron wire was wound around their core. At a later date, materials with better magnetic properties were discovered, and today more modern cables have a thin permalloy or mumetal tape wrapped helically around the conducting core. This has a high magnetic permeability. By this means, cable transmission speeds can generally be increased by a factor of 4. Some of the early cable designs were speeded up 8 or 10 times.

Adding inductance to a cable is referred to as "loading" it. Wrapping magnetic tapes or wires around a cable is too expensive for most purposes, and it is found satisfactory to insert loading coils at intervals along the cable. The inductance of and the distance between loading coils are chosen to minimize the distortion on the line and so permit high-speed transmission over it. Often the loading coil consists of a ring of powdered permalloy of high magnetic permeability wrapped around with copper wire. Such coils may be fitted at intervals of a mile or so on cable circuits. Open-wire pairs are never normally loaded because their capacitance is much smaller and because their characteristics tend to change with adverse weather conditions.

Loading will be discussed again in connection with alternating-current (ac) circuits which carry a much higher data rate than those discussed in this chapter. If the reader would like to steal a glimpse ahead to Fig. 8.6 he will see how the attenuation constant of an unloaded line varies with frequency. Loading can make the attenuation almost constant over a limited but usable range of frequencies, as shown. Figure 8.5 shows some typical loading coils on the telephone lines.

In general, for both ac and dc circuits, a lessening of distortion on the line means that a higher transmission rate can be achieved, and likelihood of errors is less.

8 AC SIGNALING AND BANDWIDTH

Light, sound, radio waves, and ac signals passing along telephone wires are all described in terms of *frequencies*. In all these means of transmission the instantaneous amplitude of the signal at a given point oscillates rapidly, just as the displacement of a plucked violin string oscillates. Rate of oscillation is referred to as the frequency and described in terms of *cycles per second,* or *hertz.*

With light we see different frequencies as different colors. Violet light has a higher frequency than green; green has a higher frequency than red. With sound the higher frequencies are heard as higher pitch. A flute makes sounds of higher frequency than a trombone. Normally, the light and sound reaching our senses do not consist of one single frequency but of many frequencies or a continuous band of frequencies all traveling together. A violin note has many harmonics higher than the basic frequency with which the violin string is vibrating. The human voice consists of a jumble of different frequencies. When we see a red light it is not one frequency but a collection of frequencies which combine to give this particular shade of red. The same is true with the electrical and radio signals of telecommunications. We will not usually be discussing one single frequency but a collection, or a band of frequencies occupying a given range.

THE SPEECH SPECTRUM The human ear can detect sounds over a range of frequencies; in other words, it can hear sounds of different pitch. A sensitive ear can hear sounds of frequencies ranging from about 30 Hz up to 20,000 Hz, though most people have a range somewhat less than this.

When we refer to a sound of a given frequency, we mean that the air is vibrating with that number of oscillations per second. To transmit this sound the microphone of a telephone converts the sound into an equivalent number of electrical oscillations per second. The telephone channels over which we wish to send data are, then, designed to transmit electrical oscillations of a range equivalent to the frequencies of the human voice, although these frequencies are often changed for transmission purposes.

In fact, the telephone circuits do not transmit the whole range of the human voice. It was found that this was unnecessary for the understanding of the speech and the recognition of the speaker. Figure 8.1 illustrates the characteristics of human speech and shows that its strength is different at different frequencies. Most of the energy is concentrated between the frequencies 300 and 3100 Hz, and each telephone channel is designed to transmit only this range. This is a decision based upon economics. It permits the maximum number of telephone conversations to be sent at one instant over the various physical media discussed in Chapter 9 while still making the human voice intelligible and the speaker recognizable.

Figure 8.1 may be described as a *spectrum diagram.* The reader should become familiar with such a form of diagram as it will be used several times subsequently in explaining the techniques used in data transmission. Figure 8.1 relates to audible frequencies. Diagrams used on later pages of this book refer to frequencies of non-audible signals. Sound spectra or electrical signal spectra are broadly equivalent to light spectra with which the reader may, perhaps, be more familiar. A light beam may be split up into several different frequencies by a spectrometer to produce a band of different colors in the same way that sunlight is split up by rain to form a rainbow. Sometimes the spectrum may contain a continuous band of color, and other times, sharp spectral lines. These spectral diagrams we

Figure 8.1   Spectrum of human speech. In order to transmit speech so that the speaker is recognizable and understandable only the range indicated need be sent.

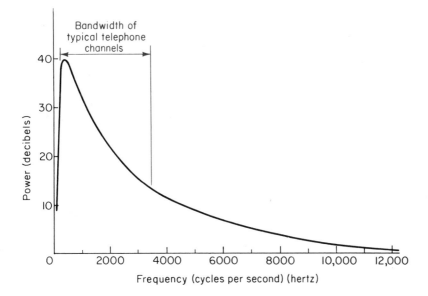

shall use for purposes of explanation do the same to the jumble of frequencies in an electrical signal, and some diagrams, such as those in Fig. 11.11, show sharp spectral lines rather than a continuous spectrum.

A spectral line (of light or electrical waves) relates to transmission at one frequency only. For a single frequency of the transmission may be represented by the equation

$$a = A \sin 2\pi ft$$

where a is the instantaneous amplitude, which is a function of time t, and A is the maximum amplitude.

We shall refer many times to this *sine wave*. It is illustrated in Fig. 8.2.

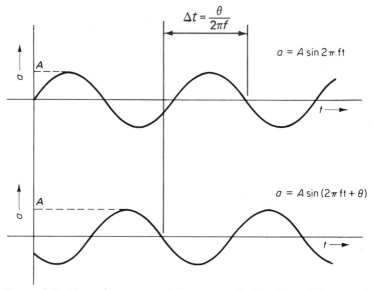

Figure 8.2 Two sine waves of frequency f with phase difference θ.

There are f peaks of the type in this diagram occurring every second—that is what we mean when we say the "frequency" is f Hz. There are f complete cycles per second, two of which are illustrated in each diagram in Fig. 8.2.

PHASE   Figure 8.2 shows two sine waves both of frequency f, which are displaced from each other in time so that their peaks do not occur simultaneously. They are therefore said to be different in *phase*. If one wave is described by $a = A \sin 2\pi ft$, then the other is described by $a = A \sin (2\pi ft + \theta)$, where θ is said to be the *phase difference*.

The time for one cycle of a sine wave, in other words, time $1/f$, is equivalent to a phase difference or angle of $\theta = 360°$ (or 2π radians). Two waves of the same frequency but differing in phase by 360° are identical. One of the methods of sending digital information that will be discussed in Chapter 11 is to transmit it as variations in phase. The maximum that the phase can be varied for this purpose is less than 360° because a 360° change would be indistinguishable from the original.

Two sine waves differing in time by Δt have a phase difference of $\theta = 2\pi f \Delta t$ radians. The time difference Δt, therefore, as indicated in Fig. 8.2, is $\theta/2\pi f$.

DECIBELS   The unit normally used for expressing signal strength in telecommunications is the *decibel*. It is also used to quote gains and losses in signal strength. Spectrum diagrams like that in Fig. 8.1 have decibels on the vertical axis. The unit measures *differences* in signal strengths, not the absolute strength of a signal, and it is a logarithmic unit, not a linear one. Both of these facts sometimes cause confusion. Box 8.1 gives the information the reader should know about decibels and similar units.

The decibel was first used as a unit referring to sound. It made sense to refer to sound levels by a logarithmic unit because the response of the human ear is proportional to the logarithm of the sound energy, not to the energy itself. If one noise sounds twice as great as another, it is not in fact twice the power, but it is approximately 2 decibels greater. The sound energy reaching your ears in the New York subway may be 10,000 times greater than in the room where you are reading this book, but it does not sound 10,000 times greater. It sounds about 40 times greater—you have to shout 40 times harder to make yourself heard to a person the same distance away. Ten thousand times the sound energy is called "40 decibels greater."

Table 8.1 shows the intensities of some common sounds measured in decibels relative to the threshold of human hearing. It makes sense to refer to *transmission* with a logarithmic unit because the signal strength falls off *logarithmically* as it passes down a cable, and so, we can refer to a constant attenuation per mile.

BANDWIDTH   While the frequency range 300 to 4000 or even 300 to 3000 is satisfactory for voice transmission, music would sound poor because it would be clipped of the higher and lower frequencies which give it its quality. To faithfully reproduce the deep notes of percussion or double bass we need to go down to 30 Hz, and to reproduce the high harmonies which make instruments sound realistic, a frequency up to 15,000, or better, 18,000, is desirable. It is toward these extremities that the high-fidelity enthusiasts trive.

AM radio transmits sound frequencies up to 5 kHz and thus it is capable of reproducing music that does not sound too distorted but is not high fidelity. FM radio can produce the whole range needed for high-fidelity reproduction. We say that AM radio uses a *bandwidth* of 5 kHz whereas FM has a bandwidth of 18.

BOX 8.1 Decibels, dBv, dBm, and nepers

DECIBELS   The unit which is normally used for expressing differences in signal strengths in telecommunications is the *decibel*. The decibel is a unit of power *ratio*. It is not an absolute unit but a unit which is employed to compare the power of two signals. Signal-to-noise ratio is normally quoted, for example, in decibels.

A decibel is equal to 10 times the logarithm (to base 10) of the power ratio:

$$\text{number of decibels} = 10 \log_{10} \frac{P_1}{P_2}$$

where P_1 is the larger power (normally) and P_2 is the smaller.

The decibel also used to be defined as the unit of attenuation caused by 1 mile of standard No. 19 gage cable at a frequency of 866 Hz, though this definition is now regarded as obsolete. 1 decibel attenuation means that a signal has dropped to 0.794 of its original

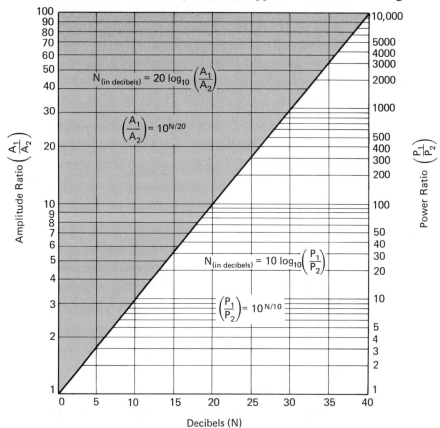

BOX 8.1 Continued

power. 1 decibel gain means that a signal has increased to 1.259 of its
original power.

Voltage and current ratios are also quoted in decibels. Power is
proportional to the square of the amplitude of a signal. A power ratio
of 100, say, is equivalent to an amplitude ratio of 10. Therefore, where
the two current levels are a_1 and a_2, or the two voltage levels are V_1 and
V_2, we have

$$\text{number of decibels} = 20 \log_{10} \frac{a_1}{a_2} \quad \text{or} \quad 20 \log_{10} \frac{V_1}{V_2}$$

Decibels are used to express such quantities as gain in amplifiers,
noise levels, losses in transmission lines, and also differences in sound
intensity. The decibel is a valuable unit for telecommunications because
losses or gains in signal strength may be added or subtracted if they
are referred to in decibels as shown in the figure below.

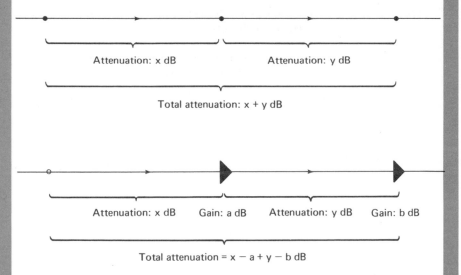

Suppose that a signal is transmitted over a line which reduces it
in power in a ratio 20 to 1. It then passes over another section of line
which reduces it in a ratio 7 to 1. The net reduction is in the ratio 140
to 1. Expressing this in decibels, the first reduction is $10 \log_{10} 20 =$
13.01 decibels, and the second reduction is $10 \log_{10} 7 = 8.45$ decibels.
The net reduction is the sum of these: 21.46 decibels ($10 \log_{10} 140 =$
21.46 decibels).

Similarly, if we say that line loss is 2 decibels per mile, then the
loss at the end of 25 miles of line is 50 decibels. We therefore need an
amplifier of gain 50 decibels to produce a signal of the original power.

BOX 8.1 Continued

The chart on p. 136 will enable the reader to quickly convert power or amplitude ratios to decibels and vice versa.

1 decibel of attenuation means that 0.79   of the input power survives.
3 decibels of attenuation means that 0.50   of the input power survives.
10 decibels of attenuation means that 0.1   of the input power survives.
20 decibels of attenuation means that 0.01   of the input power survives.
30 decibels of attenuation means that 0.001 of the input power survives.
40 decibels of attenuation means that 0.0001 of the input power survives.

dBv, dBm, and dBw

The term dBv is a measure of voltage defined by the equation $dBv = 20 \log_{10}$ [voltage (volts)/1 volt]

The term dBm is a measure of power defined by the equation $dBm = 10 \log_{10}$ [power (milliwatts)/1 milliwatt].

The term dBw is a measure of power defined by the equation $dBw = 10 \log_{10}$ [power (watts)/1 watt].

Table 8.1 The intensity of common sounds measured in decibels relative to the threshold of hearing.

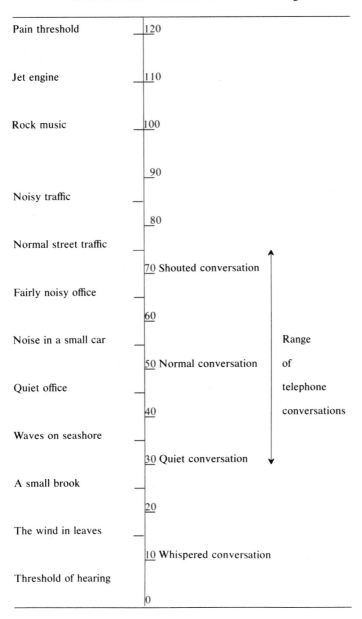

Pain threshold	120
Jet engine	110
Rock music	100
	90
Noisy traffic	
	80
Normal street traffic	
	70 Shouted conversation
Fairly noisy office	
	60
Noise in a small car	Range
	50 Normal conversation of
Quiet office	telephone
	40 conversations
Waves on seashore	
	30 Quiet conversation
A small brook	
	20
The wind in leaves	
	10 Whispered conversation
Threshold of hearing	
	0

Bandwidth means the *range* of frequencies which are transmitted. A telephone channel capable of sending signals from 300 to 3300 Hz has a bandwidth of 3 kHz.

The waves of FM radio do not actually travel at frequencies 30 to 18,000. The transmission occurs at frequencies of the order of 100 million. A similar consideration is true with AM radio and with high-frequency media used as carriers of telephone channels or data transmission.

The transmission media may work efficiently only at frequencies of, say, 70 to 150 MHz (1 million cycles per second is referred to as 1 megahertz). This high frequency must therefore in some way be made to carry the lower frequencies. Stated in another way, the low frequencies must modulate the carrier frequencies to produce a signal which can be transmitted efficiently and from which, after transmission, the lower frequencies can be recovered.

Let us suppose that a bandwidth of 4000 is to be used for voice transmission and that the carrier frequency is 30 kHz. The conversion process may change the frequency band from 0–4000 to 30,000–34,000 Hz. The bandwidth is still 4000 and will still carry the same quantity of information, be it voice or data.

The term bandwidth, therefore, says nothing about the frequency of transmission; it only indicates the size of the range of frequencies. It can be proved formally that the quantity of information a channel can carry is proportional to its bandwidth.

MULTIPLEXING   The physical cables and radio links which the common carriers use for transmitting signals (described in the next chapter) have a high bandwidth—many times higher than the 4 kHz bandwidth needed for speech. To use them economically, many speech channels are transmitted together as a single signal occupying the bandwidth of the physical facility. For efficiency the speech channels are packed together as tightly as possible, like sardines in a tin. Each speech channel occupies no more than 4 kHz of the available bandwidth.

To achieve this, each speech channel is raised in frequency to a 4 kHz slot assigned to it, as shown in Fig. 8.3. This process is referred to as *multiplexing* and is described in more detail in Chapters 12 and 13.

The word "multiplexing" means any process which *permits more than one separate signal to be transmitted simultaneously over one physical channel.* When a segment of frequency spectrum is assigned to each of several signals the process is called frequency-division multiplexing (FDM).

MODULATION   How is speech raised in frequency as in Fig. 8.3? This is carried out by a process called *modulation.*

A sine wave at the high frequency is modified by the speech to be transmitted in such a way that it "carries" it. The sine wave which is modified is called

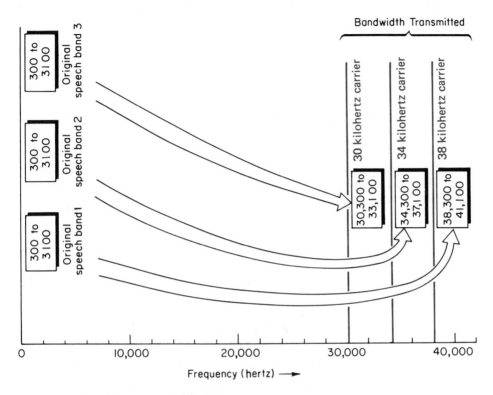

Figure 8.3   Telephone multiplexing. Each speech channel is raised in frequency to fit into a given 4 KHz bandwidth which is assigned to it.

a *carrier*. In Fig. 8.3, three carriers are shown having frequencies of 30, 34, and 38 kHz. Each is modulated by the speech band at frequencies 300 to 3100 Hz to form a block of frequencies slightly higher than the carrier frequency. The 30,000-Hz carrier is modulated to give a signal in the range 30,300 to 33,100 Hz, and so on. The signals are transmitted together in a block of frequencies ranging from 30,000 to 42,000 Hz (or larger). After transmission the signals are separated and *demodulated* to give three signals again at 300 to 3100 Hz.

AM and FM radio transmission stand for "amplitude modulation" and "frequency modulation," respectively, two techniques of carrying sounds at the frequencies to which we tune a domestic radio. "Tuning" is the process of selecting one channel from the many that are received simultaneously at slightly different frequencies. Amplitude modulation and frequency modulation, and other variations, are also used for sending data over telecommunication links. Different modulation systems have different advantages depending on the system needs, and the system planner may be faced with having to choose between modulation methods. This is discussed in Chapters 11 and 17.

BANDWIDTH OF A   A telephone call, as we shall see, may pass through
TELEPHONE CHANNEL   many multiplexing processes. The multiplexing is en-
                         gineered to precise standards, however, and the prop-
erties of the channels which the users perceive are remarkably similar.

Figure 8.4 shows the properties of a typical channel which has passed
through multiplexing equipment. It shows how the signal strength varies with frequency. There is little attenuation of the signal between 300 and 3100 Hz, but
outside those limits its strength falls off rapidly. The main energy of human speech
lies within this frequency range, as can be seen from Fig. 8.1.

The electronics of the telephone plant have deliberately chopped the signal
up in the shape of Fig. 8.4 so that it fits completely into 4 kHz slices. There is
some wastage between the slices, leaving a comfortable gap between channels.
The gap is needed to minimize interferences or "crosstalk" between channels.

LOCAL LOOPS   Multiplexing is normally used only between telephone
                         offices. If a local call is made, employing only local
loops to the telephone exchange, a larger bandwidth than that in Fig. 8.4 could be
obtained and a faster data rate could be transmitted.

Figure 8.4   This curve shows how the amplitude of a signal varies with
frequency after transmission over telephone circuits. The signal is received strongly between the frequencies of 300 and 3400 Hz. When
data is transmitted it must be manipulated with a modem to fit into those
frequencies. Unfortunately telephone company control signals for public lines are also transmitted in this band as shown in Figs. 21.3 and 21.4,
and must not be interfered with by the data transmission.

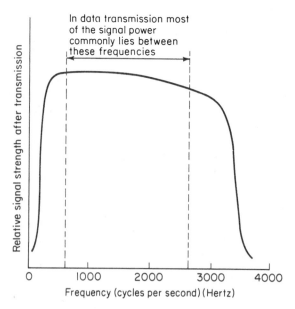

Some lengthy local loops have loading coils connected to them. These reduce attenuation but lower the potential bandwidth and bit rate that could be used. To achieve high-speed transmission over the local loops, the loading coils must be removed.

LOADING Loading, as was discussed in the previous chapter, is a means of decreasing the attenuation of a wire-pair line and holding it more nearly constant over a given frequency range. Its main purpose is to combat the effect of the capacitance between the wires. Local loops from subscriber to central office may have loading coils about every 6000 feet.

Figure 8.5 is a photograph of typical loading coils. Figure 8.6 shows the difference in attenuation between a loaded wire pair (operating at voice frequencies)

(a)
Construction of loading coils manufactured by
L. M. Ericsson, Sweden. (From L. M. Ericsson,
Sweden.)

(b)
Bell System loading coils for telephone lines on
the assembly line at the Western Electric plant.
(From AT&T.)

Figure 8.5   Loading coils.

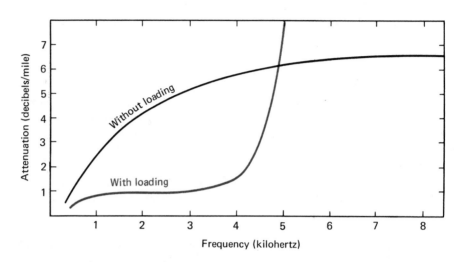

Figure 8.6 The effect of loading on a wire-pair loop using voice-frequency transmission (24-gauge cable). If frequencies much above 4 KHz are to be transmitted, the loading coils must be removed.

and the same unloaded. It will be seen that for a certain band of frequencies the attenuation on the loaded line is less than on the unloaded line and also is fairly constant. Above a certain frequency, however, the attenuation rises fast. The combination of the capacitance of the line and the inductance of the loading coil causes the line to act like a low-pass filter; that is, it transmits signals of frequencies up to a given cutoff point, and above that, attenuation increases greatly. The loading coil must be selected for the particular frequency band that is to be transmitted. It would be quite different on a line that is to carry one voice channel from a line that carries several.

If high frequencies or a high data rate are to be transmitted over a subscriber loop, the telephony loading coils must be removed. In fact frequencies up to a megahertz or so can be transmitted over a telephone loop. This is done with Picturephone transmission. To transmit such high frequencies the signal levels must be carefully coordinated to avoid interference with other services that may be using the higher frequencies on other pairs in the same cable. Generally the gain has to be equalized at short intervals. A baseband data transmission rate of 250,000 bits per second is generally attainable over unloaded local loops. This fact forms the basis for local distribution of the Bell System Dataphone Digital Service, DDS (discussed later), which provides users with bit rates up to 56,000 bits per second.

A second factor limiting high-speed data transmission over local loops is the capacitative coupling between wire pairs carrying high data rates and other wire pairs in the same cable. To avoid this problem coordination, rules are necessary to select which wires in the cable are permitted to be used for high-speed data.

The data-processing designer may have to take loading into consideration on short lines. Some computer manufacturers' specifications for short-distance line equipment state that the lines must be nonloaded. This is generally intended for lines installed by a computer user within his own premises over a distance not greater than about 8 miles.

SENDING
DATA THROUGH
TELEPHONE CHANNELS

When we send data over telephone channels, it must fit into the bandwidth shown in Fig. 8.4. Because telephone channels are so widespread, it is desirable to produce an electronic process that will enable as high a data rate as possible to be squeezed into the space shown in Fig. 8.4 but without incurring an excessive error rate.

This is achieved by another modulation process. A device called a *modem* is used. ("Modem" is a contraction of the words "modulate" and "demodulate"). Some telephone companies refer to a modem as a *data set*. Once again a sine-wave carrier is modulated. This time the carrier frequency is chosen by the equipment manufacturer and modulated so that the significant signal components are inside the speech bandwidth. It is modulated by the data to be sent in such a way that the resulting signal appropriately fills the bandwidth available.

Figure 8.8 shows some typical modems, and Fig. 8.7 illustrates their use. Modems on telephone circuits operate at speeds up to 9600 bits per second. They are commonly in use at the speeds shaded in Table 3.2

Modems are designed and manufactured both by common carriers and by manufacturers of data-processing equipment. Some data processing machines have modems built into them.

Certain baseband signals can be sent down a local telephone loop without modems. A fast bit stream can be transmitted in this way if the loop has no load coils connected to it. However, a baseband signal has a substantial part of its en-

Figure 8.7   The use of modems.

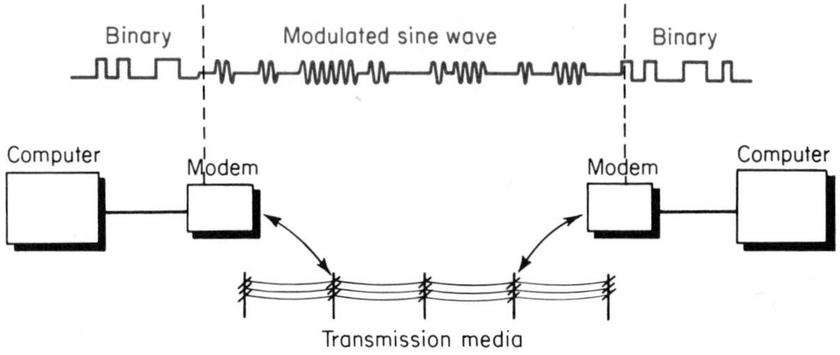

ergy at frequencies below 300 Hz, and this could not be transmitted over the channel in Fig. 8.4. Telephone loops use transformers for terminating the cable pair, and so cannot handle baseband signals which might contain passages of unchanging voltage. A baseband signal may be conditioned before transmission to ensure that it has adequate strength and no passages of unchanging voltage.

Figure 8.8   Typical modems.

The BPO Datel Modem No. 2 for transmitting up to 200 bits per second over Datel 200 lines in the United Kingdom.

The Bell System Dataphone Modem No. 202A for transmitting at speeds up to 1200 bits per second over public dial-up voice lines.

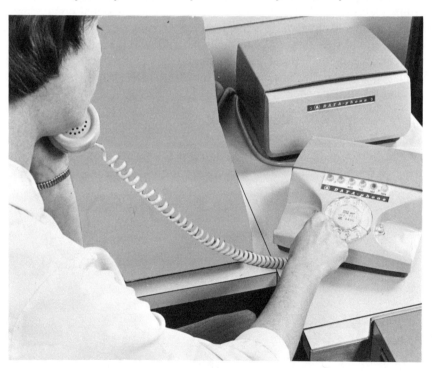

CONTROL SIGNALS   There is serious additional complexity which we will discuss in Chapter 21. On public (i.e., dial-up) telephone lines the telephone company uses some of frequencies inside the band of Fig. 8.4 for its own control signaling. The modem must be designed so that it cannot interfere with these control signals. Control signals are not used on leased lines, so the modems for public lines are either more complex than those for leased lines or else they transmit a lower bit rate, simply avoiding the signal frequencies. Glancing ahead, Figs. 21.3, and 21.4 show control signals in the voice band.

A variety of different modulation processes is possible, and these are further discussed in Chapter 11. There is much scope for ingenuity in the design of modems, and the increasing speed at which data can be sent over telephone lines is largely due to improving modem design. Often the data-processing systems designer is faced with a choice of modems he may use. A further function of the modem is to protect the common carrier lines from undesirable signals which might cause interference with other users or with the network's signaling system.

TWO SEPARATE
MODULATION
PROCESSES
     It will be seen that in their passage between data-processing machines, over analog telephone lines, the bits may undergo two quite separate modulation processes, one carried out on the subscriber's premises by the modem and the other carried out on the telephone company's premises by the multiplexing equipment. The telephone company may subject the signal to many successive multiplexing operations, which we will discuss in a later chapter.

The user can exercise control over only one of these processes — the choice of modem. What the telephone company does with the signal after it enters their lines is not controllable by the user.

THE EFFECT
OF LIMITED
BANDWIDTH
     To perceive the effect of limited bandwidth on data transmission the reader may imagine the electrical signals we discuss to be a jumble of sine waves like those in Fig. 8.2, differing in frequency, f, and amplitude, A. The spectrum diagrams we will use plot A against f and so analyze this jumble.

It was proven by Fourier that any periodic function can be represented by a sum of simple sinusoidal functions. Any function with a period T (and hence a frequency $f = 1/T$) can be considered as a sum of sine functions whose frequencies are integral multiples of f.

Fourier's theorem can be written as follows:

$$F(t) = A_o + A_1 \sin (2\pi ft + \theta_1)$$
$$+ A_2 \sin (2\pi \times 2ft + \theta_2)$$
$$+ A_3 \sin (2\pi \times 3ft + \theta_3)$$
$$\vdots$$
$$+ A_n \sin (2\pi \times nft + \theta_n) \qquad (8.1)$$
$$\vdots$$

The function in question is here represented by a series of sinusoidal components (spectral lines) of differing amplitudes at frequencies f, $2f$, $3f$, and so on. These may be referred to as "harmonics." A plucked violin emitting a "middle C" note of frequency 200 Hz has harmonics at 400, 600, 800 Hz, and so on.

In practice, as with the case of the violin string, the amplitudes, A_n, of the higher harmonics are quite small and become smaller as n becomes larger.

If we wish to transmit the original signal with absolute accuracy, we must transmit all those harmonics that are of significant magnitude and preserve their phase relationships. If we do not transmit all those harmonics, the resulting signal will then be only an approximation of the original.

The middle C violin note sounds realistic when reproduced over a hi-fi unit with a frequency range up to 18,000 Hz. It is recognizable over an AM radio with an upper frequency of 5000 Hz. If somebody played the violin to you over the telephone, the 3-Hz bandwidth would make it almost unrecognizable as a violin. And if you listened only to the first few harmonics up to, say, 600 Hz, the note would definitely not be recognizable as a violin but *would* be recognizable as "middle C."

This is the situation in data transmission also. We have a limited bandwidth available, and we want to transmit the maximum number of bits per second over it. We do not therefore transmit all the harmonics but only enough for the bits to be recognizable as such.

This is illustrated in Fig. 8.9, which shows how "bits" are likely to become distorted on an actual transmission system. Suppose that we transmit pulses such as those shown at the top of the figure. The data rate is 2000 bits per second. If we transmit the fourth harmonic, as shown at the bottom of the figure, the resulting pulse shape is reasonably close to the original. It would be much closer if we transmitted, say, the eighth harmonic. To transmit the fourth harmonic a bandwidth slightly over 8000 Hz is needed, and for the eighth harmonic, over 16,000. With a bandwidth of 4000 Hz the pulses are reasonably like the original, and with 2000 could certainly be detected as being "1" or "0" bits with good equipment. At 1000 Hz the pulses bear little resemblance to their original shape, but skillfully designed detection equipment might recover the original bits. At 500 Hz there is no hope of reconstructing the original.

Figure 8.9 does not show the effects of noise and distortion. These will further change the shape of the pulses, sometimes severely, as discussed in subsequent chapters. The probability of error in the recognition of bits will be greater with the misshapen bits transmitted over smaller bandwidths.

FOURIER ANALYSIS

To understand the functioning of data transmission equipment and modems it is useful to analyze the signals transmitted into their component frequencies.

A relatively simple piece of Fourier analysis can be applied to a repetitive bit pattern with rectangular bits, as shown in Fig. 8.10.

A bit represented by a rectangular increase in voltage, A, is transmitted

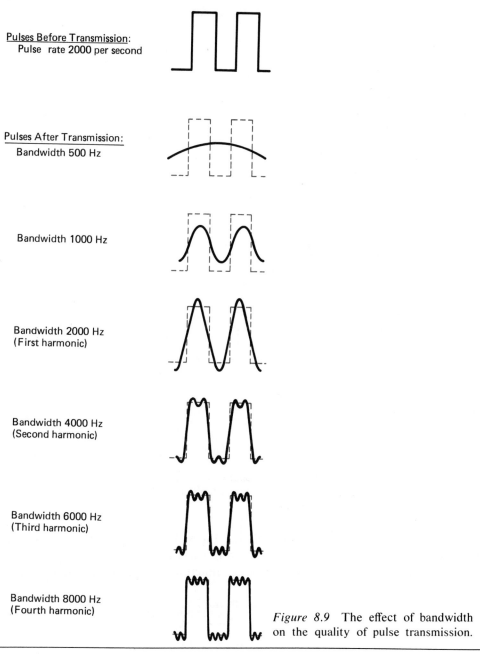

Pulses Before Transmission:
Pulse rate 2000 per second

Pulses After Transmission:
Bandwidth 500 Hz

Bandwidth 1000 Hz

Bandwidth 2000 Hz
(First harmonic)

Bandwidth 4000 Hz
(Second harmonic)

Bandwidth 6000 Hz
(Third harmonic)

Bandwidth 8000 Hz
(Fourth harmonic)

Figure 8.9 The effect of bandwidth on the quality of pulse transmission.

Figure 8.10

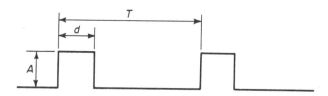

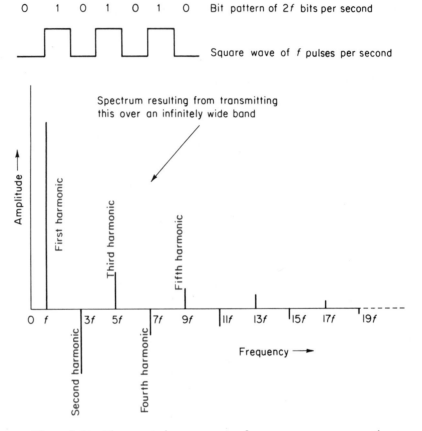

Figure 8.11   The spectral components of a square wave representing a bit pattern 0 1 0 1 0 1.... Over a limited bandwidth only the lower harmonics can be transmitted.

once every T seconds. In other words, the pattern is repeated with a frequency f, where $f = 1/T$. The duration of the bit is d seconds. It was shown by Fourier that this increase in voltage is a sinusoidal function of time, $V(t)$, as follows:

$$V(t) = \frac{Ad}{T} + \left(\frac{2Ad}{T} \frac{\sin \pi d/T}{\pi d/T}\right) \cos 2\pi ft$$

$$+ \left(\frac{2Ad}{T} \frac{\sin 2\pi d/T}{2\pi d/T}\right) \cos 2\pi \cdot 2ft$$

$$+ \left(\frac{2Ad}{T} \frac{\sin 3\pi d/T}{3\pi d/T}\right) \cos 2\pi \cdot 3ft + \cdots$$

$$+ \left(\frac{2Ad}{T} \frac{\sin n\pi d/T}{n\pi d/T}\right) \cos 2\pi \cdot nft + \cdots \qquad (8.2)$$

This consists of sinusoidal (cosine) components at frequencies f, $2f$, $3f$, and so on, up to infinity. In other words, it is a collection of sine waves superimposed on top of each other. These are referred to as "harmonics" of the basic frequency, f. The amplitudes are the terms in parentheses. For high harmonics (large n) these amplitudes are small.

For the case where $d = T/2$, the resulting square wave may be regarded as a 0 1 0 1 0 1. . . bit pattern, and the resulting spectrum is plotted in Fig. 8.11.

In Fig. 8.12 the first three harmonics are plotted: $2A/\pi$ cos $2\pi ft$, $-2A/3\pi$ cos $2\pi \cdot 3ft$, and $2A/5\pi$ cos $2\pi \cdot 5ft$. The sum of these waves is shown as the dotted line, and it will be seen that this roughly approximates to the square wave that is being transmitted.

The 0 1 0 1 0 1. . . square wave is a highly repetitive pattern, and so gives the relatively few spectral lines in Fig. 8.11, which are fundamentals of the frequency of repetition of the basic square wave. If the bit pattern is less repetitive than this, the number of spectral lines increases.

This may be seen still using the simple analysis of Eq. (8.2) by increasing the distance between the pulses. Suppose that the bits 1 0 0 0 0 are repeated so that T/d is now 5. The spectral components then become

$$\frac{2A \sin \pi/5}{\pi} \cos 2\pi ft$$

$$\frac{A \sin 2\pi/5}{\pi} \cos 2\pi \cdot 2ft$$

$$\frac{2A \sin 3\pi/5}{\pi} \cos 2\pi \cdot 3ft \qquad \text{and so on} \qquad (8.3)$$

Figure 8.12 The first three harmonics composing a bit pattern 0 1 0 1 0 1. . . . The sum of the first three harmonics only approximately represents the bit waveform transmitted.

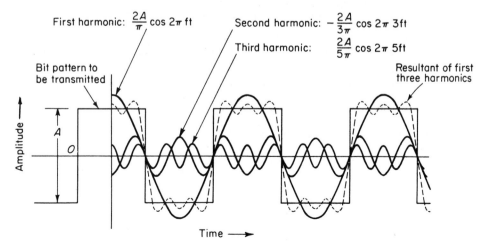

These are plotted in Fig. 8.13. Figure 8.14 shows the resulting wave shapes of such a pulse train when different bandwidths allow different numbers of spectral lines to be sent.

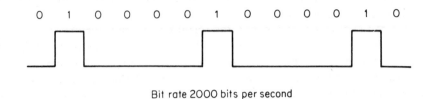

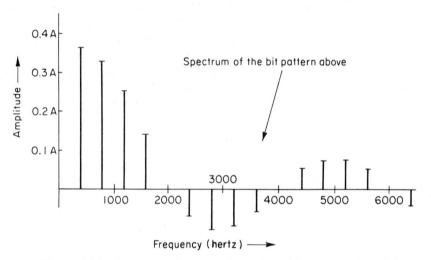

Figure 8.13 Spectrum of rectangular pulse with a separation of five times the pulse width.

If the transmission rate is 2000 bits per second, the first spectral line will be at 400 Hz, the second at 800, and so on. As is shown in Fig. 8.14, the first two spectral lines are barely sufficient to recover the original signal. When three spectral lines are transmitted, the bit pattern becomes more recognizable. A bandwidth of twice the signaling rate in bits per second gives a reasonably squared-off pulse, but the bit pattern could be recovered with less bandwidth than this.

A large number of lines occur when the pulse width, d, is one twelfth of the repetition frequency. The amplitudes of the spectral lines form an envelope shape sin x/x, where $x = n\pi d/T$ or in this case $x = n\pi/12$. Figure 8.15 is drawn, like Fig. 8.13, for a bit rate of 2000 bits per second and a bit width, d, of 1/2000 second, assuming that the presence of a pulse represents a 1 bit and its absence a 0 bit.

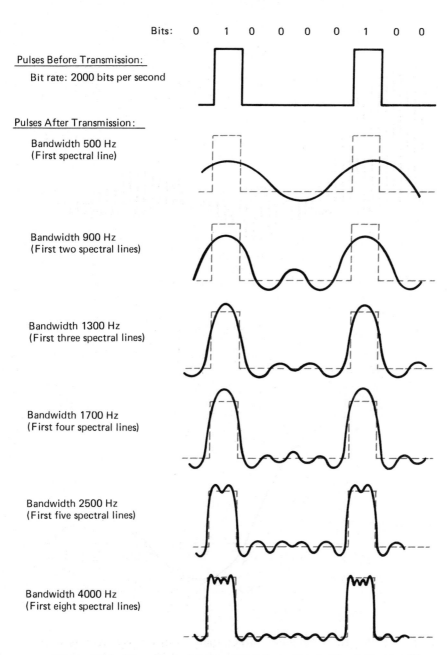

Figure 8.14 The effect of transmitting the pulses in Fig. 8.13 over differing bandwidths.

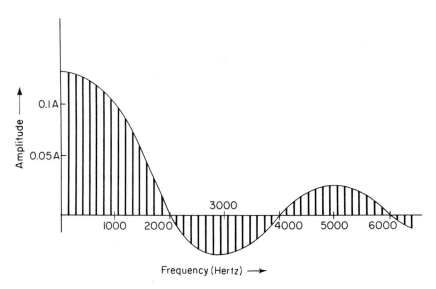

Figure 8.15 Spectrum resulting from a pulse of width *d* repeated at intervals *T*, where $T/d = 15$. Pulse width is 1/2000 second. Bit rate is 2000 bits per second.

Figure 8.16 Continuous spectrum of a solitary pulse with width 1/2000 second.

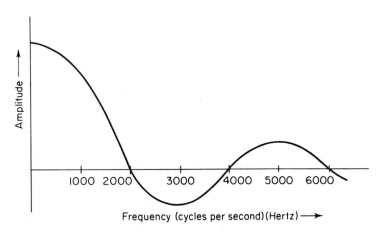

If the bit pattern is irregular rather than strictly repetitive, as in the preceding examples, the number of spectral lines will increase until for a random bit pattern, or for 1 bit on its own, the lines become merged into a continuous spectrum.

Figure 8.16 extends the ratio T/d further than in Figs. 8.15 and 8.13, making T infinite. In other words, Fig. 8.16 represents one solitary pulse, of the same width as those in Figs. 8.13 and 8.15. The spectrum is now continuous. The energy is spread throughout the spectrum rather than concentrated into single spectral lines. The zero points in the spectrum, however, still occur at the same points as in Figs. 8.15 and 8.13, and the relative amplitudes of the envelope containing the spectral lines are similar. Roughly the same bandwidth considerations still apply as those illustrated in Fig. 8.14.

9 TRANSMISSION MEDIA

As we saw in Chapter 1, during the last hundred years a series of inventions have enabled telecommunication links to be built with ever-increasing capacity. The early telegraph links carried signals at speeds up to about 30 words per minute, or about 15 bits per second. Now we are laying down cables which carry many thousands of voice channels, each with a capacity of over 9000 bits per second (with sufficiently good modems). The highest-capacity *digital* cables transmit more than a billion bits per second.

HISTORY   The ability to combine several channels in one physical link came into operation in 1874 with a scheme produced by Baudot, which permitted six users to transmit simultaneously over a single line—a dramatic improvement giving speeds up to about 90 bits per second.

In 1876, Alexander Graham Bell spoke the first sentence over his new invention, the telephone. In the years to follow, telephone lines, switchboards, and later automatic exchanges were built. "Loading," discussed in the previous chapter, was applied to telephone lines in 1899. Prior to this the longest commercial line stretched from New York to Chicago. By 1911, it was possible to speak from New York as far as Denver, a distance which today seems an amazing achievement when one realizes that amplifiers had not been invented.

In 1913, a great step forward was made when the vacuum-tube repeater was first used. A coast-to-coast service with such tubes was operating in the United States by 1915. The development of electronics followed fast, and by 1918, the first *carrier* system was in use, enabling two voice channels to be sent over a single pair of wires. The number of voice channels that can be sent over a single cable has steadily increased through the years. Coaxial cables replaced wire-pair cables for high-capacity links, and today these carry thousands of telephone channels.

In 1897, Marconi formed the Wireless Telegraph and Signal Company. In 1899, he succeeded in sending radio messages across the English Channel, and in 1901, across the Atlantic. Lodge developed means of radio tuning. Radio telegraphy grew fast.

In 1902, Fessenden developed a system for modulating radio frequencies by the human voice, but radio telephony on a commercial scale awaited the coming of vacuum-tube amplifiers and modulators. The first commercial radio stations were installed to connect two land telephone networks in 1920, between Santa Catalina Island, off California, and the mainland. By 1927 telephones in Europe and the United States were linked commercially.

Microwave radio links began to be installed after the war and today have become a major feature of telephone system. Towers, large and small, with a collection of microwave antennas, have sprung up in cities and across the countryside. The chains of microwave relays, which now span the industrialized countries of the world, carry up to 18,000 telephone channels and will probably carry more in the future.

The 1960s brought satellites, lasers, and high-speed waveguides, all of which are taking their place in today's telecommunication picture. Long-distance communication links are fast growing in capacity. As the number of circuits carried by a link increases, the cost per circuit falls. The Bell Laboratories helical waveguide system, experimentally but not yet commercially installed, carries a quarter of a million voice channels.

In this chapter we will discuss the various types of physical transmission media that are in use. In subsequent chapters we will describe in more detail how they are used for voice, television, and other signals and what is involved in sending data over them.

FREQUENCIES

The media will be contrasted by discussing the frequency at which signals are sent over them. Microwave links, for example, operate at a very high frequency, coaxial cables at a lower frequency, and wire pairs still lower. We are all familiar with the frequencies on the dial of a domestic radio. Stations on the FM band are picked up at between about 88 and 108 MHz. Stations of the AM broadcasting band in the United States are between 500 and 1600 Hz. These frequencies, along with the frequencies at which other media operate, are shown in Fig. 9.1. This is a small part of the total electromagnetic spectrum that was shown in Fig. 1.3.

What is going to be of particular interest to us is not absolute frequency of operation but the range of frequencies which can be sent over the facility. As will be discussed in Chapter 16, the quantity of data, or in general the amount of information that can be transmitted, is proportional to the bandwidth or *range* of frequencies that can be sent. In Fig. 9.1, the range of frequencies shown for microwave radio is much greater than that for FM broadcasting, for example. The former stretches from about 2000 to 12,000 MH, a range of 10,000MH. The latter

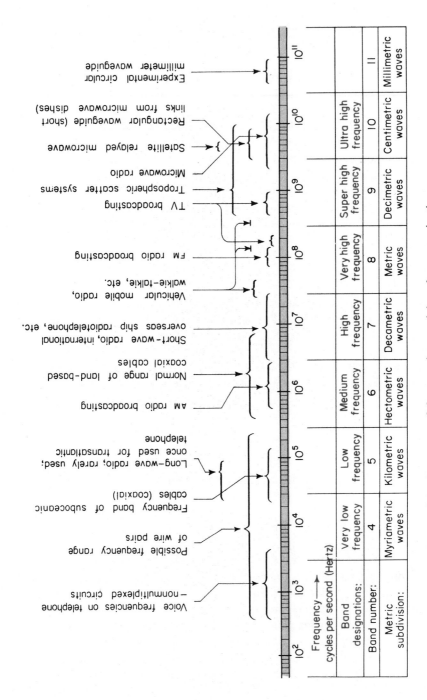

Figure 9.1 Spectrum of frequencies used in telecommunications—a small portion of the total electromagnetic spectrum shown in Fig. 1.2. *Note*: The allocation of radio frequencies to different uses is much more complicated than on this chart which has been simplified to show the main categories discussed in this book.

Band designations:	Very low frequency	Low frequency	Medium frequency	High frequency	Very high frequency	Super high frequency	Ultra high frequency	
Band number:	4	5	6	7	8	9	10	11
Metric subdivision:	Myriametric waves	Kilometric waves	Hectometric waves	Decametric waves	Metric waves	Decimetric waves	Centimetric waves	Millimetric waves

Voice frequencies on telephone — nonmultiplexed circuits

Possible frequency range of wire pairs

Frequency band of suboceanic cables (coaxial)

Long-wave radio, rarely used; once used for transatlantic telephone

AM radio broadcasting

Normal range of land-based coaxial cables

Short-wave radio, international overseas ship radiotelephone, etc.

Vehicular mobile radio, walkie-talkie, etc.

FM radio broadcasting

TV broadcasting

Tropospheric scatter systems

Microwave radio

Satellite relayed microwave

Rectangular waveguide (short links from microwave dishes)

Experimental circular millimeter waveguide

Frequency → cycles per second (Hertz)

stretches from about 80 to 150 MH, a range of about 70 MHz. Wire pairs transmit frequencies up to about 200 or 300 KH normally. One could, then, transmit far more information over microwave than over the FM broadcasting frequencies and far more over the latter than over wire pairs.

TODAY'S TRANSMISSION　　　The following are the media of main interest today.
MEDIA

1. Open-Wire Pairs

In earlier years most telephone connections were made by means of wire pairs stretched between telephone poles. The pairs of wires shown in Fig. 9.2 are suspended from insulators on the crossbars of the poles. The wires are copper, or steel coated with copper — steel for strength, copper for conductivity. At frequencies above 1000 Hz, most of the current flows on the outside "skin" of the

Figure 9.2　At the turn of the century American cities had vast numbers of open-wire pairs. Today they are found only in rural districts and are declining. (Courtesy AT&T.)

wire, in the copper coating. The wires in a pair are about 0.128 inch in diameter and spaced by about 8 to 12 inches.

A wire pair can carry telephone conversations a long way without amplification. It was on such wires, for example, that New York could speak to Denver before vacuum-tube amplifiers were invented. Today it is often desirable to send several voice channels together over the same pair of wires. This needs a higher frequency, and at higher frequencies the attenuation is greater. Therefore amplifiers are placed closer together in the line.

Wire pairs are susceptible to *crosstalk*. Electromagnetic or inductive coupling produces interference, and a conversation on one pair could become faintly audible on the next. Large separation of the adajacent pairs and periodic reversal of the wires reduce this to an almost negligible level. Weather conditions affect the loss or *attenuation* on open-wire lines. Leakage occurs at the insulators when they are wet. The electrical resistance of the wires rises with temperature, and wet and humid conditions increase the attenuation.

Open-wire pairs have now largely been replaced by cables but are still seen in rural districts and in less highly developed countries. They are fast becoming part of the romantic past and as such form the setting in songs such as "Moonlight in Vermont."

2. Wire-Pair Cables

In the wire-pair cables which have replaced open-wire pairs, the conductors are insulated and brought close together. Many of them may be packed into one cable. This would tend to increase *crosstalk* considerably. The conductors are twisted in pairs to minimize the electromagnetic interference between one pair and another. Different twist lengths are used for nearby pairs. The group of conductors is wrapped in a tough protective sheath. Cables laid in cities have many hundreds of wire pairs in one cable, as shown in Fig. 9.4. Figure 9.3 shows a smaller twisted wire-pair cable slung, for convenience, beneath the bars of telephone poles already carrying open-wire pairs. The wires in cables are much smaller than wires for open pairs. Short cables use wires about 0.015 inch in diameter. Longer cables use thicker wires up to about 0.056 inch. Because of this the resistance of the wires is higher, and the signal needs to be amplified more frequently than with open-wire pairs. The amplifiers, or "repeaters," are in manholes or attached to the poles carrying the cable at intervals along the cable route, commonly every 3 or 4 miles. Similar repeaters are used with open-wire pairs but these might be, say, 10 miles apart, or less for some systems.

Twisted wire cable, as with the other media we will discuss, can carry more than one voice channel. The frequencies of the human voice are raised to higher frequencies. Different channels are raised by different amounts, and in this way the frequency range available (shown in Fig. 9.1) is filled. This is a form of *multiplexing,* called "frequency division multiplexing," and it is discussed further in later chapters. It is common for one wire pair on a trunk route to carry 12 voice

Figure 9.3 This telephone pole crossbar was designed to carry open-wire pairs, two of which are shown. For convenience a group of twisted-wire pairs have been slung beneath the same bar. The twisted-wire pairs here have an amplifier and the open-wire pairs do not. Open-wire pairs transmit signals 30 miles or more without amplification. Twisted-wire pairs normally have amplifiers every 2 to 4 miles. (Photo by author.)

Figure 9.4 Many hundreds of wire pairs may be grouped together in a rough sheath to form a cable like this. Such cables are laid under the streets of cities, or along the roadside. They are used for subscriber loops and short trunks. (Photo by author.)

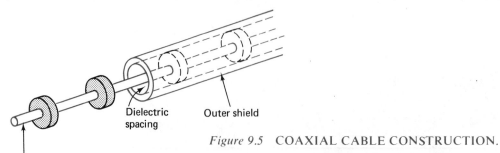

Inner conductor
carrying current
of about 0.4 to
10 megahertz

Dielectric
spacing

Outer shield

Figure 9.5 COAXIAL CABLE CONSTRUCTION.

Many of these coaxial tubes are bound
together around a central wire core for
strength. Some wire-pair conductors
are also included in the cable.

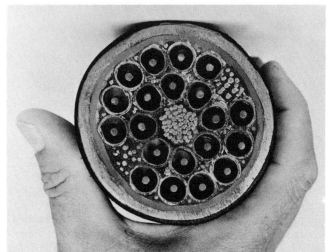

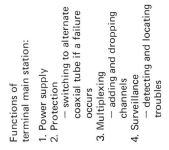

Functions of
terminal main station:

1. Power supply
2. Protection
   — switching to alternate
   coaxial tube if a failure
   occurs
3. Multiplexing
   — adding and dropping
   channels
4. Surveillance
   — detecting and locating
   troubles

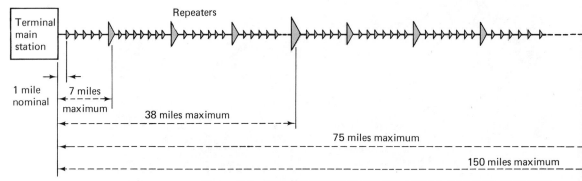

Terminal
main
station

Repeaters

1 mile
nominal

7 miles
maximum

38 miles maximum

75 miles maximum

150 miles maximum

Figure 9.6 COMPONENTS OF COAXIAL TRUNK SYSTEM

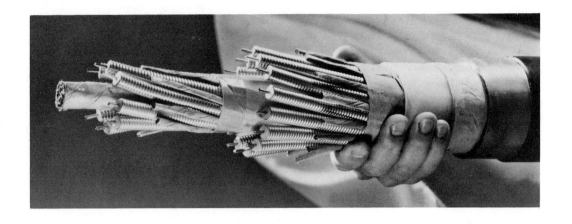

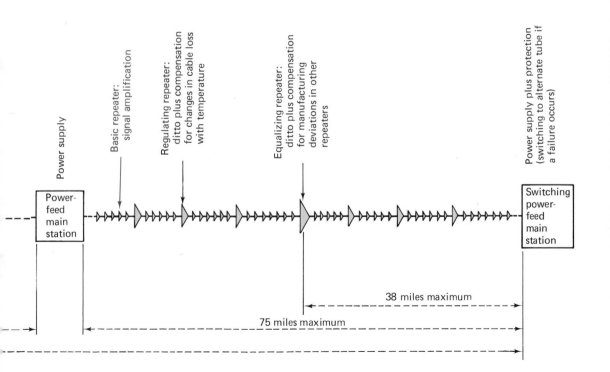

Power supply

Basic repeater:
signal amplification

Regulating repeater:
ditto plus compensation
for changes in cable loss
with temperature

Equalizing repeater:
ditto plus compensation
for manufacturing
deviations in other
repeaters

Power supply plus protection
(switching to alternate tube if
a failure occurs)

Power-
feed
main
station

Switching
power-
feed
main
station

38 miles maximum

75 miles maximum

channels simultaneously in two directions using frequencies up to about 268 kHz. Because of the desire to pack many channels into one cable, it is required that the cable operate at the highest possible frequency. Unfortunately, however, the loss of signal strength, or "attenuation," becomes great at high frequencies. Figure 10.7 is an illustration of the different signal attenuation encountered at different frequencies.

Recent developments in repeater design permit frequencies of 1 MHz to be transmitted over wire pairs. Such repeaters would be spaced at intervals of about 1 mile. This capability will permit Picturephone signals to be sent over wire pairs. Three wire pairs will be used for one unit, one for vision in each direction and one for sound in both directions.

The *capacitance* between conductors is much greater in a cable pair than in open-wire lines, because the conductors are much closer together. This has a more serious effect at high frequencies than at low frequencies. For this reason, when multiplexing was first developed it was used only on open-wire lines. The capacitance of a cable could be reduced by increasing the separation between the wires, but this would increase the cost and substantially reduce the number of wires a cable could carry.

3. Coaxial Cable

As the frequencies become higher, the current flows more on the outside edge of the wire. It uses an increasingly small cross section of the wire, and so the effective resistance of the wire increases. This is called the "skin effect." Furthermore, at higher frequencies an increasing amount of energy is lost by radiation from the wire. Nevertheless, it is desirable to transmit at as high a frequency as possible so that as many separate signals as possible can be sent over the same cable. The skin effect limits the upper frequencies.

A coaxial cable can transmit much higher frequencies than a wire pair. It consists of a hollow copper cylinder, or other cylindrical conductor, surrounding a single wire conductor. The space between the cylindrical shell and the inner conductor is filled with an insulator. This may be plastic or may be air, with supports separating the shell and the inner conductor every inch or so. A coaxial cable is shown in Fig. 9.5.

Several coaxial tubes are often bound together in one large cable as in Figs. 9.5 and 9.6. A number of twisted-wire pairs are packed in among the coaxial tubes and are usually used for control purposes. At higher frequencies there is virtually no crosstalk between the separate coaxial tubes in such a link because the current tends to flow now on the inside of the outer shell and the outside of the inner wire. Because of this shielding from noise and crosstalk, the signal can be dropped to a lower level before amplification.

A very large number of separate telephone calls can be transmitted together down a coaxial cable system. Whereas a single-wire pair commonly carries 12 or 24 voice channels, one single coaxial tube commonly carries 3600, and the highest-capacity ones carry 10,800. The highest capacity cables have 22 tubes, 10 transmitting in each direction and two spares which can be switched into operation if a failure occurs on any of the working tubes. These cables carry a total of 108,000 2-way telephone conversations.

The main reason for this higher capacity is that the signal loss, or attenuation, does not become severe until very high frequencies. Figure 10.8 illustrates this and should be compared with Fig. 10.7.

Furthermore, other forms of distortion are much less. On a wire pair carrying voice frequencies, the velocity of propagation of the signal varies widely with frequency. This is illustrated in Fig. 9.7. The lower frequencies of the human voice will arrive later than the higher frequencies. The longer the line, the greater the delay will be, and so the greater the distortion will be. This is referred to as "delay distortion" and is further discussed in Chapter 27. It can have a more serious effect when the line is used for data rather than for voice. Coaxial cable circuits give a higher velocity of propagation, which varies only very slightly with frequency, thus giving very little delay distortion. The propagation velocity along a coaxial cable at frequencies above about 4 kHz is approximately equal to the velocity of light, or if the space inside the conducting cylinder is filled with plastic, then to the velocity of electromagnetic waves in that material, which might be 25 to 45% lower than in air.

Figure 9.7 Propagation velocity of signals over loaded and unloaded wire pairs. The difference in propogation velocity at different frequencies causes *delay distortion*, discussed in Chapter 27.

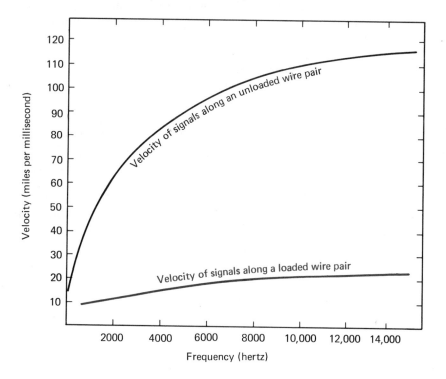

A *loaded* wire pair *gives* less distortion, but the transmission velocity is lower. It is normally about 10,000 to 20,000 miles per second at voice frequencies. The signal in a coaxial cable can thus travel at 10 times the speed of an equivalent signal in a loaded voice-frequency wire pair. This is of value because, as will be discussed in the next chapter, it often removes the need for *echo suppressors*.

The additional expense of coaxial cables is justified by the following advantages; the first is by far the most important:

1. A *much* larger number of channels can be sent over one cable.
2. Negligible crosstalk between cables.
3. Lower delay distortion and variation of amplitude with frequency.
4. Higher propagation speeds, which can remove the need for echo suppressors on many long lines, because the interval between the speech and its echo becomes very small.

Not surprisingly, many thousands of miles of coaxial cable links have been installed as intercity trunks throughout the world. The highest-capacity highways of the telecommunications world have been coaxial trunks such as that illustrated in Fig. 9.6. Today only the waveguide has a higher capacity.

4. Microwave Radio

The main contender with coaxial cable circuits for bulk transmission is microwave radio. This medium has been used more extensively than coaxial cable for the building of long-haul trunks. Like coaxial cable, microwave links carry thousands of voice channels and are in widespread use for the transmission of television. A microwave route, however, carries much less traffic than the largest coaxial routes.

Many cities of the world now have their skylines proudly dominated by a tower carrying microwave antennas. Tokyo has a tower like the Eiffel Tower, but 40 feet higher. East Berlin has one 1185 feet high. One of London's most expensive dinners can be eaten in a revolving restaurant just above the microwave antennas, and Moscow, outdoing the rest, has one 250 feet higher than the Empire State Building.

As can be seen from Fig. 9.1, microwave radio is at the high-frequency end of the radio spectrum. Unlike long-wave radio, it is not reflected by the ionosphere, and its scattering by hills and other objects is less. It needs line-of-sight transmission, and the antennas which relay it, forming chains across the country, are all on towers within sight of one another. Figure 9.8 shows typical microwave systems. Relay towers are usually spaced 20 to 30 miles apart. On a long-distance telephone conversation, or television transmission, the signal is picked up every 30 miles or so, and retransmitted.

Most major terrestrial long-distance links today are coaxial cable or microwave. Fig. 9.9 shows the layout of the Bell System long-distance highways.

A long-distance microwave circuit has fewer amplifiers than a coaxial cable link of the same length. The microwave link has amplifiers at each relay point, in other words about 30 miles apart. The coaxial cable has amplifiers every 1½ to 4 miles. A coaxial cable circuit coast to coast across the United States therefore has a thousand or so amplifiers. The equivalent microwave circuit has only 100 or so. It is a disadvantage to have too many amplifiers because a slight defect in them is cumulative. For television transmission, for example, the amplification needs to be held constant within narrow limits for parts of the signal at different frequencies. If it has to pass through a thousand amplifiers with similar characteristics, this means that each one of the thousand must be very exact indeed in this respect. This is difficult and expensive to engineer. Therefore, microwave links have come into wide use for television transmission. Fortunately, the television peak hours do not coincide with the peak usage of telephones, and so the same facility can be used by day for telephones. As will be seen in a later chapter, one television channel can carry 1200 telephone channels.

During the night hours these high-capacity links are underutilized. Furthermore, many computers stand idle on their third shift. As the night hours are at different times traveling east to west, it would make economic sense to link computers at different longitudes and fill some of the idle time on both the machine and the transmission facilities.

Unlike lower-frequency radio, microwave antennas are fixed rigidly in order to focus a beam of the narrowest angle possible on their distant associated antenna. It is common for a beam of about a 1° angle to be used, and a typical antenna size is about 10 feet across. Figure 9.10 shows the relative strengths of signals traveling in different directions from a typical antenna. Microwave radio is scattered by hills and other objects. The beams from the antennas must clear trees and the buildings; otherwise their reflections may cause echoes.

Different moisture and temperature layers can cause the beam to bend and vary in amplitude, just as we sometimes see light shimmering over a hot surface or causing minor mirages along a road surface in the sun. Occasionally these effects can cause fading. Rain can change the attenuation slightly, especially at the higher microwave frequencies, and occasionally trouble is caused by reflection from unanticipated objects such as helicopters or new skyscrapers in a city. To a limited extent automatic compensation for changes in the radio attenuation is built into the repeaters.

In addition to the long-haul trunks, many short microwave links of lower capacity are in use. The telephone companies find it convenient in some locations to use them as feeders to the main exchanges. Television companies use them for outside broadcasting. The army uses a portable microwave receiver-transmitter as a field telephone.

Privately owned microwave links have been set up by some corporations. The American Electric Power Service Corporation, for example, has more than a thousand path-miles of private microwave links used mainly for voice. Other corporations have set up shorter links. In some cases, they have been set up primarily

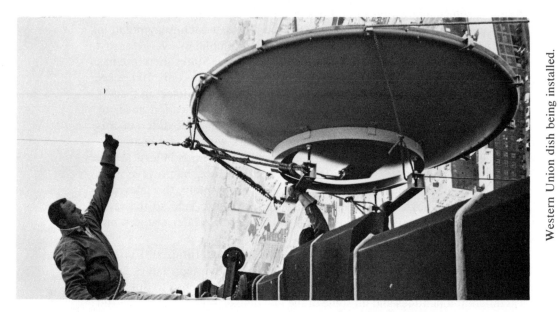

The Post Office Tower, London.

Microwave dish on a 17th century Portugese windmill.

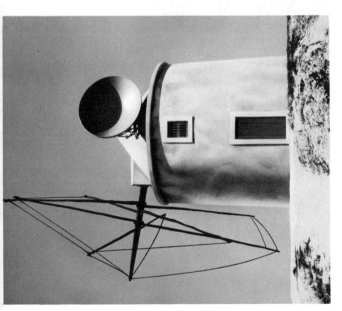

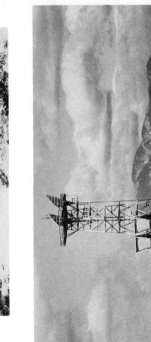

Typical AT&T long-haul microwave towers.

Figure 9.8   Microwave antennas.

169

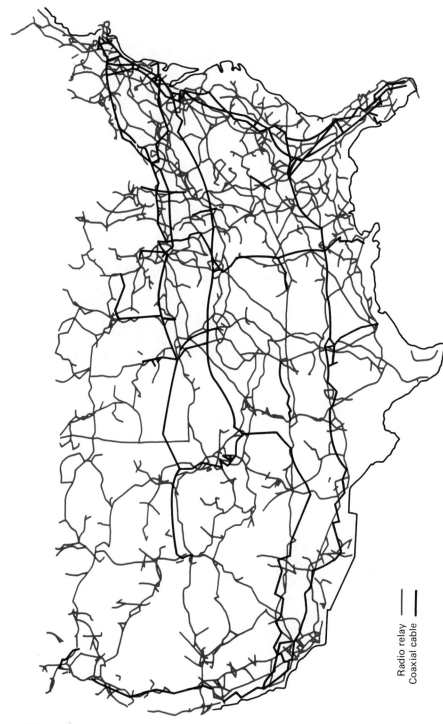

Radio relay ———
Coaxial cable ━━━

Figure 9.9   Microwave and coaxial cable trunks in the Bell System.

for data where there was a large amount of data to transmit — as, for example, when part of a programming team is remote from the computer being used.

A major problem with microwave is radio interference. Many cities are now congested with criss-crossing microwave beams. Only certain parts of the radio spectrum are allocated to the common carriers, and, unless new frequencies can be employed, further expansion of city microwave transmission is limited. This congestion seriously restricts the siting of satellite earth stations, which use the same microwave frequencies. Figure 9.11 shows the microwave congestion around New York City [1].

5. Satellites

A communication satellite provides a form of microwave relay. It is high in the sky and therefore can relay signals over long distances that would not be possible in a single link on earth, because of the curvature of the earth, mountains, and atmospheric conditions.

The first communication satellites were in relatively low-altitude orbits and consequently speeded around the earth in a few hours. This was bad because the ground antennas had to move constantly to beam signals to them, and they were only overhead for a brief period. Transatlantic television transmission was born but was confined to 5-minute sessions.

Today's communication satellites appear to remain stationary above the equator at a height of 22,300 miles. In fact they travel round the earth in exactly the time the earth takes to rotate. Small gas jets on board a satellite make occasional adjustments to its position to keep it almost stationary.

Satellites are powered by solar batteries. Like microwave links on earth they handle several thousand voice channels. Because the high-power transmission from their earth stations interferes with terrestrial microwave links, the earth stations are sited outside the cities — sometimes 50 miles away. Western Union's earth station serving the New York area, for example, has to avoid the microwave paths shown in Fig. 9.11 and so is sited 50 miles from New York in Glenwood, N.J. Figure 9.12 shows Western Union's microwave and satellite system. Figure 9.13 shows their Westar satellite, the first domestic satellite in the United States (following Canada's lead with the Anik satellites) and an earth station. A new generation of satellites will operate at a somewhat higher frequency than terrestrial microwave, and will permit satellite antennas to be sited in cities, and in general at any corporate premises. In Chapter 15 we will discuss satellites.

Communication satellites are sufficiently far away that the radio beam takes a quarter of a second to travel up to them and back. A half-second pause is thus introduced between a message transmitted in one direction and the response to it being received. Complex systems engineering will be required to integrate satellites and terrestrial links into effective communication networks.

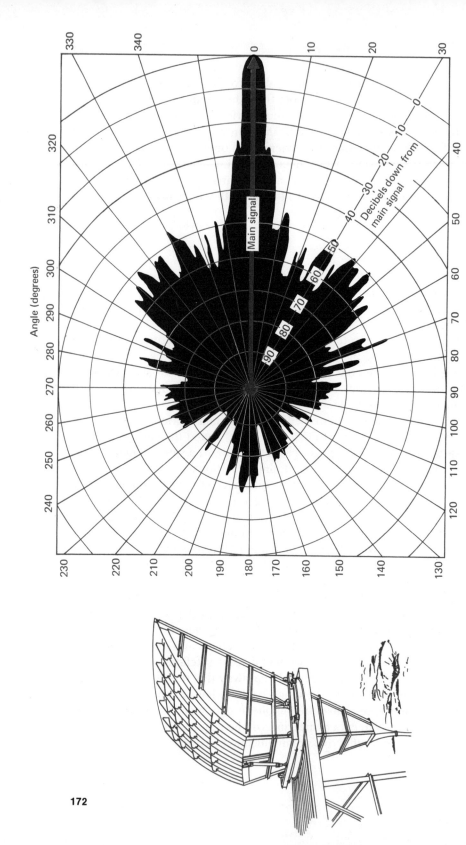

172

Figure 9.10 The relative strengths of the signal transmitted in different directions from a horn-reflector antenna. The chart is for a vertically polarized signal at 3740 MHz. The main signal is a beam which spreads out over an angle of about 3°. (Courtesy AT&T.)

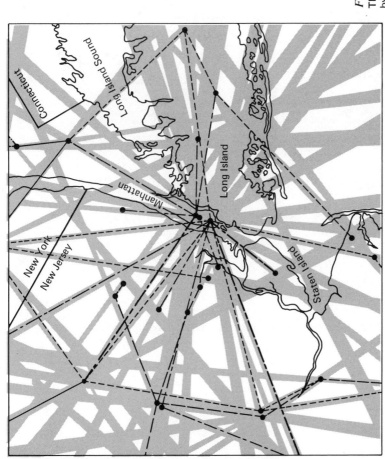

MICROWAVE INTERFERENCE

4 gigahertz ------
4 and 6 gigahertz ————
6 gigahertz —·—·—
6 and 11 gigahertz —··—··—
11 gigahertz ————

Figure 9.11 Microwave interference.
There is severe microwave congestion in most large cities, as indicated by the shaded beams in this map of the New York area. Several different frequencies are used so the routes do not produce as much mutual interference as the shaded beams suggest. Nevertheless not much expansion of microwave facilities will occur unless new frequencies are employed by the common carriers.

Todays satellites use the same frequencies as terrestrial microwave links. Consequently satellite transmitter earth stations will not be sited close to cities like New York until new satellite frequencies are used. (12/14 gigahertz).

Westar

Alaska

Hawaii

Puerto Rico

Microwave
repeater station
Satellite earth station

Figure 9.12   Western Union's microwave and satellite system. (Courtesy Western Union.)

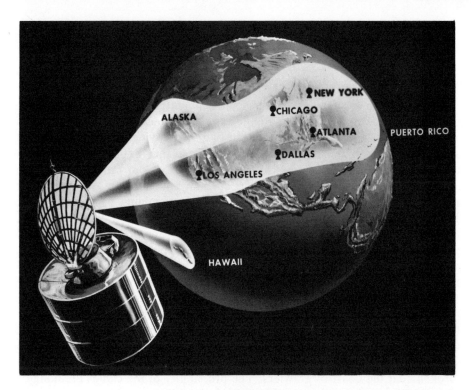

Figure 9.13 Westar satellite and earth station.

The Westar satellite and the positions of Western Union's first five earth stations.

Western Union's earth station of Glenwood, N.J., serving the New York area.

6. Submarine Cable

Satellites have spectacularly increased the potential for overseas communication. Prior to the launching of satellites the main means of communication across the sea was by cables on the ocean bed. The laying of telegraph cables across the sea was a major triumph of engineering in the 1850s, starting with an unsuccessful cable across the English Channel in 1850 and progressing to the transatlantic cable of 1858, which had a spectacular but short life. After immense difficulties and frustrations in the laying of the first transatlantic cable, it fired the public imagination when it finally worked. But the signals that trickled through it were so minute that only the most sensitive mirror galvanometer of the day could detect them. The submarine cables of that era had no repeaters. The first message, from Queen Victoria to President Buchanan, was of only 90 words but took $16\frac{1}{2}$ hours to transmit. Sixteen days later the cable failed and never worked again.

It was 1956 before the first voice cable was laid under the Atlantic. The difficulty of building underwater amplifiers of the immense reliability needed prevented this from being done earlier. The cable had vacuum-tube amplifiers every 39 nautical miles. It was 0.62 inch in diameter and surrounded by heavy armor to give it strength and protection from corrosion. Even so the cables were occasionally broken near the coast by trawlers. The probability of the cables breaking or failing in deep water has to be very low as they would be prohibitively expensive to repair.

Table 9.1 summarizes the first four generations of AT&T subocean cables. The cables are coaxial, but with a larger spacing between the inner and outer conducters than for cables on land. Figure 9.14 shows the construction of subocean cables.

Table 9.1   Four Generations of Transatlantic Cables

	First Generation 1956	Second Generation 1963	Third Generation 1969	Fourth Generation 1976
AT&T system designation	SB	SD	SF	SQ
Capacity (3 kHz voice channels)	48	140	845	4000
Top frequency on cable	164 kHz	1.1 MHz	5.9 MHz	28 MHz
Cable	2 separate armored 0.62 in. cables	1 armorless 1 in.	1 armorless $1\frac{1}{2}$ in.	1 armorless $1\frac{1}{2}$ in.
Repeaters	Vacuum tube	Vacuum tube	Transistor	Transistor
Repeater spacing (nautical miles)	39	20	10	10
Cable length (nautical miles)	2200	3500	4000	4000

Because of the long spacing between amplifiers, the upper frequency at which the cables can be operated is lower than with land cables. The capacity of the first cables was therefore only 48 voice channels, and two cables were used for two-way talking. AT&T's fourth-generation cables carry 4000 two-way voice channels in one cable.

Today's cables have their capacities increased about 90% using an ingenious technique called TASI, described in Chapter 12. The future of such cables, however, is uncertain because satellites span the oceans at much lower cost. On many of today's transatlantic calls, transmission in one direction is by satellite and the return path is by cable. This makes the overall delay in obtaining a response less than the delay of half a second that is encountered if a satellite is used for both directions. Figures 9.15 and 9.16 show the major international cables and satellite lines.

7. High Frequency Radio Telephone

Before 1956 almost all transatlantic telephone traffic was carried by high-frequency radio telephone. It is still used for international telegraphy and for telephone to ships at sea and to countries not connected by cable and with no satellite antenna.

As will be seen from Fig. 9.1 the *high-frequency* band is of much lower frequency than that of microwave. High-frequency (HF) radio transmission is reflected by the ionosphere. Because of the movement of and changes in the ionosphere, it is subject to fading, distortion, and periodic blackouts. It is used for shortwave worldwide broadcasting and can be picked up on some domestic radio sets. Long-distance HF telephone or telegraph circuits rarely form part of a computer data transmission system, except for transmitting telegraph signals from remote parts of the world. The data error rate is extremely high, and elaborate means of error detection and automatic retransmission are needed. Nevertheless, some such links have occasionally been used with success.

8. Tropospheric Scatter Circuits

The troposphere is much lower than the ionosphere and more stable. It extends up to about 6 miles. The ionosphere is above 30 miles. The troposphere scatters radio waves, and this is used for telecommunication links of up to about 600 miles where it is not possible, or economic, to construct land lines or microwave links. There are tropospheric scatter circuits over the frozen mountains of Alaska and from the United States to Nassau. There used to be one to Cuba. It is possible that other points in the Caribbean will become linked with such circuits.

Island chains such as these are typical of areas that can benefit from these circuits. The tropospheric scatter circuit is used to transmit beyond the visible horizon. The received signal is the result of a multiplicity of reflections of different paths from the troposphere.

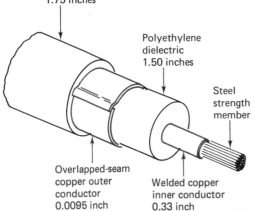

Protective polytheylene jacket
1.75 inches

Polyethylene
dielectric
1.50 inches

Steel
strength
member

Overlapped-seam
copper outer
conductor
0.0095 inch

Welded copper
inner conductor
0.33 inch

Figure 9.14 SUBOCEAN CABLES

AT&T's third generation of transatlantic telephone cable, TAT-5, from Rhode Island to Spain, carries 825 voice channels. Transistorized repeaters are inserted every ten miles and are powered by dc voltages of several thousand volts applied at the ends of the cable between the inner and outer conductors.

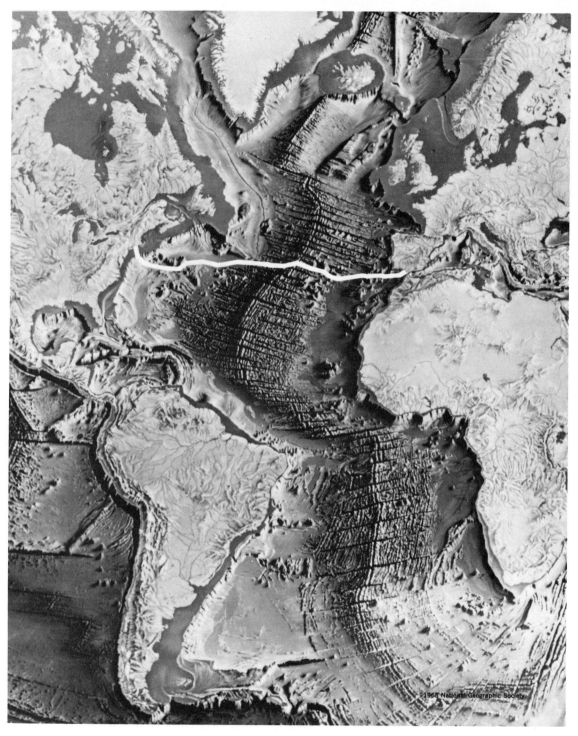

The white line is the route of TAT-5. In laying the cable, AT&T's Long Lines Department had to avoid such deep sea hazards as mountains, earthquake sites, and dangerous currents. The cable took 2 years to complete. Satellites can give transatlantic circuits of much lower cost per channel.

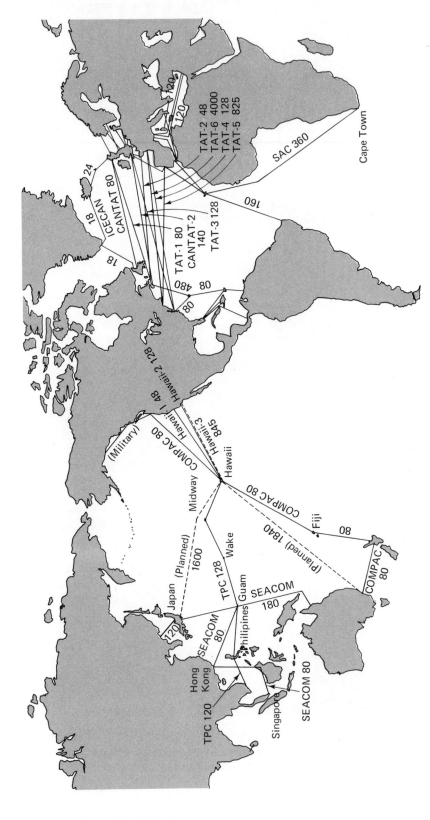

Figure 9.15   The main oceanic telephone cables. The numbers indicate the number of voice channels in a cable (without TASI).

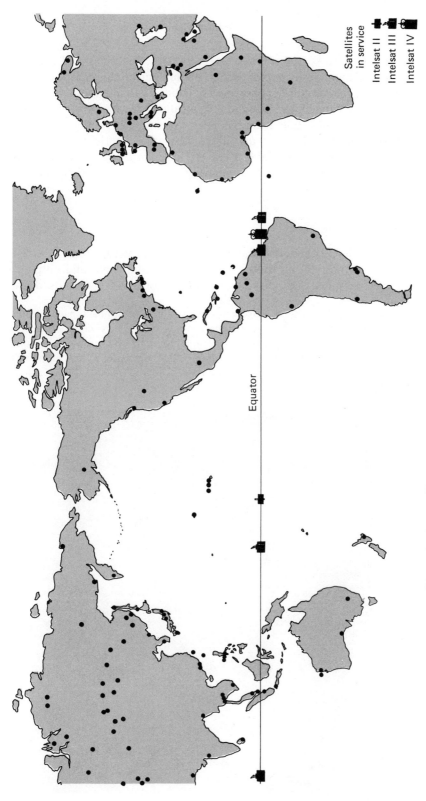

Figure 9.16 Satellites and earth stations for international telecommunications (1975). Russian satellites are not shown. Nor are domestic satellites.

Satellites
in service

Intelsat II

Intelsat III

Intelsat IV

Equator

Tropospheric scatter circuits use very large antennas and a higher transmitter power than microwave circuits. The number of channels that can be engineered into such a circuit depends on its distance. On a link of 100 miles, several hundred voice channels can be sent. A typical number on longer links is 72 channels. Over short links, television has been transmitted. Tropospheric scatter circuits are subject to fading and are affected by atmospheric conditions; however, they are much more dependable than HF circuits using the ionosphere. Their operating cost per channel mile is higher than that of the other media we have discussed.

9. Short-Distance Radio

It is possible that short-distance radio links of the type used in taxis, walkie-talkies, and so on, will be used widely for computer input and output. A small mobile radio terminal could have many applications. It could be used in the con-

Figure 9.17 The Kleinschmidt telescripter, a compact page printer designed for mobile installations such as police cars, commercial vehicles, ships, and aircraft. The terminal operates over radio telephone and has error control facilities. Should the operator leave his car, he will receive the message upon his return. (Photograph from SCM Corporation.)

trol of a railroad marshaling yard, for example, in police cars, or in aircraft. It may work in conjunction with a nearby transmitter on a device which itself is connected to a computer by landline. It may operate with the public mobile telephone service for vehicles. Figure 9.17 shows a typical terminal in a vehicle connected by mobile radio telephone links.

Unlike some of the systems discussed earlier, only a small amount of the available capacity of the medium would be made use of. A high degree of redundancy could be built in to protect the data from the considerable amount of noise that would be encountered.

10. Waveguides

The next major step forward in expanding the capacity of terrestrial systems is the *waveguide.*

A waveguide is, in essence, a metal tube down which radio waves of very high frequency travel. There are two main types of waveguide, rectangular and circular. Rectangular waveguides have been in use for some time as the feed between microwave antennas and their associated electronic equipment. It is normal to see a waveguide going up a microwave tower to the back of the dish that transmits the signal. They are not used for long-distance communication and are rarely employed for distances over a few thousand feet. They consist of a rectangular copper or brass tube, 15 inches across or smaller (Fig. 9.18). Radiation at microwave frequencies passes down this tube.

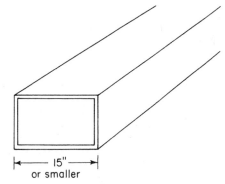

Figure 9.18　Rectangular waveguide.

15"
or smaller

Circular waveguides are pipes about 2 inches in diameter. They are constructed with precision and are capable of transmitting frequencies much higher than rectangular waveguides, or the other media discussed. Figure 9.19 shows the construction of a Bell System waveguide. This is referred to as a helical waveguide because a fine enameled copper wire is wound tightly around the inside in a helix. This is surrounded by a layer of thin glass fibers and then by a carbon layer. The whole is encased in a strong steel case and bonded to it with epoxy resin. The purpose of this construction is to attenuate undesired modes of wave propagation.

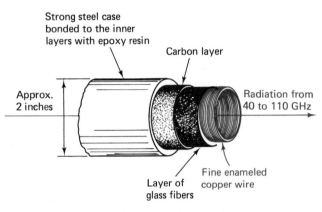

Strong steel case
bonded to the inner
layers with epoxy resin

Carbon layer

Approx.
2 inches

Radiation from
40 to 110 GHz

Fine enameled
copper wire

Layer of
glass fibers

Figure 9.19 Construction of the helical waveguide.

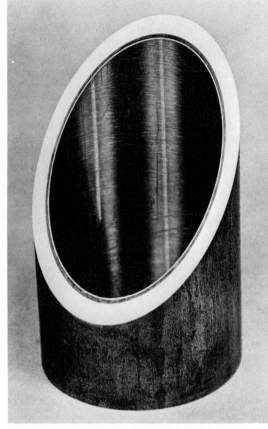

A major part of the cost of building waveguide systems is the civil engineering cost. Waveguides are not flexible and cannot go around sharp bends. Gentle bends, like a railroad track, cause only a minor loss in signal strength (Fig. 9.20) .

The loss in waveguides actually becomes less as frequency increases up to frequencies of about 100,000 MHz. This is shown in Fig. 10.9. Theoretically, it should continue to lessen indefinitely as frequency increases, though an upper limit is set by today's engineering. The AT&T WT4 waveguide system carries about 230,000 2-way telephone calls simultaneously, and it is likely that this capacity will be doubled. The first segment of this system for public use has been constructed in New Jersey. After a trial period of evaluation, and possibly modification, a nationwide system will be constructed. The WT4 system uses the pulse code modulation techniques described in Chapter 14. It transmits signals in digital form with a total bit rate of 16.4 billion bits per second. Such a system will be excellent for data transmission. The noise level encountered on the Bell System waveguide now operating is low enough for circuits of several thousand miles to be constructed, for example, coast-to-coast links in the United States.

Figure 9.20 WAVEGUIDES.

A circular waveguide under test in England. Waveguides cannot have sharp gends in them—only gentle bends like a railroad track. The civil engineering cost of laying down such a waveguide is high. (Courtesy Standard Telephone & Cables Ltd.)

WT-4, the Bell System's millimeter waveguide system, under construction in New Jersey. Here the waveguide is being inserted into a "pusher" which, in turn, slides the waveguide into its protective sheath. WT-4 will be able to transmit 230,000 two-way telephone calls simultaneously. (Courtesy AT&T.)

OPTICAL FIBERS   Substantially higher in the electromagnetic spectrum are the frequencies of light (Fig. 1.2). The early 1970s brought spectacular breakthroughs in the production of optical fibers of extraordinary transparency. Such fibers act like waveguides for the light frequencies of 10^{14} to 10^{15} Hz. These are more than 10,000 times the frequencies of microwaves in use today, and hence have the potential of carrying an extremely high bandwidth. Some of today's optical fibers have a signal attenuation as low as that of twisted copper wire-pairs or coaxial cable. Many very fine fibers are bundled together into a cable which is as flexible as today's telephone cables. Optical cables thus have the potential of being used either for new local distribution facilities which can carry high bandwidth signals, or for long-haul systems of very high capacity.

Intensive research and development is being done on the transmitters, repeaters, modulators, and other equipment that would be necessary elements of working optical fiber systems. The physical characteristics of optical fibers are such as to suggest that communications technology will continue to climb the steep curve of Fig. 1.2.

REFERENCE

1. R. Bowers and J. Frey, "Technology Assessment and Microwave Diodes," *Scientific American,* Feb. 1972.

10 ATTENUATION AND REPEATERS

The strength of the signal transmitted by any transmission medium falls off with distance. Therefore, if it is transmitted more than a few miles, it must be amplified. Most transmission systems have amplifiers at intervals to restore the signal to its former strength. The amplifier for analog signals is broadly similar to that used in a domestic radio or hi-fi unit for amplifying those signals. There are, however, other electronic circuits associated with the transmission line amplifier which correct the various forms of distortion that occur during the transmission. The amplifier and its associated circuits are referred to as a *repeater*.

The distance between the repeaters depends on the degree of *attenuation* or falling off in strength of the signal. The signal strength cannot be allowed to fall too low; otherwise it becomes increasingly difficult to distinguish it from the noise that is always present. If noise is mixed with the signal, the amplifier amplifies this as well as the signal.

TWO-WIRE AND FOUR-WIRE CIRCUITS A four-wire circuit (see Fig. 6.2) has two amplifiers in its repeaters, one for each direction of transmission, as shown in Fig. 10.1.

A two-wire circuit may have one amplifier which serves both directions. This is made possible by suitably arranged coils which act like transformers. (The terms "two-wire" and "four-wire" are used regardless of whether the transmission path actually consists of wires.) Figure 10.2 shows a simple two-wire repeater. Whichever direction the signal travels in, it is amplified and the amplified signal fed back into the line. Most two-wire repeaters use two amplifiers, one for each signal direction.

A four-wire circuit, used for long-distance trunks, may not necessarily have four actual wires. Many of them do, and these are referred to as *physical four-wire circuits*. However, two conductors can form an *equivalent four-wire circuit*. (The

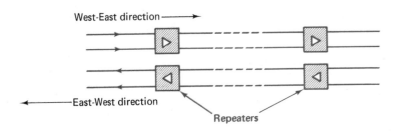

West-East direction ⟶

⟵ East-West direction

Repeaters

Figure 10.1   A four-wire circuit.

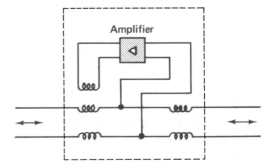

Amplifier

Figure 10.2   A simplified diagram of an amplifier on a two-wire circuit (seldom used today).

computer profession would probably call it a "virtual" four-wire circuit!) The two directions of transmission use different frequency bands. The east-west and west-east signals are changed in frequency by different amounts and can then travel on the same two physical wires without interfering with each other. Similarly, in data transmission, signals can be made to travel over two wires if their frequencies are changed so that they do not interfere with each other. The two directions are separated in frequency rather than space.

Where it is desirable to minimize the number of physical channels, two-wire or *equivalent* four-wire circuits will be used. Thus two-wire circuits are normally used for local loops. When several voice channels are packed onto one pair of wires, however, four-wire transmission is normally used, as with intertoll trunks (Fig. 6.2).

Whether they are physical four-wire or equivalent four-wire circuits depends on how important it is to minimize the number of physical paths. With open-wire pairs, for example, a physical four-wire circuit would mean doubling the number of wires stretched between the telephone poles. This is undesirable, and usually equivalent four-wire transmission is used, needing one pair of wires only. On the other hand, high-capacity intercity trunks carrying large numbers of conversations at once usually separate the directions of transmission physically. The electronic equipment is designed to pack many channels all going in the same direction into one coaxial cable or microwave facility.

HYBRID COILS   Where the two-wire line from your telephone joins the four-wire trunk, a connecting circuit is needed. An outgoing signal on the two wires is transferred to the appropriate pair of the four-wire line by this, and an incoming signal on the other pair of the four-wire line is transferred to the two-wire line.

The essence of this junction circuit is the *hybrid coil,* shown in Fig. 10.3. The signal on the two-wire line traveling west to east in the diagram is picked up by the coil entering the upper amplifier, amplified, and transmitted down the west-east half of the four-wire line. A signal traveling east to west is amplified by the lower amplifier and enters the two-wire line. It enters the hybrid coil at its center, and hence the signals induced into the uppermost part of the coil cancel out. If the hybrid coil were perfect and the balancing network precisely duplicated the line section it faces, no signal would enter the upper amplifier.

On a two-wire line, two amplifiers are often used rather than the one in Fig. 10.2. Two hybrid coils take the signals into these—one for each direction.

ECHOES   Unfortunately, the hybrid circuit is not perfect. When a signal travels east to west in Fig. 10.3 a small portion of it *does* find its way into the upper amplifier. The signals induced into the two halves of the upper coil do not cancel out exactly. The balancing network shown in Fig. 10.3 is chosen to minimize this unfortunate effect. However, the line char-

Figure 10.3   A hybrid coil circuit used to convert two-wire signals into four-wire signals and vice versa. Two such circuits are used in the transmission path shown in Fig. 6.2 at the two-wire/four-wire junctions.

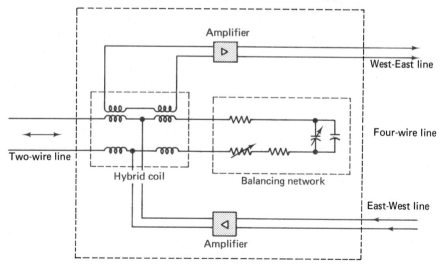

acteristics vary somewhat with ambient temperature, and different telephone handsets are connected to the network. The circuit cannot be balanced perfectly for all these conditions.

When a small fraction of the east-to-west signal goes into the upper amplifier, it is amplified and travels back west to east. There is thus an *echo* on the line. Other circuit components can cause echoes also. In fact, whenever the transmission line has a sharp change in impedance a small portion of the signal is reflected and travels back down the line.

Echoes have varying psychological effects on persons making a telephone call. The effects are sometimes not unpleasant, but sometimes they are annoying. An echo coming 0.005 second after the speaker's voice is not necessarily objectionable. It sounds rather like speaking in a bathroom. (People might even be tempted to sing.) Echoes coming a few tenths of a second after the speaker have a serious effect. Experiments show that they cause many people to stutter or repeat words. Many people speak in a disconnected manner. Some stop entirely.

ECHO SUPPRESSORS   On most of the world's telephone systems it is considered that echoes with a round-trip delay of more than about 0.045 second cannot be tolerated. Where the echo reaches the speaker after this period or longer an *echo suppressor* is employed to reduce it to a negligible level. Many circuits with less delay than this also use echo suppressors. An echo suppressor is a circuit which reduces the return transmission by inserting a large amount of loss into the return path.

The echo suppressor is activated by a circuit which detects the human voice. A *speech detector* on the east-west channel operates the echo suppressor of the west-east channel and these echoes are attenuated for the duration of the east-west speech. Likewise, a speech detector is also needed on the opposite direction of travel. If A is talking to B on the telephone and B wants to speak, B cannot be heard, except perhaps very faintly, until A stops talking. In a fast conversation with A and B interrupting one another, they may hear the echo suppressors switching on the other's voice, sometimes in midsentence. This does not usually impair conversation seriously.

With today's telephone plant only long lines need suppressors. On loaded wire-pair cables, signals travel along the line at speeds of the order of 20 miles per millisecond (see Fig. 9.7). This will vary considerably depending on the characteristics of the line and its loading coils, but roughly a signal would have to travel about 500 miles and back before its propagation time reaches 0.045 second. When coaxial cables or microwave radio are used, or if the wire pairs carry the higher frequencies used when they transmit several channels together, then the velocity of propagation will be nearer to the speed of light. Over such media the signal could travel nearly 4000 miles and back before the 0.045-second delay occurs; however, there is another cause of delay where many signals travel together down one link. Normally, they are not all going to the same destination and so must be separated for switching purposes. As will be explained in a later chapter, this involves dropping them down to voice frequencies to carry out the switching operation and then packing them together again. This introduces a small delay, and so the total delay will depend on the number of such "terminals" on the link. As a result of this we find echo suppressors in all terrestrial telephone circuits over 1500 miles in length, and often on shorter circuits. Longer alternative routes are used to bypass a busy group of trunks, and often a connection as short as 300 miles contains an echo suppressor. Care is taken to have no more than one echo suppressor switched into a connection because of the degradation they cause to the transmission of speech.

A communication satellite is 22,300 miles from the earth, and the propagation delay to and from it is therefore about 0.25 second (186,000 miles per second is the velocity of the signal to and from it). Echo suppressors are thus needed when such a satellite is used.

Echo suppressors designed for voice cannot be used when *data* are transmitted over a voice circuit. If they exist, they must be disabled. There are two main reasons for this. First, the speech detector is designed to detect *only* speech. It must be insensitive to noise, whistles, or other extraneous sounds, or the false operation could play havoc with the conversation. Digital data would not necessarily operate it satisfactorily. Second, to detect speech the circuit must listen for a short time before it operates. On a two-wire path the first syllable of a sequence can be clipped due to the echo suppressors being this slow in reversing their direction. This would not be satisfactory with data signals, and so time has to be allowed for the echo suppressors to reverse their direction of operation.

To achieve the most efficient use of a full-duplex line for data transmission it is desirable to transmit in both directions at the same time. Data signals may flow in one direction, while control signals and error retransmission requests flow in the opposite. Or, on a line with many data machines attached to it, data may be flowing from one of them to the computer at the same time as the computer is replying to another. This, however, cannot be the case on a line with an echo suppressor. Echo suppressors, unless disabled, prevent simultaneous two-way transmission over the public network.

ECHO SUPPRESSOR DISABLERS   In some countries, when dial-up telegraph transmission was introduced using the existing voice lines, echo suppressors were built which could be disabled by a specific control tone. These disablers are now being used for higher-speed data transmission, and they permit bidirectional transmission. When the echo suppressors are out of action it is possible to use either simultaneous two-way transmission or half-duplex transmission, with relatively fast reversal of transmission direction.

Typical echo suppressor disablers require that a single-frequency tone in the band 2000–2250 Hz be transmitted from either or both ends of the line. This tone must last for approximately 400 milliseconds, and there must be little or no energy at any other frequency. The echo suppressor will then stay out of action until there is no signal being transmitted from either end of the line for a period of approximately 50 milliseconds. The data machines must be designed in such a way that they do not leave the line silent for 50 milliseconds, or the echo suppressor will have to be redisabled. A "carrier" signal, explained in Chapter 13, ensures that the line does not become silent. Once the suppressor is reenabled, it again functions normally until another disabling tone is sent.

The disabling tone can be *heard* on the telephone as a continuous high-pitched whistle. It is therefore used as a signal to the person dialing that he has established the connection he desires to the computer. The girl in Fig. 8.8, for example, wishes to establish a connection between her data terminal and a computer. She dials the number of the computer—a conventional telephone number. She may briefly hear a conventional ringing tone, and then a high-pitched whistle comes down the line—referred to as the *data tone*. This whistle will disable any echo suppressors. When she hears the whistle she knows that she is in contact with the computer and she then presses the key labeled DATA on her data set. This switches the set to operate as required for data, and her terminal machine is now connected to the distant computer.

What happens to the echoes, then, if the echo suppressor is disabled? They will still be present as they would be on a line which, like the majority, has no echo suppressor. Faint echoes reach the receiving machine as unwanted noise. As will be seen in subsequent chapters, this, along with other noise, limits the maximum speed at which we can transmit. Note that it is the receiving machine we are now worried about, not the transmitting one, whereas in speech telephony it was the *talker* who was disturbed by the echoes. The echoes, in general, are not sufficiently strong to cause substantial errors in the interpretation of the data signals when data are transmitted at 1200 or 2400 bits per second over a voice line. When the speed is increased, however, more ingenious means of transmitting are required (Chapter 11), as all noise becomes more significant.

ATTENUATION CONSTANT   The attenuation of a cable is usually described in terms of an attenuation constant which is related to the decay of signal strength.

Consider a short section of cable of length Δl as shown in Figure (10.4). The voltage of the signal entering this section is V and that of the signal leaving is $V - \Delta V$. The loss of voltage is

$$-\Delta V = I R_s \, \Delta l$$

where I is the current flowing in the section and R_s is the resistance, per unit length. However, $I = V/R$, where R is the resistance relative to earth where the voltage is zero.

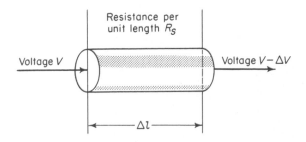

Figure 10.4

$$\therefore \frac{\Delta V}{V} = -\frac{R_s}{R} \Delta l$$

In the limiting case

$$\frac{1}{V} dV = -\frac{R_s}{R} dl$$

Integrating,

$$\log_e V = -\frac{R_s}{R} l$$

$$\therefore V = e^{-(R_s/R)l}$$

The voltage ratio between two points distance l_1 and l_2 down the line is therefore

$$\frac{V_1}{V_2} = e^{-R_s/R(l_1-l_2)}$$

In general,

$$\frac{V_1}{V_2} = e^{-\alpha(l_1-l_2)}$$

where α is a constant characteristic of the line and is called the attenuation constant.

For alternating currents one must take the capacitance and inductance of the line into consideration, and the expression for α becomes complex.

The attenuation constant is usually expressed in decibels per mile. Some typical values of attenuation constants are given in Table 10.1. Figures 10.5 to 10.7 show how the attenuation varies with frequency on wire pairs, coaxial

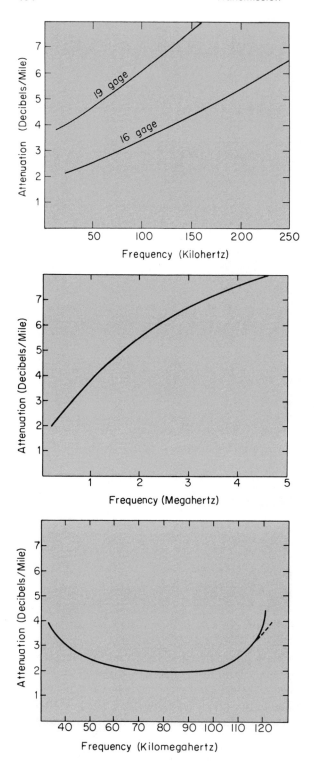

Fig. 10.5 Typical attenuation on twisted-wire-pair cables.

Fig. 10.6 Attenuation in a coaxial cable ($\frac{3}{8}$-inch outside diameter).

Fig. 10.7 Attenuation in a helical waveguide (2-inch diameter).

cable, and helical waveguide. This characteristic determines bandwidth and hence the capacity of such links.

Table 10.1   Typical attenuation constants

Transmission Medium	Frequency	Typical Attenuation Coefficient (decibels/mile)
Wire pairs on poles	1000 Hz	0.1
Twisted wires in cable, 16 gage	48 kHz	2
16 gage	140 kHz	3.5
22 gage	48 kHz	6
22 gage	140 kHz	8
Coaxial cable, $\frac{3}{8}$ inch o.d.	300 kHz	2
	2000 kHz	6
	8000 kHz	10
Transatlantic coaxial cable	160 kHz	1.4
Rectangular waveguide	5×10^9 Hz	9
Helical millimeter waveguide	5×10^{10} Hz	2

LOCATION OF REPEATERS   The repeaters are spaced at intervals sufficiently close to prevent the signal from being attenuated to a level at which it will be too small relative to the possible sources of noise. For voice transmission it is generally desirable to maintain an overall signal-to-noise ratio of 30 decibels or better.

On many good-quality circuits, when the signal has traveled a distance such that it has fallen in power by a factor of the order of 100, in other words 20 decibels, it is boosted back to its original value. Thus on a coaxial cable system with an average attenuation coefficient of 5 decibels per mile, repeaters may be installed every 4 miles. On the other hand, on open-wire pairs with an average attenuation of, say, 0.4 decibel per mile, the repeaters may be placed at distances of about 50 miles.

The higher the frequency transmitted over a wire pair or coaxial cable, the greater the attenuation. To make a wire pair or coaxial tube carry more information, higher bandwidth and hence higher frequencies are needed. Therefore more repeaters are needed to preserve the same level of attenuation. Over the years repeaters have become more reliable and less expensive. The spacing between repeaters has dropped, permitting higher frequencies to be used, and the information-carrying capacity of the cables has increased.

Figure 10.8 illustrates the use of repeaters on a 12.6-mile length of cable with an attenuation constant of 5 decibels per mile. If there were no repeaters on this line, the total drop in signal amplitude would be 63 decibels. This is too great a drop as the signal level would fall below the level of the background noise. When the signal is amplified *the noise is amplified with it,* as shown by the dotted line in Fig. 10.8. If the signal falls too low, then the ratio of signal strength to that of background noise becomes low and never improves because the two can never be separated. On the second line section of Fig. 10.8 the repeater spacing is slightly

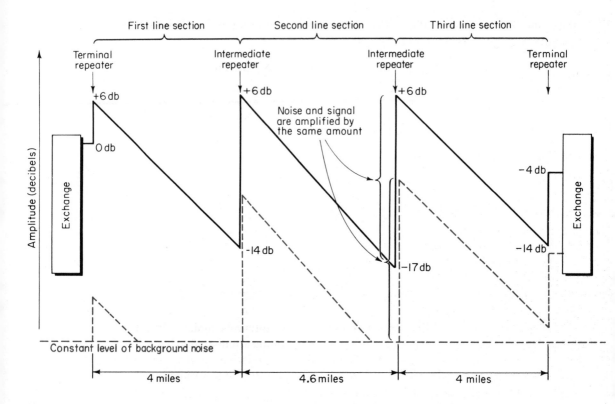

Figure 10.8 Attenuation and amplification on a line with a coefficient of 5 decibels per mile.

greater than on the first section. This allows the signal to fall slightly closer to the background noise level. This closer spacing of signal and noise remains until the signal reaches the exchange. It can be seen from Fig. 10.8 that if the repeaters had been closer together, the signal-to-noise ratio would have remained at a better figure. The one repeater interval that is greater than the others causes more than its fair share of degradation of the signal, and so it is advisable to have all the repeaters the same distance apart.

**ATTENUATION
IN RADIO
TRANSMISSION**

A signal transmitted by radio does not decay exponentially as is the case with transmission lines. Instead it obeys an inverse square law, the power falling off inversely as the square of the distance transmitted.

Suppose that a microwave antenna transmits with a power P_1, and that a similar antenna at a distance l receives the signal, the received signal having a power P_2.

Suppose that the transmitting antenna is designed so that it radiates a beam which is square in cross section and radiates in a solid angle θ by θ radians (Fig. 10.8). At distance l the cross-sectional area of the beam is $(\theta l)^2$. Let the area of the antenna be a. Then only $a/(\theta l)^2$ of the transmitted power is received:

$$\frac{P_2}{P_1} = \frac{a}{(\theta l)^2}$$

It can be shown that for an antenna with a square transmitting aperature the angle θ is given approximately by the following relation:

$$\theta = \frac{\lambda}{\sqrt{a}} \qquad (10.1)$$

where λ is the wavelength transmitted. Therefore

$$\frac{P_2}{P_1} = \left(\frac{a}{\lambda l}\right)^2 \qquad (10.2)$$

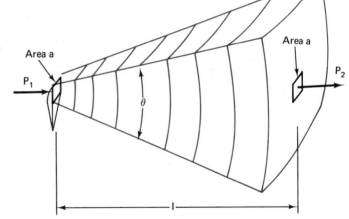

Figure 10.9 Calculation of microwave attenuation.

The amplitude ratio is therefore

$$\frac{A_2}{A_1} = \frac{a}{\lambda l} \qquad (10.3)$$

where A_1 is the transmitting amplitude and A_2 is the amplitude received. λ is inversely proportional to the frequency f of the wave:

$$\lambda f = C$$

where C is the velocity of light. We may then write (10.3) as

$$\frac{A_2}{A_1} = \frac{af}{lC} \qquad (10.4)$$

THE SPACING OF　　　Let us substitute figures into Eq. (10.4) for a typical
MICROWAVE LINKS　　　system. A frequency band in common use is 5925–
　　　　　　　　　　　　6425 MHz. The midpoint of this is $f = 6.175$ MHz. The
velocity of light is 186,000 miles per second. A typical antenna for such transmission might have an aperture 8 feet × 8 feet. Therefore, substituting, the attenuation ratio per mile is

$$\frac{(\frac{8}{5280})^2 \times 6.175 \times 10^9}{1 \cdot 86 \times 10^5} = 0.0762$$

Thus if the microwave antennas are situated 30 miles apart,

$$\frac{A_2}{A_1} = \frac{1}{30} \times 0.0762 = 0.00254$$

which is a loss of 52 decibels.

It will be seen that the signal strength does not fall off exponentially as in the
case of wire systems. It falls off much more slowly. The preceding system has an
attenuation of 32 decibels over 3 miles, 52 decibels over 30 miles, 72 decibels
over 300 miles, and so on. Contrast this with a cable system having an attenuation
of 16 decibels over 3 miles. This would suffer an attenuation of 160 decibels over
30 miles, 1600 decibels over 300 miles, and so on. The microwave systems would
thus need far fewer amplifiers, but each requires large antennas and the tower to
support them.

Equation (10.4) shows that A_2/A_1 can be improved by increasing the frequency, f. For this reason the highest frequencies practical are used in microwave
links. However, above about 10 GHz (10,000 MHz), the effect of rain and snow
in attenuating the signal becomes increasingly and prohibitively serious. Also the
refraction of the beam by different moisture and temperature layers becomes serious and can cause deep fading of the signal. Therefore microwave systems operate at frequencies between 3000 and 12,000 MHz.

A_2/A_1 can also be decreased by increasing the size of the antennas' aper-
tures, a. There is also a limit to this set by the economics of the system; however,
where for some reason the microwave antennas *have* to be a long distance apart,
the antenna size can be increased. This is sometimes the case in mountains or
links across large expanses of water. Communication via satellite uses microwave
frequencies, and here the ground antennas are very large, and specially designed.

As will be seen from Eq. (10.1) that increasing the frequency or enlarging the
transmitting antenna narrows the angle of the beam transmitted, and this is the
reason these two factors are effective in reducing the signal loss. When the beam
becomes too narrow it becomes more expensive to build antennas that will keep it
rigidly on its target 30 miles or so away. This is particularly so as microwave
dishes are fixed on the top of tall towers subject to high gales. In the typical example we have just cited, the angle of the beam is approximately 1°.

Figure 10.10   The curvature of the earth sets a limit to the spacing of microwave repeaters. If they are more than 30 miles apart and not on hills, very large and expensive towers are needed. Also, fading becomes a serious problem above 30 miles.

A major limitation on the separation between microwave relays is often, however, the curvature of the earth (Fig. 10.10). They must normally be built within line of sight of one another. Microwave antennas, even on the top of towers a few hundred feet high, often cannot be more than 30 miles apart. Sometimes large antennas are placed on mountain ridges to transmit greater distances.

Above 30 miles separation, *fading* effects become markedly more severe. This effect, combined with the others, has led to the building of microwave systems in which most of the towers are of the order of 30 miles apart. When it becomes necessary to extend microwave transmission into higher frequencies – the 10- to 18-GHz range – then fading effects are more significant. Rain causes much more severe fading and this overrides other tower-spacing considerations and causes the towers to be built much closer together.

COMPANDORS   One of the problems in the design of amplifiers and repeaters is that the signal to be amplified can vary widely in strength. Sometimes the person telephoning may shout. Another person may talk very softly with his mouth a long way from the telephone mouthpiece. After going through the long chain of amplifiers both of these signals must emerge at a reasonable volume, and noise-free.

The problem can be understood by looking at Fig.10.8. This illustrates the passage of a signal of one particular amplitude. After this has been transmitted it remains about 10 decibels above the level of the noise that was amplified with it. Some of the signal will be of greater strength and so better off than this, but some will be of lower strength. Suppose that a weak signal is 15 decibels lower than the one illustrated. When this reaches the repeaters in the diagram it will be only about the same amplitude as the noise level, or even lower. The noise will be amplified with it, and the weak signal will be drowned in noise.

To overcome this problem, it is desirable that the very strong and very weak signals be adjusted in amplitude so that they are closer together. The repeaters can then be more effective in keeping them above the noise level. This is done with a device called a *compandor*. The compandor compresses the range of signal amplitudes before transmission, and then, after transmission, restores them to their former levels. Thus there are two parts of the compandor circuits, a *compressor* at the transmitting end and an *expandor* at the receiving end.

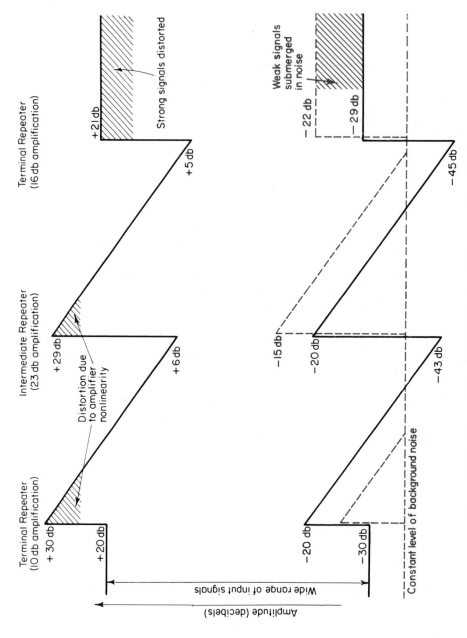

Figure 10.11 Transmission of signals of a wide range of amplitudes without a compandor.

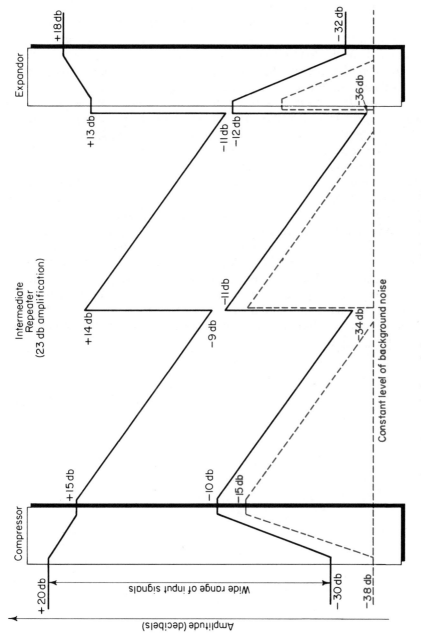

Figure 10.12 Transmission of the same range of signals, with a compandor.

There may be as much as 50 decibels difference between a loud voice input and a weak voice input. On a typical circuit this is "compressed" so that the difference is reduced to about 25 decibels. The compressor raises the level of the weak signals so that when transmitted they do not sink below the noise level, and it lowers the level of the loud signals so that they do not overload the amplifiers. The transmission of speech within this limited range of volumes gives much clearer results than if the range were greater.

Figure 10.11 illustrates the transmission of signals of a wide range of amplitudes. They are transmitted over a link which has one intermediate repeater. Figure 10.11 shows the transmission without using a compandor. The weakest signals drop below the level of the thermal noise. After amplification they are still submerged in noise. The stronger signals, on the other hand, overload the amplifier, which does not have a linear response over such a large range. When the compandor is used on these signals (Fig. 10.12), they do not fall below the noise level or overload the amplifier. The expandor lowers the amplitude of noise, and so the channel is more silent when no one is speaking.

Compandors do not, of course, improve the signal-to-noise ratio for steady power signals. The improved clarity is a result of the wide variation in human speech amplitudes. Data signals are transmitted at a steady power and so will not be improved by compandors. They may, in fact, fare somewhat worse on compandored channels than on noncompandored channels because the former may be engineered for a smaller signal-to-noise ratio, knowing that the compandor would make this acceptable for speech.

11 MODULATION AND DEMODULATION

When we wish to transmit music, or the human voice, or data signals, it often happens that their frequency spectra are unsuited to the medium we must use for transmission.

Take the case of radio broadcasting, for example. You wish to listen to a violin concerto on the radio so you tune your radio to the station on which it is playing—96.3 MHz. This is the frequency, the midpoint of the frequency band at which this communication system is then operating. Waves of this frequency travel through the atmosphere and are picked up by your radio set. However, the violins are not being played at this frequency. You would not hear them if they were. The sound of the violins is in the range 30 to 20,000 Hz. This sound must therefore be used to modulate the waves of very much higher frequency which the medium uses for transmission. The same is true in the transmission of data. The higher-frequency signal must therefore be made to "carry" the lower-frequency signal. The higher-frequency signal is "modulated" by the sound that it is to carry. The same is true with the transmission of data; the data modulate the frequencies which the transmission medium can transmit.

Modulation methods were first used almost exclusively in radio transmission. However, their use in line transmission soon followed because it was realized that communication lines had a bandwidth greater than that needed for speech. Consequently, several telephone conversations could be sent together down one telephone line. The way to do this was to change the frequencies so that several sound channels could be packed into one wider bandwidth, as was shown in Fig. 8.3. Modulation was needed to achieve this "frequency division multiplexing."

As the telephone business grew, channels of wider and wider bandwidth were developed. Coaxial cable links became available, and later microwave links. This trend to channels of greater bandwidth is still continuing. To pack more and more signals into the one physical circuit, more and more modulation is needed—modulation upon modulation, as will be described in the following chapters.

This is a problem exclusively for the common carriers. With the advent of data transmission, attention was focused on modulation, not only by the common carriers but by manufacturers of computers, office copying equipment, instrumentation, and data terminals. Modulation has become the key to using the world's communication links for sending information they were not designed to handle.

If we wish to send *data* over a telephone circuit, we must use a *carrier frequency* somewhere near the middle of the voice band and superimpose the data upon it in much the same way that the violin concerto was superimposed upon the broadcasting frequency of 96.3 MHz. The carrier selected is *modulated* with the data to be sent. The output of the modulation process needs to have a spectrum resembling the spectrum of the human voice (Fig. 8.1), or at least that truncated part of the voice spectrum which fits the telephone bandwidth. The signal has to be fitted into the shape shown in Fig. 8.4.

To make a voice circuit carry data efficiently, the modulation technique is designed to tailor the waveform to the characteristics of the channel. This tailoring process gives rise to much ingenuity. The modulation technique must be devised to maximize the quantity of data transmitted and minimize the effects of noise and distortion. Whatever bandwidth or media we use, modulation enables us to convert our data signals so that we transmit them with maximum efficiency. It matches the characteristics of the signal sent to the characteristics of the transmission medium.

THE SCOPE OF
THIS CHAPTER
There are, as we stressed before, two quite different types of modulation equipment: first, that used by the telephone companies in multiplexing voice and other signals onto higher-frequency transmission media; and second, the modems that the computer planner is concerned with to make his data travel over the voice channel.

The basic principles of modulation are the same for both of these types of equipment. The reader, however, is presumed to be more interested in data transmission, and so the main emphasis of this chapter is placed on the modulation of data rather than on modulation in the telephone plant. Amplitude modulation and frequency modulation, described below, are in regular use by the common carriers for handling telephone circuits. Phase modulation is not in common use for voice transmission. It is becoming the most favored method for digital transmission. In the following discussion, we will describe how amplitude and frequency modulation are applied to transmitting data; basically the same theory could be applied also to the transmission of voice. These are the two methods that are used for "frequency division multiplexing," described in the following chapter.

A new and different means for building telecommunication plant is coming into use which employs pulse techniques. Everything is converted to binary pulses that look like computer data. These techniques are described in a later chapter.

When we employ a *sine-wave carrier* to convey data, it has three parameters which we could modulate: its amplitude, its frequency, and its phase. There are thus three basic types of modulation in use: *amplitude modulation, frequency modulation,* and *phase modulation.* Each of these methods is in common use today. The sine-wave carrier may be represented by

$$a_c = A_c \sin (2\pi f_c t + \theta_c) \qquad (11.1)$$

where a_c is the instantaneous value of carrier voltage at time t,

A_c is the maximum amplitude of carrier voltage,

f_c is the carrier frequency, and

θ_c is the phase.

The values of A_c, f_c, or θ_c may be varied to make the wave carry information. This is illustrated in Fig. 11.1. A sinusoidal carrier wave of, say, 1500 Hz, a midfrequency of a voice channel band, is modulated to carry the information bits 0 1 0 0 0 1 0 1 1 0 0. In this simplified diagram the channel is being operated inefficiently because far more bits could be packed into the carrier oscillations shown. The tightness of this packing determines the speed of operation.

Furthermore, here we modulate the carrier by placing it in one of two possible states in each case. With amplitude modulation we could send several different amplitudes, as in the case illustrated in Fig. 11.8. Similarly, with frequency modulation, we could use several frequencies rather than just the F and F_2 shown. With phase modulation Fig. 11.1 illustrates only 180° phase changes; we could use phase changes which are multiples of 90°, giving four possible states, or 45° giving eight, and so on. Increasing the number of states of the carrier that are used increases the complexity of the decoding or demodulation circuits and considerably increases the susceptibility of the transmission to noise and distortion. If distortion can change the carrier phase by ±30°, for example, four-phase modulated signals can still be correctly detected, but not eight-phase.

We are therefore seeking a workable compromise between the quantity of data that can be packed into the transmission and the ability of the modem to decode it correctly in the presence of noise and distortion.

For correct decoding an accurate replica of the original carrier must be given to the demodulating circuit. There are a number of ways of obtaining this. In some cases it is sufficient to generate the replica independently in the demodulating equipment. A reference frequency may be generated by a high-precision quartz oscillator and used for decoding frequency modulation. It is in phase modulation that it is most difficult to obtain a reference. The demodulator can have no absolute sense of phase.

The original carrier may be reconstructed from information in the signal. This may be done by transmitting a separate tone of narrow bandwidth along with the signal, or it may possibly be obtained from the modulated signal itself. Sometimes the signal is briefly interrupted at intervals to give information about the carrier.

Let us examine the three main modulation methods in more detail.

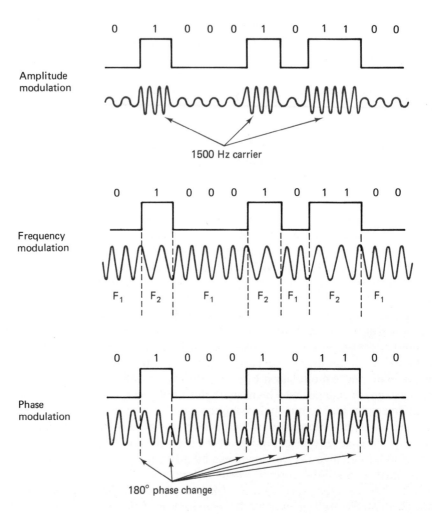

Figure 11.1   The three basic methods of modulating a sine wave carrier (a simplified diagram showing only binary signals). The amplitude, the frequency, or the phase, can be changed to carry the data.

1. AMPLITUDE MODULATION

In amplitude modulation the *amplitude* of the carrier wave is varied in accordance with the signal to be sent. In its simplest form the carrier is simply switched on and off to send 0 and 1 bits as in Fig. 11.1.

In general, the signal to be sent is multiplied by the carrier wave,

$$a_c = A_c \sin (2\pi f_c t + \theta_c)$$

This results in a signal which contains the original carrier plus two *side bands,* one higher in frequency than the carrier and the other lower. If the signal being transmitted has a frequency f_m (modulation frequency), this will give an *upper side band* with frequency $f_c + f_m$ and a *lower side band* with frequency $f_c - f_m$. It is in these two side bands that the information is carried.

Any signal that is to be sent — voice or data — can be represented by a series of sine waves using Fourier analysis. It was shown in Chapter 8 how an on or off pulse square wave such as the 1s and 0s of data transmission can be represented as a series of sine curves [Eq. (8.2)]. We are then modulating a sine-wave carrier by another sine wave or series of sine waves. This can be represented mathematically as follows:

Let us assume, for simplicity, that the phase constant of the carrier is zero. The carrier $a_c = A_c \sin 2\pi f_c t$ is to be modulated by a wave we represent by $a_m = A_m \sin 2\pi f_m t$. The resultant wave is

$$a_{mc} = (A_c + a_m) \sin 2\pi f_c t$$

$$= (A_c + A_m \sin 2\pi f_m t) \sin 2\pi f_c t$$

$$= A_c \sin 2\pi f_c t + A_m (\sin 2\pi f_m t) \sin 2\pi f_c t$$

$$= A_c \sin 2\pi f_c t + \frac{A_m}{2} \cos 2\pi (f_c - f_m)t - \frac{A_m}{2} \cos 2\pi (f_c + f_m)t$$

$$= A_c \sin 2\pi f_c t + \frac{A_m}{2} \sin \left[2\pi (f_c - f_m)t + \frac{\pi}{2} \right]$$

$$+ \frac{A_m}{2} \sin \left[2\pi (f_c + f_m)t - \frac{\pi}{2} \right] \tag{11.2}$$

This contains the three components, the carrier at frequency f_c which contains no information and the two *side bands* at frequencies $f_c - f_m$ and $f_c + f_m$ which do contain information because their amplitude is proportional to A_m. Thus:

carrier: $A_c \sin 2\pi f_c t$

lower side band: $\dfrac{A_m}{2} \sin \left[2\pi (f_c - f_m)t + \dfrac{\pi}{2} \right]$

upper side band: $\dfrac{A_m}{2} \sin \left[2\pi (f_c + f_m)t - \dfrac{\pi}{2} \right]$

A_m/A_c is referred to as the *modulation factor* or *modulation index.*

The maximum practical value of the modulation index is 1, and often it is less. If the amplitude of the modulating wave became greater than that of the carrier, giving a modulation index greater than 1, the resultant wave would have an envelope with more peaks than the modulating wave, and the original signal would not be recovered. This is shown in Fig. 11.2, and it is referred to as *overmodulation.*

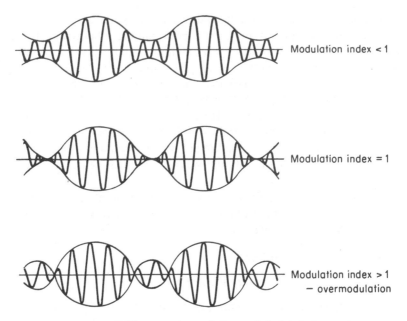

Modulation index < 1

Modulation index = 1

Modulation index > 1
— overmodulation

Figure 11.2   Different values of the modulation index A_m/A_c.

If the modulating signal consists in effect of *several* sine waves, then we shall have that number of frequencies both in the upper and lower side bands.

In voice transmission, for example, if a carrier of 60 kHz is modulated by speech filling a band of 300 to 3000Hz, the resulting transmission will be of an upper side band of 60,300 to 63,000 Hz, a lower side band of 57,000 to 59,700, and the original carrier of 60,000 (Fig. 11.3).

The Fourier series representing a square wave of frequency f_m, in other words a 1 0 1 0 1 0 bit pattern being sent at a rate of $2f_m$ bits per second (ignoring phase), is

$$\frac{4A_m}{\pi}\left(\sin 2\pi f_m t - \frac{1}{3}\sin 6\pi f_m t + \frac{1}{5}\sin 10\pi f_m t - \frac{1}{7}\sin 14\pi f_m t + \frac{1}{9}\sin 18\pi f_m t \cdots\right)$$

$$(11.3)$$

The results of using this to modulate the preceding carrier will be the following products (ignoring phase):

carrier: $A_c \sin 2\pi f_c t$

upper side band: $+\dfrac{2}{\pi} A_m \sin 2\pi(f_c + f_m)t$

$-\dfrac{2}{3\pi} A_m \sin 2\pi(f_c + 3f_m)t$

$$+ \frac{2}{5\pi} A_m \sin 2\pi (f_c + 5f_m)t$$

$$- \frac{2}{7\pi} A_m \sin 2\pi (f_c + 7f_m)t$$

$$+ \frac{2}{9\pi} A_m \sin 2\pi (f_c + 9f_m)t, \cdots, \text{etc.}$$

lower side band: $+ \dfrac{2}{\pi} A_m \sin 2\pi (f_c - f_m)t$

$$- \frac{2}{3\pi} A_m \sin 2\pi (f_c - 3f_m)t$$

$$+ \frac{2}{5\pi} A_m \sin 2\pi (f_c - 5f_m)t$$

$$- \frac{2}{7\pi} A_m \sin 2\pi (f_c - 7f_m)t$$

$$+ \frac{2}{9\pi} A_m \sin 2\pi (f_c - 9f_m)t, \cdots, \text{etc.}$$

This is illustrated in Fig. 11.4 for a modulation index of 1.

The more general case is likely to produce a spectrum with more lines on it than this. It can be shown, for example, that a regular train of square pulses, each of width d seconds and sent at a speed of S pulses per second, is represented by the following frequency lines:

$$2A_m Sd + \sum_{n=1}^{\infty} 4A_m Sd \frac{\sin \pi n Sd}{\pi n Sd} \cos 2\pi n St \qquad (11.4)$$

The lines are thus at frequencies $S, 2S, 3S, 4S, \ldots$ and are bounded by an envelope of the form $\sin x/x$, where $x = n\pi Sd$.

When these pulses modulate a sine wave of frequency f_c the resulting transmission is at frequencies $f_c + S, f_c - S, f_c + 2S, f_c - 2S, f_c + 3S$, etc., with again a large component at f_c carrying no information.

This is illustrated in Fig. 11.5, again for a modulation index of 1 and a value of Sd equal to $\frac{1}{6}$. (Interval between pulses = 6 × pulse width.)

The question now becomes apparent: *How much of the spectra like those in Figs. 11.4 and 11.5 do we in fact want to transmit?*

The carrier wave contains no information and may be suppressed. The lower side band duplicates the information that is in the upper side band, and therefore some systems transmit only one sideband. Further, not all the components shown are needed to extract the information. It could be extracted from a relatively narrow band of frequencies where the side-band amplitude is largest.

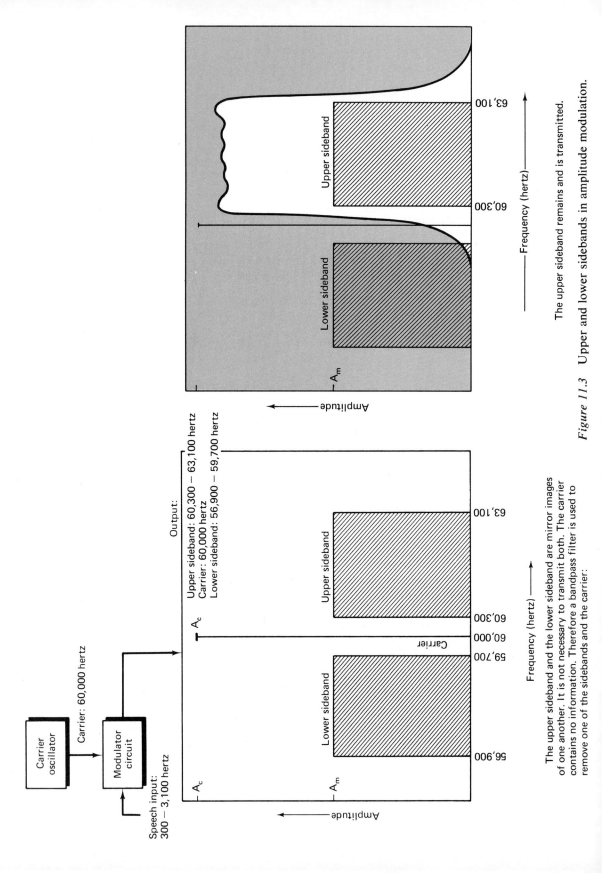

Carrier oscillator

Carrier: 60,000 hertz

Modulator circuit

Speech input: 300 — 3,100 hertz

Output:
Upper sideband: 60,300 — 63,100 hertz
Carrier: 60,000 hertz
Lower sideband: 56,900 — 59,700 hertz

A_c

A_c

A_m

Lower sideband

Carrier

Upper sideband

56,900   59,700   60,000   60,300   63,100

Frequency (hertz) —→

Amplitude ←—

The upper sideband and the lower sideband are mirror images of one another. It is not necessary to transmit both. The carrier contains no information. Therefore a bandpass filter is used to remove one of the sidebands and the carrier:

A_m

Lower sideband

Upper sideband

60,300   63,100

Frequency (hertz) —→

Amplitude ←—

The upper sideband remains and is transmitted.

Figure 11.3 Upper and lower sidebands in amplitude modulation.

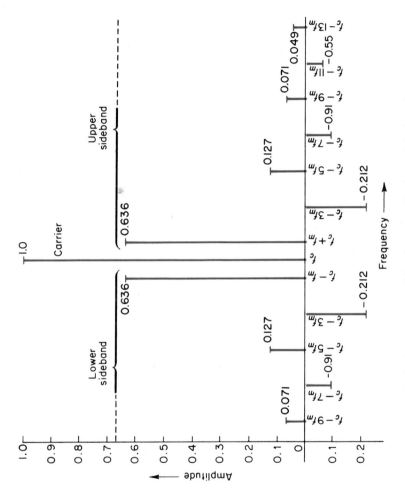

Figure 11.4 Spectrum resulting from amplitude modulation of a square wave representing bits 0 1 0 1 0 1 0 1. . . .

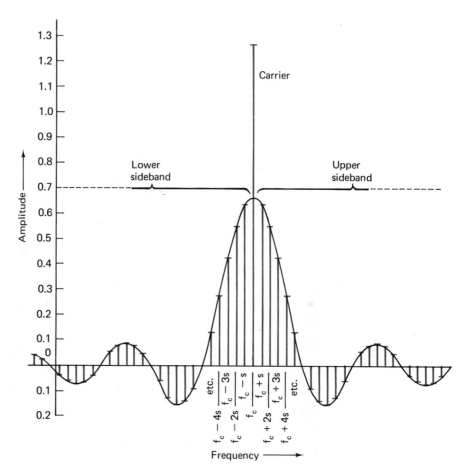

Figure 11.5 Spectrum resulting from amplitude modulation of a train of pulses. Interval between pulses is six times pulse width.

To transmit all of the spectrum shown would consume an unnecessarily wide bandwidth, or, given a fixed line or channel, it would restrict the signaling speed. Also it would need more power to transmit. As energy must be limited to values that will not be harmful to other users sharing the common facility, it is desirable to use all the power in transmitting that part of the spectrum which contains the most information content. This gives the best chance of separating the signal from the noise picked up.

The power is proportional to the square of the amplitude. Therefore, referring back to Eq. (11.2), the power transmitted in the carrier is proportional to A_c^2 and the power in each side band is proportional to $(A_m/2)^2$.

For a modulation factor of 1, $A_c = A_m$, it needs four times as much power to transmit the carrier, which contains no data, as to transmit either side band. In practice, modulation factors of less than 1 are used, and so the carrier may need six or eight times as much power as the side band.

It is common, therefore, to find *suppressed carrier amplitude modulation* in which the carrier has been removed by a filter and only the side bands transmitted and also *single-side-band amplitude modulation* in which one side band has been removed. It can be shown that the latter improves the signal-to-noise ratio over full amplitude modulation by 4 decibels or more. It also halves the bandwidth required.

DETECTION   When the modulated waveform reaches the demodulator, a "detection" process must convert it back to the original signal. For amplitude modulation there are two main types of detection: "Synchronous," "coherent," or "homodyne" detection and "envelope" detection.

Synchronous, coherent, or homodyne detection, which will be referred to below by the term "synchronous" detection, involves the use of a locally produced source of carrier which has the same frequency and phase as that bringing the received signal. The transmission is multiplied by this carrier, and this enables the signal to be extracted. Some additional components appear with it consisting of side bands centered around $2f_c$, $4f_c$, $6f_c$, and so on. These are filtered off with a low-pass filter.

The result is now close to the original but is not yet a rectangular or accurately synchronous waveform, so it may be trimmed up. If the signal is binary, the receiver interprets each pulse as being either a 1 or 0 and produces a new, clean rectangular pulse similar to the original. This may be done simply by a "slicer," which straightens up the edges of the pulse, or by a means for regenerating the waveform. In the latter case a local source of timing may be used so that a newly timed synchronous wave is generated.

Envelope detection involves rectifying and smoothing the signal so as to obtain its envelope. This is again combined with "slicing" or regenerating new pulses. After rectification, two threshold levels, as illustrated in Fig. 11.6, will tell the slicer when to start a 0 and when to start a 1.

In synchronous detection, to produce the reference wave of the same frequency and phase as the carrier, it is normally necessary to transmit some information with the signal for this purpose. Such a wave can be extracted from the carrier, and so the carrier may not be completely suppressed. It is usually partly suppressed because of its relatively high energy content. We now have a form of transmission in which one side band is suppressed, and only enough of the carrier is sent to give a reference frequency and phase for synchronous detection.

Envelope detection does not require the reference wave to be produced and so is considerably less expensive, as producing it is one of the main problems of synchronous detection. Envelope detection does, however, need both side bands and full-amplitude carrier. Suppressed side-band amplitude modulation would lead to an envelope shape which differs from that of the original signal.

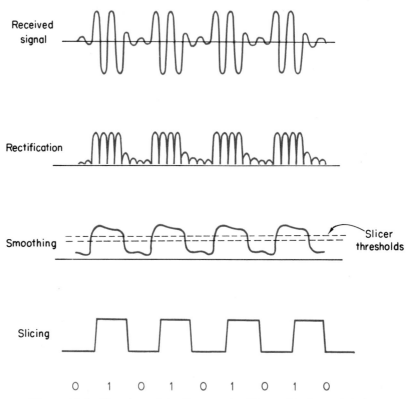

Figure 11.6 Envelope detection, used with amplitude modulation.

In choice of detection method we thus have a compromise between speed and cost. Envelope detection needs twice the bandwidth of synchronous detection because both side bands must be transmitted. Synchronous detection, however, is considerably more complex and expensive.

These various functions of an amplitude modulation modem are illustrated in Fig. 11.7.

MULTIPLE-LEVEL TRANSMISSION The preceding discussion relates to two-level amplitude modulation. The carrier is transmitted at two relative amplitudes as in Fig. 11.1. It is possible to transmit with more than two levels of amplitude. If four levels are used as in Fig. 11.8, the levels can be made to represent the bit pairs or "dibits" 00, 01, 10, and 11, respectively. This gives a lower margin for error in the threshold decision of the slicer or regenerator. Theoretically, the number of single bits carried by the signal can be twice as great; however, the susceptibility to noise is greater. The ratio of the level difference that must be detected to the noise level is substantially lower.

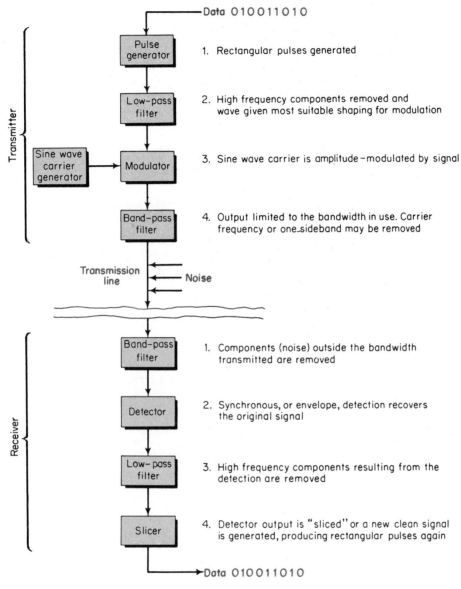

Data 010011010

Transmitter

Pulse generator	1. Rectangular pulses generated
Low-pass filter	2. High frequency components removed and wave given most suitable shaping for modulation
Sine wave carrier generator → Modulator	3. Sine wave carrier is amplitude-modulated by signal
Band-pass filter	4. Output limited to the bandwidth in use. Carrier frequency or one sideband may be removed

Transmission line ← Noise

Receiver

Band-pass filter	1. Components (noise) outside the bandwidth transmitted are removed
Detector	2. Synchronous, or envelope, detection recovers the original signal
Low-pass filter	3. High frequency components resulting from the detection are removed
Slicer	4. Detector output is "sliced" or a new clean signal is generated, producing rectangular pulses again

Data 010011010

Figure 11.7 Circuit elements used with amplitude modulation.

Figure 11.8 Amplitude modulation with four states. This gives a data rate theoretically twice as great as with two states, but much more susceptible to noise.

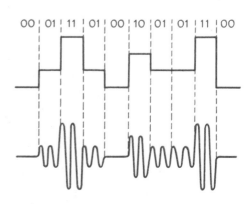

Similarly, eight levels would permit 3 bits to be carried per level and so give three times the speed of two-level transmission, but the level difference that must be detected would be still less.

Unfortunately, amplitude modulation is already vulnerable to the common types of noise. Multilevel amplitude modulation has not often been used; however, the same principle applied to phase modulation is now becoming common over good lines.

2. FREQUENCY MODULATION

When frequency modulation was developed it was used to replace amplitude modulation where better performance in the presence of impulse noise and voltage level changes was needed. The signal is transmitted at constant amplitude and so is resistant to changes in amplitude. However, a larger bandwidth is needed. Although amplitude modulation techniques are still used extensively in the common carrier's plant, frequency modulation is now being used more and more for *modems*. It has been adopted by the Bell System in the United States for their Dataphone 200 service and by the GPO in the United Kingdom for their Datel 200 and 600 service.

In frequency modulation, the frequency of the carrier wave varies in accordance with the signal to be sent. This is illustrated in a simplified form in Fig. 11.1. The frequency of the carrier assumes one value for a 1 bit and another for a 0 bit. This type of on/off modulation is sometimes called *frequency-shift keying* (FSK) *or carrier-shift keying*.

The modulation can also be a continuous analog process, the input signal being any waveform, which again we may regard as a collection of sine waves.

The unmodulated carrier, as before, may be represented by

$$a_c = A_c \sin 2\pi f_c t$$

If its frequency f_c is modulated by a sine wave of frequency f_m, we have

$$a_{cm} = A_c \sin 2\pi(f_c + \Delta f_c \sin 2\pi f_m t)t \tag{11.5}$$

where Δf_c is the maximum frequency deviation that can occur.

The ratio $\Delta f_c/f_m$ is now referred to as the *modulation index*.

In general, the spectrum resulting from frequency modulation is much more complex than the equivalent amplitude modulation, with many more components. If the modulating waveform is a simple sine wave, it can be shown that the resulting wave will contain side bands at frequencies $f_c + f_m$ and $f_c - f_m$, as before, and also at $f_c + 2f_m$, $f_c - 2f_m$, $f_c + 3f_m$, $f_c - 3f_m$, $f_c + 4f_m$, $f_c - 4f_m$, and so on. In other words, there are *an infinite number of side bands spaced at intervals equal to the modulating frequency.* The farther they are from the carrier frequency, the lower their amplitude.

It can be shown that the spectrum is as follows:

$$
\begin{aligned}
a_{cm} = A_c J_0\!\left(\frac{\Delta f_c}{f_m}\right) \sin 2\pi f_c t \\[6pt]
+ A_c J_1\!\left(\frac{\Delta f_c}{f_m}\right)[\sin 2\pi (f_c + f_m)t - \sin 2\pi (f_c - f_m)t] \\[6pt]
+ A_c J_2\!\left(\frac{\Delta f_c}{f_m}\right)[\sin 2\pi (f_c + 2f_m)t + \sin 2\pi (f_c - 2f_m)t] \\[6pt]
+ A_c J_3\!\left(\frac{\Delta f_c}{f_m}\right)[\sin 2\pi (f_c + 3f_m)t - \sin 2\pi (f_c - 3f_m)t] \\[6pt]
+ A_c J_4\!\left(\frac{\Delta f_c}{f_m}\right)[\sin 2\pi (f_c + 4f_m)t + \sin 2\pi (f_c - 4f_m)t] \\[6pt]
+ \cdots, \text{ etc.}
\end{aligned}
\tag{11.6}
$$

where $J_0(\Delta f_c/f_m)$, etc., are Bessel† functions.

The Bessel functions giving the relative amplitudes of these spectral components are shown in Fig. 11.9.

Using this diagram Fig. 11.10 was drawn for the case of a carrier of frequency 10,000 being modulated to carry a sine wave of frequency 1000. Figure 11.10 compares the spectrum for amplitude modulation with those for frequency modulation using modulation indices of 1, 2, and 5.

It will be seen that the spectrum lines carrying information are concentrated into a narrower range of frequencies when $\Delta f_c/f_m$ is small. The transmission f_m can therefore occur over narrower bandwidths.

The spectra could be considerably more complicated for actual data transmission because, as we have discussed above, the data consist in effect not of one sine wave but of many. One relatively simple case may be considered, however, and that is where data bits are sent strictly at two frequencies f_c and $f_c + \Delta f_c$. This is analogous to two amplitude modulated signals, one of frequency f_c and one of $f_c + \Delta f_c$, which fit together exactly. We may thus consider two square waves such as that in the spectrum of Fig. 11.4. If these are sent and with the same amplitude, the spectrum will contain two sets of lines such as those in Fig. 11.4, separated by Δf_c. This is drawn in Fig. 11.11 for different values of Δf_c, and here again the information is concentrated into a relatively narrow bandwidth when Δf_c is small.

†For values of Bessel functions see, for example, E. Jahnke and F. Emde, *Tables of Functions*, Dover, New York, 1945.

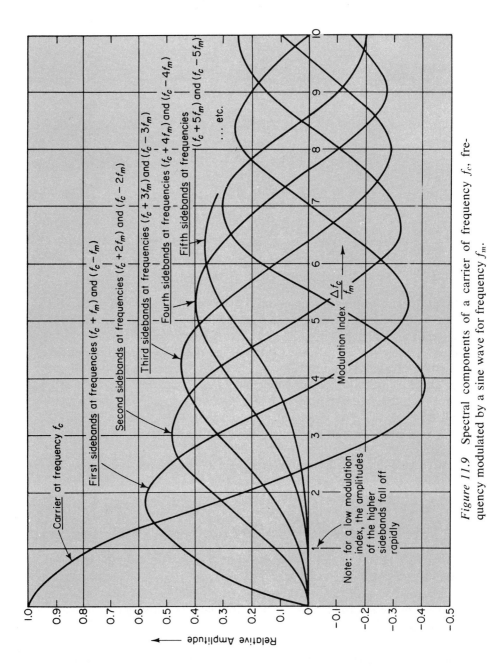

Figure 11.9 Spectral components of a carrier of frequency f_c, frequency modulated by a sine wave for frequency f_m.

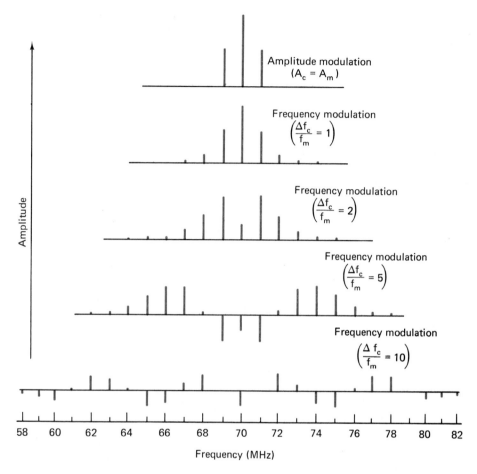

Figure 11.10 Spectra resulting from modulating a carrier of frequency 10,000 with a sine wave of frequency 1000; a comparison of amplitude modulation and frequency modulation with different modulation factors. The bottom example needs lower transmission power than the top ones, but it needs higher bandwidth. In general there is a trade-off between use of power and use of bandwidth.

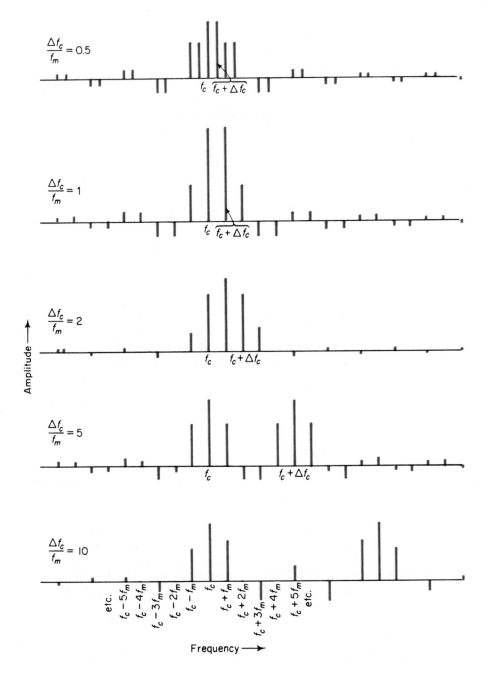

Figure 11.11 Spectra resulting from frequency modulation of a square wave representing bits 0 1 0 1 0 1 0 1. . . .

The bulk of the signal is in fact carried in the frequencies between $f_c + (\Delta f_c + f_m)$ and $f_c - (\Delta f_c + f_m)$. This may be the range transmitted, with other frequencies removed. If it is necessary to keep distortion to a very low figure, a wider bandwidth may be employed, for example,

$$f_c + (\Delta f_c + 3f_m) \quad \text{to} \quad f_c - (\Delta f_c + 3f_m)$$

It will be seen in Figs. 11.10 and 11.11 that the carrier amplitude is not so large relative to the side bands as for amplitude modulation. In some cases the carrier disappears completely. The information, however, is spread over the carrier and side-band frequencies. The carrier is therefore not suppressed in frequency modulation as it may be in amplitude modulation. However, it becomes smaller and smaller as the modulation index is increased. One set of side bands may be suppressed, as the upper and lower side bands are again mirror images of one another.

There is a trade-off between use of power and use of bandwidth. The bottom diagram of Fig. 11.10 shows a signal which can be transmitted with low power but which is spread across a wide bandwidth. The top diagram of the same figure uses more power but less bandwidth. On some transmission links plenty of power is available, and so the signal is packed tightly into a small bandwidth, thereby increasing the total information-carrying capacity. On other links power is a scarcer resource than bandwidth. On communication satellites, for example, power on board the satellite is in short supply. Therefore FM modulation is used with a fairly high modulation index, and in a typical system only a thousand 4-kHz voice channels are carried in a 36-MHz bandwidth.

Multiple-level transmission can be used in a manner similar to that with amplitude modulation, and again this packs more bits into a given bandwidth but increases the susceptibility to errors. Dibits, for example, as in Fig. 11.8, may modulate the carrier so that it uses four different frequencies. This will double the data rate, but the signal-to-noise ratio needed to achieve the same error rate will be much greater — or for the same signal strength, there may be many more errors.

DETECTION The frequency-modulated signal is transmitted at constant amplitude. The noise it encounters will occasionally change its frequency but will more commonly have amplitude modulation effects. The latter can be ignored by the detection process. To do this only a narrow amplitude slice is used for the detection. This is centered around zero amplitude. Ideally, only the instant the received wave crosses zero should be used in the detection process. In the detection circuit a device called the "limiter" converts these zero crossings into a square wave. This has then removed any amplitude distortion.

The output of the *limiter* can then be converted by different types of circuits to produce the original bit pattern. A frequency-sensitive circuit may be used to produce an amplitude variation proportional to the instantaneous frequency. Alternatively, pulses may be generated corresponding to each zero crossing and these pulses passed through a low-pass filter to produce a wave with an amplitude variation equivalent to the bit pattern transmitted. This is illustrated in Fig. 11.12. As with amplitude modulation in Fig. 11.6, a *slicer* is used to generate a rectangular waveform according to certain threshold values, and thus the modem produces a clean output.

Fig. 11.13 illustrates frequency modulation.

Figure 11.12   The detection process with frequency modulation.

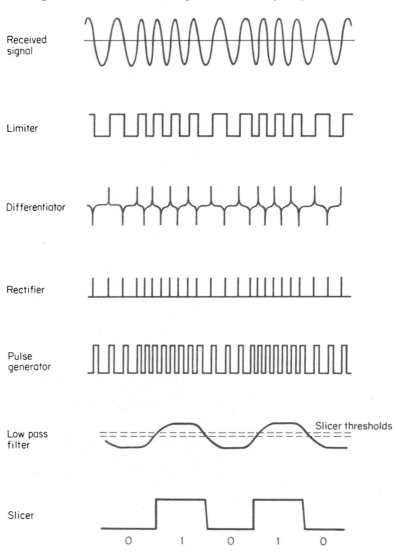

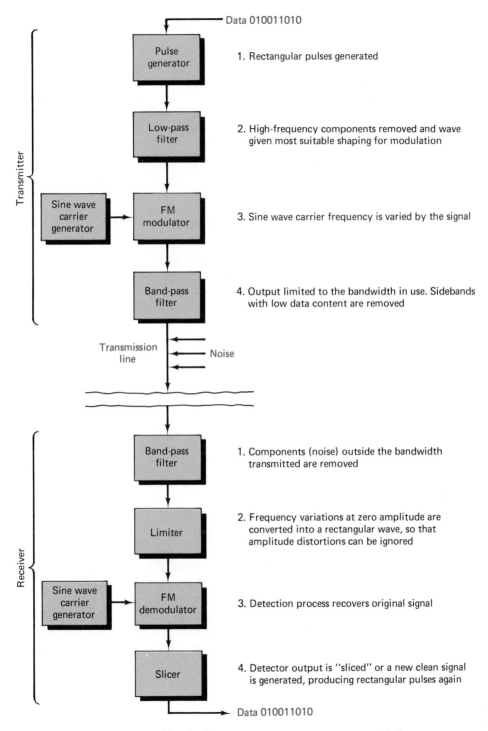

Data 010011010

Transmitter

Pulse generator	1. Rectangular pulses generated
Low-pass filter	2. High-frequency components removed and wave given most suitable shaping for modulation
Sine wave carrier generator → FM modulator	3. Sine wave carrier frequency is varied by the signal
Band-pass filter	4. Output limited to the bandwidth in use. Sidebands with low data content are removed

Transmission line ← Noise

Receiver

Band-pass filter	1. Components (noise) outside the bandwidth transmitted are removed
Limiter	2. Frequency variations at zero amplitude are converted into a rectangular wave, so that amplitude distortions can be ignored
Sine wave carrier generator → FM demodulator	3. Detection process recovers original signal
Slicer	4. Detector output is "sliced" or a new clean signal is generated, producing rectangular pulses again

Data 010011010

Figure 11.13   Circuit elements used with frequency modulation.

223

3. PHASE MODULATION

Just as frequency modulation for data transmission to a large extent replaced amplitude modulation because of its better resilience to noise, so phase modulation is now to some extent replacing the others, though at the time of writing it is still less used than amplitude and frequency modulation. It is not used for multiplexing in telephone plant except in certain isolated cases.

In phase modulation the phase of the carrier is varied in accordance with the data to be sent. A sudden phase change of $+180°$ cannot be differentiated from a change of $-180°$. Therefore the maximum range over which the phase can be varied is $\pm180°$. As small changes in phase cannot be transmitted and detected with accuracy, phase modulation is not normally used for the transmission of speech and music, for which frequency modulation and amplitude modulation are commonly used. The small range of variations can be used, however, to code the 2 bits of binary transmission, or 4 bits, 8 bits, or possibly even more when multiple-level codes are used. With four phases in use each interval carries 2 bits of information (a "dibit") and with eight phases, 3 bits.

The unmodulated carrier may, as before, be represented by

$$a_c = A_c \sin 2\pi F_c t$$

If its phase is modulated by a sine wave of frequency f_m, we have

$$a_{cm} = A_c \sin (2\pi F_c t + \Delta\theta_m \sin 2\pi f_m t) \qquad (11.7)$$

where $\Delta\theta_m$ is the maximum change in phase and is here called the *modulation index*.

The instantaneous frequency of the wave is $(1/2\pi) \times$ (the rate at which its angle is changing at that instant), in this case

$$\frac{1}{2\pi} \times \frac{d}{dt} (2\pi F_c t + \Delta\theta_m \sin 2\pi f_m t) = F_c + f_m \Delta\theta_m \cos 2\pi f_m t \qquad (11.8)$$

Thus the instantaneous frequency is F_c, the carrier frequency + a term $f_m \Delta\theta_m \cos 2\pi f_m t$. This is equivalent to frequency modulation of the carrier frequency f_m by a wave of frequency f_m.

Δf, the maximum frequency deviation is $f_m \Delta\theta_m$.

Phase modulation is thus equivalent to frequency modulation with a modulation index $f_m \Delta\theta_m / f_m = \Delta\theta_m$. (This holds only when the modulation is sinusoidal.)

Thus again the resulting wave will contain an infinite number of side bands spaced at intervals equal to the modulating frequency, i.e., side bands at frequencies $F_c \pm f_m$, $F_c \pm 2f_m$, $F_c \pm 3f_m$, and so on.

It can be shown that the spectrum is as follows:

$$A_{cm} = A_c J_0(\Delta\theta_m) \sin 2\pi F_c t$$
$$+ A_c J_1(\Delta\theta_m) \left[\sin 2\pi(F_c + f_m)t - \sin 2\pi(F_c - f_m)t \right]$$
$$+ A_c J_2(\Delta\theta_m) \left[\sin 2\pi(F_c + 2f_m)t + \sin 2\pi(F_c - 2f_m)t \right] \qquad (11.9)$$
$$+ A_c J_3(\Delta\theta_m) \left[\sin 2\pi(F_c + 3f_m)t + \sin 2\pi(F_c - 3f_m)t \right],$$
$$+ \cdots, \text{etc.}$$

where $J_0(\Delta\theta_m)$, etc., are Bessel functions, again. These are illustrated in Fig. 11.14.

Figure 11.14 Spectral components of a carrier of frequency f_c, phase modulated by a sine wave of frequency f_m.

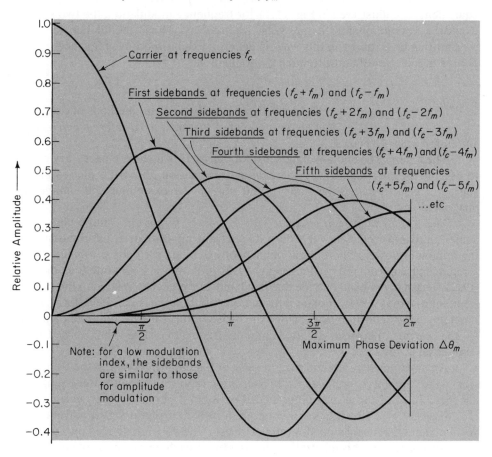

The spectra resulting from phase modulating a carrier with a sine wave would thus be similar to those for frequency modulation in Fig. 11.10. Because of the greater modulation index possible with frequency modulation, it is possible to spread the data content over a greater bandwidth than with phase modulation. Generally phase modulation uses a smaller bandwidth than frequency modulation, or, conversely, more information can be sent in a given bandwidth. The highest transmission speeds on a given bandwidth have thus often been obtained with phase modulation.

As is illustrated in Eq. (11.8) the instantaneous frequency is the derivative of the angle or phase. A sine phase variation therefore is equivalent to a cosine frequency variation. A nonsinusoidal phase variation, on the other hand, would not produce a similar frequency variation.

One relatively simple case is where a 0 bit and a 1 bit are transmitted as two identical signals differing only in that they are 180° apart in phase. This is equivalent to two amplitude modulated signals which fit together exactly. Unlike the equivalent case illustrated in Fig. 11.11 for frequency modulation, the two signals would have components at the same frequencies. Those modulation systems transmitting bit patterns in this way, in general, exhibit more resemblance to amplitude modulation than to frequency modulation.

DETECTION   There are two basically different methods of detection in phase modulated systems: *fixed-reference detection* and *differential detection*.

The receiver has no absolute sense of phase. It is therefore necessary either to use the signal in some way to generate information about the phases at the source or else to manage without it and operate by examining the *changes* in phase that occur.

The former approach needs a *fixed reference* giving the source phase. To achieve maximum efficiency it is desirable to transmit this carrier information with the minimum power. There are a number of ingenious methods of obtaining the reference phase from the carrier frequency. Alternatively, a separate tone may be sent: a very narrow band outside the data band, harmonically related to the carrier frequency so that it may contain information about the phase of the latter. Unfortunately, delay distortion may change the phase of this tone by a different amount to the carrier, because of its different frequency. This then has to be compensated for.

Again, a phase reference may be sent in bursts at intervals in the transmission. Data are often organized into words or records and separately sent and checked. Each of these may be preceded by a burst of carrier reference.

Figure 11.15 illustrates the generation of a four-phase signal and its subsequent detection using a fixed-reference phase. The data are divided into pairs of bits, called *dibits*. In this illustration the first bit of each pair is used to modulate a sine-wave carrier, and the second bit is used to modulate the same carrier delayed

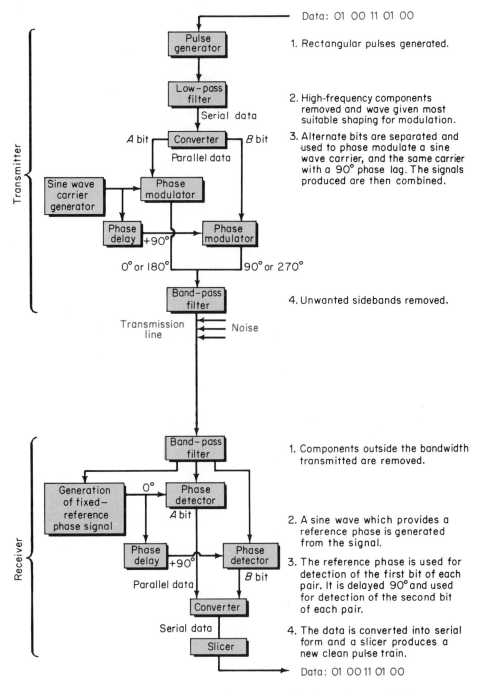

Data: 01 00 11 01 00

Transmitter

Pulse generator

1. Rectangular pulses generated.

Low-pass filter

Serial data

2. High-frequency components removed and wave given most suitable shaping for modulation.

Converter — *A* bit / *B* bit

Parallel data

3. Alternate bits are separated and used to phase modulate a sine wave carrier, and the same carrier with a 90° phase lag. The signals produced are then combined.

Sine wave carrier generator

Phase modulator

Phase delay +90°

Phase modulator

0° or 180°   90° or 270°

Band-pass filter

4. Unwanted sidebands removed.

Transmission line   Noise

Receiver

Band-pass filter

1. Components outside the bandwidth transmitted are removed.

Generation of fixed-reference phase signal   0°

Phase detector   *A* bit

2. A sine wave which provides a reference phase is generated from the signal.

Phase delay +90°

Phase detector   *B* bit

Parallel data

3. The reference phase is used for detection of the first bit of each pair. It is delayed 90° and used for detection of the second bit of each pair.

Converter

Serial data

Slicer

4. The data is converted into serial form and a slicer produces a new clean pulse train.

Data: 01 00 11 01 00

Figure 11.15   Possible circuit elements used for four-phase modulation.

227

in phase by 90°. Similarly, two sine waves are used for detection, one being 90° in phase later than the other and both having been generated by some means or other from the signal itself.

Differential detection does not attempt to generate a fixed-reference phase at the receiver. Instead, the data are coded by means of *changes* in phase. Thus in two-phase transmission a 1 bit may be coded as a +90° change in the phase of the signal and a 0 bit as a −90° change. In four-phase transmission (similarly, with eight-phase transmission) the changes might be as follows:

Bits	Phase Change
0 0	−135°
0 1	− 45°
1 1	+ 45°
1 0	+135°

The detector now merely looks for changes in phase and does not need a reference-phase signal. There is no need to have the coding start at any specific phase. If the phase of the signal slips or drifts because of interference, the system recovers without aid.

To carry out the detection, the signal received is delayed one symbol interval and compared with the signal then being received. This comparison indicates the phase change that has occurred between the symbol intervals. The phase change detected is then converted into bits, dibits, etc., as appropriate.

Because of this mechanism for delaying by one symbol interval, the speed of the transmission cannot easily be varied. Furthermore, it is difficult to use this type of detection for other than synchronous transmission in which the bits of characters or words are sent in a continuous, equally spaced stream, with no start-stop bits or gaps between characters, as in telegraphy.

12 MULTIPLEXING

All the transmission media discussed in Chapter 9 have a capacity great enough to carry more than one voice channel. In other words, their bandwidths are considerably greater than the 3 kHz needed for transmitting the human voice. At the top end of the scale the microwave and coaxial circuits carry thousands of voice channels; helical waveguide systems carry about a quarter of a million. At the lower end of the scale, each voice channel may be split into 12 or 24 telegraph channels (see Table 12.1).

Where a facility is set up, such as a chain of microwave links, which has a broad bandwidth, it is very desirable to make the maximum use of this bandwidth by making it carry as many channels as possible. The worldwide demand for communication facilities of all types is increasing at a tremendous rate, and economics and the need to conserve precious radio-frequency space demand that the common carriers devise means of increasing the capacity of their facilities. It is often desirable to construct a communication link with as wide a bandwidth as possible and then divide the bandwidth between as many users as possible. Many separate signals are *multiplexed* together so that they can travel as one signal over a high bandwidth.

In a multiplex system, two or more signals are combined so that they can be transmitted together over one physical cable or radio link. The original signals may be voice, telegraph, data, or other types of signals. The resulting combined signal is transmitted over a system with a suitably high bandwidth. When it is received it must be split up into the separate signals of which it is composed.

The word "multiplexing" is also used in other connections in data processing. For example, a "multiplexing channel" on a computer is one on which several devices can operate at the same time. Several printers, card readers, or other input/output machines operate simultaneously, and the bits which are sent to them, or received from them, are intermixed as they travel along the single channel. Also, a "multiplexer" is sometimes used as the name of a device which

Table 12.1   Frequencies used for subdividing a
voice channel into 24 telegraph
channels, each 50 baud; CCITT
Recommendation No. R. 31.

Channel Number	Lower Frequency of Band (Hz)
1	420
2	540
3	660
4	780
5	900
6	1020
7	1140
8	1260
9	1380
10	1500
11	1620
12	1740
13	1860
14	1980
15	2100
16	2220
17	2340
18	2460
19	2580
20	2700
21	2820
22	2940
23	3060
24	3180

receives, transmits, and controls data on several communication links at the same
time. The bits arrive or are transmitted at a rate which is slow compared with the
machine's scanning speed, and therefore the machine is capable of overlapping its
handling of many links. Again, the word multiplexing has also been used in connection with time-sharing systems to refer to the use of one computer handling in
real time the communication with several terminal or console operators. The computer, working at much faster speed than its human operators, switches its attention rapidly from one to another, and one operator need not know that the computer is in fact interleaving the conversations of many such people.

The word "multiplexing," in general, means the use of one facility to handle
several separate but similar operations simultaneously. In telecommunication language it means the use of one telecommunication link to handle several channels
of voice or data. Multiplexing is possible because the operations that are multiplexed take place at a considerably slower speed than the operating speed of the
facility in question. Multiplexing is the key to cost reduction in telecommunications.

As will be discussed later, communication channels are normally grouped together in "packages" which fill the bandwidth available on different types of plant. Twelve voice channels may be multiplexed together, for example, to form a signal of bandwidth 48 kHz. Much telecommunication plant is built to handle this bandwidth. Other transmission media handle a far higher bandwidth than this; five 48-kHz links may be multiplexed to feed a facility of 240-kHz bandwidth. In this case the original voice-channel signal has passed through two multiplexing processes. It may pass through more than two. Ten of the 240-kHz links may meet at a point where they are again multiplexed together to travel on a facility of a bandwidth of 2400 kHz (in fact, higher because the packing is not 100% efficient). Yet again three of these "packages" may occupy an 8-MHz bandwidth. The communication routes spanning the countries of the world thus carry large numbers of channels that were themselves grouped together stage by stage.

The original signal may go through many multiplexing stages and equivalent demultiplexing. It is thus worked upon by a variety of electronic conversion processes before it ultimately arrives at its destination, little the worse for these multiple contortions.

THREE METHODS
OF MULTIPLEXING

There are three methods of transmitting more than one signal over one path: *space-division multiplexing, frequency-division multiplexing,* and *time-division multiplexing.*

Space-division multiplexing means that more than one physical transmission path are grouped together. Wire-pair cables, for example, are constructed containing many hundreds of wire pairs (Fig. 9.4). Coaxial cables contain 20 or so tubes, as shown in Fig. 12.1, giving a high total bandwidth.

Frequency-division and *time-division multiplexing* are alternative techniques for splitting up a single physical path. The information to be transmitted may be thought of as occupying a two-dimensional continuum of frequency and

Figure 12.1 Space-division multiplexing. Twelve coaxial units are grouped with one cable. Each coaxial unit carries 1800 one-way voice channels combined by frequency-division multiplexing. The entire cable carries about 11,000 two-way voice channels, or the equivalent.

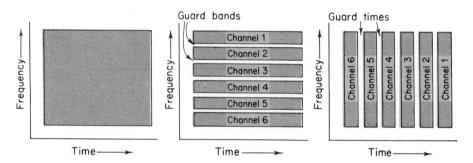

Figure 12.2 (Left) The space available for communication. (Center) Frequency-division multiplexing. (right) Time-division multiplexing.

time, as illustrated on the left of Fig. 12.2. The quantity of information that can be carried is proportional to the period of time used and to the range of frequencies, the bandwidth, used. If the quantity of information required from one channel is less than that which the physical facilities can carry, then the space available can be divided up either in frequency slices, or time slices, as in Fig. 12.2 (center and right).

In either case the engineering limitations of the devices employed prevent the slices from being packed tightly together. With frequency division, a *guard band* is needed between the frequencies used for separate channels, and with time division, a *guard time* is needed to separate the time slices. Logs cannot be sawed without some sawdust. If the guard bands or guard times were made too small, the expense of the equipment would increase out of proportion to the advantage gained.

FREQUENCY-DIVISION MULTIPLEXING

A familiar example of frequency multiplexing is radio broadcasting. The signals received by a domestic radio set contain many different programs traveling together but occupying different frequencies on the radio bandwidth. The speech and sounds from each radio station *modulate* a carrier of a frequency allocated to that station, amplitude modulation being used for AM broadcasting and frequency modulation for FM. The tuning circuits in the radio set allow one such signal to be separated from all the others.

Frequency-division multiplexing on a telephone line is basically similar. The signal, for example the human voice, is used to modulate a carrier which may have a much higher frequency. The signal thus occupies a relatively narrow bandwidth which is part of a much wider bandwidth transmitted. Other signals modify carrier frequencies that are spaced from each other by a given interval, typically 4 kHz. These modulated carriers are all amplified and transmitted together over the channel. The spacing between the carrier frequencies is slightly greater than the bandwidth needed to transmit the signal in question. This was illustrated in Fig. 8.3.

Figure 12.3 illustrates the principle of the equipment for frequency-division multiplexing. Here 12 signals each needing a bandwidth of not more than 4 kHz are combined so that they can be sent together over one physical channel.

At the sending end there are 12 modulators and at the receiving end 12 demodulators. In fact, the modulators and demodulators are combined into 12 single units to permit two-way transmission to take place. For simplicity, Fig. 12.3 illustrates only one-way transmission. The signals pass through 12 low-pass filters to remove any high-frequency components and are then used to modulate 12 separate carrier signals each 4 kHz apart. The frequencies resulting from each modulation process must be restricted to their own band. If they spill over into the bands occupied by other signals, the signals will not be separated correctly at the receiving end. As was shown in Chapter 11, the modulation process — whether it be amplitude, frequency, or phase modulation — produces components spread over a frequency range wider than that of the original signal. Frequency and phase modulation spread their products over a wide spectrum (Figs. 11.9, 11.10, and 11.14). Therefore, the outputs of each of the 12 modulators must be filtered again to stop them from interfering with each other. Band-pass filters are used to restrict each signal to the allocated 4-kHz band shown.

When the signal is received a converse process takes place. Twelve band-pass filters let through the frequencies of one signal only, as shown. These then pass into 12 demodulating circuits, and the original signal is recovered.

It is desirable that the 12 carrier frequencies used for modulation are accurately separated by 4 kHz and that those used for demodulation are *identical*. As we are packing as many separate signals as practicable into one physical channel, there is little room for error in the carrier frequencies used. For this reason the carrier frequencies may all be generated from the same source. In Fig. 12.3 a sine wave is used to produce carriers at 60, 64, 68 kHz and so on. The reference frequency which is the source of the carriers may be transmitted along with the 12 channels so that it can be used for controlling the frequencies used for demodulation at the other end, as shown.

The reference frequency transmitted is referred to as the *pilot*. Generating the carriers needed for modulation and demodulation is simplified if these are all harmonically related to the pilot.

If the carrier used for the demodulation is slightly different in frequency from that used for modulation, this results in distortion of the signal and *frequency offset*. Frequency offset is disturbing in the case of music transmission. A note and its harmonics are offset by the same amount, and consequently the harmonic is no longer a multiple of the base frequency. This results in "inharmonic" music in which the distortion is particularly noticeable.

When the modulation has taken place as described, the resulting band of 12 signals still may not be suitable for the transmission link in question. The link may be designed to carry a different range of frequencies in which case the newly

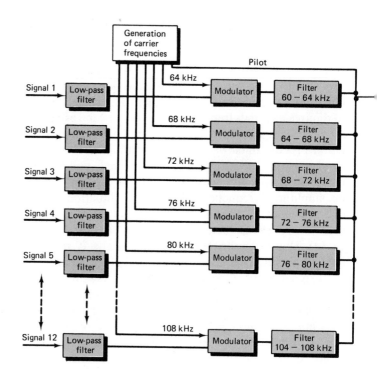

Figure 12.3 Frequency-division multiplexing. A diagram illustrating the principle.

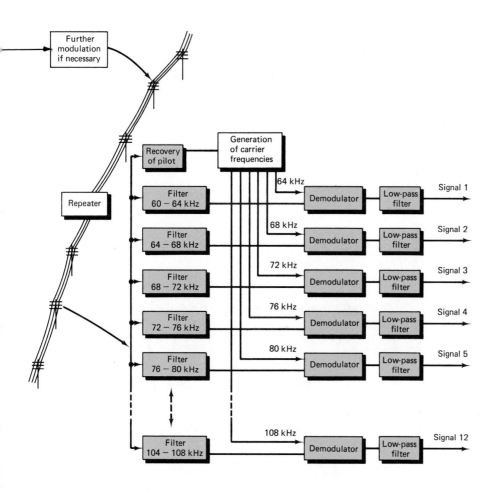

Further
modulation
if necessary

Repeater

Recovery
of pilot

Generation
of carrier
frequencies

Filter
60 – 64 kHz

64 kHz

Demodulator

Low-pass
filter

Signal 1

Filter
64 – 68 kHz

68 kHz

Demodulator

Low-pass
filter

Signal 2

Filter
68 – 72 kHz

72 kHz

Demodulator

Low-pass
filter

Signal 3

Filter
72 – 76 kHz

76 kHz

Demodulator

Low-pass
filter

Signal 4

Filter
76 – 80 kHz

80 kHz

Demodulator

Low-pass
filter

Signal 5

Filter
104 – 108 kHz

108 kHz

Demodulator

Low-pass
filter

Signal 12

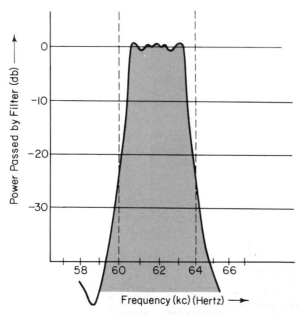

Figure 12.4 Characteristics of a filter using a 4-kilohertz bandwidth.

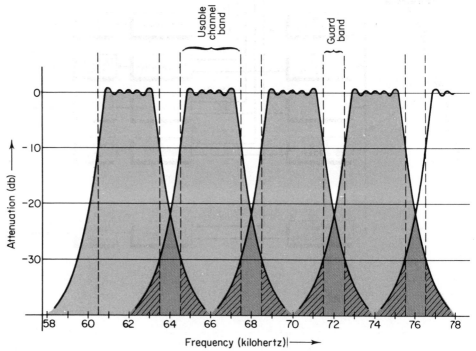

Figure 12.5 Because the filters do not have a sharp cutoff at the edges of the bands they pass, the usable channels must be separated by a guard band. Even so there is some slight cross talk as illustrated by the shading. A high-amplitude noise impulse in one channel will make itself felt in the adjacent channels.

formed band may be changed *as a unit* to the requisite frequency band. In other words, it has to undergo a further modulation process. Also it may be combined with other similar groups of signals so that several groups occupy a still greater bandwidth. We have multiplexing upon multiplexing.

To achieve efficient use of a bandwidth, one set of side bands resulting from the modulation process is usually removed. This is "suppressed side-band" modulation, as illustrated in Fig. 11.3.

Unfortunately, the filters that can be constructed at a reasonable cost do not have a sharp cutoff at the edges of the band of frequencies they pass. Instead their attenuation varies with frequency in a manner somewhat like that illustrated in Fig. 12.4. This curve could relate to the filtering of a 4-kHz band after modulation of a carrier at 64 kHz, or very slightly less than that.

When several such filtering processes are packed together the result will be as in Fig. 12.5. It will be seen that the bands the filters pass overlap in the regions of high attenuation. To minimize this, multiplexing schemes define a "guard band" as illustrated, and the system is designed so that as far as possible the required data occupy only those frequencies shown between the dashed lines. The useful bandwidth for each channel in the illustration is somewhat less than 4kHz.

Even with a guard band, there will still be some low-amplitude interference between channels, illustrated by the shaded portions of Fig. 12.5. The system must be designed so that this cross talk is of sufficiently low amplitude not to cause annoyance or, if possible, not to be discernible. The guard bands or the overlapping of the filter attenuation curves cause the total useful bandwidth to be less than the total transmission medium bandwidth. The ratio

$$\left(\frac{\text{total useful bandwidth}}{\text{total transmission medium bandwidth}} \right)$$

is sometimes called the *frequency efficiency* of the system. We may find that into a 48-kHz bandwidth we are packing 12 channels, but they have a usable bandwidth of only about 3400 Hz each. The frequency efficiency is then $(12 \times 3400)/48,000 = 0.85$. It will be seen that the frequency efficiency depends first on the width of the guard band and hence on the filter characteristics and second on the degree to which the modulation products are dispersed in frequency. If, for example, we transmitted both sets of sidebands, or used frequency modulation with a modulation index that scatters the products over a wide frequency range, we would be increasing the bandwidth needed for each channel and so lowering the frequency efficiency of the multiplexing process.

To carry out multiplexing the transmission medium used must be linear. Amplitude-frequency distortion or phase delay on the medium as a whole will have a serious effect on the individual channels transmitted. Equalizers are applied to the overall transmission rather than to individual channels. The amplifiers and, if a higher level of multiplexing is carried out, the modulators which operate on the group of channels must be made with a good linearity.

Even though the group transmission is carefully equalized the effect of filters on individual channels is to produce some amplitude-frequency distortion over the channel band. This can be seen from Fig. 12.5.

The economics of multiplexing arise partly from the fact that one repeater amplifies many channels. In Fig. 12.3 all frequencies from 60 to 108 kHz are amplified together. If no multiplexing were used, a separate repeater would be needed for each channel. This saves expense especially on long-distance transmission on which many repeaters would be used.

Figure 12.3 shows, schematically, transmission in one direction only. In actual systems two-way transmission is necessary. At each end of the link modulators and demodulators would be needed, and these may have some circuitry in common, particularly the oscillator. A trunk carrying multiple signals in this way is usually a four-wire line (see Fig. 6.2). It may be an equivalent four-wire which actually employs two wires. The frequencies below 60 kHz in Fig. 12.3 may be used for transmission in the opposite direction. The two directions of transmission are thus multiplexed together to travel over the same two wires at different frequencies.

TIME-DIVISION MULTIPLEXING

The major alternative to frequency-division multiplexing is *time-division multiplexing*. Here the time available is divided up into small slots, and each of these is occupied by a piece of one of the signals to be sent. The multiplexing apparatus scans the input signals in a round-robin fashion. Only one signal occupies the channel at one instant. It is thus quite different from frequency multiplexing in which all the signals are sent at the same time but each occupies a different band.

Time-division multiplexing may be thought of as being like the action of a commutator. Consider the commutator sketched in Fig. 12.6. The mechanically driven arm of this device might be used to sample the output of eight instruments. Providing the values of the voltages from the instruments are not varying too rapidly compared with the rotation time of the arm, the individual inputs can be reconstructed from the composite signal. Such a device is used in telemetering. To separate the signals when they are received, a commutator similar to that illustrated might be used but with the input and output reversed. The receiving commutator must be exactly synchronized with the transmitting commutator. The time multiplexing devices we meet in telecommunications today are normally electronic and of much higher speed but in principle are similar to the commutator.

Whereas frequency-division multiplexing fits naturally into the world of analog signals, time-division multiplexing has long been used with *digital* signals. In many types of equipment the bits from different bit streams are interleaved so that they can travel together over a single physical path. The multiplexor channel of a computer, for example, interleaves the bit streams of the devices attached to it.

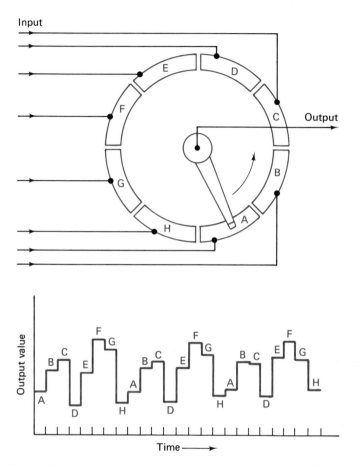

Figure 12.6 A commutator can be used as a simple form of time-division multiplexing.

If analog signals are to be time-division-multiplexed, they must be *sampled*. The sampling could be done by a series of accurately timed switches or sampling gates, shown schematically in Fig. 12.7. Here the signals are first put through low-pass filters which restrict them to 4 kHz or somewhat below this. They are then sampled by passing through a series of gates which open for a brief period of time. At the receiving terminal, the reverse process takes place. The signal is amplified by a wideband power amplifier which raises it to the level of the original signal. It then again passes through a series of accurately timed sampling gates. These separate the signals. A low-pass filter on each channel then integrates the samples, and the original is recovered, or at least that part of it below about 4 kHz.

There is clearly a problem in this. The sampling gates must open and shut at *exactly* the right instants in time. The gates at the receiving end must *exactly* correspond to those at the transmitting end and must take into consideration any

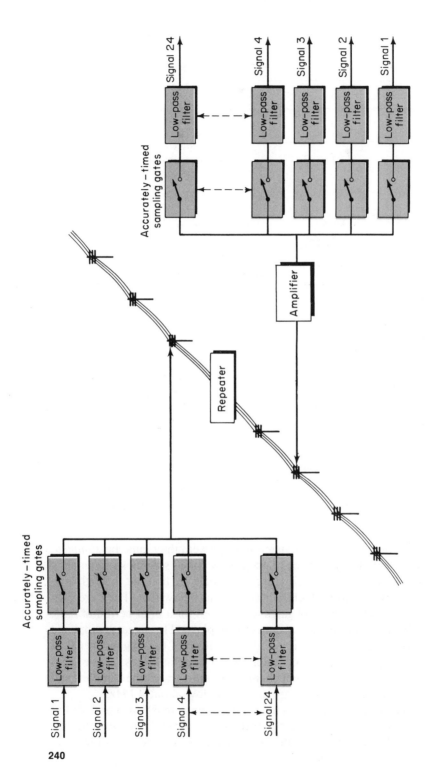

Figure 12.7 Time-division multiplexing of sampled analog signals. In practice the signals are usually encoded digitally before transmission as discussed in Chapter 14.

transmission delay. Furthermore, because the pulses tend to spread out slightly as they travel down the line, the samples cannot be packed tightly together. A *guard time* is needed between them. In addition to this, it is necessary to establish synchronization so that the receiving terminal can identify which pulse is which.

In Chapter 14, we will discuss how time-division multiplexing is used on telephone circuits.

TASI

An ingenious technique is used on some of today's very long-distance, frequency-division multiplexed lines for packing extra voice conversations into these expensive links. This is called *time assignment speech interpolation,* or TASI.

On a link carrying a conversation, both parties do not normally speak at once and for a small portion of the total connection time (usually about 10%) nobody is speaking. The long-distance link is normally a four-wire line, and so any one-way path is in use only about 45% of the total time. In other words, for 100 talkers only about 45 on average will be talking simultaneously. TASI aims to use less than 100 channels to carry these 45 voices. There is a spread about this average, and so it cannot suffice with only 45 channels. There is a possibility that most of the talkers at one instant will be talking in the same direction. As the number of talkers becomes large, however, the ratio of talkers to channels required becomes close to 1/0.45.

The TASI equipment is designed to detect a user's speech and assign him a channel in milliseconds after he begins to speak. An almost undetectable amount of his first syllable is lost. He retains the channel until he stops speaking; a short time after then, if the channel is needed for another talker, it will be taken away from him. If the traffic volume is high, the speaker's path may be snatched away as he pauses in midsentence, and when he utters his next words a different path is given to him. If, on the other hand, there are few users, he will retain his path. TASI can sometimes be heard in operation by a user if he knows what to listen for. On the transatlantic cable there are relatively few paths, and a caller during a peak period can detect a slight change in the noise amplification, which indicates that a channel has been given to his called party in the nick of time to catch his next word. There is a small but finite chance that a TASI circuit will be unable to find enough free paths and that for a very brief period of time words will be lost.

The cost of TASI switching equipment is high, but less expensive than the long channels which it saves. Data should not be transmitted with conventional modems over TASI circuits.

13 TRUNKS AND WIDEBAND FACILITIES

The main telecommunication highways of the world carry many different signals simultaneously, and these signals are gathered together into standardized groupings. The composition of a grouping is precisely defined in terms of the number of channels, the frequencies used, the carrier frequencies, and the multiplexing techniques employed to form the grouping.

A standardized grouping may travel over a variety of different physical facilities – differently structured coaxial cable routes, microwave routes, satellites, and so on. It is necessary that the structures be standardized so that groupings of signals may pass from one type of transmission facility to another, and may pass from one common carrier to another, without having to demultiplex the grouping into its constituent channels and remultiplex them. To a major extent, international standardization has been achieved, but some of the North American standards are different from the CCITT standards adopted by much of the rest of the world.

BELL SYSTEM GROUPS   Figure 13.1 shows the main blocks of channels which form the standards for North American carriers. At the bottom of the diagram is the telegraph channel. Twelve 150-Hz telegraph channels are derived from one voice channel, or 24 50-bit-per-second telex channels can be derived from one voice channel (see Table 12.1). Twelve voice channels can be multipled together to form a *channel group*, sometimes called simply a *group*. This was the highest level of multiplexing in the 1930s; since then larger and larger blocks have come into use:

Five channel groups are combined to form one *supergroup*.

Ten supergroups are combined to form one *mastergroup*.

Six mastergroups are combined to form one *jumbogroup*.

242

In the 1970s three jumbo groups have been combined to form one *jumbogroup multiplex* (Ma Bell ran out of superlatives!). The jumbo multiplex group is used on the highest capacity coaxial cable link, the L5 carrier, and contains 10,800 voice channels.

The L5 carrier, like other cables, also uses space-division multiplexing. It contains 22 coaxial tubes: 10 carry signals in one direction and 10 in the opposite

Figure 13.1   The hierarchy of frequency-division multiplex groups used in the Bell System.

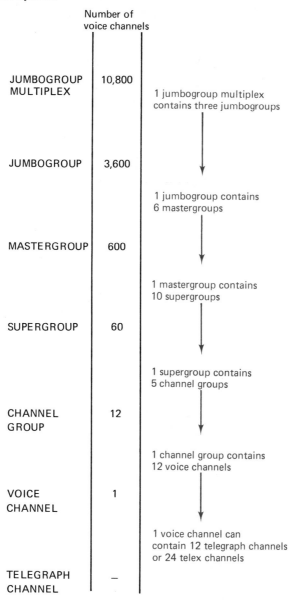

Number of voice channels

JUMBOGROUP MULTIPLEX	10,800
	1 jumbogroup multiplex contains three jumbogroups
JUMBOGROUP	3,600
	1 jumbogroup contains 6 mastergroups
MASTERGROUP	600
	1 mastergroup contains 10 supergroups
SUPERGROUP	60
	1 supergroup contains 5 channel groups
CHANNEL GROUP	12
	1 channel group contains 12 voice channels
VOICE CHANNEL	1
	1 voice channel can contain 12 telegraph channels or 24 telex channels
TELEGRAPH CHANNEL	—

direction, and 2 are spares which are used if one or two of the others fail. Each tube carries a multiplex jumbo group of 10,800 voice channels. The cable thus carries a total of 108,000 voice channels in both directions.

In addition to the main groupings shown in Fig. 13.1, some Bell System facilities transmit a group of three mastergroups.

CCITT STANDARD   The CCITT international standards are the same as
GROUPS   the North American ones for the supergroup and
                         channel group (which CCITT calls simply a "group")
but differ for higher-capacity groups, as shown in Fig. 13.2. Consequently a mastergroup or higher cannot travel directly from American to European telephone networks. They must be demultiplexed to the supergroup level.

Figure 13.2 CCITT and North American standards for multiplex groups are different except at the *channelgroup* and *supergroup* levels.

Number of voice channels	CCITT Standard	Bell System standard
12	Group	Channel group (sometimes called "Group")
60	Supergroup	Supergroup
300	Mastergroup	
600		Mastergroup
900	Supermaster group (sometimes called "Mastergroup" or "Hypergroup")	
1800		Mastergroup multiplex
3600		Jumbogroup
10,800		Jumbogroup multiplex

FREQUENCIES USED   Figure 13.3 shows the frequencies used for the Bell
System groups.

The voice channel might be regarded as the first building block. It occupies a
bandwidth of about 300 to 3100 Hz. Telegraph signals multiplexed into the voice-
channel frequencies can travel over voice channels everywhere.

The next building block, the *channel group,* enables 12 voice channels to
travel as a unit in a band of 60 to 108 kHz. This frequency band is used throughout most of the world as a standard. The band 12 to 60 kHz is also used. Many of
the world's *wire-pair* cables and their associated plant are designed for 60 to 108
kHz. Without this standardization, international telephone transmission would
have been much more expensive.

The next building block is the *supergroup* containing 5 channel groups and
occupying frequencies of 312 to 552 kHz. The telephone companies thus manufacture another subdivision of transmission facilites to carry this band of $5 \times 12 =$
60 telephone channels. The frequencies 312 to 552 kHz are generally agreed upon
internationally. However, several different pilot frequencies (the carrier frequency used for modulation) are used by different countries.

The next step up the bandwidth is the mastergroup. In North America, 10
supergroups are multiplexed together to form 1 mastergroup, which carries 10×5
$\times 12 = 600$ voice channels. This is suitable for the bandwidth available on a coaxial cable or microwave link.

Television transmission requires the bandwidth of two AT&T mastergroups. Consequently AT&T used the *mastergroup multiplex* block of frequencies shown in the center of Fig. 13.3.

The Bell L3 coaxial cable system was originally built to carry either three
mastergroups and a supergroup (total of 1860 voice channels) or one mastergroup,
a supergroup, and a television channel. Today television is transmitted separately
to achieve the higher transmission quality needed for color. Microwave facilities
similarly carried the mastergroup multiplex block. The massive growth of commercial television in the United States has been a spur to building of such wideband facilities to carry the programs across the nation. Fortunately, television
peak broadcasting periods do not coincide with the time of day when people are
using their telephones the most, and so the same bandwidth is used for television
in the evening and telephone by day.

The larger-capacity and more recent coaxial cable systems needed the
jumbo group (in the L4 carrier system) and then the *jumbogroup multiplex* (in the
L5 carrier system). Figure 13.4 shows the frequencies used on the L5 carrier system.

The first coaxial cable system was installed by AT&T in 1941. Figure 13.5
shows the growth in capacity of coaxial systems since then—a growth both in
space-division multiplexing (number of tubes per cable) and in frequency-division
multiplexing (higher-capacity groups).

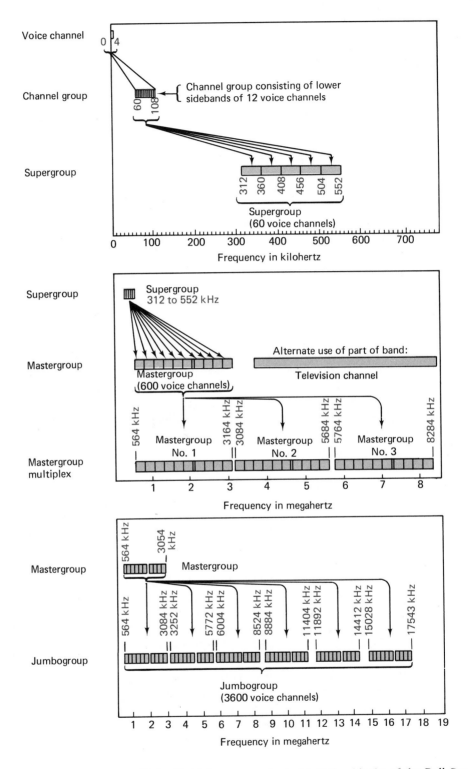

Figure 13.3 Multiplex groups: logical building blocks of the Bell System.

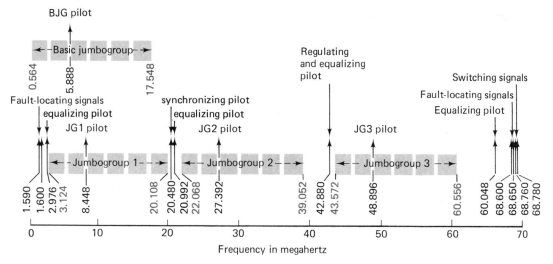

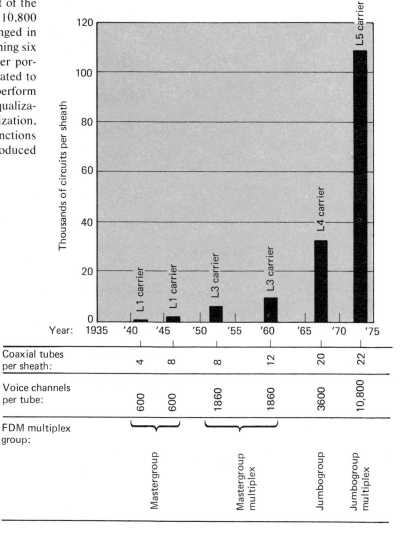

Figure 13.4 (top) Frequencies used on the L5 carrier system. Most of the L5 spectrum is occupied by 10,800 voice channels. These are arranged in three jumbogroups, each containing six 600-channel mastergroups. Other portions of the spectrum are dedicated to test and control signals that perform such functions as regulation, equalization, fault location, synchronization, and protection switching—functions critical to reliable service. (Reproduced from Reference 13.)

Figure 13.5 (right) Growth of AT&T coaxial cable systems—a 225-fold increase in the capacity of a cable since 1941 and a 10-fold drop in cost per voice channel. More coaxial tubes per cable and higher multiplex groups. (Source AT&T.)

Year:	1935	'40	'45	'50	'55	'60	'65	'70	'75
Coaxial tubes per sheath:		4	8	8		12		20	22
Voice channels per tube:		600	600	1860		1860		3600	10,800
FDM multiplex group:		Mastergroup		Mastergroup multiplex				Jumbogroup	Jumbogroup multiplex

Table 13.1 The North American transmission systems

	ATT Name of Transmission System	Main Application	Today's Usage	Analog (A) or Digital (D)	Reference (see end of chapter)	Frequency Band Used	No. of Voice Circuits Per Wire Pair / Per Tube	No. of Voice Circuits Per Cable System
Open-wire pair	O Carrier	Short haul	Declining	A	5	Up to 200 kHz	4–24	
Twisted-wire pair	K Carrier	Short haul	Medium	A	2	Up to 300 kHz	12	
	N1 Carrier	Short haul	Declining	A	3	Up to 300 kHz	12	
	N2 Carrier	Short haul	Medium	A	4	Up to 300 kHz	12	
	N3 Carrier	Short haul	Large	A		Up to 300 kHz	24	
	T1 Carrier	Up to 50 miles	Large	D	14		24 (1.544 million bits/sec)	
	T2 Carrier	Up to 500 miles	New	D	15		96 (6.3 million bits/sec)	
Coaxial cable	L1 Carrier	Long haul	Declining	A	10	Up to 3 MHz	600	1,800
	L3 Carrier	Long haul	Large	A	12	Up to 10 MHz	1,800	9,000
	L4 Carrier	Long haul	Large	A	13	Up to 20 MHz	3,600	32,400
	L5 Carrier	Long haul	Becoming large	A			10,800	108,000
	T4 Carrier	Long haul or Short haul	New	D			274 million bits/sec/channel	

(Twisted-wire pair: Many wire pairs in one cable.)

248

	Carrier/Satellite				Frequency	No. of Voice Circuits Per Radio Channel	Typical no. of Voice Circuits Per Route
Microwave radio	TD-2 Carrier	Long haul	Large	A	6	3.7–4.2 GHz	600–1200	12,000
	TD-3 Carrier	Long haul	Large	A		3.7–4.2 GHz	1200	12,000
	TH-1 Carrier	Long haul	Large	A	11	5.925–6.425 GHz	1800	10,800
	TH-3 Carrier	Long haul	Large	A	11	5.925–6.425 GHz	1800	10,800
	TM-1 Carrier	Short haul	Low	A	9	5.925–6.425 GHz	600–900	3,600
	TJ Carrier	Short haul	Low	A	7	10.7–11.7 GHz	600	1,800
	TL-1 Carrier	Short haul	Medium	A	8	10.7–11.7 GHz	240	720
	TL-2 Carrier	Short haul	Medium	A	9	10.7–11.7 GHz	600–900	2,700

						Total Digital Bit Rate	No. of Voice Circuits Per Waveguide
Waveguide	WT-4 Carrier	Long haul	Experimental	D			1.6×10^{10}	234,000

	Satellite Name				Frequency	No. of Voice Circuits Per Transponder	No. of Transponders Per Satellite
Satellite	INTELSAT IV	International	Largest international facility	A		5.925–6.425 up and 3.7–4.2 down	600	12
	ANIK	Long haul	Low	A		5.925–6.425 up and 3.7–4.2 down	1200	12
	WESTAR	Long haul	Low	A		5.925–6.425 up and 3.7–4.2 down	1200	12
	RCA SATCOM	Long haul	Low	A		5.925–6.425 up and 3.7–4.2 down	1200	24

249

BELL TRANSMISSION   AT&T gives letter designations to its various trans-
SYSTEMS   mission systems. The coaxial systems mentioned
                         above are referred to as the L carrier systems L1, L3,
L4, and, the latest, L5. Microwave transmission systems are referred to as TD,
TH, TM, TJ, and TL carriers. Twisted wire-pair systems are called K and N car-
riers. Open wire-pair systems are called O carrier systems.

They have widely differing physical characteristics but all carry the standard
blocks of channels referred to in this chapter. Making the structures of the groups
of channels independent of the physical structures of transmission systems is an
architectural concept which has served the common carriers well. (It is a little like
separating logical and physical structures in a computer system.)

Table 13.1 summarizes the main transmission systems of North America,
and Fig. 13.6 illustrates the relative cost per circuit on different systems.

So far we have discussed *analog* transmission systems, using frequency-
division multiplexing. Some of the most interesting systems for the computer
industry are the T carrier systems listed in Table 13.1 (T1, T2, and T5), which are
digital rather than analog. We will discuss those in the next chapter.

Figure 13.6   Relative cost per circuit mile of various carrier systems.
(Courtesy AT&T.)

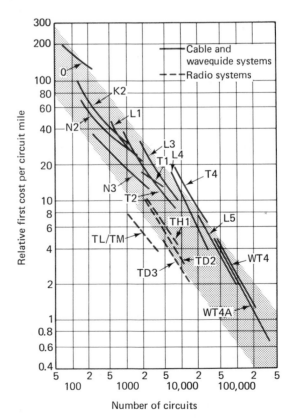

**LINKING OF
DIFFERENT FACILITIES**
Figure 13.7 shows three small towns and illustrates the types of facilities that may link them. The towns may be many hundreds of miles apart.

Suppose first that subscriber A at the botton left of the figure calls a subscriber in the same town. Both are connected to the same local switching office on dedicated two-wire subscriber loops. The telephone conversation undergoes no multiplexing. The signals travel a voice frequency.

Now consider what happens when C telephones D, both of them in town 2. These two are not connected to the same switching office. The call must therefore be routed from one central office to another on an interoffice trunk. Most of the time this trunk route will be handling several calls from people in that town talking to each other. Therefore, multiplexing equipment could be used. The interoffice trunk might be a number of wire pairs each, carrying a channel group of 12 voice channels, or it could be a number of nonmultiplexed pairs in a cable if that is more economical.

When A telephones C the situation is more complex as there is no direct connection between the local control offices to which they are attached. This is often the case even within one town. A large town with several central offices would have an additional exchange called a "tandem office" for switching lines between central offices. In Fig. 13.7 A's call will travel to the town where C lives over the short-wave microwave system connecting the towns. This is connected in both of these towns to a toll office. A's call goes over his local loop to his local central office, and this routes it to the toll office on a toll-connecting trunk (or terminal trunk). Like the interoffice trunk, this trunk may be one of a channel group, and it travels with two other mastergroups between towns 1 and 2. The toll office of the receiving town demultiplexes it and sends it to the appropriate central office on a toll-connecting trunk in a channel group. From there is goes at voice frequency on to C via his subscriber's loop.

When A telephones E in town 3 there is another stage in the process because the link to E's town goes over two long-haul coaxial cable systems, an L3 carrier, carrying a mastergroup multiplex, and an L5 carrier, carrying a jumbogroup multiplex. The call then goes over a microwave relay to the toll office of town 3.

The same complex route would be taken if A were using a computer terminal for communicating with a computer in E's town. Suppose that A has a computer terminal. He dials the distant computer to use a program in its files. The computer sends a message, which is displayed on the terminal, telling him to identify himself. The data that travel in the ensuing work undergo the same multiplex processes as a voice conversation would. The data, however, may be more seriously affected by delay distortion and other distortions on the route (discussed later). The design of the modem is the key to tailoring the signal so that it can travel over this network without errors other than those caused by unavoidable noise.

At the switching offices which interconnect different trunk systems, the sig-

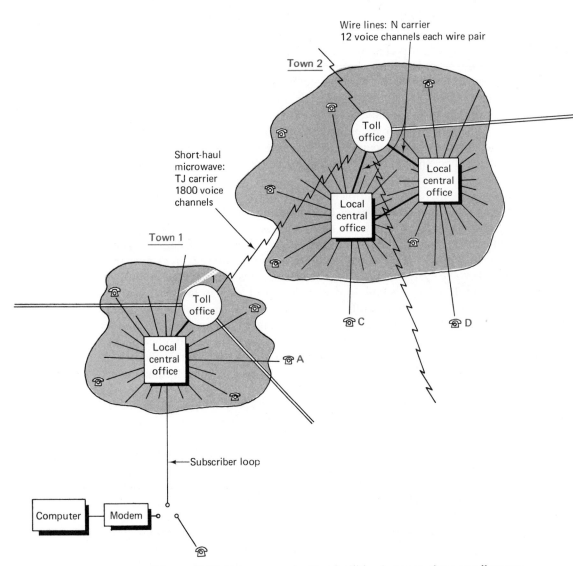

Figure 13.7. Telecommunication facilities between three small towns many miles apart.

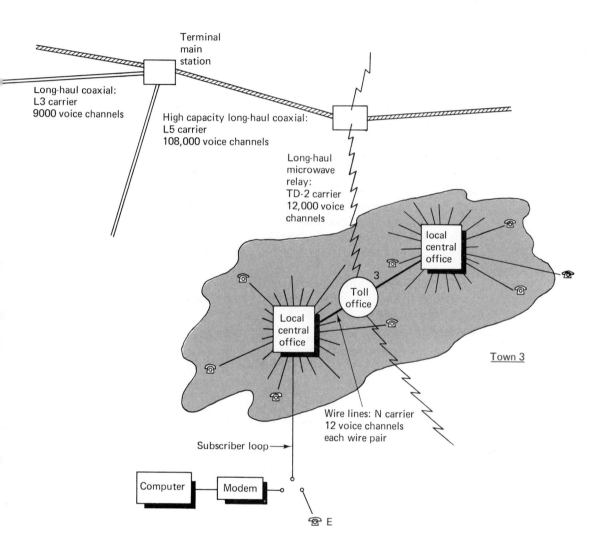

Terminal
main
station

Long-haul coaxial:
L3 carrier
9000 voice channels

High capacity long-haul coaxial:
L5 carrier
108,000 voice channels

Long-haul
microwave
relay:
TD-2 carrier
12,000 voice
channels

local
central
office

3

Toll
office

Local
central
office

Town 3

Wire lines: N carrier
12 voice channels
each wire pair

Subscriber loop →

Computer

Modem

☎ E

253

nals must be demultiplexed to lower-level groups or to a voice channel because different signals have different destinations.

RELATIVE   A long-distance telephone call thus goes through many
ATTENUATION   stages of multiplexing and demultiplexing. It is a tri-
                      umph of modern electronics that the telephone voice
can pass through all these manipulations and emerge unharmed. The curve of relative signal strength against frequency remains approximately the curve of Fig.
13.8, and hence a telephone voice from the other side of the nation sounds much
the same as it would from across the street.

The reader might wonder next time he makes a long-distance call at the multiple contortions his voice is undergoing.

A local loop with loading coils is designed, like a trunk circuit, to give low
attenuation up to about 3100 Hz and to allow the signal strength to fall off at
higher frequencies. Many loops today, however, do not employ loading. On unloaded loops of several miles the signal strength falls off more rapidly than on a
trunk circuit as shown by the dashed line in Fig. 13.8. Nevertheless higher frequencies can be transmitted over an unloaded loop than over a telephone trunk
circuit derived by multiplexing. 56,000 bits per second can be transmitted over
such loops, but not over FDM telephone trunk circuits.

Figure 13.8   After all of the electronic manipulations the voice channel
still follows this curve approximately for variation of signal strength
with frequency.

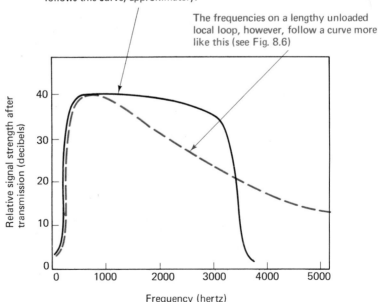

After all of the electronic manipulations the
attenuation of the telephone trunk frequencies
follows this curve, approximately.

The frequencies on a lengthy unloaded
local loop, however, follow a curve more
like this (see Fig. 8.6)

Relative signal strength after transmission (decibels)

40

30

20

10

0

0   1000   2000   3000   4000   5000

Frequency (hertz)

MICROWAVE TRUNKS The frequencies allocated for radio transmission of common carrier signals are much higher than those of the mastergroups and jumbogroups which are used. Table 13.2 lists the microwave transmission frequencies. The bands most commonly used are the 3.7–4.2 and 5.925–6.425 GHz bands. These bands each have a bandwidth of 500 MHz — much greater than the 3-MHz bandwidth of the mastergroup or 18-MHz bandwidth of the jumbogroup. Several channels each containing a high-level multiplexed block are therefore transmitted over one microwave route. The TH systems, for example, have six active radio channels (with additional backup channels in case of failure), and each channel carries a mastergroup multiplex block. The route capacity is thus 10,800 voice channels.

Table 13.2 Microwave frequencies allocated by the FCC for common carrier use in the United States

Band (MHz)	Bandwidth (MHz)
2110–2130	20
2160–2180	20
3700–4200	500
5925–6425	500
10,700–11,700	1000

The mastergroup multiplex block occupies frequencies up to 9 MHz. It is frequency-modulated as a unit onto a carrier of "intermediate frequency," about 60 to 80 MHz. This band is shifted again to the correct transmitting frequency in the gigahertz band and amplified. A waveguide carries it up to the antenna, where it is radiated.

A television channel used to replace two of the mastergroups in any of the mastergroup multiplex blocks. Now higher quality television trunks are used. Figure 13.9 shows microwave terminals.

WIDEBAND CHANNELS Most common carriers lease their channel groups and supergroups, as well as leasing individual channels, to provide a wideband service. Three grades of wideband services in the United States are listed in Table 13.3.

Table 13.3

	Equivalent Bandwidth (kHz)	Equivalent Number of Voice Channels
Series 8000 channel	48	12
Type 5700 channel	240	60
Type 5800 channel	1000	240

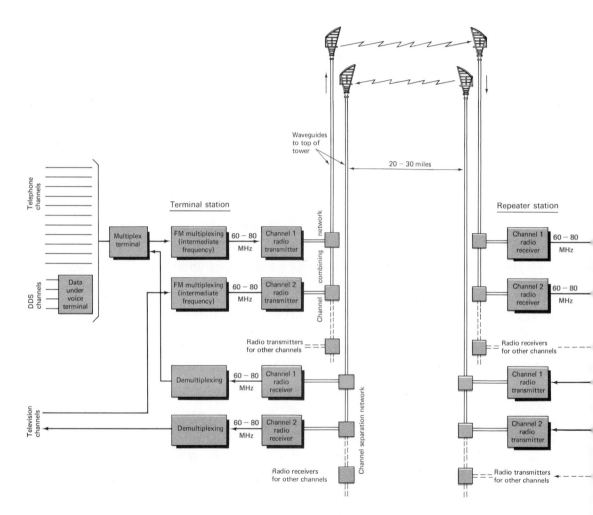

Figure 13.9 The equipment used in a microwave trunk route.

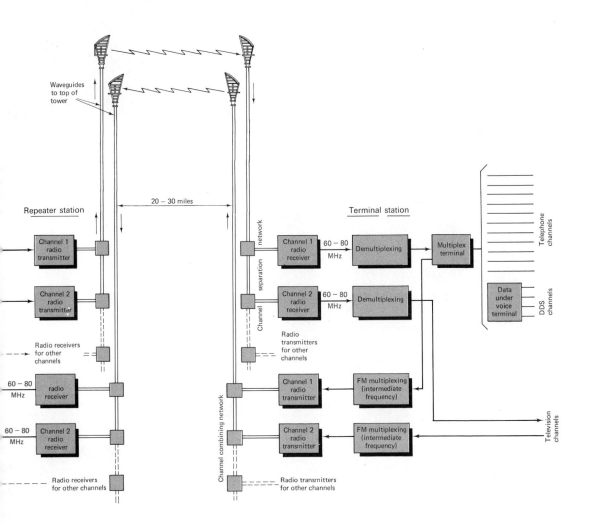

Waveguides
to top of
tower

20 – 30 miles

Repeater station

Terminal station

Channel separation network

Channel combining network

Channel 1
radio
transmitter

Channel 2
radio
transmitter

Radio receivers
for other
channels

60 – 80
MHz

radio
receiver

60 – 80
MHz

Channel 2
radio
receiver

Radio receivers
for other channels

Channel 1
radio
receiver

60 – 80
MHz

Demultiplexing

Multiplex
terminal

Telephone
channels

Channel 2
radio
receiver

60 – 80
MHz

Demultiplexing

Data
under
voice
terminal

DDS
channels

Radio
transmitters
for other
channels

Channel 1
radio
transmitter

FM multiplexing
(intermediate
frequency)

Channel 2
radio
transmitter

FM multiplexing
(intermediate
frequency)

Television
channels

Radio transmitters
for other channels

257

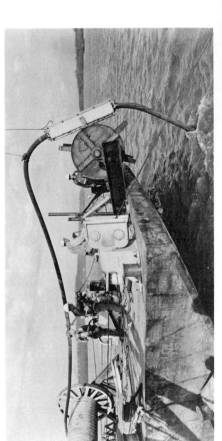

Figure 13.10 The laying down of the L5 carrier system – 108,000 telephone circuits. (Courtesy AT&T.)

The series 8000 channel is thus equivalent to one channel group, the type 5700 channel to a supergroup, and the type 5800 to four supergroups.

When an organization leases a wideband channel such as these, it will also lease appropriate channel terminals. A variety of terminating equipment is possible. The channels may be used as wideband channels or may be subdivided by the user into private channels of low bandwidth such as telephone, teletypewriter, or data transmission channels. The channel group, for example, can have terminating equipment from AT&T which will enable it to be used for one of the following:

1. Twelve telephone channels between two points.
2. Forty-eight 150-baud channels between two points.
3. One hundred forty-four teletypewriter channels between two points.
4. Equivalent combinations of the above.
5. Data transmission at 40,800 bits per second, plus one telephone channel for coordination purposes.
6. Two-level facsimile signals in the frequency range of approximately 29 to 44 kHz plus one telephone channel for coordination purposes.
7. Two-level facsimile signals requiring up to 50,000 bits per second plus one telephone channel.
8. A channel of bandwidth up to 20 kHz of high quality, i.e., only minor deviations in gain and delay characteristics.

Similarly, a supergroup can have terminating equipment that will enable it to be used for one of the following:

1. Seven-level magnetic tape transmission at approximately 105,000 bits per second plus a control channel and a telephone channel for coordination purposes.
2. Two-level facsimile requiring up to 250,000 bits per second plus four channels of teletypewriter grade for control and coordination.
3. A channel of bandwidth up to 100 kHz having only minor deviations in gain and delay characteristics.

A mastergroup or a standard television channel is capable of transmitting data at rates up to several million bits per second, but standard tariff offerings are not (yet) available for such uses.

Providing wideband transmission facilities can be made difficult by the fact that most installed wideband systems link only the toll offices. It is usually the case that, as in Fig. 13.6, there are no facilities wider than the channel group on the subscriber side of the toll office and no facilities wider than the voice channel linking the subscriber to his local central office. Special arrangements, therefore, have to be made. The subscriber loop can carry a bandwidth much higher than

Figure 13.11 Some radio towers carry many different microwave systems.

that of the voice channel if the loading coils are removed. For Picturephone transmission it is made to carry 1 MHz (after careful adjustment).

For a subscriber to obtain digital transmission at rates higher than those of voice-channel modems, the techniques discussed in the next chapter will become increasingly important.

REFERENCES

1. J. L. Hysko, W. T. Rea, and L. C. Roberts, "A Carrier Telegraph System for Short Haul Application," *Bell System Technical Journal,* 31 (July 1952), pp. 666–687.

2. R. S. Caruthers, "The Type N-1 Carrier Telephone System: Objectives and Transmission Features," *Bell System Technical Journal*, 30 (Jan. 1951), pp. 1–32.

3. R. C. Boyd and F. J. Herr, "The N2 Carrier Terminal—Objectives and Analysis," *Bell System Technical Journal*, 44 (May–June 1965), pp. 731–759.

4. G. W. Bleisch and C. W. Irby, "N3 Carrier System: Objectives and Transmission Features," *Bell System Technical Journal*, 45 (July–Aug. 1966), pp. 767–799.

5. R. D. Fracassi, "Type-O Carrier: System Description," *Bell Laboratories Record*, 32 (June 1954), pp. 215–220.

6. A. A. Roetken, K. D. Smith, and R. W. Friis, "The TD-2 Microwave Radio Relay System," *Bell System Technical Journal*, 30 (Oct. 1951), pp. 1041–1077.

7. J. Gammie and S. D. Hathaway, "The TJ Radio Relay System," *Bell System Technical Journal*, 39 (July 1960), pp. 821–877.

8. S. D. Hathaway, D. D. Sagaser, and J. A. Wood, "The TL Radio Relay System," *Bell System Technical Journal*, 42 (Sept. 1963), pp. 2297–2353.

9. R. W. Friis, J. J. Jansen, R. M. Jensen, and H. T. King, "The TM-1/TL-2 Short-Haul Microwave Systems," *Bell System Technical Journal*, 45 (Jan. 1966), pp. 1–95.

10. R. D. Ehrbar, C. H. Elmendorf, R. H. Klie, and A. J. Grossman, "The L3 Coaxial System," *Bell System Technical Journal*, 32 (July 1953), pp. 781–1005.

11. J. P. Kinzer and J. F. Laidig, "Engineering Aspects of the TH Microwave Relay System," *Bell System Technical Journal*, 40 (Nov. 1961), pp. 1459–1494.

12. R. H. Klie and R. E. Mosher, "The L-4 Coaxial Cable System," *Bell Laboratories Record*, 45 (July–Aug. 1967), pp. 211–217.

13. *Bell Laboratories Record*, 51 (Nov. 1973). Special Issue on the L-5 System.

14. K. E. Fultz and D. B. Penick, "The T-1 Carrier System," *Bell System Technical Journal*, 44 (Sept. 1965), pp. 1405–1452.

15. J. H. Bobsin and L. E. Forman, "The T-2 Digital Line," *Bell Laboratories Record*, 51 (Sept. 1973).

14 PCM: A REVOLUTION BEGINS

For half a century, telecommunications has been dominated by analog transmission and frequency-division multiplexing. Multibillions of dollars are tied up in such equipment.

However, if the telecommunications industry were to start again today building the world's transmission links, frequency-division multiplexing would have limited use, and a different technology would be dominant. Trunks would transmit a *digital* bit stream, not analog. Most links except local loops would be incapable of transmitting analog signals without conversion to digital form. The technique used today to transmit analog signals in digital form is called *pulse code modulation* (PCM). The telephone voice becomes a bit stream looking like computer data, and to combine separate signals *time-division multiplexing,* not frequency-division, is used.

Whether a telecommunication link is converted to digital technology is determined by economic considerations. When an entire nation's telephone system is considered, vast sums of money are involved. The links which can benefit most profitably are converted first. With many links it is still cheaper to use analog transmission. However, the costs of digital transmission are dropping. It will be a decade before the majority of the Bell System is digital because the vast investment in analog plant cannot be abandoned.

COMPUTER DATA The swing to PCM is good news for the computer industry. If digital bit streams form the basis of telecommunications, then computer data will no longer need to be converted to analog form for transmission. In other words, there will be no need for modems at each terminal. This can be seen today in AT&T's Dataphone Digital Service (discussed in Chapter 18), which performs end-to-end digital transmission. The user has a *service unit* instead of a modem, which contains a buffer from which the data are transmitted.

On the other hand, analog information such as the sound of the human voice needs to be converted to digital form before it can be transmitted over digital links. This conversion is done with a device which in a sense is the converse of a modem. It is called a *codec*. Just as "modem" is a contraction of the words <u>modu</u>late and <u>demodulate</u>, so "codec" is a contraction of the words <u>code</u> and <u>decode</u>.

Ironically, instead of the computer industry having to convert its data with a modem to travel over the telephone lines in an analog form, the telephone industry will have to convert its analog signals with a codec to travel over digital lines.

A telephone call when digitized by the telephone company requires 64,000 bits per second for transmission. This rate is much higher than the 4800 or 9600 bits per second at which data travel over analog telephone lines. The balance of cost between telephone transmission and data transmission is thus swinging substantially in favor of data.

ADVANTAGES
OF DIGITAL
TRANSMISSION

What are the advantages of transmitting the telephone voice in digital form?

Oddly enough one of the major advantages existed on early teletype links, but when several teletype channels were multiplexed into one voice circuit the advantage was lost. Now the changing economics are bringing it back. With analog transmission such as that discussed in the previous chapter, whenever the signal is amplified, *the noise is amplified with it*. As the signal passes through its many amplifying stations, so the noise is cumulative. With digital transmission, however, each repeater station *regenerates* the pulses. New, clean pulses are reconstructed and sent on to the next repeater where another cleaning-up process takes place. And so the pulse train can travel through a dispersive noisy medium, but, instead of becoming more and more distorted until eventually parts are unrecognizable, it is repeatedly reconstructed and thus remains impervious to most of the corrosion of the medium. Of course, an exceptionally large noise impulse may destroy one or more pulses so that they cannot be reconstructed by the repeater stations.

A major disadvantage of digital transmission would appear to be that much greater bandwidth is required. However, because the signal is reconstructed at frequent intervals down the line, it can tolerate much more battering than if it had to travel a long distance without reconstruction. It can survive traveling over a channel with a high level of distortion and with a poor noise-to-signal ratio. There is a trade-off between bandwidth and noise-to-signal ratio. A given transmission link such as a pair of wires can be operated at a higher bandwidth, i.e., higher frequencies can be transmitted over it, but the resulting signal will be more distorted and the noise-to-signal ratio will be higher. The trick that makes digital transmission worthwhile is to *reconstruct* the signal sufficiently often that it survives the bad distortion. A high bit rate can then be transmitted.

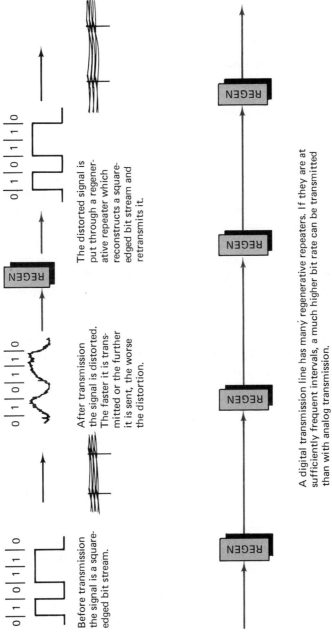

0 | 1 | 0 | 1 | 1 | 0

Before transmission the signal is a square-edged bit stream.

0 | 1 | 0 | 1 | 1 | 0

After transmission the signal is distorted. The faster it is transmitted or the further it is sent, the worse the distortion.

REGEN

0 | 1 | 0 | 1 | 1 | 0

The distorted signal is put through a regenerative repeater which reconstructs a square-edged bit stream and retransmits it.

REGEN   REGEN   REGEN   REGEN

A digital transmission line has many regenerative repeaters. If they are at sufficiently frequent intervals, a much higher bit rate can be transmitted than with analog transmission.

Figure 14.1

265

Consider a telephone wire pair under the streets of a city. With analog transmission it can carry a *channel group* — 12 voice channels. Now suppose that we transmit digitally over the same wire pair. The digital signal becomes distorted as it is transmitted, as shown in Fig. 14.1. We catch it before it becomes too distorted to recognize whether a bit is 0 or 1. The bit stream is then reconstructed, retimed, and retransmitted. The faster the bit stream is transmitted, the greater will be the distortion and the closer the spacing of repeaters necessary to reconstruct the signal correctly. If, for the sake of argument, repeaters reconstructed the signal every 100 feet, then a *very* high bit rate could be transmitted. The wire pair would be handling a high bandwidth. The distortion would be severe, but the bit stream would get through. Clearly, repeaters could not be spaced that close together. The cost would be too high and the wires are not accessible. How closely can they reasonably be spaced? There is a manhole or other access to the wires about every 6000 feet. With that spacing, today's telephone wire pairs can be made to carry 2.048 million bits per second, which is equivalent to 32 telephone speech channels.

In North America 1.544 million bits per second (24 speech channels) are used with conservative engineering and 6.3 million bps (96 speech channels) with modern lines carefully adjusted. The CCITT has standards for 1.544 million bps (24 speech channels) and 2.048 million bps (30 speech channels with two channels for control purposes).

One coaxial tube such as those in Fig. 9.5 has been used to transmit 274 million bps. The capacity of a 22-tube cable is then 2.74 billion bps. The WT4 waveguide shown in Fig. 9.19 is designed to carry 60 channels of 274 million bps, giving a total capacity of 16.4 billion bps.

ECONOMIC FACTORS   The economic circumstances favoring digital transmission stem from several factors.

First, the cost of digital circuitry produced in large quantities is dropping fast due to LSI (large-scale integration) technology. Telecommunication networks employ large quantities of any circuit used and so can benefit from the mass-production economics of LSI.

Second, as we have seen, it is becoming economical to build channels of higher bandwidth. Satellites, waveguides, and optical fibers promise much higher bandwidths. A high level of multiplexing is needed to make use of high-capacity channels. When many calls are packed in a high bandwidth the cost of a voice-channel mile drops greatly, but the cost of the increasing number of multiplexing and routing operations becomes high. Frequency-division multiplexing uses fairly expensive circuit components such as filters. When thousands of telephone conversations travel together over coaxial cable or microwave links, they must be demultiplexed, switched, and then multiplexed together again at each switching point. While there is great economy of scale in the *transmission,* there is not in this multiplexing and switching operation. As the channel capacities increase, so the multiplexing and switching costs assume a greater and greater proportion of the total network cost.

Digital circuitry, on the other hand, is dropping in cost at a high rate. With the maturing of LSI techniques, it will drop even more. Where digital rather than analog transmission is used, this increasingly low-cost circuitry handles the multiplexing and switching. The telecommunication networks becomes in some aspects like a vast digital computer.

Third, the use of digital transmission makes it possible to operate on links with a high noise-to-signal ratio. On links with a wide-ranging trade-off between noise-to-signal ratio and bandwidth, a somewhat larger number of voice channels can be derived by PCM techniques then by analog techniques. This is the case on today's wire-pair telephone lines and satellite channels, for example.

Fourth, digitization of voice channels can be done with more complex techniques than PCM, which will permit the speech to be encoded into fewer bits.

Last, an additional economic factor is the rapidly increasing use of data transmission. Although data transmission still employs only a small proportion of the total bandwidth in use, it is increasing much more rapidly than other uses of the telecommunication networks. Data can be transmitted over digital voice links with a total equipment cost that can be as low as a tenth of that for transmission over analog links with modems.

In terms of the immediate economics of today's common carriers, pressed for capacity, digital transmission is appealing for short-distance links because with relatively low-cost electronics it can susbtantially increase the capacity of existing wire pairs. This is particularly important in the congested city streets.

An important long-term advantage is the fact that all signals — voice, television, facsimile, and data — become a stream of similar-looking pulses. Consequently, they will not interfere with one another and will not make differing demands on the engineering of the channels. In an analog signal format, television and data are much more demanding in the fidelity of transmission than speech and create more interference when transmitted with other signals. Eventually, perhaps, there will be an integrated network in which all signals travel together digitally.

PULSE AMPLITUDE
MODULATION

To convert an analog signal such as speech into a pulse train, a circuit must sample it at periodic intervals. The simplest form of sampling produces pulses, the amplitude of which is proportional to the amplitude of the signal of the sampling instant (see Fig. 14.2). This process is called *pulse amplitude modulation* (PAM).

Compare the PAM illustration in Fig. 14.2 with that for amplitude modulation of a sine-wave carrier in Fig. 11.2. Envelope detection can be used for demodulating the PAM signal in much the same way as that described in Chapter 11 for amplitude modulation.

The pulses of Fig. 14.2 still carry their information in an analog form; the amplitude of the pulse is continuously varible. If the pulse train were transmitted over a long distance and subjected to distortion, it might not be possible to re-

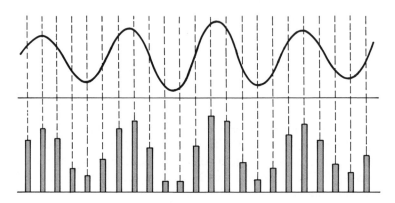

Figure 14.2   Pulse amplitude modulation (PAM).

construct the original pulses. To avoid this, a second process is employed which converts the PAM pulses into unique sets of equal amplitude pulses, in other words into a binary bit stream. The receiving equipment then detects only whether a bit is 0 or 1; it detects the presence or absence of a pulse, not its size.

As we shall see in Part III of this book the PAM pulses themselves are used in certain switching equipment in which the switching is done by electronically controlling the flow of PAM pulses.

PULSE CODE   The amplitude of the PAM pulse can assume an in-
MODULATION   finite number of possible values ranging from zero to
                      the maximum.

It is normal with pulse modulation to transmit not an infinitely finely divided range of values but a limited set of specific discrete values. The input signal is *quantized*. This process is illustrated schematically in Fig. 14.3. Here the signal amplitude can be represented by any one of the eight values shown. The amplitude of the pulses will therefore be one of these eight values. An inaccuracy is introduced by rounding values in a computation. Figure 14.3 shows only eight possible values of the pulse amplitude. If there were more values, the "rounding error" would be less. In systems in actual use today, 128 pulse amplitudes are used—or 127 to be exact, for the zero amplitude is not transmitted.

After a signal has been quantized and samples taken at specific points, as in Fig. 14.3, the result can be coded. If the pulses in the figure are coded in binary, as shown, 3 bits are needed to represent the eight possible amplitudes of each sample. A more accurate sampling with 128 quantized levels would need 7 bits to represent each sample. In general, if there were N quantized levels, $\log_2 N$ bits would be needed per sample.

The process producing the binary pulse train is referred to as *pulse code modulation* (PCM). The resulting train of pulses passes through frequent repeater stations that reconstruct the pulse train and is impervious to most types of tele-communication distortion other than major noise impulses or signal interruptions. The mere presence or absence of a pulse can be recognized easily even when distortion is present, whereas determination of pulse magnitude would be more prone to error.

On the other hand, the original voice signal can never be reproduced exactly, because of the quantizing errors. This deviation from the original signal is sometimes referred to an *quantizing noise*. It is of known magnitude and can be reduced, at the expense of bandwidth, by increasing the number of sampling levels; 128 levels, needing 7 bits per sample, are enough to produce telephone channels having a signal-to-noise ratio comparable to that achieved on today's analog channels.

Figure 14.3 Pulse code modulation (PCM).

1 The signal is first "quantized" or made to occupy a discrete set of values

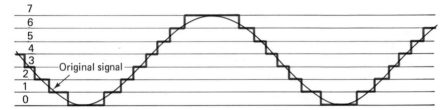

2 It is then sampled at specific points. The PAM signal that results can be coded for pulse code transmission

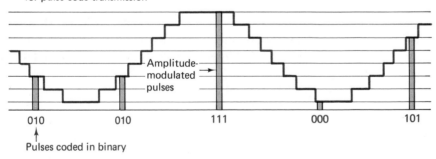

3 The coded pulse is transmitted in a binary form

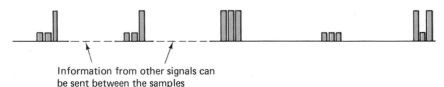

HOW MANY The pulses illustrated in Figs. 14.2 and 14.3 are sam-
SAMPLES ARE pling the input at a limited number of points in time.
NEEDED? The question therefore arises, How often do we need
to sample the signal in order to be able to reconstruct it
satisfactorily from the samples? The less frequently we can sample it, the lower
the number of pulses we have to transmit in order to send the information, or,
conversely, the more information we can transmit over a given bandwidth.

Any signal can be considered as being a collection of different frequencies,
but the bandwidth limitation on it imposes an upper limit to these frequencies.
When listening to a violin, you hear several frequencies at the same time, the
higher ones being referred to as "harmonics." You hear no frequencies higher
than 20,000 Hz, however, because that is the upper limit of the human ear. (The
ear has a limited bandwidth like any other channel.) When listening to a full or-
chestra, you are still hearing a collection of sounds of different frequencies, although now the pattern is much more complex. Similarly, other signals that we
transmit are composed of a jumble of frequencies. A digital signal can be analyzed
by Fourier analysis into its component frequencies.

It can be shown mathematically that *if the signal is limited so that the highest frequency it contains is W Hz, then a pulse train of 2W pulses per second is
sufficient to carry it and allow it to be completely reconstructed.*

The human voice, therefore, if limited to frequencies below 4000 Hz, can be
carried by a pulse train of 8000 PAM pulses per second. The original voice
sounds, below 4000 Hz, can then be *completely* reconstructed.

Similarly, 40,000 samples per second could carry hi-fi music and allow complete reproduction. (If samples themselves were digitized, as with PCM, the reproduction would not be quite perfect because of the quantizing error.)

Table 14.1 shows the bandwidth needed for four types of signals for human
perception, plus the digital bit rate used or planned for their transmission
with PCM.

In telephone transmission, the frequency range encoded in PCM is somewhat less than 200 to 3500 Hz; 8000 samples per second are used. Each sample is
digitized using 7 bits so that $2^7 = 128$ different volume levels can be distinguished.

Table 14.1   Bandwidths and equivalent PCM bit rate for
typical signals

Type of Signal	Analog Bandwidth Used (kHz)	Number of Bits Per Sample	Digital Bit Rate Used or Needed (1000 bps)
Telephone voice	4	7	$4 \times 2 \times 7 = 56$
High-fidelity music	20	10	$20 \times 2 \times 10 = 400$
Picturephone	1000	3	$1000 \times 2 \times 3 = 6000$
Color television	4600	10	$4600 \times 2 \times 10 = 92,000$

This gives $7 \times 8000 = 56,000$. High-fidelity music with five times this frequency range would need five times as many samples per second, and, to achieve subtle reproduction, more bits per sample.

MULTIPLEXING   As noted, $4000 \times 2 \times 7 = 56,000$ bps are needed to carry a telephone voice. However, all the transmission facilities that this bit stream is likely to be sent over can carry a much higher bit rate than this.

It is therefore desirable to send many telephone signals over one physical path. This is done by interleaving the "samples" that are transmitted, i.e., *time-division multiplexing*. If four voice signals are to be carried over one pair of wires, the samples are intermixed as follows:

<blockquote>
sample from speech channel 1

sample from speech channel 2

sample from speech channel 3

sample from speech channel 4

sample from speech channel 1

sample from speech channel 2

sample from speech channel 3

.

.

.
</blockquote>

This is illustrated in Fig. 14.4. By sampling the signals at the appropriate instants in time, a train of PAM pulses is obtained; these pulses are then digitally encoded. For simplicity, only a 4-bit code is shown in the diagram. Each PAM pulse is encoded as 4 bits. The result is a series of "frames," each of 16 bits. Each frame contains one sample of each signal.

To decode the signal, it is necessary to be sure where each frame begins. The signals can be reconstructed with this knowledge. The first 4 bits relate to speech channel 1, the second 4 to channel 2, and so on. A synchronization pattern must also be sent in order to know where each frame begins. This, in practice, can be done by the addition of 1 bit per frame. The added bits, when examined alone, form a unique bit pattern that must be recognized to establish the framing.

This process takes place at electronic speeds in computerlike circuits. The circuit components for digital multiplexing of this type are much lower in cost than those for frequency-division multiplexing. In the latter process, the range of frequencies available for transmission is divided up into smaller ranges, each of which carries one signal.

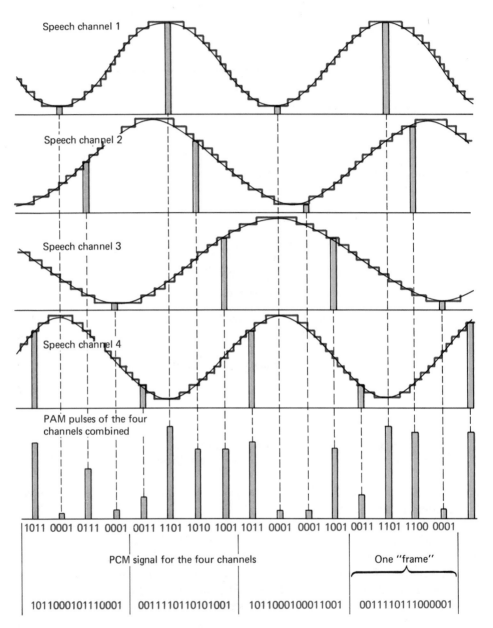

Figure 14.4 A simplified picture of time-division multiplexing with PCM transmission. The reader may extend this, in his mind, to 24 channels and 128 sample levels (7 bits per sample).

COMPANDING   If the signal being transmitted were of low amplitude, then the procedure illustrated in Fig. 14.3 would not, of itself, be so satisfactory. The quantizing noise, still the same absolute magnitude, would now be larger relative to the signal magnitude. The quantizing error is a function of interval between levels and not of the signal amplitude; therefore the signal-to-quantizing noise ratio is lower for smaller signals. For this reason a *compandor* is normally used.

A compandor is a device which, as discussed in Chapter 10, compresses the higher-amplitude parts of a signal before modulation and expands them back to normal again after demodulation. Preferential treatment is therefore given to the weaker parts of a signal. The weaker signals traverse more quantum steps than they would otherwise, and so the quantizing error is less. This is done at the expense of the higher-amplitude parts of the signal for these cover less quantum steps.

The process is illustrated in Fig. 14.5. The effect of companding to move the possible sampling levels closer together at the lower-amplitude signal values is sketched on the right-hand side of Fig. 14.5, which shows the quantizing of a weak signal and a strong signal. The right-hand side of the diagram is with companding, and the left-hand side without. It will be seen that on the left-hand side the ratio of signal strength to quantizing error is poor for the weak signal. On the right-hand side it is better. Furthermore, the strong signal is not impaired greatly by the use of the compandor. In practice, the PAM pulses are companded, and one equipment serves all the channels which are being multiplexed together.

Different telephone administrations must use the same companding rules if

Figure 14.5 With a compandor the quantization of the weak signal gives more separate values, and therefore a better ratio of signal to quantizing noise. Standard sets of compounding rules are specified in the CCITT Recommendation G.711[1].

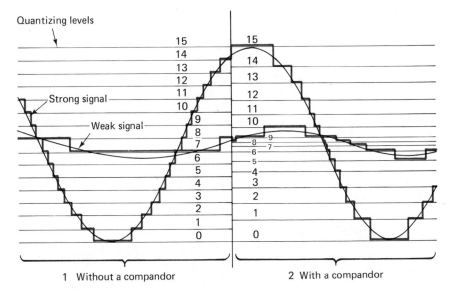

they are to interchange signals. The CCITT specifies companding rules for PCM transmission, referred to as A-law and μ-law companding, in CCITT Recommendation 9.711 [1].

REGENERATIVE REPEATERS   The main reason high bit rates can be achieved on wire-pair circuits using pulse code modulation is that repeaters are placed at sufficiently frequent intervals to reconstruct the signal.

The Bell System places them at intervals of 6000 feet, which is the spacing of loading coils employed when the wires are used for analog transmission; the repeaters replace the loading coils.

Figure 14.6   A high speed PCM repeater. It receives the waveform at the top and reconstructs from it the waveform at the bottom.

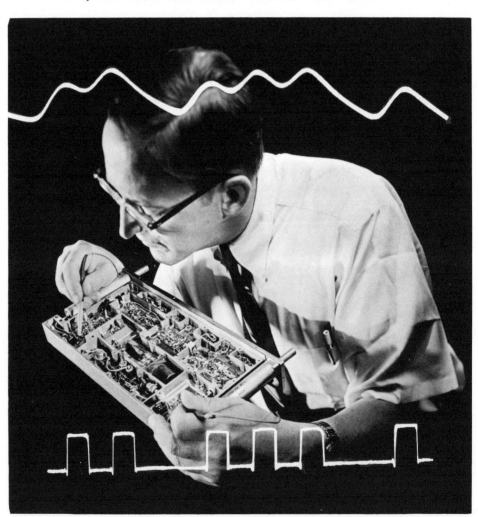

A regenerative repeater has to perform the following three functions, sometimes referred to as the three "R"s: reshaping, retiming, and regeneration. When a pulse arrives at the repeater, it is attenuated and distorted. It must first pass through a preamplifier and equalizer to reshape it for the detection process. A timing recovery circuit provides a signal to sample the pulse at the optimum point and decide whether it is a one or zero bit. The timing circuit controls the regeneration of the outgoing pulse and ensures that it is sent at the correct time and is of the correct width.

Figure 14.6 shows a PCM system repeater.

THE BELL SYSTEM   The most widely used transmission system at present
T1 CARRIER   with time-division multiplexing is the *Bell System T1 carrier*. This carrier uses wire pairs with digital repeaters spaced 6000 feet apart to carry 1.544 million bps. Into this bit stream 24 speech channels are encoded, using pulse code modulation and time-division multiplexing.

The T1 carrier is used for short-haul transmission over distances of up to 50 miles. It has been highly successful, and more than 70 million voice-channel miles of it are in operation. Most readers of this book have talked over a digital telephone channel without knowing it.

The Bell T1 PCM System uses 7 bits for coding each sample. The system is designed to transmit voice frequencies up to 4kHz, and therefore 8000 samples per second are needed, and 8000 frames per second travel down the line. Each frame, then, takes 125 microseconds. A frame is illustrated in Fig. 14.7. It contains 8 bits for each channel. The eighth bits form a bit stream for each speech channel which contains network signaling and routing information, for example to establish a connection and to terminate a call. There are a total of 193 bits in each frame, giving $193 \times 8000 = 1.544$ million bps.

The last bit in the frame, the 193rd bit, is used for establishing and maintaining synchronization. The sequence of these 193 bits from separate frames is examined by the logic of the receiving terminal. If this sequence does not follow a given coded pattern, then the terminal detects that synchronization has been lost. If synchronization does slip, then the bits examined will in fact be bits from the channels—probably speech bits—and will not exhibit the required pattern. There is a chance that these bits will accidentally form a pattern being sought. The synchronization pattern must therefore be chosen so that it is unlikely that it will occur by chance. If the 193rd bit was made to be always a 1 or always a 0, this could occur by chance in the voice signal. It was found than an alternating bit pattern, 0 1 0 1 0 1 . . . , never occurs for long in any bit position. Such a pattern would imply a 4-kHz component in the signal, and the input filters used would not pass this. Therefore the 193rd bit transmitted is made alternately a 1 and a 0. The receiving terminal inspects it to ensure that this 1 0 1 0 1 0 . . . pattern is present. If it is not, then it examines the other bit positions that are 193 bits apart until a

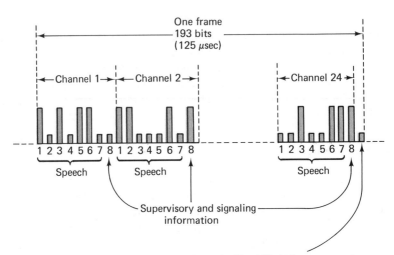

Framing code. The 193rd bits on successive
frames follow a pattern which is checked to
ensure synchronization has been maintained.
If synchronization is lost this pattern is
scanned for to re-establish it. The pattern used
is 010101. . . repeated.

Figure 14.7 The bit structure of a North American PCM transmission
link operating at 1.544 million bits per second (T1 Carrier). The above
frame is repeated 8000 times per second, thus giving 8000 samples per
second on each channel plus an 8000 bps bit stream for control signaling.
The CCITT Recommendation for 1.544 bps PCM is slightly different
(see Fig. 21.7) [2].

1 0 1 0 1 0 . . . pattern is found. It then assumes that these are the framing pulses.

This scheme works very well with speech transmission. If synchronization
is lost, the framing circuit takes 0.4 to 6 milliseconds to detect the fact. The time
required to reframe will be about 50 milliseconds at worst if all the other 192 positions are examined, but normally the time will be much less, depending on how far
out of synchronization it is. This is quite acceptable on a speech channel. It is
more of a nuisance when data are sent over the channel and would necessitate the
retransmission of blocks of data. Retransmission is required on most data transmission, however, as a means of correcting errors that are caused by noise on the
line and detected with error-detecting codes.

Data-processing specialists who have had reason to complain that the
printer on the other end of a telephone line is slow can be encouraged by the
thought that over a PCM telephone line they could print at about 4000 lines per
minute, or fill a very large screen full of data in a second. To do so, however, the
high-capacity link would have to go into their premises. At present it commonly
ends at the local telephone office.

HIGHER-CAPACITY PCM   The T1 carrier is only the beginning. A hierarchy of interlinking digital channels is planned for the Bell System. The next step up is the T2 carrier, which takes the signals from four T1 carriers, or, alternatively carries one Picturephone signal, and operates at 6.3 million bps. Millions of T2 voice-channel miles are operational. Higher still are the other T carriers, which will carry hundreds of megabits per second over broadband transmission facilities.

Furthermore, by coding the telephone voice in a more complex way than that of the T1 or T2 carriers the voice can be carried with half the bit rate.

Higher-capacity digital carriers are discussed in the sequel to this book, *Future Developments in Telecommunications* (2nd edn.).

Figure 14.8   CCITT Recommendation for the structure of PCM channels for transmission at 2.048 million bits/sec [3]. 30 speech channels of 64000 bps are derived, each with a signaling channel of 500 bps.

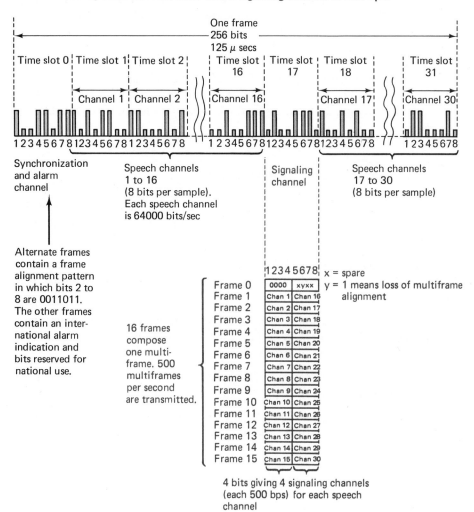

CCITT   The CCITT has made two recommendations for PCM
RECOMMENDATIONS   transmission, one for transmission at the T1 carrier
                          speed of 1.544 million bps [2] and one for transmission
at 2.048 million bps [3], which can be achieved over most telephone wire pairs.

As is often the case, the CCITT recommendation for 1.544-million-bps
transmission is slightly different from the North American standard set by AT&T.
It employs a 193-bit frame with 8 bits per channel as in Fig. 14.6, but the frame
alignment bit is the first bit, not the 193rd bit as in Fig. 14.6, and it carries a different synchronization pattern. Twelve such frames are grouped together to form
one *multiframe*.

If separate signaling is provided for each channel, *two* signaling bit streams
are derived from the 8th bits and only every sixth frame contains signaling bits, as
shown in Fig. 21.7. This gives a smaller bit rate for signaling but leaves 8 bits per
channel in 5/6 of the frames for carrying speech or information.

Figure 14.8 shows the CCITT 2.048-million-bps recommendation, which
most of the world outside North America is starting to use for PCM transmission.
In this, 16 frames of 256 bits each form a multiframe. There are 32 8-bit time slots
in each frame, giving 30 speech channels of 64,000 bps each plus one synchronization and alarm channel and one signaling channel, which is submultiplexed to
give four 500-bps signaling channels for each speech channel.

The difference between the CCITT and North American standards will prevent the world becoming linked with standard digital channels on satellites and
other media.

REFERENCES

1. *CCITT Recommendation A. 711* (on compandors), *Green Book*, Vol. III:
   Line Transmission, International Telecommunications Union, Geneva,
   1973.

2. *CCITT Recommendation A.733* (on PCM multiplex equipment operating at
   1544 KB/S), *Green Book,* Vol. III: Line Transmission, International Telecommunications Union, Geneva, 1973.

3. *CCITT Recommendation A.732* (on PCM multiplex equipment operating at
   2048 KB/S), *Green Book,* Vol. III: Line Transmission, International Telecommunications Union, Geneva, 1973.

4. J. S. Mayo, "A Bipolar Repeater for Pulse Code Signals," *Bell System Technical Journal,* 41 (Jan. 1962), pp. 25–98.

5. J. S. Mayo, "Experimental 224 MG/S PCM Terminals," *Bell System Technical Journal,* 44 (Nov. 1965), pp. 1813–1842.

6. I. Dorros, J. M. Sipress, and F. D. Waldhauer, "An Experimental 224 MG/S Digital Repeatered Line," *Bell System Technical Journal*, 45 (Sept. 1966), pp. 993–1044.

7. C. G. Davies, "An Experimental Pulse Code Modulation System for Short-Haul Trunks," *Bell System Technical Journal*, 41 (Jan. 1962), pp. 1–24.

8. M. R. Aaron, "PCM Transmission in the Exchange Plant," *Bell System Technical Journal*, 41 (Jan. 1962), pp. 99–142.

9. R. H. Shennum and J. R. Gray, "Performance Limitation of a Practical PCM Terminal," *Bell System Technical Journal*, 41 (Jan. 1962), pp. 143–172.

10. H. Mann, H. M. Straube, and C. P. Villars, "A Companded Coder for an Experimental PCM Terminal," *Bell System Technical Journal*, 41 (Jan. 1962), pp. 173–226.

11. K. E. Fultz and D. B. Penick, "The T1 Carrier System," *Bell System Technical Journal*, 44 (Sept. 1965), pp. 1405–1452.

12. J. F. Travis and R. E. Yeager, "Wideband Data on T1 Carrier," *Bell System Technical Journal*, 44 (Oct. 1965), pp. 1567–1604.

13. J. O. Edson and H. H. Henning, "Broadband Codes for an Experimental 244 MB/S PCM Terminal," *Bell System Technical Journal*, 44 (Oct. 1965), pp. 1887–1940.

14. F. J. Witt, "An Experimental 224 MB/S Digital Multiplexer – Using Pulse Code Stuffing Synchronization," *Bell System Technical Journal*, 44 (Nov. 1965), pp. 1843–1886.

15 COMMUNICATION SATELLITES

Perhaps the most exciting and most rapidly evolving of all the developments in transmission is the communications satellite.

On April 6, 1965, the world's first commercial satellite, Early Bird, rocketed into the evening sky at Cape Kennedy. The success of the transmission experiments that followed this was spectacular. Before long earth stations were being built around the world, and new and more powerful satellites were on the drawing board.

Originally the satellites were intended primarily to span the oceans. As has been discussed earlier, the laying of subocean telephone cables is expensive, and these cables have a bandwidth too low to carry live television. Early Bird alone increased transatlantic telephone capacity by more than one third. The television pictures were excellent. Pope Paul VI was soon seen live from the Vatican in the United States, and President Johnson was seen all over Europe. A Pacific satellite followed in 1967.

The satellites were soon used with computer systems. Pan American successfully hooked up their real-time Panamac system to Early Bird, so that reservation agents in Rome or London could smile at their customers while conversing with a computer in New York via satellite. IBM linked its laboratories on opposite sides of the Atlantic and was soon talking about a far-flung information system using satellite-transmitted data responses for its World Trade operations.

Four generations of Comsat satellites were launched in 6 years: INTELSAT I (named Early Bird), II, III, and IV. INTELSAT IV was followed in North America by the Canadian ANIK satellites (ANIK means "brother" in Eskimo) and then by various private industry satellites and in Europe by the French-German Symphonie satellite. The first U.S. domestic satellite was the WESTAR satellite launched in 1974 by Western Union.

TRANSPONDERS   A communication satellite (Fig. 15.1) is, in essence, a
                          microwave relay in the sky. It receives microwave sig-
nals in a given frequency band and retransmits them at a different frequency. It
must use a different frequency for retransmission; otherwise the powerful transmitted signal would interfere with the weak incoming signal. The equipment which
receives a signal, amplifies it, changes its frequency, and retransmits it is called a
transponder.

Most satellites have more than one transponder. The bandwidth handled by
a transponder has differed from one satellite design to another, but most contem-
porary satellites (e.g., INTELSAT IV, ANIK, and Western Union's WESTAR)
have transponders with a bandwidth of 36 MHz. How this bandwidth is utilized
depends on the earth station equipment. The WESTAR satellites, which are typical, may be used to carry any of the following:

1. One color television channel with program sound.

2. 1200 voice channels.

3. A data rate of 50 Mbps.

4. The center 24 MHz of each band may relay either
   (a) 16 channels of 1.544 Mbps, or
   (b) 400 channels of 64,000 bps, or
   (c) 600 channels of 40,000 bps.

The WESTAR satellites each have 12 such transponders, two of which are
spares used to back up the other 10 in case of failure. Future satellites will have a
larger number of transponders.

EARTH STATIONS   A satellite earth station consists of a large dish, such as
                            those in Fig. 15.2, which points at the satellite in basi-
cally the same way that an earthbound microwave relay dish points at the next
tower in the chain. The earth station antenna, however, is larger, giving a narrower beam angle.

The first earth stations were massive. The earth station at Andover, Maine,
originally built for AT&T's Telstar satellite and then used with Early Bird and its
successors, has a dome 18 stories high housing a huge steerable horn-shaped
antenna weighing 380 tons, with electronics cooled by liquid helium. Many of
today's earth stations, such as that in Fig. 15.2, are small enough to be erected
quickly in a parking lot behind a factory or office building. Dishes of 20 feet or less
are used, whereas the large Comsat earth stations use 100-foot dishes. Initially,
earth stations were owned only by the common carriers (and military). Now the
small earth stations are owned or leased by private industry and access common
carrier satellites.

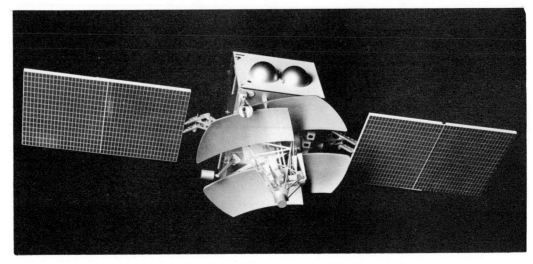

The RCA SATCOM satellite. 24 transponders of 36 GHz.

NASA's ATS-6 satellite with a 30-foot umbrella antenna which broadcasts to small low-cost earth stations.

Figure 15.1

Figure 15.2 A 10-meter diameter antenna used by the American Satellite Corporation for transmitting the *Wall Street Journal* in facsimile form to remote printing plants. The WESTAR satellite is used. The *Journal* pages are received on page-size photographic film. One page takes 3–5 minutes of transmission time.

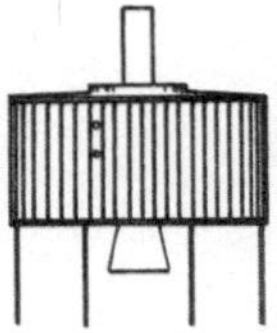

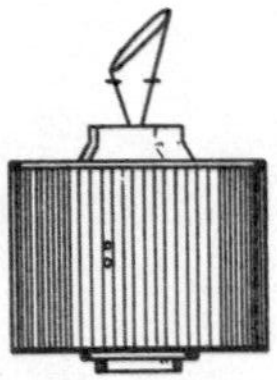

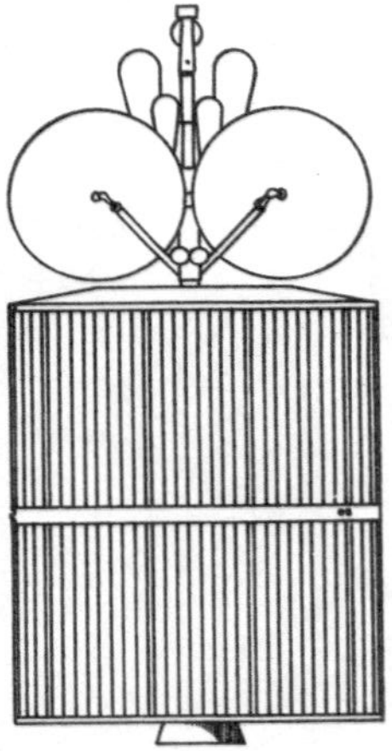

Name	Intelsat I (Early Bird)	Intelsat II	Intelsat III	Intelsat IV	Being designed Intelsat V
Year of launch	1965	1967	1968	1971	≈ 1979
Diameter	28 inches	56 inches	56 inches	93 inches	600 inch sails
Height	23 inches	26 inches	78 inches	111 inches	264 inches
Weight in orbit	85 lbs	192 lbs	322 lbs	1547 lbs	3200 lbs
Number of antennas	1	1	1	3	6
Primary power (watts)	40	75	120	400	1000
No. of transponders	2	1	2	12	27
Bandwidth of transponder	25 MHz	130 MHz	225 MHz	36 MHz	
Cost of satellite	$3.6 million	$3.5 million	$4.5 million	$14 million	≈ $25 million
Cost of launch	$4.6 million	$4.6 million	$6 million	$20 million	≈ $23 million
Design lifetime	1.5 years	3 years	5 years	7 years	10 years
Total cost per year	$5.47 million	$2.70 million	$1.90 million	$4.85 million	≈ $4.8 million
Maximum No. of voice circuits	240	240	1200	6000	≈ 24,000
Cost/voice circuit/year	$23,000	$11,000	$1600	$810	≈ $200

Figure 15.3 The INTELSAT satellites .

While most earth stations simply transmit and receive the telecommunication signal with a fixed antenna, at least one must carry out the additional function of *controlling* the satellite. Western Union's earth stations, for example, are unmanned, with the exception of one at Glenwood, N.J. which monitors both the satellites and the other earth stations.

During a satellite launch the Glenwood station has the critical function of maneuvering the satellite into position once it has separated from its launch vehicle. For 7 years or more after the launch it must maintain the satellite in its correct position by occasionally firing small gas jets on board the satellite. It can send commands to the satellite to turn transponders on and off (to save power), to switch to redundant and backup equipment, and to control the charging of batteries and positioning of antennas.

THE DROPPING COST OF SATELLITE CHANNELS   The first four generations of INTELSAT satellites carried increasing numbers of channels and had progressively longer design lives, as shown in Fig. 15.3. Consequently the cost per voice channel per year dropped dramatically. The process will continue with INTELSAT IV and V.

The bottom line of Fig. 15.3 shows the drop in cost per satellite voice channel per year. Fig. 15.4 plots the trend. The figure shown is the *investment* cost of the satellite and its launch. The cost to a subscriber will be much higher because it must include the earth station and links to it, and must take into consideration the fact that the average channel utilization may be low.

The satellites and their launch costs are referred to as the *space segment* of satellite communications. The comment is sometimes made among systems planners that the space segment costs are dropping to such a low level that overall system costs will be dominated by the organization of the ground facilities.

The cost of an earth station, however, has dropped much more spectacularly than that of a satellite. The first Comsat earth stations cost more than $10 million. (The first Bell System earth stations for TELSTAR cost *much* more.) Earth stations have dropped in cost until now a powerful transmit/receive facility such as that in Fig. 15.2 can be purchased for about $100,000. Receive-only facilities are a fraction of this cost. At the same time the traffic that can be handled by an earth station is increasing as satellite capacity increases. Combining these two trends we find that the investment cost per channel per earth station is dropping, as shown in Fig. 15.5.

The **total** *earth segment* costs are *not* dropping because to provide increased accessibility to the satellites many earth stations are being built. Prior to 1973 the United States had only a handful of earth stations. Now many corporations are setting up their own satellite antennas.

There is a trade-off between the cost of the satellite and the cost of its earth station. If the satellite has a large antenna and considerable power, smaller earth stations can be used. If the satellite makes higher use of its bandwidth allocation, the cost per channel will be lower. There is a limit to bandwidth utilization, and so the main effect of increasing satellite cost will be to reduce earth antenna size and cost.

As the earth facilities drop in cost, more antennas will be constructed and more traffic will be sent, making it economical to use more powerful satellites, which will make the earth facilities drop further in cost.

Figure 15.4   Satellite costs per circuit are dropping rapidly.

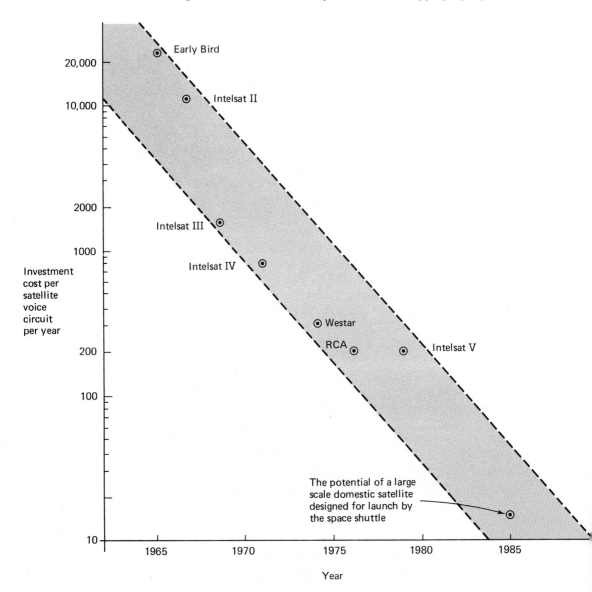

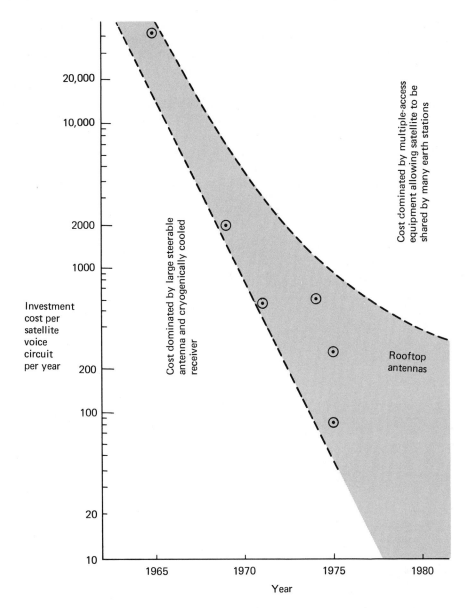

Figure 15.5   Earth station costs per circuit are also dropping rapidly.

SATELLITE ORBITS   Modern communications satellites have orbits very
different from their experimental predecessors such as
AT&T's Telstar satellites and RCA's Relay satellites. The latter traveled rapidly
around the earth at a relatively low height. The Telstar satellites had highly ellipti-

cal orbits, Telstar I from about 600 to 3800 miles and Telstar II from 600 to 6200 miles. The apogee of the elipse was positioned so that the satellite was within line-of-sight of certain stations for as long as possible. As with early manned orbital flights and most other satellites launched in the first decade of space flight, they traveled around the earth in a few hours: Telstat 1, 2 hours and 38 minutes, and Telstar 11, 3 hours and 45 minutes. Herein lay their disadvantage for telecommunications; they were within line of sight of the tracking station for only a brief period of time, often less than half an hour. The Russians also use elliptical orbits for their Molniya communication satellites, but their orbits are larger so that the satellites are within sight for longer periods.

Figure 15.6 plots the time a satellite takes to travel around the earth versus its height. The orbit at a height of 22,300 miles is special in that a satellite in that

Figure 15.6   Rotation times of earth satellites in circular orbits.

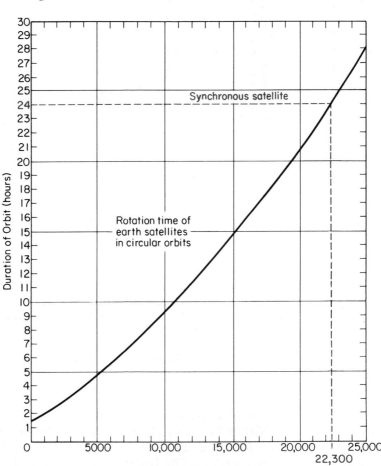

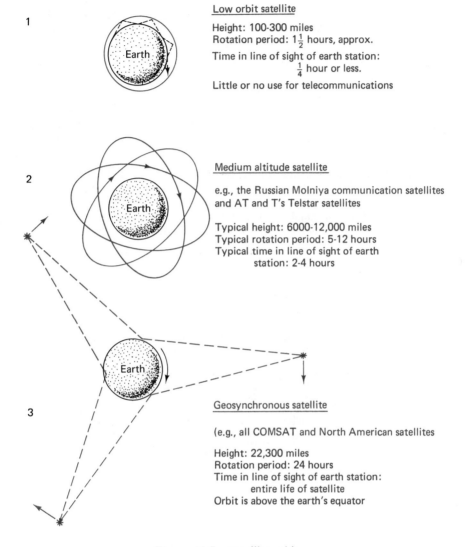

1

Low orbit satellite

Height: 100-300 miles
Rotation period: $1\frac{1}{2}$ hours, approx.

Time in line of sight of earth station:
              $\frac{1}{4}$ hour or less.

Little or no use for telecommunications

2

Medium altitude satellite

e.g., the Russian Molniya communication satellites
and AT and T's Telstar satellites

Typical height: 6000-12,000 miles
Typical rotation period: 5-12 hours
Typical time in line of sight of earth
              station: 2-4 hours

3

Geosynchronous satellite

(e.g., all COMSAT and North American satellites)

Height: 22,300 miles
Rotation period: 24 hours
Time in line of sight of earth station:
              entire life of satellite
Orbit is above the earth's equator

Figure 15.7   Satellite orbits.

orbit takes exactly 24 hours to travel around the earth—the earth's rotation time. If its orbit is over the equator and it travels in the same direction as the earth's surface, then it appears to hang stationary over one point on earth. This orbit is called a *geosynchronous* orbit. The apparently stationary satellite is called a geosynchronous satellite.

      The INTELSAT satellites hang in the sky stationary over the Atlantic and Pacific. The U.S. domestic satellites hang over South America or the Pacific west of Equador. Figure 15.7 shows satellite orbits. It is clear now that the right place for a communications satellite is in geosynchronous orbit.

As shown at the bottom of Fig. 15.7, three geosynchronous satellites can cover the entire earth with the exception of almost unpopulated regions close to the poles. An advantage of using satellites in such a high orbit is that they cover a large portion of the earth. Figure 15.9 shows the maximum spacing between earth stations for different satellite heights, assuming that 5° is the minimum angle of elevation of the ground station antennas.

The placement of a satellite in a synchronous orbit needs high precision spacemanship. The launch vehicle first places it into a lengthy elliptical orbit with the highest part of the ellipse about 22,300 miles from earth. This orbit is then measured as exactly as possible and the satellite orientation adjusted so that it will be in precisely the right attitude for the next step. When the satellite is at the farthest end of its ellipse, traveling approximately at right angles to the earth's radius,

Figure 15.8 The number of available orbital slots for satellites covering North America is limited, but can be greatly increased by using satellites of higher frequencies.

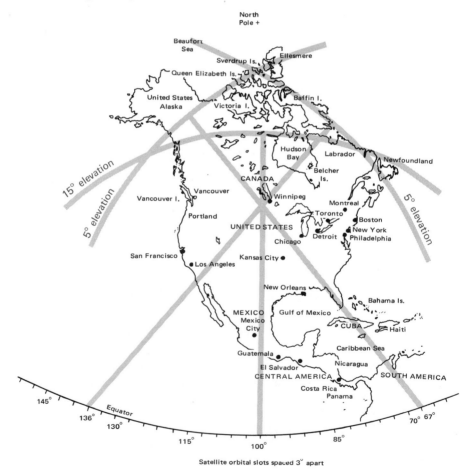

Satellite orbital slots spaced 3° apart

a motor is fired at precisely the right instant to put the satellite in a circular orbit around the earth. The satellite's velocity is then adjusted to synchronize with the earth's rotation, and its attitude is swung so that its antenna points in the right direction.

During the launching of the first INTELSAT II satellite, the "apogee" motor, which should change the elliptical orbit into the circular one, terminated its 16-second thrust prematurely and left the satellite plunging through space on a large elliptical nonsynchronous orbit. Comsat, however, managed to use it. Following its unplanned journey through space with their big antennas they succeeded in transmitting the first live color TV between Hawaii and the American mainland. It was also used for commercial telephone circuits during those periods when its wanderings brought it within line-of-sight of suitable earth stations.

Once a synchronous orbit is achieved the satellite will need periodic adjustment to keep it where it is needed. Solar gravitation will pull it slowly out of the equatorial orbit and irregularities in the earth's gravitational field will make it drift along the equator over a period of months. Solar radiation pressure tends to make the satellite drift. Similarly, its attitude will change slightly, and after a long period its antennas may not point correctly toward earth. To compensate for this, the satellite is equipped with small gas jets which can be fired on commands from earth to make minor adjustments to its velocity and attitude. The jets can be operated in short squirts by signals from an earth station, to nudge the satellite into the required position to keep its orbit "stationary." The orbital position may be adjusted every few weeks. The *attitude* of some satellites is stabilized automatically with on-board control equipment.

ORBITAL POSITIONS   Satellites cannot be too closely spaced in orbit; otherwise the up-link microwave beams for adjacent satellites interfere with one another. Five-degree spacing between satellites has been accepted practice; however, it appears feasible to operate satellites using today's frequencies with 3° or 4° spacing. Higher frequency satellites can be spaced more closely.

Figure 15.8 shows the orbital slots that could serve North America with 3° spacing. 5° elevation of the earth station antenna is about the minimum that can be used. Figure 15.8 shows two arcs showing the positions of antennas with 5° elevation for satellites at the outer limits of North America. The satellite at 67° longitude cannot reach Alaska. There is more freedom in positioning the earth station if not less than 15° elevation of the antenna is needed; Figure 15.8 shows the locus of 15° elevation antennas for a satellite at 100° longitude. It is desirable that the elevation of the earth station antenna be greater than 25° if possible because for a smaller angle the beam passes through enough earth atmosphere and rain to cause much signal loss. Earth stations of higher elevation can be less expensive because they have to achieve less signal gain.

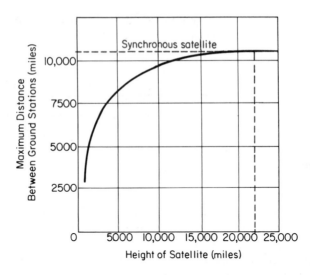

Figure 15.9   Maximum separation of earth satellite stations.

ANTENNAS IN SPACE   Satellite antennas, like all microwave antennas, are directional. Those on the early satellites were not highly directional, and transmitted most of their signal into empty space. Later antennas pointed towards the earth as a whole because intercontinental transmissions were needed. At the geosynchronous orbit the earth subtends an angle of 17°, and the antennas distributed the transmitted energy over this angle.

If the antenna covers a smaller angle, then the signal strength received on earth from a satellite of a given power is greater. INTELSAT IV carries two narrower-angle antennas subtending 4.5° as well as its 17° earth-coverage antenna. A domestic satellite designed for a country such as Canada or Brazil requires an antenna which focuses on that country.

The NASA ATS 6 satellite has an antenna large enough to transmit narrow beams like search-lights 0.25° in width. This antenna is an umbrella-like structure 30 feet in diameter which unfolds in space. The beam, referred to as a *spot beam*, can illuminate an area about the size of England. The satellite has multiple antenna feeds underneath the umbrella, each of which bounces its transmission off the umbrella to form a spot beam traveling in a given direction. Similarly when receiving, the umbrella focuses the signal from the area of the spot beam on to the antenna feed.

There is a trend toward satellites with multiple narrow-beam transmissions. Not only do they give a higher effective radiated power, but also the same frequency can be reused several times for different portions of the earth. The satellite can therefore contain more transponders for a given fixed bandwidth. Different satellites can also reuse the same frequency, thus conserving the valuable spectrum space. If a satellite transmits many spot beams it is desirable that there should be on-board equipment which can switch signals from one beam to another.

The early satellites with earth-coverage antennas were stabilized about one axis. Satellites transmitting narrow beams require three-dimensional stabilization. To achieve stabilization, most satellites are spun so that they act as their own gyroscope. WESTAR, for example, spins at 100 revolutions per minute. If multiple directional antennas are used, the antenna array must be despun so that it remains stationary relative to the earth. Alternatively, the satellite can contain a gyroscope for stabilization. The use of narrow beams places severe requirements on the systems that stabilize satellites. Typical satellites have their axis held steady to $\pm 0.1°$. Their axis can be adjusted by commands from earth, but also on-board control equipment senses the edge of the earth and holds the satellite steady.

Also the satellite position in orbit must be maintained so that it does not drift by more than $\pm 0.1°$. This permits the use of cheaper nonsteerable earth station antennas.

DELAY   A disadvantage of satellite transmission is that a delay occurs because the signal has to travel far into space and back. The signal propagation time is about 270 milliseconds and varies slightly with the earth station locations.

The bad effects of this delay have been much exaggerated by organizations which operate long-distance terrestrial links. The claim is frequently heard that the delay is psychologically harmful in telephone conversations and renders satellite links useless for interactive data transmission.

A telephone user may wait for the reply of the person he is talking to for an extra 540 milliseconds if the call goes via satellite in both directions. He certainly notices this delay but very quickly becomes used to it if he makes many satellite calls. Assessment of the psychological effect should not be based on the first satellite call a person makes. The delay seems less harmful to a person who is accustomed to it than the effects of TASI (Chapter 12) on heavily loaded circuits, which sometimes deletes the first spoken syllables. Transcontinental callers sometimes confuse the effects of TASI with the effect of the satellite delay.

While a telephone user can learn to ignore one or two 270-millisecond delays in a conversational response, four such delays (1080 milliseconds) may strain his tolerance. It is therefore desirable that the switching of calls should be organized so that no connection contains two or more round trips by satellite. Where satellites supplement the terrestrial toll telephone network the switching can usually be organized to limit the delay to 270 milliseconds. A transatlantic call, for example, often goes one way by satellite and the other way by submarine cable.

In interactive data transmission via satellite a terminal user will experience a constant increase in response time of about 540 milliseconds. A systems designer has to take this into consideration in designing the overall system response time. In many interactive systems it is desirable that the mean response time not be greater that 2 seconds. This is achieved satisfactorily on many interactive systems

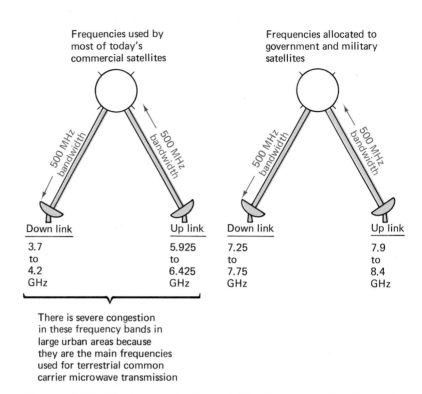

Figure 15.10 shows:

Frequencies used by most of today's commercial satellites			Frequencies allocated to government and military satellites	
Down link		**Up link**	**Down link**	**Up link**
3.7 to 4.2 GHz		5.925 to 6.425 GHz	7.25 to 7.75 GHz	7.9 to 8.4 GHz

There is severe congestion in these frequency bands in large urban areas because they are the main frequencies used for terrestrial common carrier microwave transmission

Figure 15.10 The communication satellite frequency allocations of 250 MHz bandwidth or more, below 40 GHz.

using satellites today. Line-control procedures must be selected which are appropriate to satellite channels. It would be inappropriate, for example, for a computer to "poll" devices at the other end of a satellite link, asking them one by one if they have any data to send. Polling is efficient only when the propagation and turn-around time is low. In general, procedures which require protocol signals to travel to and fro before the data message is transmitted are to be avoided on satellite links.

The comment is sometimes made that satellite channels are inappropriate for fast interactive computing with short messages. A more correct comment would be that some commonly-used terrestrial protocols are inappropriate over satellite channels. Various control procedures are in use, which *are* efficient for interactive systems using satellite channels [1].

FREQUENCY ALLOCATION

Figure 15.10 shows the frequencies allocated to satellite use. A different frequency band is allocated to the up-link and down-link. The frequencies are referred to with phrases such as the *4/6-GHz band*, the *11/14-GHz band*, and the *20/30-GHz band*, the first number in each case referring to the down-link band and the second number to the up-link band.

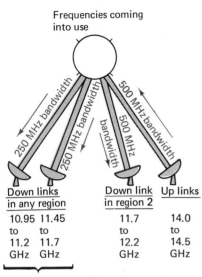

Frequencies coming
into use

Frequencies allocated
but not yet used because
more development work is
needed

250 MHz bandwidth
250 MHz bandwidth
500 MHz bandwidth
500 MHz bandwidth
500 MHz bandwidth

2000 MHz bandwidth
1500 MHz bandwidth
1500 MHz bandwidth
2000 MHz bandwidth
1500 MHz bandwidth

Down links in any region	Down link in region 2	Up links	Up link	Down links
10.95 to 11.2 GHz	11.45 to 11.7 GHz	11.7 to 12.2 GHz	14.0 to 14.5 GHz	17.7 to 19.7 GHz

Down links in any region		Down link in region 2	Up links	Up link		Down links	
10.95 to 11.2 GHz	11.45 to 11.7 GHz	11.7 to 12.2 GHz	14.0 to 14.5 GHz	17.7 to 19.7 GHz	19.7 to 21.2 GHz	27.5 to 29.5 GHz	29.5 to 31.0 GHz

These frequencies are also allocated to mobile radio and terrestrial common carrier microwave, but are not congested

(These frequencies are allocated to the western hemisphere, North and South America, only)

Frequencies allocated for communication satellites
at the 1971 WARC (World Administrative Radio Conference)

Most of today's commercial satellites use the 4/6-GHz band—the frequencies shown in the leftmost diagram. These are the main frequencies used by the terrestrial microwave common carriers (see Table 13.2). The bandwidth of 500 MHz can be split up in similar ways to those described in the previous three chapters. A supergroup, mastergroup, or higher group can be taken directly from a terrestrial microwave or coaxial trunk to the satellite transmitter as shown in Fig. 15.11. Similarly, a T1 carrier or other PCM bit stream can be placed directly on the satellite link.

**MICROWAVE
INTERFERENCE**
Unfortunately, because satellite and terrestrial microwave links use the same frequencies, there is a serious problem of radio interference. Four types of interference are theoretically possible:

1. Transmission from the earth station interferes with the terrestrial link receiver.

2. Terrestrial link transmission interferes with reception from the satellite.

3. Transmission from the satellite interferes with the terrestrial link receiver.

4. Terrestrial link transmission is received by satellite.

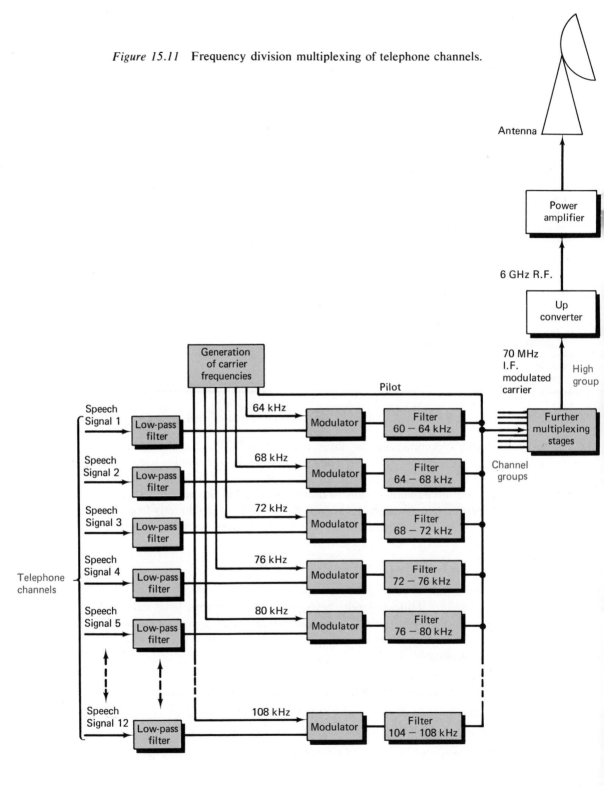

Figure 15.11 Frequency division multiplexing of telephone channels.

First multiplexing stage
as in Fig. 12.3

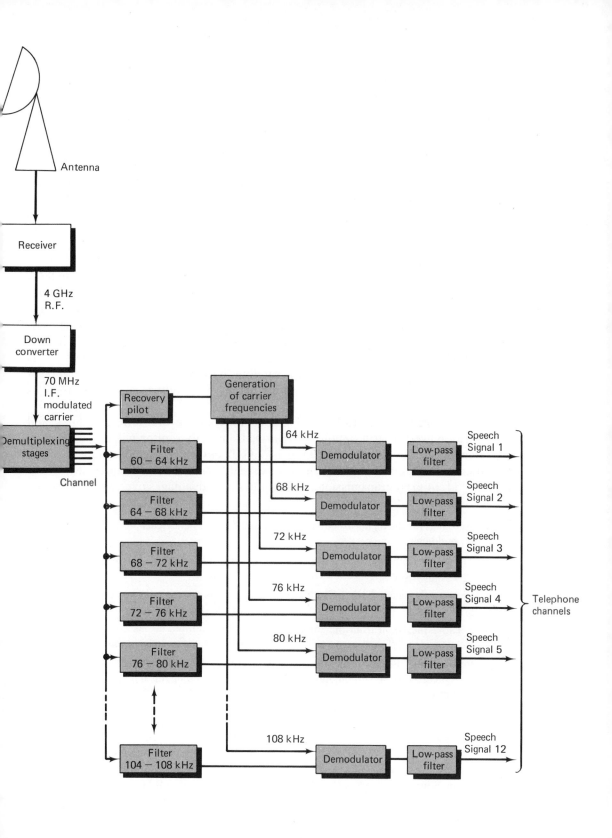

Final demultiplexing stage
as in Fig. 12.3

The first is by far the most serious. An earth station must transmit a powerful signal to compensate for the vast distance and the low gain of the satellite-receiving equipment. The dish-shaped antenna transmits a highly directional beam toward the satellite, but nevertheless some of the signal spills in other directions (as in Fig. 9.9) and may interfere with a microwave receiver. The earth station transmitter must therefore not be too close to a microwave antenna.

The second of the above types of interference is the next most serious. To avoid it, an earth station should not be located close to a terrestrial microwave path so that part of the terrestrial beam shines into the receiving antenna.

Because of the already serious microwave congestion in the cities (Fig. 9.10), earth stations using the 4/6-GHz band cannot be located in many urban areas. In large cities they often have to be 50 or more miles away.

To avoid the interference problems it seems highly desirable that satellites should use their own frequency bands. When this is done, small satellite earth stations can be on the rooftops of city buildings. The 11/14-GHz and 20/30-GHz bands were allocated with this intent. Unfortunately radiation at 11/14 GHz suffers more absorption by clouds, rain, and the atmosphere than radiation at 4/6 GHz, and 20/30 GHz is still worse. Because of the absorption, higher gain is needed at these frequencies. Several satellites have been designed for operation at 11/14 GHz, but 20/20 GHz, with its higher bandwidths, is still regarded as experimental.

The up-link of 14 to 14.5 GHz does not compete with terrestrial links, and hence a powerful earth station transmitting from a city rooftop will cause no interference. The down-links at 10.95 to 11.7 GHz *can* be interfered with, and so the antenna must not be in the path of terrestrial beams at these frequencies.

11/14-GHz transmission has several other advantages over 4/6-GHz transmission:

1. Antennas for the higher frequency can be smaller and hence cheaper and easy to install on buildings.

2. The higher frequency tends to spread out less. Beams tend to be narrower, and so more such beams can crisscross a city without interference.

3. Because the beam from the satellite is narrower, its energy is more concentrated, and greater energy tends to be received by an antenna of a given size.

4. Because the beam from the satellite is narrower, multiple beams could use the same frequency to different areas on earth, i.e., a form of space-division multiplexing.

5. Because the beam to the satellite is narrower, more satellites can occupy the geosynchrous orbit without interference, thereby extending this uniquely valuable earth resource.

ECLIPSES   A satellite is prone to two types of eclipses. First, when the earth's shadow passes across the satellite its solar batteries stop operating. These eclipses last from minutes to slightly more than an hour on 43 consecutive nights in spring and fall. Less commonly, the moon's shadow passes across the satellite, like a solar eclipse on earth. The satellite can carry backup batteries or storage cells to ensure continuous operation if the need for continuous operation justifies carrying the extra weight.

A more serious form of outage occurs when the sun passes directly behind the satellite. The sun, being of such a high temperature, is an extremely powerful noise source and so blots out transmission from the satellite. This outage lasts about 10 minutes on 5 consecutive days twice a year. The only way to achieve continuous transmission is to have two satellites and switch channels to the non-eclipsed one before the eclipse begins. Most satellites *are* duplicated in orbit, not only to provide protection from eclipses but also protection if one satellite should fail.

Another brief random form of outage is sometimes caused by airplanes flying through the satellite beam.

UNIQUE PROPERTIES OF SATELLITE LINKS   A satellite channel is often used simply as a substitute for a point-to-point terrestrial channel. However, it has certain properties which are quite different from conventional telecommunications. It should not be regarded as merely a cable in the sky. Different types of design are needed, especially in computer systems, to take advantage of satellite properties and avoid the potential disadvantages.

A satellite channel is unique in the following respects:

1. There is a 270-millisecond propagation delay.

2. Transmission cost is independent of distance. A link from Washington to Baltimore costs the same as a link from Washington to Vancouver. A computer center can be placed anywhere within range of a satellite without affecting transmission costs. It is becoming economical to centralize many computing operations. In an international organization worldwide links can be similar in cost to national links if the regulatory authorities so permit.

3. Very high bandwidths or bit rates are available to the users if they can have an antenna at their premises, or radio-link to an antenna, thereby avoiding local loops.

4. A signal sent to a satellite is transmitted to all receivers within range of the satellite antenna. The satellite *broadcasts* information unlike a terrestrial link.

5. Because of the broadcast property, dynamic assignment of channels is possible between geographically dispersed users. This can give economies, especially with data transmission on a scale not possible with terrestrial links, but needs new forms of transmission control.

6. Because of the broadcast property, security procedures must be taken seriously.

7. A transmitting station can receive its own transmission and hence monitor whether the satellite has transmitted it correctly. This fact can be utilized in "contention" forms of transmission control in which two transmissions from different locations might coincide on the same channel and destroy each other [1].

DIGITAL   A 36-MHz transponder on the WESTAR satellite can
TRANSMISSION   be used to relay 50 million bps. With advanced modem
                         design 72 million bps have been relayed. At this bit
rate digital encoding and transmission of voice channels give a higher satellite
throughput than analog voice. Most of the recently installed satellite equipment
used PCM transmission of voice or digitized voice with some form of compaction
differential PCM or delta modulation. Straight PCM uses 64,000 bps for one
voice channel. Operating compaction schemes give 32,000 bps.

Digital TV compression has been used to compress color television down to
30 Mbps, thus increasing a transponder's TV channel capacity from 1 to 2.

While such digital techniques can increase the capacity of a satellite, alternatively they can be used to decrease the cost of the earth stations. Earth stations
transmitting a lower bit rate can employ a smaller antenna and less expensive
equipment.

The future of satellites clearly lies in digital technology, and it emphasizes,
once again, the intermingling of telecommunications and computer technologies.
Satellite data rates begin to make computer data transmission look inexpensive, at
least when compared with established common carrier costs. On the other hand,
to take advantage of the unique properties of satellites, computer control of the
assignment and switching of channels is essential. It is a complex problem to allocate the satellite capacity to all its potential geographically scattered users so as to
maximize the usefulness of that capacity. We will return to this subject again at
the end of Part III of the book.

Sooner or later the old established common carriers will have to come to
grips with the fact that satellite trunks can be constructed at a fraction of the cost
of terrestrial trunks. Several billion dollars per year are being spent on expanding
the terrestrial trunk networks.

REFERENCES

1. For example, see the techniques developed at the University of Hawaii described in the following papers:

   a. Norman Abramson, "Packet Switching with Satellites," *AFIPS Conference Pro-
      ceedings, National Computer Conference,* 82 (1973), pp. 695–702.

   b. Leonard Kleinrock and Simon Lam "Packet Switching in a Slotted Satellite Chan-
      nel," *AFIPS Conference Proceedings, National Computer Conference,* 42 (1973),
      pp. 703–710.

   c. Lawrence G. Roberts, "Dynamic Allocation of Satellite Capacity Through Packet
      Reservation" *AFIPS Conference Proceedings, National Computer Conference,* 42
      (1973), pp. 711–716.

d. Franklin F. Kuo and Richard D. Binder, "Computer-Communications by Radio and Satellite: THE ALOHA SYSTEM," *ALOHA SYSTEM Technical Report B73-4*, University of Hawaii, Honolulu, Aug. 1973. Also published by the International Advanced Study Institute on Computer Communication Networks, University of Sussex, Sept. 9–15, 1973.

16 THE MAXIMUM CAPACITY OF A CHANNEL

The capacity of a channel may be described as the maximum rate at which information can be sent over it without error, and this, for data transmission purposes, may be measured in bits per second.

The rate at which we can send data over a channel is proportional to the bandwidth of the channel. Hartley [1] proved in 1928 that a given bandwidth $\times$ time is required to transmit a given quantity of information. This can be pictured by imagining a phonograph record with data recorded on it in the dots and dashes of Morse code. If we double the speed of play of the record, we halve the time needed to relay this coded data. Doubling the speed of the record doubles the frequencies of the sound, as well as the bandwidth used. Similarly, we may increase the record speed and relay the data very fast providing there is some way of interpreting the squeaks that result. When we exceed a certain speed the sounds will no longer be audible because we have exceeded the frequencies detectable by the human ear. The human ear has a limited bandwidth.

In 1924 and 1928, Nyquist [2] published papers also concerned with the capacity of a noiseless channel. He showed that if one sends $2W$ different voltage values (or other symbols) per second, these can be carried by a signal with no frequencies greater than W. If frequencies greater than W are sent, they are *redundant,* unnecessary for the reconstruction of the series of signal values at the receiver.

In other words, a bandwidth W can carry $2W$ separate voltage values per second. If one is sending binary signals as in the telegraph signaling discussed in Chapter 7, the sending voltage has one of two separate values. One can therefore send $2W$ bps. However, if one sends two bits simultaneously by having *four* possible voltage levels at any one instant, then the $2W$ voltage values per second are used to code $4W$ bps. Eight alternative voltage values at one instant can be used to code 3 bits and achieve a signaling rate of $6W$ bps.

In general, n bits can be sent at any one instant by using one of 2^n possible signal levels. Therefore with 2^n possible and distinguishable signal levels, a signaling rate of $2nW$ bps can be transmitted through a channel with W Hz of bandwidth.

If L is the number of signaling levels,

$$2n = L$$

$$\therefore n = \log_2 L$$

Therefore the channel capacity, C, in absence of noise, is given by

$$C = 2W \log_2 L$$

The question thus arises, How many signaling levels can be transmitted and be separately distinguishable at the receiver? Noise and distortion on the line, fluctuations in the attenuation, and a limit on the signal power which can be used obviously restrict this.

BAUDS   The speed of a transmission line is often quoted as being a certain number of "bauds." This term is usually used to relate to the signaling speed actually used on a line, not to the capacity of the line. It refers to *the number of times the line condition changes per second.* If the line condition represents the presence or absence of 1 bit, then the signaling speed in bauds is the same as bits per second. If, however, the line can be in one of four possible states, that is $L = 4$, then one line condition represents a "dibit," that is, 2 bits instead of 1 ($n = 2$). x bauds will then be the same as $2x$ bps. The Bell 201 data set, for example, transmits data in dibits; each pair of bits is coded as one of four possible combinations. If the signals are coded into eight possible states, then one line condition represents 3 bits. One baud equals 3 bps, and so on.

The reader should note that the term "bauds" is sometimes taken to mean "bits per second." While this is true with many lines because they use two-state signaling, it is not true in general. For any line not using two-state signaling, it is wrong. *The term bauds has sometimes therefore proven confusing, and we have avoided using it in most of this book.*

SIGNALING ON A   Shannon [3], 20 years after Nyquist, proved again
CHANNEL WITH NOISE   mathematically that a channel has a finite maximum capacity. He discussed a continuous channel as well as one transmitting discrete values. His work relates first to a noiseless channel and then, more interestingly in our case, to a channel with noise.

Shannon proved that if signals are sent with a signal power S over a channel perturbed by white noise (random, i.e., Gaussian, fluctuations) of power N, then the capacity of the channel in bits per second is

$$C = W \log_2 \left(1 + \frac{S}{N} \right) \tag{16.1}$$

where W is the bandwidth of the channel.

This formula gives the maximum signaling rate over a communication channel in terms of three parameters which are known or measurable. Shannon's law is one of the most fundamental laws of telecommunications. According to this law, the maximum number of data bits that can be sent over a channel in time T seconds is

$$WT \log_2 \left(1 + \frac{S}{N} \right)$$

Shannon's law relates to transmitting bits the sequence of which is *completely unpredictable*. As will be discussed later in this chapter, if there were some way of anticipating a nonrandom bit sequence, the transmission rate might be increased. However, for an unpredictable bit sequence Shannon's proof shows that there is *no possible way* of exceeding this quantity of information for these channel parameters. An engineer can design very ingenious modulation techniques and elaborate coding systems, but try as he might he will never send more than this number of bits over the channel, unless he increases either the bandwidth available or the signal-to-noise ratio.

Let us suppose that a certain section of telephone line is known to have a signal-to-noise ratio of 20 decibels. In other words, the noise power from the line is $\frac{1}{100}$ of the signal power transmitted. We wish to use this line to transmit data, and the bandwidth available is 2600 Hz. Using these in Eq. (16.1), we find that the capacity of the line is

$$C = 2600 \log_2 \left(1 + \tfrac{100}{1} \right)$$

$$\log_2 x = (\log_2 10)(\log_{10} x) = 3.32 \log_{10} x$$

$$\therefore C = 2600 \times 3.32x \log_{10} 101 = 17,301$$

The *maximum possible* rate at which data could be transmitted over this voice line is thus about 17,300 bps.

If the signal-to-noise ratio were 30 decibels — a more typical figure — then we would have

$$C = 2600 \log_2 \left(1 + \tfrac{1000}{1} \right) = 25,900 \text{ bps}$$

The only way to improve upon this is to make fundamental changes in the construction of the line which would increase the transmitting power of the amplifiers, increase the bandwidth, or reduce the noise. This could be done by placing the amplifiers closer together. There is nothing that can be done by way of ingenious *terminal* equipment design that would give a bit rate higher than these figures. If the common carrier provides a line of given characteristics, Shannon's law gives the maximum transmission rate that we could ever achieve.

Ingenious inventors occasionally propose a scheme that would do better than Eq. (16.1). However, so fundamental is Shannon's law that any such plausible scheme should be treated with the same attitude as inventions for perpetual motion machines. Somewhere in the scheme there is a flaw, and one can say with assurance that it will not work.

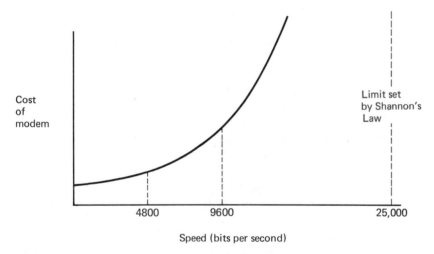

Speed (bits per second)

Figure 16.1   The cost of modems increases with transmission speed for a line of given characteristics. The cost becomes very high before it approaches the limit set by Shannon's Law. The figures on the above chart relate to a typical telephone line.

Systems used in practice on voice lines work at speeds very much lower than those above. It is common to find transmission schemes operating at 1200 and 2400 bps on such lines. Sometimes the part of the bandwidth that can actually be used for transmitting is less than 2600 Hz because of signaling considerations. But even on a good-quality line with the full bandwidth available, speeds used are not normally greater than 9600 bps. Part of the reason for this is that as one approaches closer to the Shannon maximum, the encoding necessary becomes very complex with longer and longer word lengths and therefore more and more delay in encoding and decoding. The cost of the modem increases with the speed of transmission, as shown in Fig. 16.1.

WHITE GAUSSIAN   The noise referred to in Shannon's equation is Gaus-
NOISE   sian noise. This means that the amplitude of the noise
signal varies around a certain level in a purely *random*
fashion with respect to time. The amplitude of the noise signal follows a *Gaussian*
distribution. Shannon's equation was proved using this assumption.

This may seem a somewhat unjustified assumption to make. How do we
know that the noise amplitude follows a Gaussian distribution? The noise I hear in
my New York apartment certainly does not follow any such pattern. It consists of
the television next door, a rhythmic banging from the plumbing, and the wail of
police cars screaming through the night. However, if I open my window and listen
to the sounds from far away, the hum of the big city, this infinite jumble of diminutive sounds does approximate that of Gaussian distribution.

In telecommunication circuits there will certainly be much noise that is not
Gaussian, such as the clicks and whistles we sometimes hear on the telephone.
However, in all electronic circuitry there is a steady continuing background of
random noise. This is sometimes referred to as *thermal* noise.

The atoms and molecules of all substances vibrate constantly in a minute
motion which causes the sensation of heat. The higher the temperature, the
greater this vibration. As the atoms vibrate they send out electromagnetic waves,
and as there are many atoms, we have a chaotic jumble of electromagnetic waves
of all frequencies. The electrons in electrical conductors move in a similar random
fashion. These motions form the ultimate unavoidable noise background to all
electronic processes. It is a continuous Guassian noise like the hum of a distant
city.

We thus have to send data signals against this background of a ceaseless random variation in amplitude, usually of low intensity. When one can hear it, it
sounds like a hiss. If the volume of an FM radio is turned up full when there is no
program being received, one can hear the hiss of this noise.

This Gaussian noise in electronic circuits is referred to as *white noise*. It is
called "white" because it contains all spectral frequencies equally on average just
as white light contains all the colors of the rainbow equally. It sounds like a hiss,
different from the hum of diminutive sounds from a distant city because these,
although Gaussian, are not "white" because they contain more low frequencies
than high frequencies. High frequencies are absorbed more than low frequencies
as sound travels through the air and is reflected off buildings and off the ground.

Figure 16.2 illustrates Gaussian noise.

If the electronic equipment were perfect, and *perfectly* insulated from external interference, there would still be white noise. Any electrical conductor
is a source of white noise of power, N, where

$$N = kTW \tag{16.2}$$

where T is the temperature in degrees Kelvin,
     W is the bandwidth, and
     k is Boltzmann's constant: 1.37×10^{-23} joule per degree.

Figure 16.2   Thermal (random, Gaussian) noise.

The power of the white noise is thus proportional to the absolute temperature. This fact is of little value on earthbound communication links but is important in receiving very long-distance transmissions from space. The transmitter can operate at much reduced power in the icy depths of space when shielded from sunlight. The first communication satellite earth stations used electronics cooled by liquid helium.

Thermal noise, caused by the random motion of electrons, is the inescapable minimum. Practical circuitry in amplifiers and modulators will typically generate noise, however, which exceeds thermal noise by some 5 to 25 decibels. Large numbers of such devices are encountered on a long circuit, and so the Gaussian noise at the end of the channel exceeds thermal noise by many decibels. Signal-to-noise ratios in well-maintained telecommunications channels will seldom be better than 30 decibels.

The power of thermal noise, unlike some other types of noise discussed in the next chapter, is proportional to bandwidth. For a given transmitting power, therefore, as will be seen from Eqs. (16.1) and (16.2), *the channel capacity is not quite proportional to bandwidth*. Doubling the bandwidth of the channel does not quite double its capacity. A 40,000 Hz channel will not carry quite as much data as ten 4000 Hz channels. Figure 16.3 sketches the relationship. An infinite bandwidth would carry a finite, maximum number of bits per second.

**THE TRANSMISSION
RATES ACHIEVED
IN PRACTICE**

Making the best use of the bandwidth available is, in fact, to a large extent dependent on the design of the modems, and herein lies scope for much ingenuity. Assuming only white noise, the signaling speed at a given bandwidth can be evaluated theoretically for each type of modulation. This can be established as a function of signal-to-noise ratio and probable error rates. [4] Expressions for the maximum signaling rate can be obtained which, as would be expected, give lower rates than the Shannon formula.

Figure 16.4 plots the Shannon formula, showing the speed per cycle of bandwidth in terms of signal-to-noise ratio on telephone channels. The shaded area be-

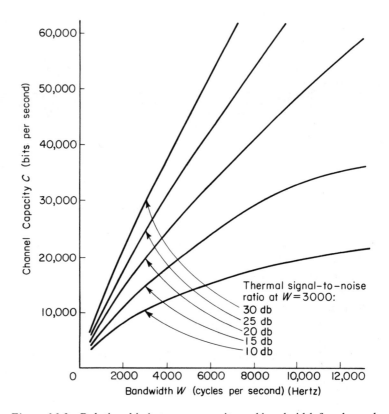

Figure 16.3 Relationship between capacity and bandwidth for channels with differing levels of Gaussian noise.

low this curve gives an indication of the speeds that are achieved in practice to-day. It will be seen that they are very much less than the Shannon limit. In fact, if one half of the Shannon limit were achieved, it would be regarded as a fast and sophisticated system.

Figure 16.4 shows the maximum signaling speeds with two common types of modem, assuming that they are operating in the presence of Gaussian noise with an error rate of 1 bit in 10^4. This is a somewhat larger error rate than is acceptable in practice, and hence these speeds are not found on actual systems. The figures on these dashed curves refer to number of states the signal can be in at any one instant. A binary signal, for example, can be in one of *two* states. However, if there are four possible voltage levels rather than two, then these can represent 2 bits at one instant instead of only 1. Eight levels can represent 3 bits. 2^n levels can represent n bits at one instant. However, as the number of voltage levels that must be distinguished increase, their spacing decreases, and the randomly fluctuating noise amplitudes are more prone to do damage.

It is clear that one way to increase the capacity of a channel is to raise the

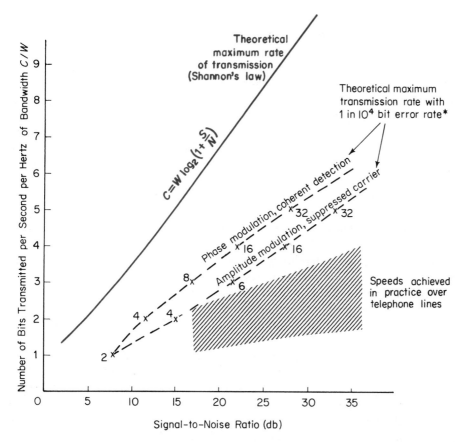

Figure 16.4 The maximum data transmission speeds achieved in practice today are considerably lower than the Shannon theoretical maximum. These speeds fall into the shaded area above. The dotted lines are
the theoretical maximum speeds with two common types of modulation
(Chapter 11), in the presence of Gaussian noise, and calculated for an
error rate of 1 error bit in every 10,000 transmitted. (Curves reproduced
from *Data Transmission* by Bennett and Davey, McGraw Hill, 1965.)

signal-to-noise ratio. As we have seen, the level of white noise is determined by
natural phenomena which are beyond our control. Other types of noise *can* be
controlled, although not entirely, and many measures are taken to do this. However, there is a level below which N cannot be pushed. If the distance between the
repeaters is great, then the signal strength falls substantially, whereas the noise
level generated at all points of the line remains the same. Too great a repeater
separation gives too low an S/N ratio. The telephone companies can thus improve
S/N by increasing the number of repeaters, though beyond a certain point it is not
economical to do this.

How about increasing S, then? For the common carriers this is also largely a question of economics. Beyond a certain level it becomes expensive to increase the signaling power. For the telephone subscriber there is a limit to the signal strength he is allowed to transmit. In the United States he may connect his own devices directly to the telephone channel, but he must go through the DAA isolation transformer which limits the signal strength that can be transmitted.

More than a hundred years ago the communications industry learned a bitter lesson about making S too large. After some of the toughest financial wrangling of the nineteenth century and more than a year of heartbreaking failures on the high seas, the first transatlantic telegraph cable was laid — a magnificently impressive feat for its era. It worked very slowly, taking half a day to transmit data that we can send over today's cable in a second. However, the press headlines were sensational beyond precedent. Dr. Whitehouse, a telecommunication scientist in England, decided that higher voltages were needed. His colleague, Dr. William Thomson (Lord Kelvin), disagreed. Whitehouse, however, insisted on using large induction coils which he had built. The line insulation gradually broke down, and the 2500 tons of cable became useless. It was 8 years before another cable was laid. One American newspaper claimed that the cable had been a hoax, and an English writer "proved" that it had never been laid at all.

**DIGITAL
CHANNELS**
As we commented in previous chapters, the electronics on a wire-pair line can be changed so that it becomes a high-speed digital channel rather than an analog channel. Digital regenerative repeaters are used instead of analog amplifiers, and the bit rate transmitted is much higher than if all the analog telephone channels on the wire transmitted at the Shannon limit. Does transmission of 2.048 million bps over a telephone wire pair violate Shannon's law?

The answer is no. What is happening on PCM channels is that higher frequencies are being transmitted; a higher bandwidth is being used. This gives a *much* worse signal-to-noise ratio, but two factors compensate for the deficiency. First, the repeaters are closer together so the signal does not sink into the background noise. Second, only binary bits are sent, so only the presence or absence of a bit need be detected — not the range of amplitudes needed in analog transmission.

The signal-to-noise ratio *is* worse, but the bandwidth is higher, and Shannon's equation shows that while capacity increases proportionally to bandwidth, it increases only as the log of the signal-to-noise ratio. A 10-fold increase in bandwidth compensates for a 1000-fold decrease in signal-to-noise ratio.

On other channels also, increasing the frequency used at the expense of signal-to-noise ratio can be a good trade-off. Very high bit rates can be transmitted over coaxial cable and helical waveguide channels. On established microwave channels, however, the bandwidth cannot be increased because of the frequency allocation, and the spacing between the repeaters cannot be decreased because of the tower construction. The best that can be done is to squeeze 1.5 million bits per second under the mastergroup using "Data Under Voice."

SATELLITE   On satellite channels the noise-to-signal ratio is of con-
CHANNELS   cern partly because of the immense transmission dis-
                      tances and partly because of the fact that power is a
scarce resource on the satellite.

The satellite designer with a limited power supply can use it either to increase the signal-to-noise ratio to increase the total effective bandwidth by using closer channel spacing or more transponders and directional antennas. Shannon's equation shows that much more capacity can be bought by increasing the effective bandwidth than by increasing the transmitting power of a transponder.

On the other hand, more power used for reception and transmission on the satellite makes possible lower-cost earth stations. As there may be a large number of earth stations and an already adequate satellite capacity, this can be the better trade-off.

ENTROPY   The classic Shannon formula relates to data in which
                     when 1 bit or character is sent we have no clue as to
what the next bit or character will be. This is normally the case with the computer techniques in use today. It is either not possible, or if it is possible the machine makes no attempt, to estimate the probability of the next bit being 1 or the next character being A or 6.

This is not true with human communication. Here there is often a good chance of guessing what the next letter or next word will be. If I send the characters E L E P H A N, a human recipient will expect the next character to be a T. When a telegram begins "CHRISTMAS COMES BUT ONCE A," it is reasonable to guess that the next word will be "YEAR."

If we can guess what the next bit, character, or word will be, the information is to some degree redundant, and so we may be able to devise a coding scheme which enables us to send more information than the channel capacity would otherwise allow.

The same applies to noise. We have discussed only white noise, in which we have no clue as to what the instantaneous noise amplitude at any point in time will be. We know that it is likely to lie in a certain range, with a Gaussian probability distribution. There may be other noise situations, however, in which we could make a more knowledgeable estimate of the likelihood of a certain noise pattern.

This informed guesswork constitutes an important part of the mathematical discipline of *information theory*. In Shannon's classic paper "The Mathematical Theory of Communication," long before he reached his memorable $C = W \log_2 (1 + S/N)$ formula, he had discussed the question of choice and uncertainty in the coding of information. To do this he used the concept of *entropy*.

Entropy is a measure of uncertainty or randomness. It is a concept which physicists have used for some time, largely in connection with thermodynamics, and it is intriguing to find it turning up again in a quite different discipline.

In the physical world the degree of randomness constantly increases. A parallel beam of light striking a wall is scattered in millions of random directions. Heat in a machine, organized into areas of high temperature and areas of low temperature tends to flow from the high to the low temperature. Similarly, on a communication line, voltages organized into square-edged 1 and 0 pulses become distorted, and if not regenerated by external means, they eventually become so distorted that they are inseparable from the noise background. In other words, entropy, the measure of randomness, increases. This appears to be one of the most fundamental laws of the universe, and it is sometimes quoted as an alternative way of stating the second law of thermodynamics.

The temporary strivings of an intelligent agent can increase the degree of order in a small corner of the physical world, but left to its own devices, nature restores the chaos and entropy increases. All seems condemned in the long run to approximate a state akin to Gaussian noise.

The entropy associated with a message is, in effect, a measure of the uncertainty of what is to follow in the message. If the next symbol to be sent consists of 6 bits, then the entropy associated with that symbol can vary from 0 bits to 6 bits. If it is certain, before it is sent, what each of the bits in the symbol will be, then the entropy associated with that symbol is zero. If it is completely uncertain what it will be — in other words, each bit has an equal probability of either being a 0 or 1 — then the entropy is 6 bits per symbol.

The 6-bit symbol can have $2^6 = 64$ possible different states. Let us number these 1 to 64, and let the probability of the symbol in the ith state be P_i. The 64 states are the only possible states; therefore,

$$P_1 + P_2 + P_3 + \cdots + P_{64} = 1$$

The entropy, H, of such a symbol is defined as the sum of the factors $P_i \log_2 P_i$ for each possible state the symbol can be in:

$$H = (P_1 \log_2 P_1 + P_2 \log_2 P_2 + P_3 \log_2 P_3 + \cdots + P_{64} \log_2 P_{64})$$

When it is certain what the value of the symbol will be, then one of the values of P is 1, and all the others are zero. Therefore,

$$H = -\log_2 1 = 0$$

The entropy is zero because there is no uncertainty as to what the symbol will be.

It can be shown that the maximum value of the preceding expression occurs when $P = P_2 = P_3 = \cdots P_{64} = \frac{1}{64}$. In other words, there is an equal probability of the occurrence of any state. Then

$$H = -64 \times \tfrac{1}{64} \log_2 \tfrac{1}{64} = 6$$

In general, if the symbols consist of x bits each of which could be a 1 or 0, then the entropy will range from zero to x.

For a symbol, word, or message of n possible states, the entropy is defined similarly as

$$H = -\sum_{i=1}^{n} P_i \log_2 P_i$$

and this ranges from a minimum of zero to a maximum of $\log_2 n$.

As an example of this, consider the throwing of a die. If it is a normal six-sided die, this gives an equal probability of producing any number from 1 to 6. The entropy associated with the throw is then

$$-(6 \times \tfrac{1}{6} \log_2 \tfrac{1}{6}) = 2.58$$

If, however, three of the sides are the same—let us say that three sides read 1, 2, and 3, respectively, and the other three read 4—then the entropy would be

$$-(3 \times \tfrac{1}{6} \log_2 \tfrac{1}{6} + \tfrac{1}{2} \log_2 \tfrac{1}{2}) = 1.79$$

Similarly, with data transmission, if the symbols, words, or messages contain some measure of redundancy or predictability, then the entropy associated with them will be less than if they were random.

Shannon [5] having defined entropy in this way, went on to prove that *for a noiseless channel with some discrete method of signalling such that its maximum capacity is C bps, the maximum rate at which data can be transmitted is C/H.*

Shannon went on to show that for a continuous signal the entropy of a Gaussian distribution is

$$H = W \log_2 2\pi eS \text{ bits per second}$$

where W is the bandwidth used in Hz, and S is the average power of the signal.

If N is the average white noise power and $H_{(N)}$ is the entropy of this noise,

$$H_{(N)} = W \log_2 2\pi eN$$

If the signal and noise on a continuous channel together form a Gaussian ensemble of average power $S + N$, then the entropy of this $H_{(S + N)}$ is

$$H_{(S + N)} = W \log_2 2\pi e(S + N)$$

Shannon showed that the maximum capacity of such a channel is given by

$$C = H_{(S + N)} - H_{(N)}$$
$$= W[\log_2 2\pi e(S + N) - \log_2 2\pi eN]$$

And so the equation $C = W \log_2 (S + N)/N$ is obtained.

We thus see that the Shannon formula $C = W \log_2 (S + N)/N$ applies to bit patterns which are completely unpredictable and to noise which follows a Gaussian distribution. If the bit pattern is to some extent predictable, a coding scheme might, theoretically, be devised which removes the predictability or redundancy. The message would therefore be sent in fewer bits and the original bit pattern restored on reception.

Similarly, if the noise has less entropy than Gaussian noise, a theoretically higher bit rate would be possible.

The noise that is encountered in practice often has short sharp spikes of much greater amplitude than the white-noise background. These destroy or add bits and result in an error rate much higher than would be expected from a theory assuming constant Gaussian noise.

Information theory sets a maximum on what we might expect to achieve over given channels. In a sense it provides a target for the engineers to work toward. It goes on to produce theorems about the nature of signal coding and error detection and correction codes which might aid in achieving the maximum transmission rate over a given bandwidth.

REFERENCES

1. R. V. L. Hartley, "Theory of Information," *Bell System Technical Journal,* 7 (1928).

2. H. Nyquist, "Certain Factors Affecting Telegraph Speed," *Transactions A.I.E.E.* (1924), and "Certain Topics in Telegraph Transmission Theory," *Transactions A.I.E.E.* (1928).

3. Shannon, Claude E., "Mathematical Theory of Communication." *Bell System Technical Journal* (July and October, 1948).

4. W. R. Bennett and J. R. Davy, *Data Transmission,* McGraw-Hill, New York, 1965.

5. Theorem 9 in Shannon's "Mathematical Theory of Communication," *Bell System Technical Journal,* 27 (July and Oct., 1948), p. 401.

17 LINE TERMINATION EQUIPMENT FOR DATA TRANSMISSION

We have now seen that the signal we send down a communication line can travel to its destination by a wide variety of possible means. When it disappears into our wall plaster on its telephone or telegraph wires we are not necessarily sure how it is going to travel. It may go in solitude on a wire circuit, or it may be huddled with hundreds of other signals on coaxial cable or microwave. It may race 25,000 miles into space to be beamed back by a satellite. When a time-sharing terminal chatters back after a pause of a second or so its operator does not know how the data were transmitted.

Whatever way the signal has traveled, though, for a telephone trunk the curve of attenuation against frequency is approximately that in Fig. 17.1. Telephone channels are engineered to this specification whether they are satellite

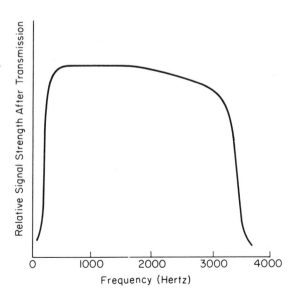

Figure 17.1 Transmission characteristics of a telephone trunk circuit. (*Note:* This does not show the signaling used on switched telephone circuits, discussed in Chapter 21.)

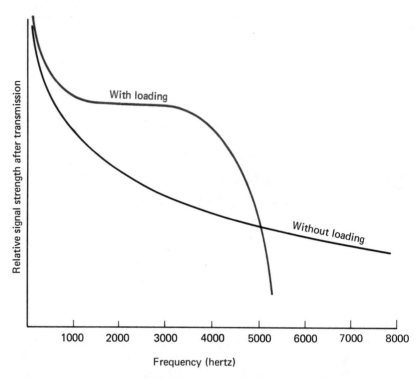

Figure 17.2 Typical transmission characteristics of a local telephone loop. A dc signal can be transmitted without modulation. A fairly high bit rate can be sent in baseband form if the loading coils are removed.

Figure 17.3 The bottom baseband signal in this diagram can be sent on local telephone loops without repeaters (Fig. 17.2) but not over trunk telephone circuits (Fig. 17.1). For trunk transmission it has to be converted using a modem.

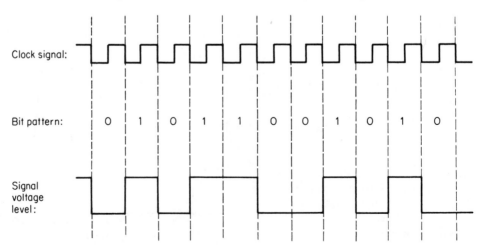

channels or open-wire pairs singing into the wind between poles. It is therefore into this frequency range that we must fit our signal. The primary function of the modem is to convert a binary data signal into a set of frequencies that fit into transmission space of Fig. 17.1.

LOCAL LOOPS   A local telephone loop, or any channel on which frequency-division multiplexing is not employed, has properties different from those of Fig. 17.1. Figure 17.2 shows typical local loop characteristics. If the loop has no loading coils (see Figs. 8.5 and 8.6), it can transmit higher frequencies or a higher bit rate. AT&T transmits 56,000 bps without modems on local telephone loops as part of the Digital Dataphone Service (DDS), and to achieve this speed the *load coils,* if they are present, must be bypassed. Most local loops today do not have load coils.

The channel in Fig. 17.1 cannot pass dc signals. The typical output of a data-processing machine is a set of rectangular pulses such as that in Fig. 17.3. Such pulses cannot be sent down a telephone *trunk* circuit in their original ("baseband") form. They must be converted to fit as efficiently as possible into the transmission space of Fig. 17.1.

The pulses of Fig. 17.3 can be sent down local loops in baseband form. However, if they are transmitted too far or too fast, they become excessively distorted—much more so on a loaded than an unloaded loop. On a loaded loop a baseband signal of 200 bps can be sent several miles without modems. On an unloaded loop a signal of 56,000 bps can be sent several miles, as in DDS. If the load coils are replaced with regenerative repeaters, the local loop can potentially handle a bit rate of several million bits per second, as on PCM transmission circuits.

Usually today the line termination equipment is designed not only for local loop transmission but for transmission to anywhere on the telephone network. Therefore the signal must be tailored to the shape of Fig. 17.1, and transmission is limited to bit rates of less than about 10,000 bps. This limitation makes clear the need for, and indicates the potential of, switched data networks, possibly employing the PCM trunks.

IN-PLANT AND   Many computer users have privately owned lines on
OUT-PLANT LINES   their own premises linking terminals to computers.
                        The terms "in-plant" and "out-plant" are used. *Out-plant* refers to common carrier lines connecting separate premises. *In-plant* refers to lines or systems within one building or complex of buildings. In-plant lines are usually copper wire pairs; they are sometimes coaxial cable. They may be installed by the firm's own engineers or by telecommunication companies.

Often no modems are used on in-plant lines. They may employ baseband digital signaling at speeds ranging from telegraph speeds to several million bits per second. For high speeds digital repeaters which regenerate the bit stream are used every thousand feet or so.

FULL-DUPLEX VERSUS Over a given physical line, the terminal equipment
HALF-DUPLEX may be designed so that it can transmit either in both
directions at once—full-duplex transmission—or else
in either direction but not both at the same time—half-duplex. It is possible also to
have simplex transmission, which can take place in one direction only, but this is
rarely used for data transmission as there is no easy way of controlling the flow of
data or requesting retransmission after errors or failures. All four-wire facilities
are capable of full-duplex working (though sometimes one finds them used in a
half-duplex manner). Some two-wire facilities can operate only in a half-duplex
mode, though over many two-wire links full-duplex operation is possible.

A terminal or a computer line adapter will work somewhat differently
depending on which of these possibilities is used. Where full-duplex transmission
is employed it may be used either to send data streams in both directions at the
same time or to send data in one direction and control signals in the other. The
control signals would govern the flow of data and would be used for error control.
Data at the transmitting end would be held until the receiving end indicated that
they had been received correctly. Control signals would ensure that no two terminals transmit at once on a line with many terminals and would organize the
sequence of transmission.

As we commented earlier, simultaneous transmission in two directions can
be obtained on a two-wire line by using two separate frequency bands. One is used
for transmission in one direction and the other for the opposite direction. By keeping the signals strictly separated in frequency, they can be prevented from interfering with each other.

The two bands may not be of the same bandwidth. A much larger channel
capacity is needed for sending data than for sending the return signals which control the flow of data. If, therefore, data are to be sent in one direction only, the
majority of the line bandwidth can be used for data. Some modems provide a data
channel of several thousand bits per second and a return channel of 75 bps or less.
Both channels can reverse their direction of transmission simultaneously.

Many full-duplex modems provide a full-speed data channel in each
direction.

FUNCTIONS OF LINE
TERMINATION The equipment at the ends of a telephone line on the
EQUIPMENT user's premises carries out several functions:

1. It must handle the initial setting up of the connection ("handshaking").

2. It must transmit and receive.

3. It must convert the digital signals into a form suitable for transmission and must convert
   them back again after transmission.

4. It may be able to initiate a call by automatically sending "dialing" pulses or their equivalent and accepting the response to them.

5. It should protect the transmission facilities.

6. It should detect any data errors which occur and take action to correct them.

7. It should detect transmission or equipment failures and diagnose where they occurred so that action can be taken.

All these functions *could* be performed by the modem or line terminal equipment which constitutes an interface between the line and the data-processing machines which use the line. Often, however, the modem (data set) carries out only the first three functions. Dialing the call may be done by the operator. The protection of the line is done in the United States by the data access arrangement (DAA), discussed below. The error detection is done by the data-processing equipment. And automatic diagnosis of failures is too often not done at all. The most economical and efficient approach is for all these functions to be carried out in an integrated fashion "under the cover" of the terminal or other data-processing machine.

There is a wide variety of different modems.

The main criteria for choosing between different types of design are

1. Speed of transmission.

2. Cost.

3. Numbers of errors in transmissions.

4. Line turnaround time.

The relative importance of these four requirements varies from one application to another.

MODEM STANDARDS   Most countries other than the United States do not have data access arrangements, and consequently users of the telephone network do not have the same freedom in attaching devices to it. It is desirable that independent organizations be able to design and manufacture modems and data-processing equipment with modems under the cover. To permit this, various standards exist for modem design. The standards permit modems of different manufacturers to communicate with one another.

It is desirable that modem standards be internationally accepted, and they should permit international transmission. CCITT modem standards have achieved this objective but not yet at the highest transmission speeds. The standards used are listed in Box 17.1.

A STANDARD   As an example of a modem standard we will discuss
4800-bps MODEM   *CCITT Recommendation V.27* [1] for a 4800-bps modem.

This modem is intended for synchronous transmission on leased telephone circuits. It is capable of operating in half-duplex fashion or full-duplex, meaning

BOX 17.1   CCITT recommendations for modem standards [1]

The interface between the modem and the data-processing machine

CCITT Recommendation

V.24   Definition of interchange circuits between modems and data terminal equipment
V.25   Automatic calling and/or answering equipment and disabling of echo suppressors

---

Modem standards for use on the switched telephone network

CCITT Recommendation

V.21   200-baud modem
V.23   600/1200-baud modem
V.26b   2400/1200-bps modem
V.30   Parallel data transmission modems
V.22   Standardization of data-signaling rates
V.25   Automatic calling and/or answering equipment and disabling echo suppressors
V.15   Use of acoustical coupling

Modem standards for use on leased telephone lines

CCITT Recommendation

V.26   2400-bps modem
V.27   4800-bps modem
V.22b   Standardization of data-signaling rates
V.35   Data transmission at 48,000 bps using 60–108-kHz (channel group) circuits

---

that a backward channel of up to 75 bps can transmit simultaneously with the forward transmission of 4800 bps and that the direction of both of these channels can be reversed.

The 4800 bps signal modulates a single carrier of 1800 ±1 Hz. It is assumed that there may be a frequency drift of up to ±6 Hz in the transmission; hence the receiver must be able to accept errors of ±7 Hz in the received frequencies.

The modulation rate is 1600 bauds (i.e., 1600 separate line conditions per second). The data are divided into groups of three consecutive bits (tribits). Each tribit is transmitted as one change in line condition, thus giving 4800 bps. The bit rate and hence the modulation rate are held constant to ±0.01%.

Each tribit is encoded as a phase change relative to the phase of the immediately preceding tribit. The encoding is as follows:

Tribit	Phase Change (%)
001	0
000	45
010	90
011	135
111	180
110	225
100	270
101	315

The receiving modem detects these phase changes and converts them into the appropriate bits.

On a bad-quality line it is sometimes difficult to detect phase changes to the required level of accuracy. During periods of noisy transmission it would be of value to change to modulating *dibits* with 90% phase differences rather than *tribits* with 45% differences. Dibits, using the same modulation rate, would give a signal of 3200 bps. Unfortunately, neither this nor 1600 bps are CCITT-recommended speeds; therefore the standard modem has no fallback transmission rate, although this is desirable technically.

A data stream 0 0 1 0 0 1 0 0 1... could result in no phase changes, and this could result in loss of synchronization between the transmitting and receiving modems. Certain other repetitive bit patterns might also cause problems. To avoid the transmission of repetitive bit patterns the bit stream is scrambled before modulation and unscrambled by the receiving modem. The procedure for scrambling is specified in the CCITT recommendation.

The reverse channel of up to 75 bps is an optional feature of the modem. The reverse channel is organized like a standard voice-frequency telegraph channel with frequency-shift modulation: 390 Hz represents a 1 bit or *mark*; 450 Hz represents a 0 bit or *space*. The use for this reverse channel is to permit control signals, especially error control signals, to be sent simultaneously with the transmission of data in the opposite direction.

ECHO SUPPRESSORS   Echos bounce back from points in a telephone line at which its impedence changes. If the line is sufficiently long that the echo of a talker would reach him 45 milliseconds or more after he speaks, the echo must be removed. The receive path is attenuated so that a talker cannot hear the echos.

A device which suppresses echos would also suppress data, and so when full-duplex transmission is used the echo suppressors must be disabled. The user's modem can disable the echo suppressors by transmitting a single-frequency tone in the band 2010–2240 Hz for at least 400 milliseconds. No other signal or tone should be transmitted during this disabling operation. The echo suppressor will remain disabled if a data signal (or other signal) is transmitted within 100 milliseconds of the disabling tone being removed. If the signal transmitted ceases for 100 milliseconds or more, the echo suppressor will come into operation again.

While echo suppressors must be disabled for full-duplex transmission, they are often left in operation for half-duplex transmission. They do, however, take many milliseconds to reverse their direction of operation. For this reason, when a modem stops receiving, it should not transmit immediately but should wait about 100 milliseconds if there are echo suppressors on the line. This *line turnaround* time is a nuisance with some types of line-control procedure, for example, with a line on which many terminals are *polled*.

DATA ACCESS ARRANGEMENT

Customer-owned and independently designed data-handling machines can be connected directly to the telephone network of North America via a direct access arrangement (DAA). In many cases the DAA is no longer needed if the equipment conforms to certain standards. The DAA is a small plastic box, shown in Fig. 17.4, which protects the telephone network from high voltages or signals of too great an amplitude. The user rents it for a few dollars per month. The DAA is, in essence, an isolation transformer. It attenuates the customer signal if it is too strong but introduces negligible other distortions into the signal. It is wired with a conventional telephone set, as shown in Fig. 17.4, and the telephone set is used for establishing dialed calls. It has a test circuit and test switch on it so that the telephone connection can be tested from the telephone company office.

A different DAA is used for data-handling machines which automatically originate and answer calls.

GETTING STARTED

Some data transmission links need an operator at each end of the link in order to establish the connection. Many machines, however, are capable of *unattended operation,* and so, when called, the modem must turn itself and its associated data-handling machine on without human intervention.

Figure 17.5 illustrates the opening sequence of events on a typical start-stop system. The frequencies quoted in this diagram are those of AT&T's type 103A data set. This machine uses frequency-shift keying on two separate frequency bands, one for each direction of transmission. In this way it gives full-duplex transmission over switched public voice lines at low speed. It transmits in a binary fashion, sending either a MARK or a SPACE frequency ("1" or "0"). These fre-

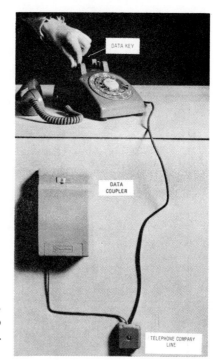

Figure 17.4 The Direct Access Arrangement, DAA, gives users in North America freedom to chines design *foreign attachment* data handling ma- and connect them to the telephone network.

NOTE: Devices of approved design can now be attached directly to the telephone wires without a DAA.

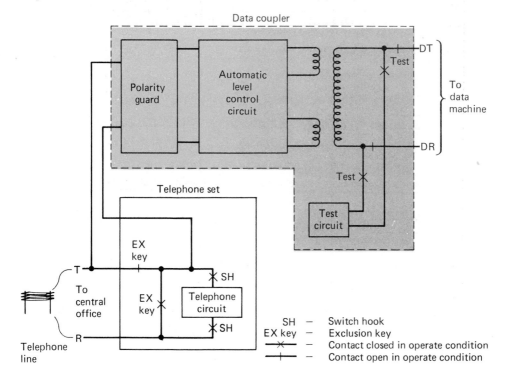

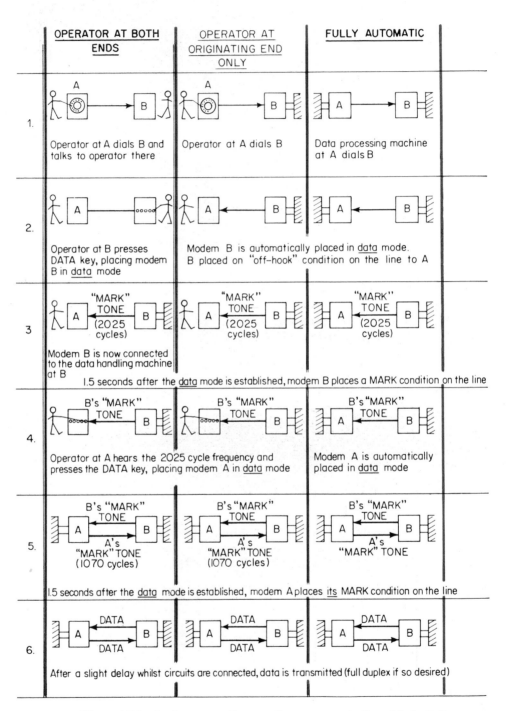

Figure 17.5 Setting up a call, manually or automatically, with the Bell type 103A data set.

Figure 17.6 The interface between the modem and the data processing equipment is a standard that is well adhered to.

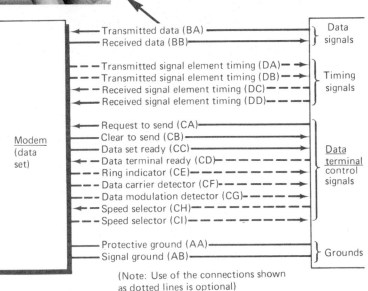

Modem (data set)

Transmitted data (BA)
Received data (BB)
} Data signals

Transmitted signal element timing (DA)
Transmitted signal element timing (DB)
Received signal element timing (DC)
Received signal element timing (DD)
} Timing signals

Request to send (CA)
Clear to send (CB)
Data set ready (CC)
Data terminal ready (CD)
Ring indicator (CE)
Data carrier detector (CF)
Data modulation detector (CG)
Speed selector (CH)
Speed selector (CI)
} Data terminal control signals

Protective ground (AA)
Signal ground (AB)
} Grounds

(Note: Use of the connections shown as dotted lines is optional)

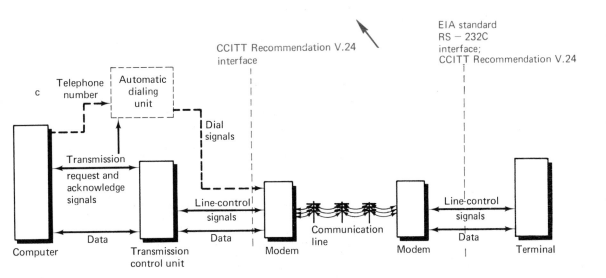

EIA standard RS – 232C interface; CCITT Recommendation V.24

CCITT Recommendation V.24 interface

c

Telephone number

Automatic dialing unit

Dial signals

Transmission request and acknowledge signals

Line-control signals

Data

Computer

Transmission control unit

Modem

Communication line

Modem

Line-control signals

Data

Terminal

BOX 17.2 Circuits commonly used in the interface between a modem and a data terminal†

	CCITT Circuit No.

1. Data signals

   (a) *Transmitted data (to the modem).* Data generated by the terminal for transmission. — 103

   (b) *Received data (to the terminal).* Data received by the modem for the terminal. — 104

2. Timing signals

   (a) *Transmitter signal element timing.* Two connections are defined. One sends signal element timing information from the transmitting terminal to its modem. The other sends timing information from the transmitting modem to its terminal. — 113, 114

   (b) *Receiver signal element timing.* Two connections are defined. One sends signal element timing information from the *receiving* terminal to its modem. The other sends timing information from the receiving modem to its terminal. — 115, 128

   The timing signal connections are optional. A modem for start-stop transmission does not use them.

3. Control signals

   (a) *Request to send (to the modem).* Signals on this connection are generated by the transmitting terminal when it wishes to transmit. The modem's carrier signal is transmitted during the ON condition of this connection. (With half-duplex operation, the OFF condition of this connection holds the modem in the receive-data state.) — 105

   (b) *Clear to send (to the terminal).* Signals on this connection are generated by the transmitting modem to indicate that it is prepared to transmit data. They are a response to the Request to Send signal from the transmitting device. (With full-duplex operation the modem is in the transmit state at all times.) — 106

   (c) *Data set ready (to the terminal).* Signals on this connection are generated by the local modem to indicate to the transmitting machine that it is ready to operate. — 107

†*CCITT Recommendation V.24* and *EIA Standard RS-232-C.* The use of the interface is shown in Fig. 17.6.

BOX 17.2 Continued

	CCITT Circuit No.

(The following control signals are optional.)

(d) *Data terminal ready (to the modem).* When the terminal sends the ON condition on this connection it causes the modem to be connected to the communication line. The OFF condition causes it to be disconnected, in order to terminate a call or free the line for a different use. — 108

(e) *Ring indicator (to the terminal).* A signal on the connection informs the terminal that the modem is receiving a ringing signal from a remote location. — 125

(f) *Data carrier detector (to the terminal).* A signal on this connection indicates to the terminal that the carrier (the sine wave that carries the signal) is being received. If the carrier is lost because of a fault condition on the line, the terminal will be notified by an OFF condition in this connection. — 109

(g) *Data modulation detector (to the terminal).* An ON condition on this connection informs the terminal that the signal is being demodulated correctly by the modem. When the quality of demodulation drops below a certain threshold the terminal may take corrective action such as requesting retransmission or requesting that a lower transmission rate be used. — 110

(h) *Speed selector.* There are two speed selector connections, one to the modem and one to the terminal. Using them, the transmission rate may be changed. — 111, 112

4. Grounds

(a) *Protective ground.* Attached to the machine frame and possibly to external grounds. — 101

(b) *Signal ground.* Establishes the common ground reference potential for the circuits. — 102

Table 17.1 A complete list of the circuits in the CCITT V24 interface between modem and data machine.

Circuit No.	Function	Direction
Grounds		
101	Protective ground or earth	
102	Signal ground or common return	
Data		
103	Transmitted data	→
104	Received data	←
118	Transmitted backward channel data	→
119	Received backward channel data	←
Control		
105	Request to send	→
106	Ready for sending	←
107	Data set ready	←
108/1	Connect data set to line	→
108/2	Data terminal ready	→
109	Data channel received line signal detector	←
110	Signal quality detector	←
111	Data-signaling rate selector	→
112	Data-signaling rate selector	←
116	Select standby	→
117	Standby indicator	←
120	Transmit backward channel line signal	→
121	Backward channel ready	←
122	Backward channel received line signal detector	←
123	Backward channel signal quality detector	←
124	Select frequency groups	→
125	Calling indicator	←
126	Select transmit frequency	→
127	Select receive frequency	→
129	Request to receive	→
130	Transmit backward tone	→
132	Return to nondata mode	→
133	Ready for receiving	→
134	Received data present	←
191	Transmitted voice answer	→
192	Received voice answer	←
Timing		
113	Transmitter signal element timing	→
114	Transmitter signal element timing	←
115	Receiver signal element timing	←
128	Receiver signal element timing	←
131	Received character timing	←
Automatic calling		
201	Signal ground or common return	
202	Call request	→
203	Data line occupied	←
204	Distant station connected	←
205	Abandon call	←
206	Digit signal (2^0) — 4 bits giving the number dialed	→
207	Digit signal (2^1)	→
208	Digit signal (2^2)	→
209	Digit signal (2^3)	→
210	Present next digit	←
211	Digit present	→
212	Protective ground or earth	
213	Power indication	←

Key: From the terminal ⟶
     To the terminal ⟵

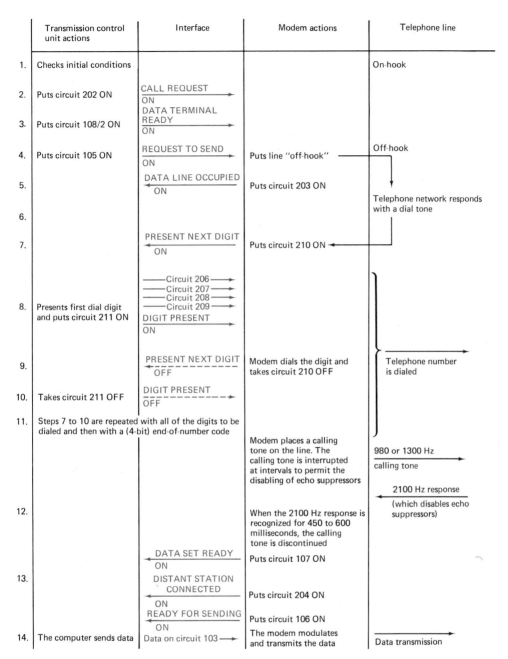

	Transmission control unit actions	Interface	Modem actions	Telephone line
1.	Checks initial conditions			On-hook
2.	Puts circuit 202 ON	CALL REQUEST ON →		
3.	Puts circuit 108/2 ON	DATA TERMINAL READY ON →		
4.	Puts circuit 105 ON	REQUEST TO SEND ON →	Puts line "off-hook"	Off-hook
5.		DATA LINE OCCUPIED ON ←	Puts circuit 203 ON	
6.				Telephone network responds with a dial tone
7.		PRESENT NEXT DIGIT ON ←	Puts circuit 210 ON ←	
8.	Presents first dial digit and puts circuit 211 ON	──Circuit 206→ ──Circuit 207→ ──Circuit 208→ ──Circuit 209→ DIGIT PRESENT ON →		
9.		PRESENT NEXT DIGIT OFF ←	Modem dials the digit and takes circuit 210 OFF	Telephone number is dialed
10.	Takes circuit 211 OFF	DIGIT PRESENT OFF →		
11.	Steps 7 to 10 are repeated with all of the digits to be dialed and then with a (4-bit) end-of-number code		Modem places a calling tone on the line. The calling tone is interrupted at intervals to permit the disabling of echo suppressors	980 or 1300 Hz calling tone → 2100 Hz response ← (which disables echo suppressors)
12.			When the 2100 Hz response is recognized for 450 to 600 milliseconds, the calling tone is discontinued	
13.		DATA SET READY ON ← DISTANT STATION CONNECTED ON ← READY FOR SENDING ON ←	Puts circuit 107 ON Puts circuit 204 ON Puts circuit 106 ON	
14.	The computer sends data	Data on circuit 103 →	The modem modulates and transmits the data	Data transmission →

Figure 17.7 The sequence of events when a transmission control unit initiates the automatic dialing of a call on the telephone network and transmits data when the connection is made. This illustration shows the use of the interface circuits and frequencies conforming to CCITT Recommendation V.25. If the call is not successfully completed, i.e. the 2100 Hz tone is not received a given number of seconds after dialing (between 10 and 40 seconds), then the modem puts circuit 205 (ABANDON CALL) on. The circuit numbers are those in Table 17.1.

329

quencies are audible, and if an earphone were connected, the operator would be able to hear the whistle of data rushing long the line.

The first step in establishing the connection is to dial the distant machine. This might be done either by the originating operator or in some cases automatically by the originating machine. When the connection is made, the answering data set must be placed in its "data mode" so that it is ready to receive and transmit data. This may be done automatically, or it may be done by the answering operator pressing the DATA key on that data set. It may be done entirely nonautomatically by two operators talking to each other and agreeing to press the DATA key.

One and a half seconds after the establishment of *data* mode, the answering set places its MARK frequency on the line. If the originating call came from an operator, she will hear this as a high-pitched whistle or *"data tone."* She will then press the DATA key on her set. She can then no longer hear the *data tone* because she has switched the set from *talk* to *data* mode and the telephone is no longer connected. If the call was not originated by an operator, the data set will be automatically placed in *data* mode. The pause of 1.5 seconds before answering is used to avoid interfering with certain tone-signaling actions that are used on the telephone network.

A similar pause of 1.5 seconds occurs after the originating set is placed in data mode, and it then places its MARK frequency on the line (a different frequency because these modems are designed for full-duplex operation). Any echo suppressor disablers on the line are told by this signal that this is a data transmission, so they stop the action of their associated echo suppressor. The echo suppressor normally permits transmission in one direction at a time only, but here simultaneous two-way transmission is needed, so the echo suppressors are automatically disabled.

Both sets, once they receive the other set's MARK frequency, place their transmitting circuits under control of their data-handling machine. Thus the connection for full-duplex data transmission has been set up.

INTERFACE BETWEEN MODEMS AND MACHINES   It is very important that the interface between the modem and the various data transmission machines be standardized, so that one modem can be replaced by another and so terminals and transmission control units can be interchanged. A standard interface between modems and business machines is specified in *CCITT Recommendation V.24* and is adhered to closely in practice. The *American Electronic Industry Association Standard RS-232-B* is the same. It is illustrated in Fig. 17.6 and Box 17.2.

A terminal or data-processing machine plugs into the modem usually with a 25-pin plug. The circuits in the plug are specified in the above standards. Figure 17.6 shows a typical group of such circuits. Other circuits can be used optionally at the discretion of the equipment designer.

CCITT Recommendation V.24 lists more circuits than those in Fig. 17.6. Some are for channels in the reverse direction. Some are to permit an automatic dialing unit (shown in Fig. 17.6) to set up a dialed connection. Table 17.1 gives a complete list of the circuits in the V.24 standard.

CCITT Recommendation V.25 gives standards for automatic calling. Figure 17.7 shows the sequence of events when a data-processing machine with an automatic calling unit sets up a call and starts to transmit (using the CCITT standard).

PARALLEL DATA
TRANSMISSION

Most data are transmitted serial by character, serial by bit. In some systems, however, data are sent a character at a time, and the components of each character are transmitted in parallel. Parallel transmission is usually employed to provide inexpensive terminals, often for data collection systems in which many terminals are needed and the data flow is in one direction.

The cheapest form of data terminal is a simple push-button telephone. Each button pressed transmits a pair of frequencies which travel well within the telephone bandwidth of Fig. 17.1. The computer receiving the data may respond to the telephone user either with tones or with voice answerback.

Figure 17.8   Parallel transmission using the Bell 400 series data sets. Inexpensive data collection terminals use these frequencies without employing a data set.

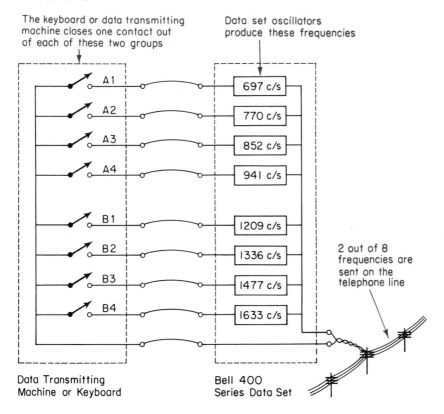

The keyboard or data transmitting machine closes one contact out of each of these two groups

Data set oscillators produce these frequencies

A1   697 c/s
A2   770 c/s
A3   852 c/s
A4   941 c/s

B1   1209 c/s
B2   1336 c/s
B3   1477 c/s
B4   1633 c/s

2 out of 8 frequencies are sent on the telephone line

Data Transmitting Machine or Keyboard

Bell 400 Series Data Set

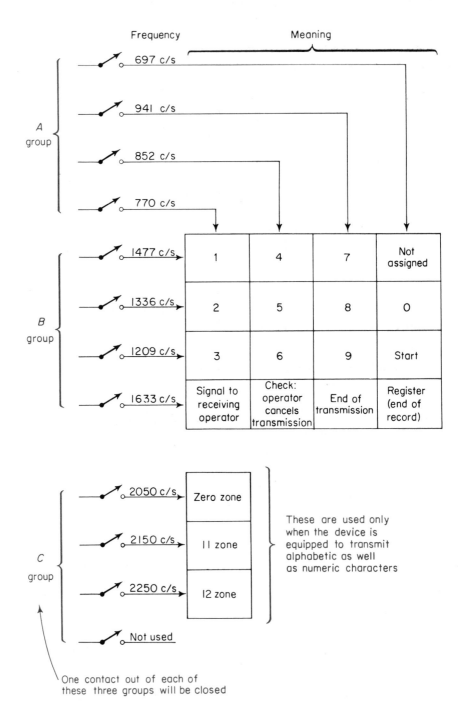

Figure 17.9 An example of coding using 3 out of 12 parallel multifrequency transmissions.

The same frequencies can be used for other types of terminals. The Bell System 400 series of data sets uses the Touchtone telephone frequencies, as shown in Fig. 17.8, and a data transmission device operating at 10 characters per second, or less, may be connected. Other devices generate the same frequencies themselves without employing a data set.

A code is used in which two frequencies out of the eight possible are transmitted at any one instant. This gives 16 possible combinations that can be transmitted. It gives some measure of transmission error detection, in that a fault causing only one or more than two frequencies to be received will be noted as an error, but this is not comprehensive error detection as in other codes.

For alphabetic transmission a third group of frequencies is added. Characters are encoded as in Fig. 17.9 on some devices, the third group of frequencies being used for the "zones" of a punched-card code. The "*12* zone" combined with a digit gives letters *A* to *I*, the "*11* zone" gives *J* to *R*, and the "*0* zone" gives *S* to *Z* and one special character.

CCITT PARALLEL TRANSMISSION

CCITT Recommendation V.30 is for parallel transmission over the public telephone network.

Telephone networks use signaling schemes which employ frequencies inside the telephone bandwidth, and these frequencies must be avoided by multifrequency transmission devices. The frequencies of Figs. 17.8 and 17.9 are suitable for North America, but unfortunately telephone networks in different countries employ different signaling frequencies. The CCITT recommendation avoids all these different national signaling frequencies with the result that a closely bunched group of 12 frequencies is employed as shown in Fig. 17.10. These frequencies exclude the use of a normal push-button telephone set.

The 12 frequencies are grouped as follows:

Group A	920 Hz	1000 Hz	1080 Hz	1160 Hz
Group B	1320 Hz	1400 Hz	1480 Hz	1560 Hz
Group C	1720 Hz	1800 Hz	1880 Hz	1960 Hz

The following systems are desired using the 12 frequencies:

1. Transmitting 16-character combinations.

2. Transmitting 64-character combinations.

3. Transmitting 256-character combinations.

For a 16-character system, groups 1 and 2 are used; for a 64-character system, all three groups are used. For a 256-character system, each character is split into two half-characters, and two groups of frequencies are used.

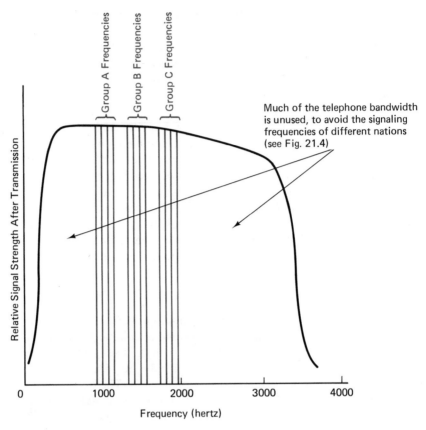

Figure 17.10 The 12 frequencies recommended by the CCITT for parallel data transmission from low-speed devices.

A backward channel can be used in conjunction with the above frequencies, at 420 Hz. This can be used for either audible signals or amplitude-modulated digital signals at a rate up to 5 bauds. The CCITT recommends that speeds up to 40 characters per second be used with the 16-character system if two of the frequencies are used for a separate binary timing channel. If there is no such timing channel, an intercharacter rest condition is recommended, with a speed up to 20 characters per second. Speeds of up to 20 characters per second are recommended with the 64- and 256-character systems

ACOUSTICAL COUPLING   Where audible tones are used, there is no need to have, as with the above machines, a direct wire connection to the communication line. Instead, acoustical coupling can be used, in which the signals are converted into sound which fits well inside the telephone earpiece. They are then converted back into data signals. Acoustical coupling is somewhat less efficient that direct coupling. It is used for transmitting between relatively slow machines such as typewriterlike terminals.

334

In a typical acoustical-coupling device the telephone handpiece fits in a special cradle, as shown in Fig. 17.11. An advantage of acoustical coupling is that the terminal can easily be made portable. A small terminal could be made to transmit to a computer from a public call box.

Although there is no electrical connection to the telephone lines, it is still possible for acoustically coupled machines to interfere with the public network's signaling, as with a directly coupled device. Also severe crosstalk can be caused on a multiplexed link by the transmission of a continuous frequency, as with a repetitive data pattern. It is therefore desirable that the coupling device randomize the signal before it is sent, though not all devices do so today.

The CCITT recommends that acoustical couplers not be used for permanent installations.

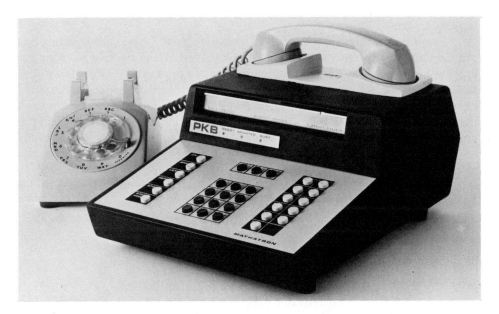

Figure 17.11   A typical example of a device with an accoustical coupler. Accoustical couplers permit a terminal to transmit over a telephone line without a direct electrical connection. Data is converted into sounds that are transmitted via the telephone. The telephone company does not have to install these couplers. The above device is Mathatron's PKB electronic calculator.

HIGH-SPEED LINE TERMINATION   Lines with a bandwidth higher than that of a telephone channel can be leased. Commonly the bandwidth of a *channel* group (48 kHz) is leased—equivalent to 12 telephone channels. Sometimes higher bandwidth circuits are used. Modems are available from various manufacturers for wideband transmission.

CCITT Recommendation V.35 is for a 48,000 bps modem operating in the 60- to 104-kHz band. Suppressed carrier AM modulation of a 100-kHz carrier is recommended.

With the spread of PCM and digital circuits it is likely that many high-data-rate links in the future will operate over digital lines and hence not need modems. Many high-speed terminals will, however, be linked to the digital trunks via copper wire-pair local telephone loops. On these, either wideband modems or *line drivers* which transmit a digital baseband signal are used.

The approach of some common carriers is to provide their customers with an interface unit which can contain different facilities depending upon the nature of the transmission link. Sometimes it contains a modem; sometimes it contains a digital baseband signaling unit. The customer does not necessarily need to know what it contains. He needs to know merely how it interfaces with his data processing equipment.

For wideband transmission AT&T provides its *303-type data station* which can contain either a modem or a digital line driver. It can operate over various types of local and toll facilities including analog, T1, and even Picturephone lines. It can be used to transmit data synchronously at rates from 19.2 kbps to 460.8 kbps, or can transmit nonsynchronous data with minimum signal element durations in the range from 52 to 2 microseconds. It provides a full duplex data circuit plus a voice channel for coordination purposes. The equipment can permit a customer to use either wideband data transmission or multiple voice circuits, and to switch between these at his discretion.

When such equipment transmits high-speed data over local telephone loops, special conditioning of the local loops is sometimes used. This conditioning employs wideband amplifiers with equalizers to provide a flat response over the required bandwidth.

DDS

The Bell System Dataphone Digital Service (DDS) provides a digital channel and hence requires no modems. The digital channel operates at speeds of 2400, 4800, 9600, or 56,000 bps in a full-duplex or half-duplex fashion.

Instead of a modem, the user is provided with a Bell *Data Service Unit*. The data processing machine passes data to, and receives data from, this unit. The unit performs the following functions:

1. It samples the data it receives synchronously.

2. It recovers the timing of the bits.

3. It transmits the bits in an appropriately coded form.

4. It generates and recognizes control signals.

5. It formats the data and control signals as required. Any combination of data bits can be sent.

6. It protects the telephone network from any damaging voltages or signals.

7. It provides remote loopback so that end-to-end testing can be performed on the link.

8. It equalizes the line (discussed in Chapter 27) to improve its transmission properties.

The data service unit is designed so that the user's terminal or data-processing equipment connects to it with the same plug that would be used with a modem on an analog line. For transmission at 2400, 4800, or 9600 bps this is a 25-pin plug with lines conforming to the EIA Standard RS-232-C interface. With 56,000 bps transmission it is a 34-pin plug conforming to *CCITT Recommendation V.35* for wideband modems.

The circuits used in the latter case are

Circuit No.	Function
101	Protective ground or earth
102	Signal ground or common return
103	Transmitted data
104	Received data
105	Request to send
106	Ready for sending
107	Data set ready
109	Data channel receive line signal detector
114	Transmitter signal element timing
115	Receiver signal element timing

SPC Southern Pacific Communications operates the Datran fast-connect switched data network. It has to provide its subscribers with both a transmission interface and the means of setting up the switched call.

Sometimes the customer is close enough to the SPC digital facilities to be connected to them by a wire-pair local pair without modems. In this case a *line driver* is used, transmitting a digital baseband signal. Often, however, the customer is many miles from the SPC digital trunks and must be connected to them via a *modem* on a leased telephone line. The line driver and modem are designed so that they look the same (Fig. 17.12) and have the same interface to the customer's equipment—an EIA Standard RS-232-C interface as with DDS. The maximum speed of the modem is 9600 bps. Customers can transmit at higher speeds, such as 56,000 bps, only if they are close enough to the digital trunks to be connected by a wire-pair loop without amplifiers.

The customer equipment also contains a dialing unit shown in Fig. 17.12, which permits user dialing, automatic dialing, and testing of the circuit.

INTERCONNECTION   SPC and other companies need to connect customers
to their facilities over wire-pair telephone circuits. The
telephone company specifies three tariffed grades of wire-pair circuit for this purpose: Type W1 which transmits digitally at 2400 or 4800 bps, Type W2 (9600 bps),
and Type W3 (56,000 bps). Box 17.3 gives details of these digital inter-connection
facilities.

> **BOX 17.3**   Tariffed wire-pair facilities (local loops)
> which telephone companies in the U.S.A.
> provide for digital interconnection.
> Over these facilities data at speeds of
> 2400, 4800, 9600 and 56000 bps can be
> transmitted

Type W1 (2.4 and 4.8 kbps)

A.   All load coils and build out capacitors will be removed and bridged tap
will not exceed a cumulative tap length of 6000 feet.
B.   The actual measured 2.4 kHz insertion loss between 135 ohm terminations will be 34 db or less.
C.   The noise between 40 Hz and 30 kHz measured between 135 ohm terminations will not exceed −45 dbm background noise and 7 counts per 15
minutes impulse noise at a threshold of −29 dbm.

Type W2 (9.6 kbps)

A.   All load coils and build out capacitors will be removed and bridged tap
will not exceed a cumulative tap length of 6000 feet.
B.   The actual measured 4.8 kHz insertion loss between 135 ohm terminations will be 34 db or less.
C.   The noise between 40 Hz and 30 kHz measured between 135 ohm terminations will not exceed −53 dbm background noise and 7 counts per 15
minutes impulse noise at a threshold of −37 dbm.

Type W3 (56 kbps)

A.   All load coils and build out capacitors will be removed and bridged tap
will not exceed an individual tap length of 200 feet or a cumulative tap
length of 2500 feet.
B.   The actual measured 28 kHz insertion loss between 135 ohm terminations will be 34 db or less.
C.   The noise between 40 Hz and 30 kHz measured between 135 ohm terminations will not exceed −56 dbm background noise and 7 counts per 15
minutes impulse noise at a threshold of −40 dbm.

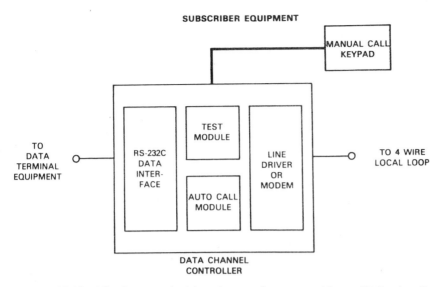

Figure 17.12 The data terminal interface equipment used by an SPC subscriber.

The customer keypad and test module.

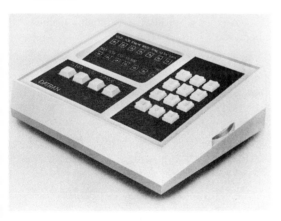

The modem or line driver (they have the same cover).

REFERENCES

1. The CCITT recommendations referred to in this chapter are summarized in Box 17.1 and are published in *The CCITT Green Book*, Vol. VIII: Data Transmission, published by the International Telecommunications Union, Geneva, 1973.

2. The following are useful Bell System manuals relating to line termination. All can be requested from AT&T with the "PUB" number shown:

   a. Bell System Technical Reference: "Data Communications Using the Switched Telecommunications Network." — PUB 41005

   b. Bell System Technical Reference: "1969–70 Switched Telecommunications Network Connection Survey (Reprints of *Bell System Technical Journal* Articles) — April 1971". — PUB 41007

   c. Bell System Technical Reference: "Analog Parameters Affecting Voiceband Data Transmission — Description of Parameters — October 1971". — PUB 41008

   d. Bell System Technical Reference: "Transmission Parameters Affecting Data Transmission — Measuring Techniques — January 1972". — PUB 41009

   e. Bell System Technical Reference: "Acoustic and Inductive Coupling for Data and Voice Transmission — October 1972". — PUB 41803

   f. Bell System Technical Reference: "Data Auxiliary Set 801A (Automatic Calling Unit) Interface Specification — March 1964". — PUB 41601

   g. Bell System Technical Reference: "Data Auxiliary Set 801C (Automatic Calling Unit) Interface Specification — September 1965". — PUB 41602

   h. Bell System Technical Reference: "Data Access Arrangement CDT for Manual Originating and Answering Terminals — May 1973". — PUB 41801

   i. Bell System Technical Reference: "Data Couplers CBS and CBT for Automatic Terminals — May 1974". — PUB 41802

18 TARIFFS

The services which a common carrier offers to the public and their prices are described in tariffs. A tariff is a document which, in the United States, is required by the regulating bodies who control the carriers. The U.S. Federal Communications Commission must eventually approve all interstate facilities, and similar state commissions control those within state boundaries. By law, all tariffs must be registered with these bodies. In most other countries, the telecommunication facilities are set up by government bodies and thus are directly under their control.

In this chapter we will summarize telecommunications facilities as the user sees them and as described in the tariffs. The illustrations are taken from the offering of the carriers in the United States and Britain. Those of other countries are broadly similar. Countries with less well-developed data transmission have less to offer in the way of wideband (high-capacity) facilities.

In the United States the subject of communication rates has become very complex. The amount and structure of charges differ from one state to another. In most other countries, the rates for more conventional channels remain relatively straightforward; however, often the carriers are government organizations, who are not obliged to publish tariffs for all their facilities. The price for less common channels, such as broadband, may have to be obtained by a special request to the carrier. In general it is desirable, when designing a system, that the organization in question be called in to quote a price for the facilities needed.

CATEGORIES OF LINE   Table 18.1 lists the main types of leased and public communication links in order of increasing speed. The speeds have been listed in terms of the number of data bits per second that may be sent over the line. Communication lines fall into one of three categories of speed:

1. *Subvoice grade.* Lines designed for telegraph and similar machines transmitting at speeds ranging, in the United States, from 45 to 150 bps. Some countries have lines of higher speed than their telegraph facilities but still much slower than the capacity of voice lines. Britain, for example, has its Datel 200 service operating at 200 bps. Most industrialized countries outside North America have a similar 200-bps service (a CCITT standard). All these lines are today commonly obtained by subdividing telephone channels.

2. *Voice grade.* At present, telephone channels normally transmit at speeds from 600 to 9600 bps. Dial-up telephone lines are commonly used for speeds from 1200 to 4800 bps today. A speed of 9600 is possible but requires elaborate modem design and powerful facilities for error correction. Telephone organizations in some other countries have not yet permitted the use of such high speeds over their telephone lines. In many countries, 1200 or 2400 bps is still the maximum.

3. *Wideband.* Wideband lines give speeds much higher than voice channels, using facilities that carry many simultaneous telephone calls. Speeds up to about 500,000 bps are in use, and higher bit rates are possible if required.

All these line types may be channeled over a variety of different physical facilities. This chapter and indeed the tariffs themselves normally say nothing about the medium used for transmission. It might very well be wire, coaxial cable, microwave radio, or even satellite. The transmission over different media is organized in such a way that the channels obtained have largely the same properties— same capacity, same noise level, and same error rate. The user generally cannot tell whether he is using a microwave link, coaxial cable, or pairs of open wires stretched between telephone poles. Only satellite transmission requires different data-handling equipment, because of the transmission delay and because the satellite can be used as a broadcast rather than point-to-point facility.

SWITCHED VERSUS LEASED LINES   The next important parameter about the lines is whether they are public switched lines or not.

Voice lines and telegraph lines can either be switched through public exchanges (central offices) or permanently connected. Facilities for switching broadband channels are in operation in some countries, although most broadband channels today are permanent connections.

When you dial a friend and talk to him on the telephone, you speak over a line connected by means of the public exchanges. This line, referred to as a "public" or "switched" line, could be used for the transmission of data. Alternatively, a "private" or "leased" line could be connected permanently or semipermanently between the transmitting machines. The private line may be connected via the local switching office, but it would not be connected to the switchgear and signaling devices of that office. An interoffice private connection would use the same physical links as the switched circuits. It would not, however, have to carry the signaling that is needed on a switched line (discussed in Chapter 21).

Just as you can either dial a telephone connection or have it permanently wired, so it is with other types of lines. Telegraph lines, for example, which have a much lower speed of transmission than is possible over voice lines, may be permanently connected or may be dialed like a telephone line via a switched public network. Telex is such a network; it exists throughout most of the world, permitting transmission at 50 bps. Some countries have a switched public network, operating at a somewhat higher speed than telex but at less speed than telephone lines. In the United States, the TWX network gives speeds up to 150 bps. TWX lines can be connected to telex lines for overseas calls. Also, certain countries are building up a switched network for very high-speed (wideband) connections. In the United States, Western Union installed the first sections of a system in which a user can indicate *in his dialing* what capacity link he needs.

CATEGORIES   Telephone switching is designed primarily for con-
OF SWITCHING   necting people to people, not machines to machines.
                          As indicated in Box 6.1 machine-to-machine communi-
nication has different requirements. One of the differences is in the switching. Telephone switching is slow, typically taking between 10 and 30 seconds to complete a call (see Table 29.1). If connections were made rapidly—at the speed of electronic switching machines—the communication links would be used differently.

New networks specifically designed for data transmission are now operating, which have high-speed switching. For these networks the tariffs can be different from telephone tariffs in that they permit a much shorter interconnection and charge a proportionately low price for it. The minimum billing interval on the U.S. telephone network is one minute. The minimum billing interval on the SPC system, a *specialized common carrier* switched data network, is one second. Consequently if a user has sporadic data messages to send, as a terminal user might when using a distant computer interactively, he will obtain cheaper transmission by employing multiple short switched data connections.

Networks which switch circuits electronically in a fraction of a second are referred to as *fast-connect networks*.

An alternative to the switching of circuits—the interconnection of electrical paths—is a network which interleaves the data from different users on fixed paths. Short messages can then be transmitted economically because many users share the facility. The interleaving can be done by small network computers which briefly store the messages from different users and then transmit them on shared transmission paths. This can be done by a concentrator network or a *packet-switching* network (discussed in Chapter 24). Networks which store and forward messages in this way can have another type of tariff structure in which the user is charged *by the amount of data transmitted* rather than *by the duration of the connection*.

Table 18.1 Main types of communication lines available

1. PUBLIC (DIAL-UP) LINES

	Bit Rate (bps)		Bandwidth (kHz)	Type of Line			Half-Duplex or Full-Duplex
				United States		United Kingdom	
	Fixed	Dependent on Modem		ATT	Western Union		
Subvoice grade	45 (6 5-bit characters per second)	Up to 45			TTY-TWX and CPT-TWX CE-TWX		HDX / HDX or FDX
	50				Telex	Telex	HDX
	110 (10 characters per second)	Up to 150 / Up to 200			TTY-TWX and CPT-TWX CE-TWX	Datel 200	HDX / HDX or FDX / FDX
The public telephone network		Up to 600 / Up to 1200				Datel 600 / Datel 600	FDX / FDX
		600 to 4800 (certain modems achieve higher speeds.)	3 (not all freely usable because of network signaling)	Public network		Public telephone network	HDX or FDX
Switched wideband networks	600 / 1200 & 2400 / 4800† / 9600† / 38,400†	(Other speeds will be achievable with other modems.)	2 / 4 / 8† / 16† / 48†		BEX (certain cities only)		FDX
	50,000	Up to 50,000		Dataphone 50 (few cities only)			FDX

†Planned but not yet available.

Table 18.1 Cont.

2. LEASED LINES

Bit Rate (bps) Fixed	Dependent on Modem	Bandwidth (kHz)	United States — ATT	United States — Western Union	United Kingdom	Half-Duplex or Full-Duplex
Subvoice grade						
50	Up to 45		1004			HDX/FDX
	Up to 55				Tariff H	FDX
100			1002			HDX/FDX
	Up to 75		1005			HDX/FDX
					Datel 100	FDX
200	Up to 150		1006	1006		HDX/FDX
	Up to 180					HDX/FDX
					Datel 200	FDX
Voice-grade lines						
	Up to 600	3			Datel 600	FDX
	Up to 1200	3			Datel 600	FDX
	600 to 10,500 (For the higher rates, conditioning is needed.)	3	3002 (C1, C2, C4, and C5 conditioning)		(Datel 2000 refers to conditioned or high-quality voice lines.)	HDX/FDX
Wideband						
2400 / 4800 / 9600			DDS			FDX
19,200		24	8803		Special quotation	FDX
40,800		48	8801			FDX
50,000		48	8801			FDX
230,400		240	5700 (originally TELPAK C) / 5800 (originally TELPAK D) (5700 & 5800 tariffs all provide "bundles" of smaller bandwidth lines.)			FDX
		1000 (approx.)				
56,000			DDS			FDX

In some data systems there is no urgency to transmit the messages in a short time. This is the case, for example, with message delivery systems. Such systems can maximize the line utilization by filing messages at the switching points until the required circuit is free, waiting if necessary until off-peak periods. This technique is referred to as *message switching* (discussed in Chapter 23).

The next section of the book discusses switching techniques. Let us note here that there are four fundamentally different types of switching which lead to different tariff structures:

   i conventional telephone switching
  ii fast-connect circuit switching
 iii packet switching
 iv message switching

PROS AND CONS   Leased voice lines have certain advantages for data
OF LEASED LINES   transmission over switched telephone connections.
                       We mentioned some of them earlier. Let us summarize
these advantages.

1. If it is to be used for more than a given number of hours per day, the leased line is less expensive than the switched line. If it is used for only an hour or so per day, then it is more expensive. The break-even point depends on the actual charges, which in turn depend on the mileage of the circuit, but it is likely to be of the order of several hours per day. This factor is clearly an important consideration in designing a data transmission network.

2. Because the leased line is permanently connected there need be no delay associated with switching times. Leased lines are therefore better than telephone switching systems for applications requiring *fast* access to a distant computer. The same argument does not apply to fast-connect or packet-switched systems.

3. Private lines can be specially treated or "conditioned" to compensate for the distortion that is encountered on them. The common carriers charge extra for conditioning. In this way the number of data errors can be reduced, or, alternatively, a higher transmission rate can be made possible. The switched connection cannot be conditioned beforehand in the same way, because it is not known what path the circuit will take. Dialing at one time is likely to set up a quite different physical path from that obtained by dialing at another time, and there are a large number of possible paths. Modems now exist that condition dynamically and adjust to whatever connection they are used on. These devices enable higher speeds to be obtained over switched circuits, but they are expensive.

4. Switched voice lines usually carry signaling within the bandwidth that would be used for data (at frequencies such as those shown in Figs. 21.3 and 21.4). Data transmission machines must be designed so that the form in which the data are sent cannot interfere with the common carrier's signaling. With some machines, this operation also makes the capacity available for data transmission somewhat less than that over a private voice line.

5. The leased line may be less perturbed by noise and distortion than the switched line. The switching gear can cause impulse noise that results in errors in data. This is a third factor that contributes to a lower error rate for a given transmission speed on private lines.

There are certain advantages which switched lines can have over leased lines, as follows:

1. If the terminal or terminals at a location have only a low usage, used switched lines will give a low overall cost.

2. The ability to access multiple distant machines using a switched network gives great flexibility. Many different machines offering different services and with different data bases may be dialed by a terminal user.

3. If a leased line fails, its user may be cut off from the facilities it connects to. With a switched system the user or using machine can redial and may obtain an alternate path to the facilities.

4. If a computer is overloaded or under repair, its user might be able to dial an alternative computer.

5. Simplicity. Leased-line systems often become complex because of the techniques used, such as polling, concentrators, and multipoint lines, which allow separate users to share the line.

Sometimes the dialing is done over a firm's own leased tie lines.

LINE CONDITIONING   As has been mentioned, private leased voice lines can be *conditioned* so that they have better properties for data transmission. Tariffs specify maximum levels for certain types of distortion. *An additional charge is made by most carriers for lines that are conditioned.*

AT&T has two types of conditioning referred to as **C** and **D** conditioning. There are five categories of **C** conditioning, C1 through C5, and two categories of **D** conditioning. A line ideal for data transmission would have an equal drop in signal voltage for all frequencies transmitted. Also, all frequencies would have the same propagation time. This is not so in practice. Different frequencies suffer different attenuation and different signal delay. **C** conditioning attempts to equalize the attenuation and delay at different frequencies. Standards are laid down in the tariffs for the measure of equalization that must be achieved. The signal attenuation and delay at different frequencies must lie within certain limits for each type of conditioning. The higher the conditioning number, the narrower are the limits. The result of the conditioning is that a higher data speed can be obtained over that line, given suitable line termination equipment (modem).

Types C1 and C2 conditioning are applicable to point-to-point and multipoint lines. Type C4 is available only on two-point, three-point, and four-point lines. Type C5 conditioning can be applied only to point-to-point lines.

D conditioning controls the signal-to-noise ratio and harmonic distortion. It is intended for voice-grade lines operating at 9600 bps. As **C** and **D** conditioning control different types of distortion, both may be used on the same circuit.

Conditioned is explained more fully in Chapter 27 and Box 27.1.

TARIFFS INCLUDING DATA SETS　　　Some common carrier tariffs are designed especially for data transmission, and the cost of the link must include the common carrier modems (data sets) needed for efficient transmission. The tariff will then state the speed or speeds in bits per second at which the link will transmit.

An example of this is the Bell System Dataphone service. A data set (Fig. 8.8) is provided which connects to the data-processing machine. A call may be dialed by an operator with the data set switched to "Voice." When she has established the call she hears a characteristic whistling tone from the data set at the dialed machine, and she then switches the data set to "Data." The international Datel services are designed for this form of operation. CCITT standards define the modem interfaces as discussed in the previous chapter.

TARIFFS FOR WIDEBAND LINES AND BUNDLES　　　The North American common carriers offer several tariffs for leased wideband lines. Some of these can be subdivided by the carrier into "bundles" of lower bandwidth. Some can be subdivided into channels for voice transmission, telephotograph, teletypewriter, control, signaling, facsimile, or data. With some tariffs the user pays a lower price for the bundles than for the individual channels.

The word TELPAK was formerly used for the "bulk" communication services offered by the telephone companies and Western Union. The word has now been eliminated from the tariffs but is still found in much literature. What used to be TELPAK C is now called a type 5700 line, and what used to be TELPAK D is now type 5800. Both can provide a wideband channel or a bundle of lesser channels.

The TELPAK customer pays a monthly change based upon the capacity of the communications channel he selects, the number of airline miles between locations, and the type and quantity of channel terminals. He has use of this channel on a full-time basis.

Originally there were four sizes of TELPAK channels: TELPAK A, B, C, and D. However, in 1964 the Federal Communications Commission ruled that rates for TELPAK A (12 voice circuits) and TELPAK B (24 voice circuits) were discriminatory in that a large user could obtain a group of channels at lower cost per channel than a small user, who could not take advantage of the bulk rates. In 1967 the TELPAK A and B offerings were eliminated.

The Type 5700 line has a base capacity of 60 voice channels (full-duplex).

The Type 5800 line has a base capacity of 240 voice channels (full-duplex).

Each voice channel in these lines can itself be subdivided into one of the following:

1. Twelve teletype channels, half- or full-duplex (75 bps).

2. Six class D channels, half- or full-duplex (180 bps).

3. Four AT&T type 1006 channels, half- or full-duplex (150 bps).

There cannot be mixtures of these channel types in a voice channel. The type 5700 line can transmit data at speeds up to 230,400 bps; the type 5800 line has a potential transmission rate much higher. Line termination equipment is provided with these links, and each link has a separate voice channel for coordination purposes.

The TELPAK channels thus serve two purposes. First, they provide a wideband channel over which data can be sent at a much higher rate than over a voice channel. Second, they provide a means of offering groups of voice or subvoice lines at reduced rates—a kind of discount for bulk buying.

Suppose that a company requires a 50,000 bps link between two cities, together with 23 voice channels and 14 teletypewriter channels, or perhaps 30 voice channels and no teletypewriter links. Then it would be likely to use the type 5700 tariff. In leasing these facilities, it would have some unused capacity. If it wishes, it can make use of this capacity at no extra charge for mileage, although there would be a terminal charge.

Government agencies and certain firms in the same business whose rates and charges are regulated by the government (e.g., airlines and railroads) may *share* bundled services. Airlines, for example, pool their needs for voice and teletypewriter channels. An intercompany organization purchases the bundled services and then apportions the channels to individual airlines. Most of the lines channeling passenger reservations to a distant office where bookings can be made are type 5700 or 5800 lines and so also are the lines carrying data between terminals in those offices and a distant reservations computer. There has been some demand to extend these shared facilities to other types of organizations that could benefit from them by sharing, but this is not permissible as yet.

TELPAK originally was proposed as an interstate service, but since then it has become generally available intrastate as well.

Although not a TELPAK offering, series 8000 is another "bulk" communications service in the United States that offers wideband transmission of high-speed data, or facsimile, at rates up to 50,000 bps; the customer has the alternative of using the channel for voice communication up to a maximum of 12 circuits. A type 8801 link, part of this series, provides a data link at speeds up to 50,000 bps with appropriate terminating data sets and a voice channel for coordination. A type 8803 link provides a data link with a fixed speed of 19,200 bps and leaves a remaining capacity that can be used either for a second simultaneous 19,200-bps

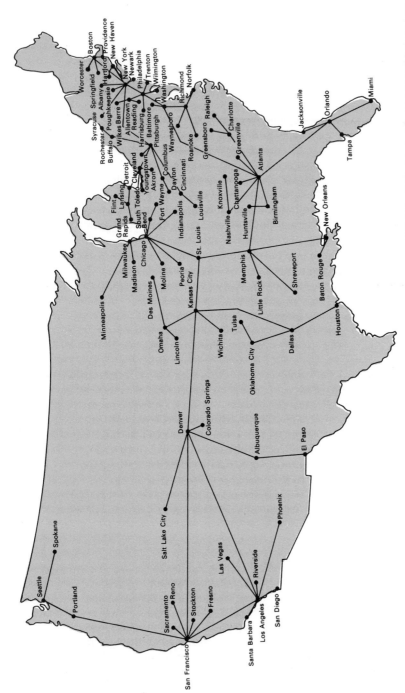

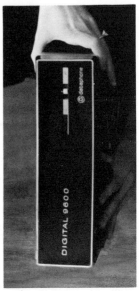

Figure 18.1 AT&T's planned DDS network giving point-to-point lines between the cities shown which transmit data at speeds of 2400, 4800, 9600, or 56,000 bits per second. These links are digital, not analog. Hence, they do not need a modem, but employ the service unit (right) which permits computers or terminals to be connected to the DDS line. (Pictures courtesy AT&T.)

channel or for up to five voice channels. These links must connect only two cities. The separate channels cannot terminate at intermediate locations.

Most countries outside North America also offer tariffs similar to the series 8000, and in most locations quotations for higher speeds can be obtained on request. Obtaining a wideband link in many such countries can be a slow process. This is particularly so if the termination is required in a small town or rural area rather than in a city to which such links already exist. Undoubtedly, as the demand for such facilities increases, the service of the common carriers in providing them will improve.

DDS   Particularly important for computer systems in the United States is ATT's Dataphone Digital Service (DDS). DDS provides point-to-point digital links operating at speeds of 2.4, 4.8, 9.6, or 56 thousand bps. Unlike the other AT&T tariffs, the DDS lines are digital and so cannot transmit analog signals such as voice. Because they are digital lines, they do not need a modem but instead employ a *service unit* which connects the subscriber's computer or terminal to the DDS line. The error rate on DDS lines should be better than that on voice lines because they are designed for digital transmission. They may also be more reliable because the service units are designed to permit end-to-end testing from a central AT&T location.

DDS service is offered only in certain cities. The planned DDS network is shown in Fig. 18.1

As we have discussed, the Bell System is in process of swinging from analog to digital transmission for many telephone trunks. The DDS service is a by-product of this process of voice digitization. The T1 and T2 carrier, and Data Under Voice (DUV) circuits, discussed earlier, are utilized to create the DDS network shown in Fig. 18.1. These digital trunks could provide data circuits at speeds up to 1.344 million bits per second, but DDS at present has an upper speed of 56,000 bps so that it can be provided without modification of existing local loops.

SPC   Southern Pacific Communications, a specialized common carrier, also offers digital circuits (which need no modems). Unlike AT&T's DDS service, the SPC circuits can be switched. The user can have a leased channel permanently connected, or a switched channel which he pays for on a usage basis.

SPC uses a microwave route shown in Fig. 2.4 on which the carriers are digitally modulated by a very high data rate. This bit stream is submultiplexed into streams of 1.344 million bits per second, compatible with the T1 carrier. Users can obtain channels derived from this at speeds up to 1.344 Mbps, thus:

asynchronous channels up to 1.2 Kbps
synchronous channels of: 2.4 Kbps
                              4.8 Kbps
                              9.6 Kbps
                             19.2 Kbps
                              56   Kbps
                             112   Kbps
                             224   Kbps
                     and 1344   Kbps

The channels of 56 Kbps and above are not initially switchable.

The use of digital channels such as DDS or SPC can have several advantages over using telephone channels:

1. Low cost
2. No modems
3. Higher speeds
4. Lower error rate
5. Higher availability because of end-to-end monitoring

SPC provides lower cost transmission than DDS for many traffic mixes. Because it is a switched network with a minimum billable connect time of 1 second, it can give much lower charges for very short connections. For example, 5 seconds of transmission on a 500-mile dialed SPC circuit costs 1 cent. 5 seconds of transmission on a dialed telephone circuit of the same distance costs 49 cents — the same as one minute. Fig. 18.2 illustrates those cost tradeoffs.

Unfortunately these new digital channels are available only in certain locations. Often a potential user cannot obtain a local channel of such a network going into his premises. His alternative is to use a telephone connection to the nearest access point of the digital network. When this is done, however, the above advantages are lost. If he has a telephone connection at both ends of the otherwise digital link he needs a total of four modems whereas he would have only needed two if he had used a telephone line all the way. And he is restricted to telephone speeds.

It can be expected that the digital networks will grow and become more widespread so that increasing numbers of users will be able to connect to them directly.

VALUE-ADDED
COMMON CARRIERS

Value-added common carriers charge by the message, or by the packet. As with fast-connect circuit-switched networks, the resulting price will be much lower than with dialed or leased telephone lines if the quantity of data sent is low. In a sense the value-added carriers are enabling the small user to obtain some of the economies of scale that a large user obtains with leased lines.

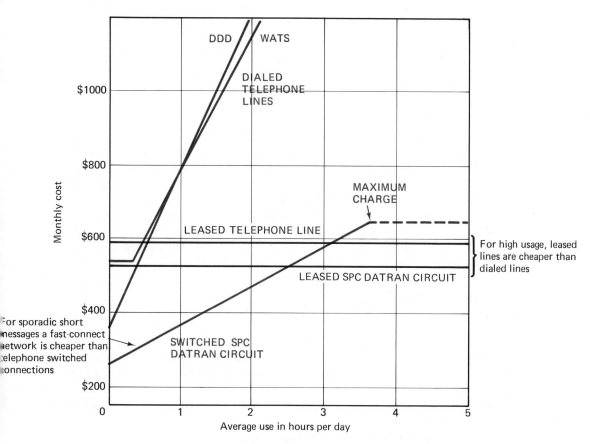

Figure 18.2   An example of data transmission charges: 4800 bps on a 500-mile connection. The prices will change but the principles should be noted. A fast-connect network is cheaper for multiple short connections. A leased line is cheaper when used many hours per day.

PRIVATE SYSTEMS   Some private lines are wholly owned by their users rather than leased. Users are generally prohibited from installing their own cables across public highways, and most privately installed communication links are wholly within a user's premises—for example, within a factory, office building, or laboratory. Railroads have their own communication links along their tracks. Some companies have private point-to-point microwave transmission links or other radio links. Recently infrared and optical links have been used for the transmission of data; line-of-sight links can be established at low cost, capable of carrying up to several million bits per second. Such links require no license, as do private microwave links. They could be used in cities for transmission between rooftops. Their main drawback is that the link can be put out of operation for a brief period by rain downpours of abnormal intensity—and for longer periods by thick fog. Relatively inexpensive millimeterwave radio equipment is also coming into use for short line-of-sight links. Small dishes often mounted on rooftops transmit at frequencies around 35 GHz and have the potential of carrying a very high bandwidth over distances of a few miles.

SWITCHED PRIVATE SYSTEMS

Many firms have private lines that are switched with private exchanges. In this way they build their own switched system. The lines that are used can be either privately-owned or leased from common carriers. The reason for designing a switched private system is either to lower the total telephone bill or else to have a switched system of higher bandwidth than the telephone system.

Telephone companies facilitate the building of private switched networks within corporations (corporate tie-line networks) by providing *common control switching arrangements* (CCSA). Such an arrangement uses switching equipment at telephone company exchanges to switch calls in a private leased network. All stations connected to the private network may call one another without using the public toll facilities.

MEASURED-USE SERVICE

When a leased line forms part of a system the transmission costs over it are independent of usage. When the toll telephone network is employed the caller is charged according to usage. Several data tariffs exist in which the user is charged on a measured-use basis. The following are measured-use services in the U.S.:

1. *Toll telephone.* The public telephone service.
2. *Dataphone.* An AT&T service using modems (data sets) on conventional public telephone lines.
3. *Dataphone 50.* An AT&T service to a few cities giving 48kHz dial-up channels capable of transmitting up to 50,000 bps.
4. *WATS (Wide Area Telephone Service).* An AT&T reduced-rate, bulk-billing, long-distance, telephone service in which the U.S. is divided into billing zones. (Tariffed for either measured-time or full-period charges).
5. *Hot Line.* A Western Union service providing fixed point-to-point telephone-grade connections on a measured-time basis.
6. *Broadband Exchange Service.* A Western Union service in which the user can dial the bandwidth he requires. So far only 2 kHz and 4 kHz are dialable and only between certain cities.
7. *Telex.* The international 50 bps telegraph service widely used over much of the world.

8. *TWX (Teletypewriter Exchange Service)*. A Western Union 150 bps teletypewriter service with dial-up connections like telex.

9. *TCS (Telex Computer Services)*. A message-switching service (Chapter 23) for Western Union telex subscribers, offering telex-to-TWX conversion.

10. *SPC*. Dial-up digital channels of various speeds on the SPC Datran network (leased-line or measured-time tariffs).

11. *TELENET*. A packet-switching network (Chapter 24) designed to provide interconnection among computers and terminals in a fraction of a second, and charge by the packet transmitted.

TELEX   Telex is a worldwide switched, public teleprinter system. It operates at 66 words per minute (50 bps) and uses the CCITT Alphabet No. 2 (see Fig. 3.6). It is operated in the United States by Western Union. Any teleprinter on the system can dial any other teleprinter in that country, and telex machines can be connected internationally without speed or code conversion. The United States can dial Canada and Mexico directly, but operator intervention is needed when dialing to other countries. Some countries permit the telex facilities to be used for other forms of dial-up data transmission. Each telex call is billed on a time and distance basis.

Each subscriber has an individual line and his own number, as with the conventional telephone service. His teleprinter is fitted with a dial, like a telephone, with which he can dial other subscribers. The teleprinter used may or may not have paper-tape equipment also. The teleprinter can be unattended. When a message is sent to an unattended teleprinter, it will switch itself on, print the message, and then switch itself off. Fig. 18.3 shows two typical telex installations.

TELETYPEWRITER EXCHANGE SERVICE   Western Union also operates a service purchased from AT&T that is competitive with telex. Again, each subscriber has a dial-up teletypewriter with his own number listed in a nationwide directory. This service is called the Teletypewriter Exchange Service (TWX), and it uses the telephone circuits combined with several TWX channels so that they can be sent over one voice channel. The combining or "multiplexing" is done at the local switching office, where the dc signals are changed to equivalent bursts of appropriate frequencies. The link between the local switching office and the subscriber is often a conventional telephone line, and in this case the teletypewriter needs a data set to convert the dc signals to appropriate frequencies in the voice range.

Other manufacturers' data transmission equipment can be connected to TWX lines and can transmit at speeds up to 150 bps, half- or full-duplex. This

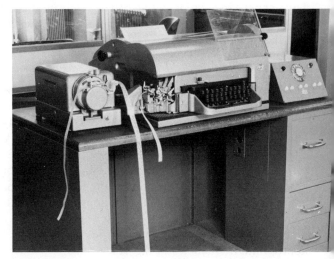

Figure 18.3 Typical telex installations. Many countries have this dial-up telegraph service operating at 50 bits per second (66 words per minute). Such telex machines have been incorporated in computer systems. (Courtesy General Post Office, London.)

process requires a special terminal arrangement at additional cost. Three types of access lines to the TWX network exist:

1. *TTY-TWX.* This is an access line with a teletypewriter provided by the common carrier. The speeds of transmission are either 6 characters per second in Murray code, a version of CCITT Alphabet No. 2, or 10 characters per second in Data Interchange Code (DIC).

2. *CPT-TWX.* CPT stands for "customer provided terminal"; to this access line the customer can attach any device operating with one of the preceding two speeds and codes and adhering to normal TWX line control. The device could be a computer with an appropriate adapter on its input/output channel.

3. *CE-TWX (formerly called "TWX Prime").* CE stands for "customer equipment." This can now be any device and is not restricted to a specific code or character speed. Two TWX subsystems are accessible, one operating at speeds up to 45 bps and the other up to 150. A CE-TWX terminal can communicate only with another CE-TWX terminal.

TWX directories listing TTY-TWX and CPT-TWX subscribers are published.

SUMMARY   Box 18.1 summarizes the main categories of communication line types and tariffs.

BOX 18.1 Categories of communication line and tariff

Types of Link	Comments
Digital	Designed for digital transmission. No modem required. Are code-sensitive in some cases.
Analog link	Transmits a continuous range of frequencies like a voice line. Modem required.
Switched public	Cheaper if usage is low. Switched telephone lines are universally available. It is necessary to avoid signaling frequencies on public telephone lines.
Leased (sometimes called "private")	Cheaper than public lines if usage is high. May have lower error rate. Higher speeds possible on leased telephone lines than switched ones because (1) conditioning is possible and (2) no signaling to avoid.

BOX 18.1 Continued

Leased with private switching | May give the lowest cost. Combines the advantages of leased lines with the flexibility of switching. Public switched wideband lines may not be available; hence private switched wideband networks are built.

Private (non-common carrier) | Often permitted only within a subscriber's premises. See next item.

Private (Non-Common-Carrier) Links:

In-plant | Very high bit rates achievable using coaxial cables or PCM on wire pairs.

Microwave radio | Permissible in special cases for point-to-point links.

Millimeterwave radio | Small antennas transmitting high bandwidth signals over a line-of-sight path up to a few miles.

Shortwave or VHF radio | Used for transmission to and from moving vehicles or people.

Optical or infrared | Used for short links—e.g., between city rooftops—at high bit rates (250,000 bps, typical). No license required. Put out of action by thick fog or *very* intense rain.

Speeds

Subvoice grade | Usually refers to speeds below 600 bps.

Voice grade | Usually refers to analog voice lines using modems of speeds from 600 to 10,500 bps.

Wideband | Speeds above those of voice lines, most commonly 19,200, 40,800, 50,000, 56,000, and 240,000.

(For a detailed list, see Table 18.1)

Mode of Operation

Simplex | Transmission in one direction only. Not normally used in data transmission except for telemetry or space applications.

Half-duplex | One direction or the other; not both at once.

Full-duplex | Transmission in both directions simultaneously. On the Bell System, costs 10% more than half-duplex and can give disproportionately higher throughput if the terminal is designed to take advantage of it.

Note: These terms sometimes describe the limitation of a machine rather than the limitation of the line it is attached to.

BOX 18.1 Continued

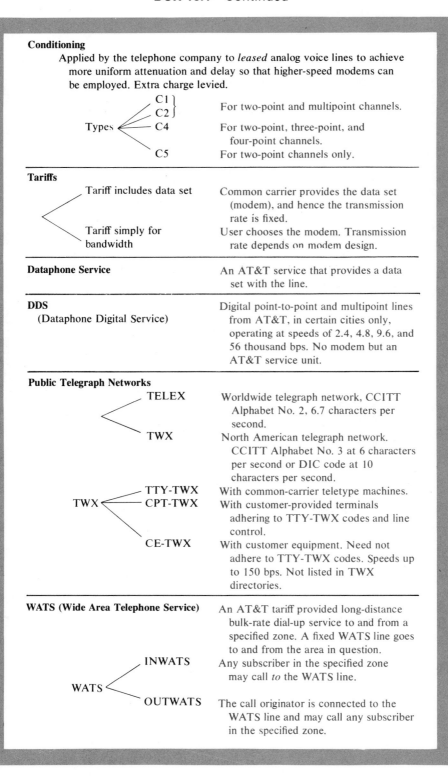

Conditioning

Applied by the telephone company to *leased* analog voice lines to achieve more uniform attenuation and delay so that higher-speed modems can be employed. Extra charge levied.

Types
- C1 ⎫
- C2 ⎬ For two-point and multipoint channels.
- C4   For two-point, three-point, and four-point channels.
- C5   For two-point channels only.

Tariffs

Tariff includes data set — Common carrier provides the data set (modem), and hence the transmission rate is fixed.

Tariff simply for bandwidth — User chooses the modem. Transmission rate depends on modem design.

Dataphone Service — An AT&T service that provides a data set with the line.

DDS
(Dataphone Digital Service) — Digital point-to-point and multipoint lines from AT&T, in certain cities only, operating at speeds of 2.4, 4.8, 9.6, and 56 thousand bps. No modem but an AT&T service unit.

Public Telegraph Networks

TELEX — Worldwide telegraph network, CCITT Alphabet No. 2, 6.7 characters per second.

TWX — North American telegraph network. CCITT Alphabet No. 3 at 6 characters per second or DIC code at 10 characters per second.

TWX
- TTY-TWX — With common-carrier teletype machines.
- CPT-TWX — With customer-provided terminals adhering to TTY-TWX codes and line control.
- CE-TWX — With customer equipment. Need not adhere to TTY-TWX codes. Speeds up to 150 bps. Not listed in TWX directories.

WATS (Wide Area Telephone Service) — An AT&T tariff provided long-distance bulk-rate dial-up service to and from a specified zone. A fixed WATS line goes to and from the area in question.

WATS
- INWATS — Any subscriber in the specified zone may call *to* the WATS line.
- OUTWATS — The call originator is connected to the WATS line and may call any subscriber in the specified zone.

BOX 18.1 Continued

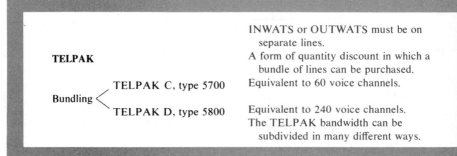

Table 18.2 Types of measured-use service. (Charge by usage rather than charging a fixed price per line.)

	Available	Channel Characteristics	Circuit-switched	Message-switched	Packet-switched
Toll telephone	All cities	Voice channel	✔		
Dataphone	All cities	Voice channel with data set	✔		
Dataphone 50	A few cities	48 KHz; Up to 50 bps.	✔		
WATS	All cities	Voice channel	✔		
Hot Line	Most cities	Voice channel between fixed points	Not switched		
Broadband Exchange	A few cities	2 KHz or 4 KHz	✔		
Telex	All cities	50 bps	✔		
TWX	All cities	Up to 150 bps	✔		
TCS	All cities	50 bps		✔	
SPC Datadial	A few cities	Synchronous data from 2.4 to 1344 Kbps.	✔		
		Asynchronous data up to 1.2 Kbps	✔		
TELENET	A few cities	Fast response time data channel connectable to machines of most speeds			✔

PART **III** **SWITCHING**

19 TELEPHONE SWITCHING

There are almost 300 million telephones in the world. Approximately half of them are in North America, but the rest of the world is fast rising in affluence and loquacity. Almost all the world's telephones must be capable of being interconnected on to another. Clearly, they cannot be linked directly because this would need $n(n-1)/2$ links, where n is the number of points to be interconnected — in other words about 4.5×10^{16} interconnections. Consequently a vast network of switching centers has grown up. Around 40–45% of the total costs of the North American telecommunications network are in switching facilities.

Switching centers are of two types: those which form part of the public network, and those which are private, installed for the use of one organization or building. The latter are called *private branch exchanges* and are discussed in Chapter 22. In this chapter we will discuss switching centers in the public networks.

Any switching office can be either *manual, automatic,* or *semiautomatic.* "Semiautomatic" means that a human operator receives the call and then dials it on automatic equipment. The relative cost of automatic switching is falling, and manual public offices are rare in advanced countries.

The theory and practice of switching has become established largely in an era prior to that of the interactive computer system. Much of it has evolved to meet the needs of telephone traffic. We will comment in later chapters that the needs of the burgeoning computer industry are different from those of telephone users, and new forms of switching are needed for machines talking to machines. A whole new body of theory and practice is evolving related to data transmission switching as opposed to the switching of *continuous* signals such as telephone and television. In this chapter we will discuss telephone network switching.

NETWORK ROUTING
STRATEGY

In a telecommunications network there is a trade-off between transmission and switching. If fewer switching centers are used, more transmission links will be needed. In a large network, such as the public telephone system, the trade-off is exceedingly complex and is constantly changing as the usage grows, subscribers alter their behavior, and equipment costs change. It is therefore necessary to have a network *architecture* which can accommodate continuous change in both equipment and topology.

Selecting the network structure and layout is a trade-off between minimizing the overall cost of the network and giving a suitably low probability of blocking (i.e., failing to establish a connection because the trunks are busy). The network must be designed with alternative routing so that calls can be routed to their destination when some parts of the trunk group have failed. The network must be adaptable so that it can accommodate sudden shifts in traffic patterns, such as telephone calls from resort areas in summertime, weekend traffic, Christmas traffic, and so on.

The facilities for alternative routing give high versatility but introduce switching complexities. Where alternative routes are sought over a long distance on many links, a number of difficulties could arise. It might be possible for a call to be routed in circles or to be routed over such a complex path that quality suffers severely. A logical scheme is needed that avoids unnecessary complications. The scheme employs a *hierarchy* of switching offices.

HIERARCHY OF
SWITCHING OFFICES

Figures 19.1 and 19.2 illustrate the trade-off between switching and transmission.

There are two main types of public network offices—*local central offices* and *toll offices*—illustrated in Fig. 13.6. Subscribers are connected on local loops to their nearby telephone exchange, called a *local control office* or *end office*. In the United States there are about 20,000 end offices. To interconnect these directly would require $n(n-1)/2$ or about 2×10^8 trunks. Not only would this be prohibitively expensive, but also the average utilization of the trunks would be very low. To lessen the number of trunks, further levels of switching are used. The switching offices used to interconnect the end offices are called *toll offices*.

There are four levels of toll offices, referred to as class 1, 2, 3, and 4 offices. This gives five levels of switching in total, the end office being referred to in the hierarchy as a "class 5" office. There are 10 class 1 offices in the U.S. telephone network, and these are interconnected by $n(n-1)/2 = 45$ high-capacity trunks.

Figure 19.3 shows the hierarchy of offices used in the United States. They have the following names:

Class 5: end office (local central office)

Class 4: toll center

Class 3: primary center

Class 2: sectional center

Class 1: regional center

The symbols used to represent these switching offices in Figs. 19.2 and 19.3 are standard symbols employed by AT&T.

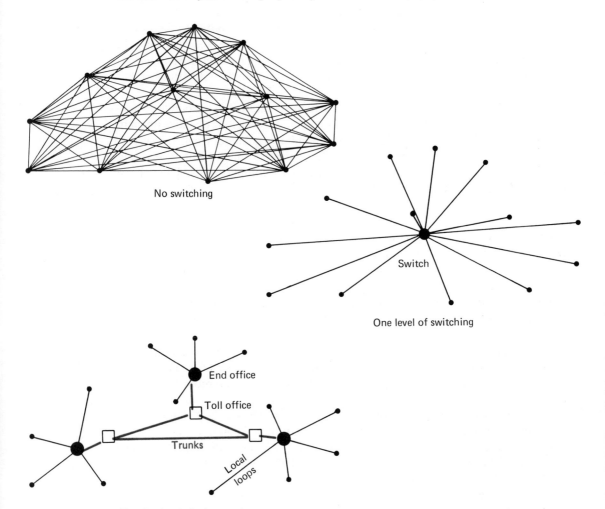

No switching

Switch

One level of switching

End office

Toll office

Trunks

Local loops

Two level switched network

Figure 19.1 There is a trade-off between switching and transmission.

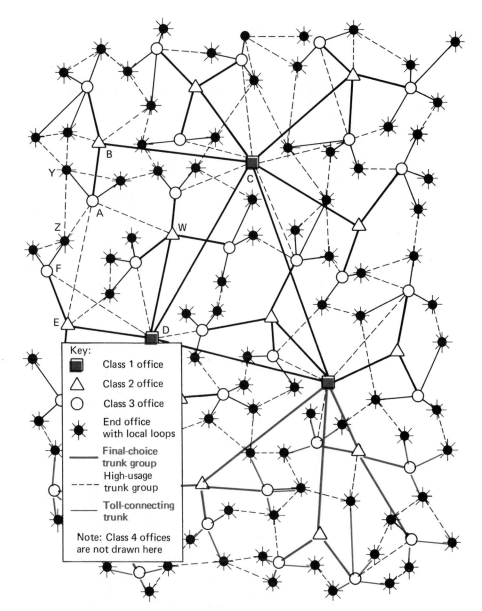

Figure 19.2 A network with four levels of switching. A call placed from subscriber Y to subscriber Z might be routed in one of the following ways (listing in the sequence in which they are attempted):

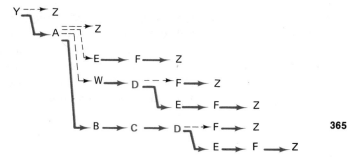

365

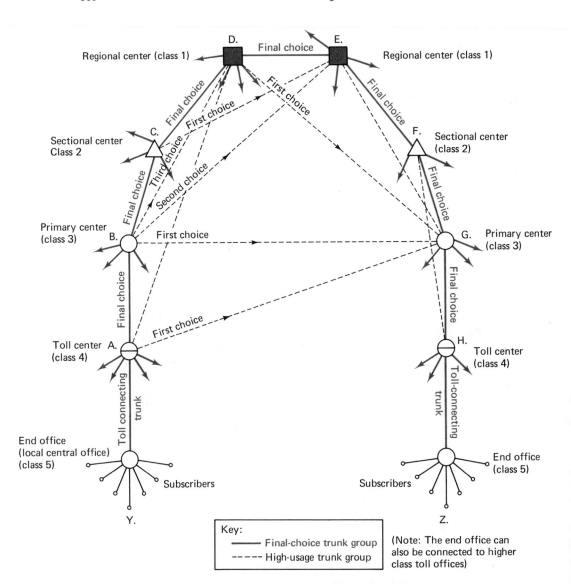

Figure 19.3 The five classes of switching office of the Bell System, showing the choice of paths when subscriber Y dials subscriber Z.

A call between distant telephones could be routed through the hierarchy as follows: telephone → end office → toll center → primary center → sectional center → regional center → regional center → sectional center → primary center → toll center → end office → telephone. Most calls, however, are not routed up and down the hierarchy following the solid lines of Fig. 19.2 or 19.3. In fact this route is used only as a last resort. It is referred to as a *final choice* path.

Suppose that subscriber Y, attached to the end office at the bottom left of Fig. 19.3, dials subscriber Z, at the bottom right. Y's end office switches the call to the nearby toll office, A, which in this case is a class 4 office. It could have been any other class. This office has a trunk to a distant class 3 office, G, in the locality of the subscriber being called, Z. This is the best choice of line for this call, and so office A tries to switch the call onto that line. That line, however, is not a very high capacity trunk. It may well turn out to be occupied. If so, A tries its second best choice, a line direct to the class 1 office, D. If this also is full, it will switch the call to its local primary office, B.

B also has a range of choices. The best route would be the line direct to G. The longest route that the call can travel is the nine-segment route A-B-C-D-E-F-G-H. It is unlucky if it does not find a shorter path than this. The longest path for any call is to travel to the nearest regional office (class 1) and from there to the regional office in the neighborhood of the recipient. This path, however, is one in which the lines are built with generous capacity.

The dotted lines in Figs. 19.2 and 19.3 are referred to as high-usage trunks. They are the first choices in routing the calls. Their layout and capacity are selected so that they are likely to be highly utilized. Their high utilization, while economically desirable, means that they will sometimes by busy when needed. If all the dotted-line paths are busy, the switching offices resort to the hierarchical final choice trunks (solid lines in Figs. 19.2 and 19.3), which are engineered to have a lower utilization and hence a low probability of being busy when needed.

The higher the trunk utilization, the more economical the network, but the higher the probability of blocking. This dilemma is overcome by dividing the paths into high-utilization first-choice trunks and lower-utilization final-choice trunks. The final-choice trunks are arranged in a tree-structured network with a class 1 offices at the top of the tree. This subdivision gives the best compromise between low cost and good grade of service. The tree structure to one class 1 office forms one numbering zone for the allocation of telephone numbers.

The network structure illustrated is made possible by automatic long-distance dialing, referred to as *direct-distance dialing*, DDD in North America, and *subscriber trunk dialing*, STD in British parlance. If the exchanges were not automatic, the caller would not tolerate going through eight toll offices and two end offices. The call would take too long to establish if all these routings had to be selected by operators pushing plugs into switchboards. The automatic exchanges, therefore, increase the probability of obtaining a line to the distant party. They increase the efficiency with which the high capacity trunks are utilized. Even so, occasionally it will not be possible to obtain a line. The lines from the local central office to the toll office may be busy, for example. In some areas the caller can detect this when dialing as he may hear a "busy" signal before he has dialed all his digits.

If Y talks to Z at different times, his conversation will probably travel over different paths. It may travel over different physical facilities, one day over microwave and the next over wire-pair cable. He may occasionally obtain a connection with much noise on the line for some reason. Occasionally, people say, "I've got a very bad line. Stay there and I'll redial you." On redialing the noise may have disappeared because now a different line has, by chance, been selected. The same is true with data transmission. If a user at a terminal dials a distant computer, he may possibly obtain a "bad" line. A red light on the terminal will indicate a succession of transmission errors, or, worse, if the error-detection procedures are not adequate, the terminal may print incorrect characters or the computer perform unexpected antics. The same is true for batch transmission. The check total may regularly disagree. Here again the user may dial another line. He may send for an engineer, who often cannot help except by redialing himself. Even leased lines

Figure 19.4 The three classes of switching office used for international traffic, employing a routing strategy similar to the hierarchy of Fig. 19.3.

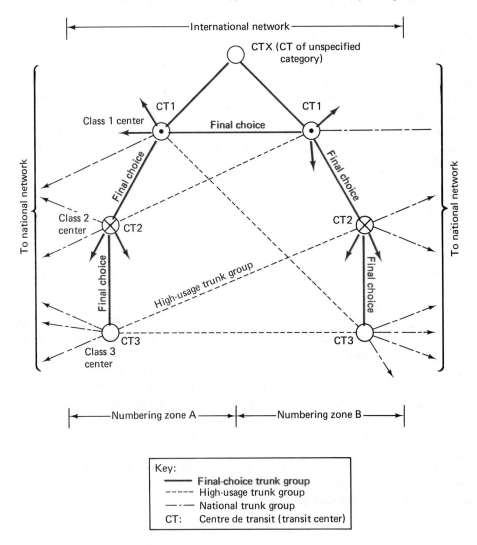

Key:
———— Final-choice trunk group
----- High-usage trunk group
—·— National trunk group
CT: Centre de transit (transit center)

may be routed over different paths by the common carrier and may give different performance on different days.

INTERNATIONAL SWITCHING
Automatic, or in some countries semiautomatic, switching is rapidly being extended to international calls. A hierarchical structure is desirable for interlinking individual national networks into a worldwide system. A three-level structure has been proposed for international routing, as shown in Fig. 19.4, with switching centers designated CT 3 (Centre de Transit), CT 2, and, the highest level, CT 1. The symbols used to designate these switching offices in Fig. 19.4 are standard symbols employed by the CCITT.

CITY SWITCHING OFFICES
In cities a separate level of switching is used employing a *tandem office*. The majority of city calls are for other subcribers in the same city but are often connected to a different end office. A large city has so many end offices that it is not economical to have trunk groups connecting all of them. If a city has 100 end offices, $n(n-1)/2 = 4950$ trunk groups would be needed to interlink them directly. Instead the end offices are interconnected via a *tandem* office, as shown in Fig. 19.5.

Calls to a different end office in the city are routed via the tandem office; calls to a different city are routed via a toll office, which is one of the hierarchy in Fig. 19.3.

CONCENTRATORS
If cities A and B each have 500,000 telephone subscribers who might call each other, there are not 500,000 voice channels between the cities. There may be only 600, and this is adequate because only very rarely would more than 600 of these subscribers want to talk to each other at once. It is clearly possible to apply the same principle on the subscriber side of the local central office.

Suppose that 40 two-wire telephone lines go from the exchange to a small community some distance away. The community has 40 telephone subscribers or possibly more than that if some have party lines. The telephone company's statistics indicate that it will be rare that more than 40 lines are in use at any one time. If a small remote switching unit could be used in the community so that only 10 lines went from there to the exchange, then the cost of 30 lines would have been saved.

This is occasionally done. The remote switching device is referred to as a concentrator.

The use of a concentrator can save money, especially if the lines in question are long. However, switching done remotely in this way is rather more expensive than the same switches in the local central office, and by lessening the flexibility with which alternative lines can be substituted, it increases the probability that the caller will not obtain a circuit when he wants one.

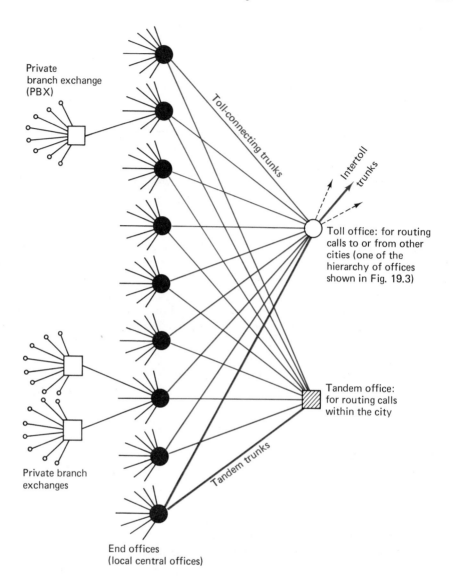

Figure 19.5   The types of telephone exchanges within a city.

NUMBERING PLAN   In North America, and some of the rest of the world,
                            the last four digits of a telephone number represent a
telephone connection to an end office. Each end office can thus handle up to
10,000 telephones, or more if they are grouped into modules of 10,000. The preceding three digits in North America are a code for the end office. As there are
more than 20,000 end offices, a further code is needed for long-distance dialing—a
three-digit area code.

The call may therefore be first directed to a high-level toll office by the area code, then directed to an end office by the following three digits, and finally to a subscriber's telephone by the last four digits. Often high-usage trunks bypass the high-level toll office, connecting the call directly to a lower level.

Each country has a *national numbering plan* for its own network, and most of them differ from that of North America. Occasionally several countries are included in one integrated numbering plan, as with the United States and Canada. In addition there exists an *international numbering plan,* administered by the CCITT.

To dial an international number, first a prefix is dialed which indicates that the call is going outside the country. Then a code is dialed for a destination country, and last the number in that country's national numbering plan is dialed. International switching is made more complicated by the fact that subscriber numbers and national numbers differ in length from one area to another; thus:

Subscriber Number		National Number		International Number		
12.34.56	in Brussels	2	.12.34.56	32	.2	.12.34.56
12.34.56	in Dusseldorf	211	.12.34.56	49	.211	.12.34.56
21.34.56	in Nice	93	.21.34.56	33	.93	.21.34.56
246.4200	in New York	212	.246.4200	1	.212	.246.4200
1234	in Perranporth	87257	.1234	44	.87257	.1234
248.4567	in London	1	.248.4564	44	.1	.248.4564

National Number: Trunk Code, Subscriber Code

International Number: Country Code, Trunk Code, Subscriber Code

To dial the above number in New York from other countries, a subscriber dials an international prefix followed by the number:

from Switzerland:	00	1	212	246	4200
from Belgium:	91	1	212	246	4200

International Prefix (indicates that the call is going outside the country)

Country Code (selects a country)

Area Code (selects a toll office region)

End Office Code (selects a local central office)

Local Loop Code (selects a subscriber)

Not all switching offices are equipped to recognize international prefixes, so not all subscribers can dial international calls. When international calls are dialed between operators a *language digit* is used in the number to indicate the language of the calling party. If an operator is called upon for assistance, the call can then be routed to an operator who speaks that language (when possible). The language digit follows the country code.

TYPES OF EXCHANGES   There have been several eras of switching equipment. Until after World War I all switching was done manually, the connections being established by operators plugging cords with jack plugs in appropriate line terminations. Some private branch exchanges are still manual today — exchanges in many hotels, for example. So are many public exchanges in developing countries and many offices for establishing international connections.

Manual exchanges gave way to automatic electromechanical operation. The subscriber dials the number he wants, and the connection is made by banks of automatic equipment consisting of relays and mechanical switches. Some automatic exchanges have grown to amazing size. In a large-city exchange one can be surrounded by bank upon bank of relays and switches, from floor to ceiling. The multiple clicks as calls are connected are so numerous that the sound approximates to the hiss of white noise, and one can wonder: Is this what the computer

Figure 19.6   The frame room in a large Bell System central office.

would have looked like if we had not invented electronics? Figure 19.6 shows a large electromechanical switching office.

Electromechanical exchanges appear crude, slow, and unreliable when compared to computer circuitry, and the era of electromechanical switching is giving way—reluctantly in places—to computerized switching. Electronic switching struggled into existence during the 1960s, the early exchanges giving many problems. With the knowledge gained from these and with improved hardware, computerized exchanges became an economic proposition by the start of the 1970s. Now there are many computerized exchanges both private and public, and their cost is dropping rapidly with the growth of miniaturized mass-produced logic circuitry. They are discussed in the following two chapters.

FUNCTIONS OF A SWITCHING OFFICE

The *basic* functions performed by a switching office are the same whether it is manual, electromechanical, or electronic. The basic stages that a call must go through are as follows:

1. When the subscriber picks up his telephone the office *must detect that service is needed.* In an automatic office, the dialing tone is switched to that line, and the mechanism waits for the subscriber to dial. In a manual office, a lamp lights adjacent to the jack for that line. The operator has a set of cords which have, in effect, two ends with a jack plug on each end. She picks a cord, plugs it into the line in question, flicks a switch so that she can talk on that line, and says "Number please."

2. The requested telephone number must now be used to *set up an interconnection path.* In the manual office, the operator simply listens to the number requested and plugs her cord into the appropriate output line. She may have the line to the telephone in question terminating on her plugboard. However, if it is a large exchange, she will not, and so she will have to route the call to another operator who has. A large office has many operators who communicate in this way.

   In the automatic office, the number is received as a train of pulses from a rotary dial or train of frequency-pairs from a push-button telephone. These signals cause the equipment to set up a path through the exchange to the appropriate outgoing line.

3. The manual switchboard has a number of interconnecting cords smaller than the total number of subscriber connections. If there are several operators, the number of lines between them is also limited. The operator, in effect, *concentrates* the lines into the set of interconnecting cords and lines in a manner similar to the remote concentrator discussed above.

   The large automatic exchange does the same. A central office with 10,000 lines would need approximately 50 million interconnections if there were to be a unique path between every input and every possible output. The switches therefore concentrate the calls into a limited number of paths through the exchange. Also, the office will be connected to a limited number of trunks, and again the incoming calls are concentrated onto these outgoing lines.

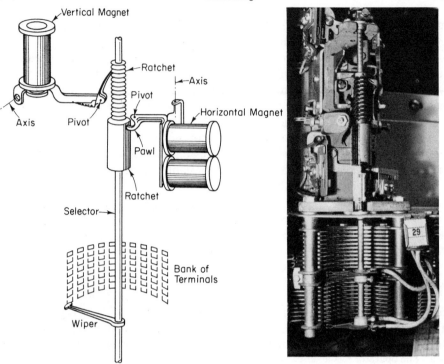

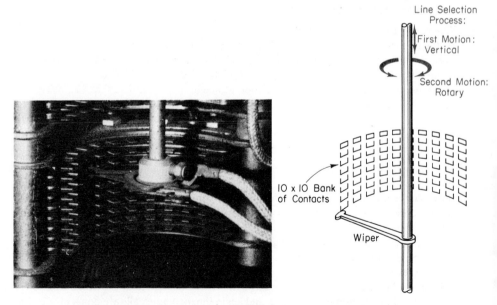

Figure 19.7 A Strowger switch. The first dialed digit reaching the switch causes vertical motion, selecting the appropriate wafer of contacts. The second digit reaching the switch causes horizontal rotary motion, selecting a contact.

Figure 19.7 (cont.)   A Strowger switch central office.

The calls, having been concentrated onto the available interconnections, find their way to the requisite part of the exchange and are *switched to the appropriate output lines*. In the manual exchange they reach the appropriate operator, who again selects a cord for the call. When the first operator tells her the number asked for, she plugs the call into the appropriate jack. It has now reached its output line.

4. The required outgoing line might be busy. It is necessary to *detect a busy condition and to notify the caller of it*. Similarly, as there are only a limited number of paths through the exchange, the exchange itself may not be capable of making the connection and again the caller must be notified. If the electromechanical system is unable to make a connection, it will switch a busy signal onto the caller's line. If the call is being connected manually, the operator will observe whether the output line jack is occupied, or if she is connecting with a trunk of line to another operator. she will make a busy test to see whether it is occupied. She informs the caller verbally.

5. It is necessary to *make the telephone of the called person ring*. The terminating automatic exchange sends a ringing signal down the line when the connection is made. In a manual office the act of plugging the cord into the jack of the outgoing line causes the signal to be sent, either to the called person's telephone or to an outgoing trunk or to the outgoing switchboard where the call will next be handled. In private branch exchanges the operator might have to operate a switch to send the ringing signal down the line.

6. The telephone of the called party has now rung. The *ringing signal must be removed from the line,* manually or automatically, when that person answers. If the telephone goes on ringing unanswered and the caller does not put his telephone down, the operator may speak to him and tell him that the party does not answer. The automatic exchange may, after a respectable wait, disconnect the call.

7. When the call is successfully established and completed, the parties put their telephones down and *the circuit must then be disconnected*. The exchange circuitry detects that the telephones are back on their rests. The automatic exchange disconnects the circuit, freeing the interconnection paths. The manual exchange has lamps lit by the jack plugs in question to inform the operator or operators that the call is over. They then pull the plugs out. If a caller fails to replace his handset on its rest, the automatic exchange may send a signal to his telephone in an attempt to make a loud enough noise to alert him.

8. Last, *the caller usually has to be charged*. In the manual exchange, the operator can write down the cost of the call. In an automatic exchange there must be some mechanical way of recording the number of calls each subscriber makes and the duration and distance of trunk calls. Some exchanges have counters for each subscriber which must be read periodically. Other exchanges use automatic punch cards or paper tape. Computerized exchanges produce magnetic tape for processing on another computer.

STEP-BY-STEP SWITCHING

The earliest automatic exchanges used a *step-by-step* switch which was invented and installed by an American mortician Strowger, who vas afraid that the town telephone operator was routing too much of his potent il business to the competition. Figure 19.7 shows a *Strowger switch*. Throughc it the world today there are still more Strowger switch exchanges being installed than any other type.

A Strowger switch responds to the train of on/off pulses that are produced by a rotary telephone dial. As the dial rotates back counterclockwise to its stop, it sends a train of dc pulses down the line, as illustrated in Fig. 19.8. Each digit sends off a different number of pulses. As each on/off pulse is received it moves the Strowger switch one position.

One Strowger switch can connect an incoming wire to one of 100 outgoing wires. The 100 outgoing contacts are arranged in 10 wafers of 10. The moving wiper contact which can complete a circuit with any of these has two possible dimensions of movement. It can slide up and down to select the required wafer, and having positioned itself at the right wafer, it can rotate to select the right contact. It is operated by an address of two digits. The first controls the vertical positioning, and the second the rotary movement.

The early Strowger switch exchanges (and many exchanges still being installed today) operate in a step-by-step fashion, illustrated in Fig. 19.9. When a call is dialed the first operation is to hunt for a line to a free Strowger switch. Suppose that the subscriber then dials the digits 45, as shown in Fig. 19.8. The four pulses of the first digit cause the contact to move vertically. As the telephone dial rotates back beneath the finger of the caller, the four pulses it sends activate an electromagnet which operates the vertical lift pawl of the switch. As each pulse arrives the pawl lifts a rack and moves the contact with its shaft up by one step. Dialing the digit 4 causes the contact to position itself at the fourth level.

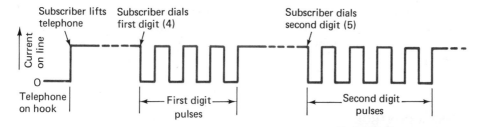

Figure 19.8 A rotary dial telephone generates a series of on/off pulses which cause the central office or PBX switching equipment to operate. (Touchtone dialing is illustrated in Fig. 21.1.)

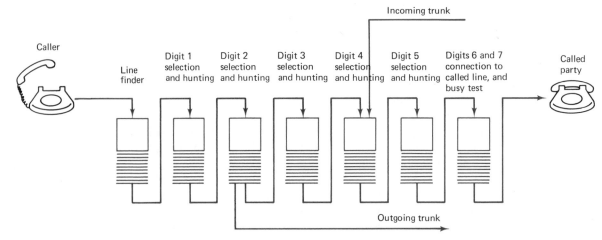

Figure 19.9 Step-by-step switching of a seven-digit number using Strowger switches (Fig. 19.7).

The Strowger switch then rotates horizontally, hunting for a connection to another free Strowger switch. This next switch accepts the next digit, 5, and again nudges its contact pulse by pulse to the fifth level and then searches the fifth level for a free connection.

The last Strowger switch in the series in Fig. 19.9 accepts the final two digits and is connected to 100 outgoing lines. When an outgoing line is finally found a test is made to see whether the line is busy, and if so, a busy signal is switched to the caller's line.

COMMON CONTROL   Step-by-step switching is rather like a cook executing a recipe one line at a time. When action is taken on the first digit the system has no knowledge of what the subsequent digits will be. In a kitchen where many different meals are prepared it would be disadvantageous if all the cooks operated from their recipes one line at a time. The cooks could operate more efficiently if they knew the whole recipe and planned its completion.

A major advance in the organization of switching offices was the introduction of *common control*. With common control the cooks read the whole recipe before they start cooking. The dialed signals are not regarded as the means of operating the switches, but rather they are regarded as information. They are stored in a memory, and the system then works from the memory to set up the connection. Logic circuitry is designed so that the digits stored in the memory are *translated* into instructions for efficiently setting up the call. The memory and decoders are common to the entire office.

Step-by-step switching has several disadvantages which disappear with *common control* switching. First, the whole system of switches is tied up during the entire length of the call. Once the switching logic is separated from the connection path, the switching logic is tied up only during the setup period. Second, in a step-by-step office the switch configurations are constrained by the telephone numbers that are assigned to that office. Third, available telephone numbers are constrained by the switch terminals that happen to be free. When subscribers move or add lines it is often necessary to change their telephone numbers. Fourth, the logic available for analyzing the telephone number is limited. Most step-by-step offices cannot detect in time the difference between a long distance number and a local number, so the subscriber has to dial "1" or some identifying code before the long-distance number.

Common control offices were first designed using Strowger switches and then, more efficiently, with cross-bar switches. Figure 19.10 shows a crossbar switch. Like other switches it connects the input wire to one of a number of possible output wires. It consists, in essence, of sets of horizontal and vertical bars which can be moved by electromagnets in such a way that they cause relaylike contacts to be closed at their coordinate intersection. By suitable arrangement of these, a path can be selected through the exchange. Crossbar switches are faster in operation than Strowger switches and less prone to create impulse noise.

Another form of exchange common in Europe and South America uses motor-driven rotary switches. Rotary switches with up to 500 three-wire outputs are used.

Figure 19.11 shows the organization of an office employing common control logic. It could be a crossbar office. The switching network is quite separate from the memory and the logic that is used. When a subscriber lifts his handset a *register* is allocated for that subscriber's use, and the subscriber is given a dial tone. The subscriber dials or keys a telephone number, and it is stored in the register. When the complete number has been received it is passed to *translater* logic, which converts it into a code which will control the switching. This code is given to a *marker* circuit the function of which is to select the speech path and control the switching network. It finds a connecting path through the switch network by examining all the stages in parallel and then closes the switches to connect this path. It checks that the connection is good before serving the next call. If the connection has not been made satisfactorily, it will try again.

When the subscriber dials a trunk call, the translater will determine what routing code is to be transmitted and pass this to a *sender* circuit. The sender stores the information and transmits it on the outgoing trunk when it has been selected.

If all calls are to be served by relatively few markers, the switching and logic

(a) An L. M. Ericsson cross-bar exchange in Europe.

(d) The select magnets.

Figure 19.10 The sequence of operation of a cross-bar switch is as follows: One of the two select magnets associated with the desired horizontal "select" bar is activated. The select bar moves to one of its two positions, operating mechanical interlocks of that level. The hold magnet moves the vertical "hold" bar. Relay-like contacts at the point of intersection of the horizontal and vertical bars are thus closed. Two select magnets not associated with the same select bar may be activated together. The vertical bar then connects the contacts on the two horizontal bars.

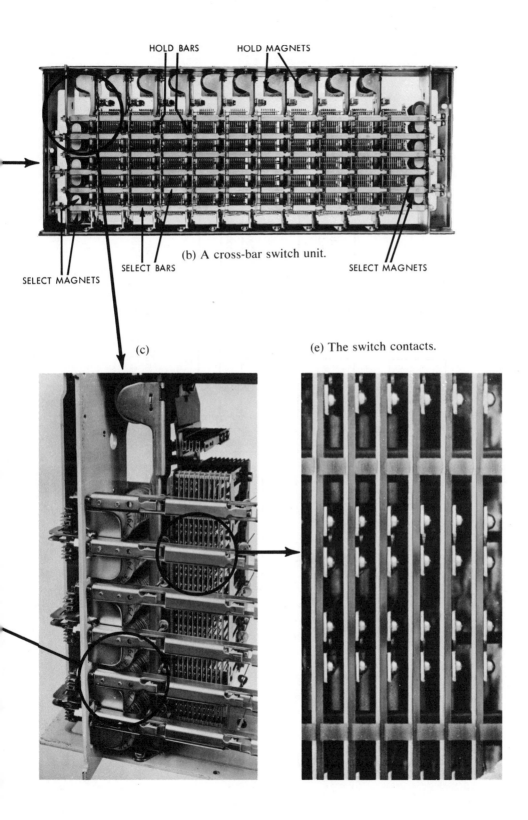

HOLD BARS   HOLD MAGNETS

(b) A cross-bar switch unit.

SELECT MAGNETS

SELECT BARS   SELECT MAGNETS

(c)

(e) The switch contacts.

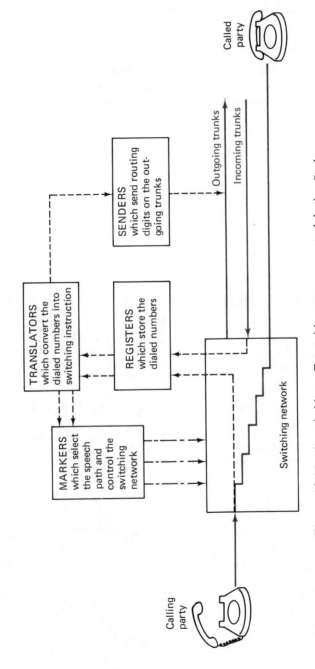

Figure 19.11 A switching office with common control logic. Such switching may employ cross-bar switches (Fig. 19.10). Strowger switch common control systems are also in use.

must operate fast. Crossbar switches are preferable to Strowger switches because of their greater speed. Better still would be electronic switching. The functions described are indeed better carried out by a computer, as discussed in the next chapter.

MULTIPLE STAGES   A simple form of switch would be a two-dimensional
OF SWITCHING   array of crosspoints like that at the top of Fig. 19.12. A
single crossbar switch is such an array—often a 10 × 10 matrix. If n lines are to be connected by a two-dimensional array, there must be n^2 crosspoints ($2n^2$ for two-wire circuits). A telephone exchange may switch several thousand lines, and this would require millions of crosspoints. Therefore, to lower the cost, several stages of switching are employed within one exchange, in much the same way that several levels of exchanges are employed within a country.

The second part of Fig. 19.12 shows a two-stage switch. Two-stage switches, however, give a substantial probability of blocking; i.e., a probability of there not being a free path. The probability of blocking can be made very low (but not zero) if a third stage is added, as in the third part of Fig. 19.12. Most telephone exchanges in fact use four stages, as in the last part of Fig. 19.12. The diagram shows 1000 lines entering a switch with a total of 40,000 crosspoints. The 1000 lines entering a switch with a total of 40,000 crosspoints. The 1000 lines would have needed 1 million crosspoints if a single-stage switch were used. Calculations with typical traffic intensities show that the probability of blocking in such a switch would be considerably less than 0.001. For more than 1000 lines the total number of crosspoints can be further reduced by employing nonsquare arrays, for example, 10 × 30 arrays. in which 30 lines are concentrated onto 10 paths.

To connect any one call there are many possible paths that could be set up through an exchange like that in Fig. 19.12. Algorithms exist for selecting the paths so as to minimize the average probability of blocking for all calls. In a computerized exchange, such algorithms can be run under stored program control.

Even with multiple stages of switching a phenomenal quantity of wiring is needed at back of electromechanical switching frames, as illustrated in Fig. 19.13.

A cross-point matrix

Input lines

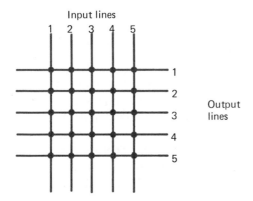

1

Output lines

A 5 x 5 matrix of crosspoints. To interconnect many lines with such a matrix would need an extremely large number of cross points; therefore switching uses several stages, as below.

2

A 2-stage switch

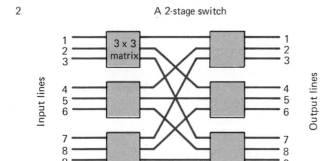

Input lines

Output lines

In a 2-stage switch, blocking can occur. For example if input 1 is connected to output 2, then input 2 and 3 cannot be connected to outputs 1 or 3.

3

A 3-stage switch

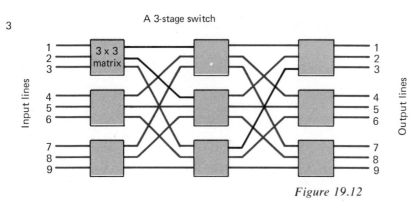

Input lines

Output lines

In a 3-stage switch blocking can still occur but is much less probable than in a 2-stage switch. If the black lines in this figure are connected, for example, input 3 cannot be connected to outputs 1, 2 or 3.

Figure 19.12

384

4

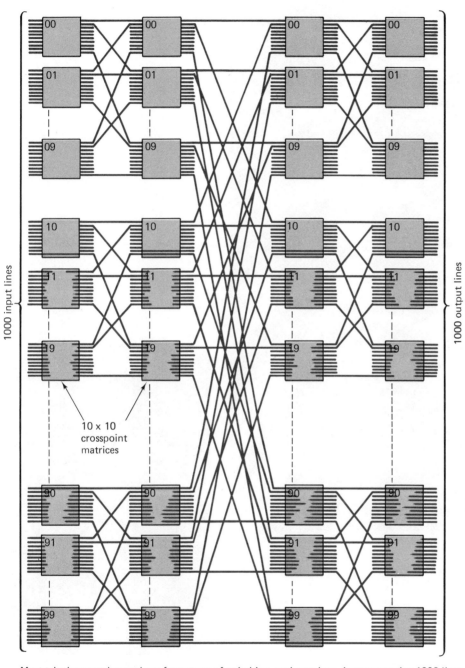

10 x 10
crosspoint
matrices

1000 input lines

1000 output lines

Most telephone exchanges have four stages of switching, as shown here, interconnecting 1000 lines

Figure 19.13 Wiring at the back of electromechanical switching frames.

20 COMPUTER-CONTROLLED CIRCUIT SWITCHING

When one observes and studies the immense complexity of a large electromechanical switching office it seems clear that a computer-program-controlled operation would bring a great improvement. For many years engineers worked on ideas for applying electronics and computer techniques to switching problems. In recent years this research has paid off, and several countries are replacing electromechanical central offices with computer-controlled switching systems. Telecommunication networks of the future will be gigantic organisms under computer control.

The first major use of electronic switching came with the Bell System's No. 1 ESS (Electronic Switching System). This system, shown in Fig. 20.1, is designed to replace, at an economic cost, central offices of a wide range of different sizes. It can handle from 10,000 to about 70,000 lines, with a maximum capacity of 100,000 calls in the busy hour [1]. For remoter areas with smaller numbers of subscribers, No. 2 ESS handles from 1000 to 10,000 lines [2]. These systems are designed for mass productions. AT&T is installing No. 1 ESS central offices at the rate of about one per day. By 1980 the Bell System will have more than 2000 No. 1 ESS offices and more than 600 No. 2 ESS offices. Although these figures represent a vast capital investment, most of the Bell System will retain its electromechanical switching for a long time—there are more than 20,000 central offices.

In the rest of the world a variety of stored-program offices are being installed and developed, including the L. M. Ericsson AKE systems from Sweden [3], the Siemens System IV from Germany, Northern Electric systems from Canada, ITT's Metaconta system marketed worldwide [4], Japan's DEX-2 system [5], the Pericles system in France, and the TXE 4 exchange in Britain. In addition to computerized central offices there are many computerized private branch exchanges, which some manufacturers see as a more lucrative market because PBX's are not a "natural monopoly."

Figure 20.1   Bell System ESS No. 1,
a computerized central office. Computers of extremely high reliability carry
out the line switching and associated
operations. (Courtesy AT&T.)

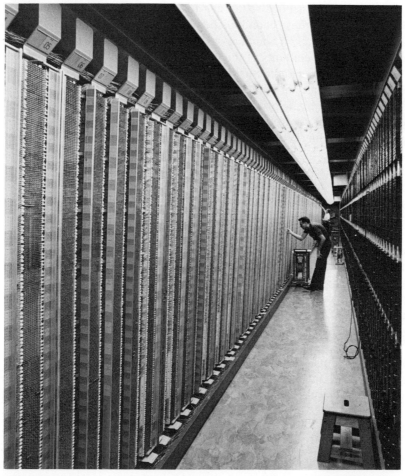

Electronic exchanges need not use stored-program computers. Several small exchanges use fixed-wired electronics. The registers, senders, translators, and markers in Fig. 19.11, for example, could be implemented electronically while the switchings network could retain a crossbar facility. The TXE 2 system [6], of which more than 100 have been installed by the British Post Office, is designed for an exchange with only 200 to 2000 lines; the TXE 3 system handles small exchanges with more than 2000 lines. There are several other noncomputerized electronic exchanges.

As the cost of computers continues to drop, computerized exchanges will become increasingly attractive. The spread of minicomputers will make the switching of small numbers of lines more economical.

ADVANTAGES The economic objectives of the No. 1 ESS are achieved merely in the replacement of existing central offices and an increased capacity in existing buildings. This advance in switching technology comes at an appropriate time, for many of the early central offices need replacement. No. 1 ESS needs a fraction of the floor space of its electromechanical equivalent; therefore, in some cases, the large expansion in switching capacity that is needed today can take place on existing premises.

In addition to these economic benefits, the flexibility of switching under program control has many advantages. Alterations in the exchange can be made by a change in memory contents of the computers and thus can be made very quickly in many cases. The manufacturing of the equipment is capable of more standardization because many of the variables that exist on electromechanical equipment are now variables in the program rather than in the hardware. An example of the power of program control was seen when a "centrex" service was added to the system for providing centralized PBX facilities (see Chapter 22). This new service was incorporated by an addition to the *programming*. Growth of the exchange is made easier, for every No. 1 ESS can handle up to 70,000 lines. New subscriber numbers can be added quickly. In addition, the accounting procedures and traffic-measuring procedures can be fully automated. Automatic features also aid considerably in maintenance.

The computer program can make possible a variety of new features not on today's electromechanical exchanges—features that are attractive to the subscribers. It would be possible to set up conference calls with several parties joining in, simply by dialing. One could dial an extension in one's own home if it were equipped with more than one telephone. Subscribers could use "abbreviated dialing," which would enable them to dial certain selected seven-digit numbers using a two-digit code. Also, various methods could be provided to allow the subscriber to have his calls automatically directed to another telephone. The subscriber could, for example, if he went to a friend's house for dinner, dial a specific code and then his friend's telephone number. Thereafter all calls could be directed to this number automatically. When he returned home, he would cancel the rerouting

BOX 20.1 An AT&T advertisement for ESS facilities

by dialing another code. Alternatively, he could inform the telephone company of the times and numbers for rerouting. Box 20.1 shows an AT&T advertisement proclaiming ESS capabilities.

Some systems, including the System AKE stored-program exchange manufactured by L. M. Ericsson in Sweden, have an automatic call transfer. Incoming calls may be transferred to another specified number, either when the subscriber is busy or when he does not answer. Again, if you dial a subscriber on this system and he is busy, you may then dial a code that instructs the computer to dial you as soon as he is free and then automatically redial him. The need for redialing is dispensed with. System AKE also has an alarm-clock service. You dial a four-digit time at which you wish to be awakened by the computer. Subscribers may dial directly to private automatic branch exchange extension telephones without operator assistance. System AKE can be set to gather a variety of statistical information about subscribers' calling habits. The system can list particulars of incoming or outgoing calls on particular numbers. It has a facility for recording malicious calls. If a subscriber receives a malicious call, he does not replace his handset for a time, and the system then records his number and the number of the calling party.

Other new facilities are planned. Such is the flexibility of operating under programmed control.

BASIC PRINCIPLES   The essential elements of a computerized switching office are shown in Fig. 20.2. The incoming lines all enter some form of switching network [1] which could be built from crossbar switches, reed relays, or solid-state switches with no moving parts. The controls for setting up different paths through the switching network are operated by the computer [2].

There is a mechanism for constantly scanning the activity on the lines [3] and informing the computer program of significant events. The program must be informed when you pick up your telephone handset; it must read the number you dial and must be informed when you replace the receiver after a call. Similarly, it must detect and interpret the signals from other switching offices that arrive on the trunks.

There are various signals that it must send down the lines, such as "busy" signals ("engaged" signals in British parlance), signals to make telephones ring, dialed-number signals, and on-hook/off-hook signals, which it sends to other switching offices. The signaling equipment [4] might be regarded as one of the computer output units in this type of system.

Information about calls must be gathered for billing purposes, and details about the uses of the network will be filed for statistical analysis. These will be recorded by the computer [7].

Figure 20.2   Basic elements of a computerized switching system.

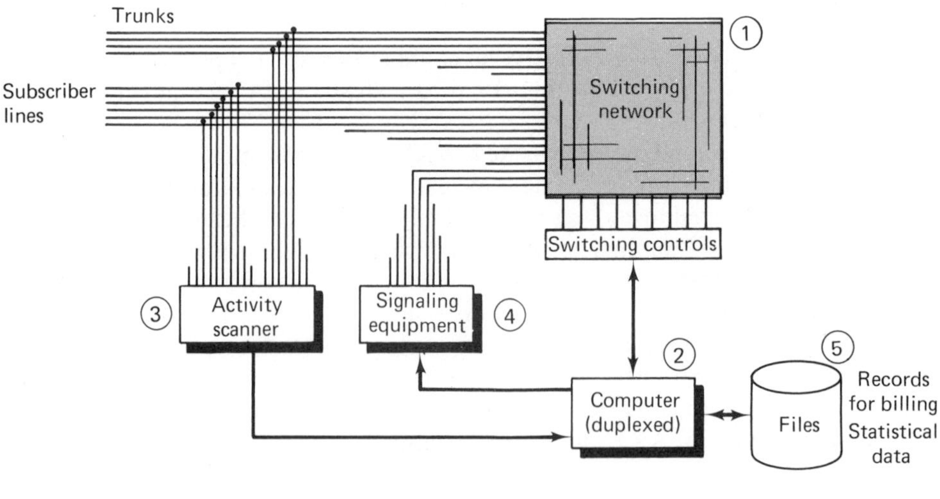

**ORGANIZATION OF
NO. 1 ESS**
Figure 20.3 shows how these principles are applied to the Bell System's No. 1 ESS.

The subscriber lines (1) and trunks (3) enter a switching network (2) in which the interconnection paths between lines or between lines and trunks can be set up. The switches used are reed switches, called *ferreeds*. Magnetic contacts encapsulated in glass, free from dust and corrosion, are opened and closed by windings outside the glass. They are arranged in pluggable units containing an 8×8 array of switches.

The switches for making the appropriate line connections are opened and closed under control of the program in the computers (7).

The switching network is organized into frames, and each frame has its own switching controller (5). The controller sets up the appropriate line paths in the frame it controls. Both the switching controllers and the computer are duplicated for reliability reasons, as is much of the remainder of the system. When one such component fails, its counterpart takes over.

The computers have two types of memory unit. Semipermanent memory units (9) contain the programs and fixed data about the lines. This is "read-only" during normal operations and so cannot be accidentally overwritten by a program error or hardware failure. Temporary memory units (8) contain information about the calls in process and other information that is changing during normal operation. Both types of memory unit and their contents are duplicated.

The computer knows the status of the lines, trunks, and signal receivers by means of a continuing scanning operation, carried out by the line scanners (4), which are also duplicated. After each scan, the computer writes information about the line status in the temporary store.

Each line and trunk is equipped with a sensing device called a ferrod (13), which can detect telephone on-hook and off-hook conditions. The ferrod is a current-sensing device consisting, in essence, of a ferrite rod with two identically wound solenoid coils through which the line current flows. Two other loops of wire are threaded through two holes in the ferrite rod; these are used for sensing whether or not line current is flowing. The scanner sends an interrogate pulse down one loop and detects whether there is a corresponsing readout pulse down the other loop. When no dc current is flowing in the line, there will be a strong readout pulse, but when current is flowing, the pulse will be suppressed. The scanners then scan these ferrods, and the information they obtain about line status is read into the temporary storage by the computer.

The scanners are provided in modules that scan 1024 lines. A scan takes place every 100 milliseconds, in which each subscriber line is examined to detect call originations. If an origination is detected, that line is connected to a device which interprets the dialing signal, either dial pulses or Touchtone. During this period, a line is sampled every 100 milliseconds. Once the call is connected, the line is again scanned every 100 milliseconds, the purpose now being to detect when the call is terminated. When the caller hangs up, the line is disconnected by the computer.

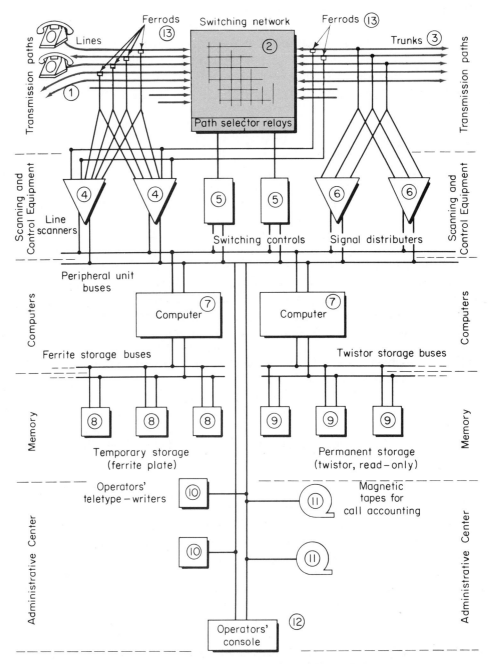

Figure 20.3 Basic configuration of No. 1 ESS.

As the scanning takes place, the results from the ferrod sensors are sent to the computer, 16 at a time, and compared with the results of the last scan, which are stored in temporary storage. If there is a change, then the computer branches to the appropriate program to handle it.

As discussed in the next chapter, it is necessary to send a variety of signals on trunk circuits. No. 1 ESS uses a signal distributor (6), again duplicated, for this purpose. Under computer program control, this distributor connects the appropriate signals to the trunks as required.

At the bottom of Fig. 20.3, are the devices used by the operating personnel. These devices include alarms, line test facilities, displays, and control circuitry for maintenance; a teletypewriter (10), duplicated, allows the operating personnel to communicate with the computer. A magnetic tape unit (11), duplicated, writes on tape all data related to billing customer calls. This information is then processed on an off-line computer. Finally, there is a device for changing the information in the semipermanent storage.

THE COMPUTERS   The engineering of the ESS computers is considerably different from that of machines familiar in today's data-processing world. The major factor leading to this difference is the need for extreme system reliability. A design objective was that the sum of the periods of overall system failure should not exceed 2 hours over its 40-year life [7]. Furthermore, the system should be able to be left unattended for long periods of time. These criteria are far beyond the standards of today's conventional data processing.

Although system outage is regarded as a catastrophe, individual errors are not. Individual errors in conventional data processing may be very serious, but in an exchange they are merely a nuisance. The subscriber may redial his call. One does not want such errors to occur often, but certainly subscribers do have to re-dial calls occasionally on today's electromechanical plant. When their dialing fails to reach its destination most subscribers blame themselves rather than the machinery the first time it happens — thinking they must have misdialed. This facet of dialer psychology is useful in planning the error and recovery procedures in a computerized exchange.

The computer components have therefore been chosen with breakdown-free operation as the main criterion rather than speed or other factors. All system components other than those concerned with a single line are duplicated, including the data buses (shown in Fig. 20.3). The computer is organized and its components designed so that faults can be found and corrected as quickly as possible. When a fault develops, it must be corrected rapidly to minimize the probability of a duplex failure. During normal operation one computer is "on-line" carrying out the switching work; the other is in active standby. Both the standby and the on-line machine are automatically monitored continuously so that any fault is signaled as soon as it occurs. When a fault occurs in any of the duplicated components of the system, a switchover occurs to the alternative unit automatically so that the telephone actions are interrupted for only a short duration.

The instruction word length is 44 bits, but 7 of these are used for automatic error detection and correction. The circuitry automatically corrects single-bit errors and detects double-bit errors. In addition to this checking, the system contains special circuitry that compares the execution of instructions in the two computers. If a difference is detected, the engineers are notified. Also, all transfers to and from the scanners and other peripheral devices are checked.

The circuitry of the computer is designed very conservatively to give the maximum reliability. It is considerably slower than conventional data-processing

Figure 20.4 The master control center of a No. 1 ESS office. The control center permits operating personnel to communicate with and control the system externally. It consists of a teletypewriter for sending and receiving communications, trouble alarm devices, a memory program and writer, and an accounting tape recorder. (Courtesy AT&T.)

machines. The cycle time of the computer is 5.5 microseconds. The storage units are of unusual design, again planned to maximize reliability. The majority of the storage is of read-only nature, and program malfunctions cannot damage it.

To handle the peak-hour traffic of a busy office with 65,000 lines, it was estimated that the system could not spend more than 5000 machine cycles per telephone call [8]. To meet this requirement, it was necessary to design an instruction set with powerful instructions that carried out several functions simultaneously. In addition to general-purpose data-processing instructions needed to make the machine flexible enough to handle a variety of possible future demands, it also contains special instructions for reading the scanners and performing operations on their information, as well as for sending output to the switching-control equipment and the signal generators. The machine thus has a mix of general-purpose and special-purpose instructions.

Some input/output programs have to be executed at strictly determined times—for example, when a subscriber is dialing, the dial pulses are sampled every 10 milliseconds. This action is governed by a clock that interrupts the program being executed, so that control is transferred to the appropriate priority program. When the condition that caused the interrupt has been dealt with, control is returned to the program that was interrupted.

As on more conventional computers, a variety of conditions can cause an interrupt, including the detection of an error by the checking circuitry. The computers have three levels of priority for normal error-free processing. Interrupts can normally occur only at 5-millisecond clock intervals; they snatch control away from lower-level programs, giving control to a higher-priority program. When the higher-priority work is completed, the computer returns to the program that was interrupted. There are another seven levels of priority, however, for error conditions. These seven interrupts can occur at any instant when a fault is detected. They trigger a variety of procedures designed to give automatic switchover to a fault-free configuration. Duplicate units can be switched in times comparable to the cycle time of the unit. Thus for most of the failures that occur, there will be no break in the processing of the majority of the calls. An abnormally large proportion of the programming in this system is concerned with reliability—more than half of all the instructions.

The programming proved to be a major problem in the installation of ESS 1. The number of instructions grew much larger than anticipated, and this fact gave rise to difficulties in handling the intended volume. There were delays and schedule slippages. For a time, some of the installed systems had troublesome bugs in them. But all this is hardly surprising to anyone who has lived through the installation of equally complex real-time systems.

Perhaps the most unconventional feature of the computers is the design of the storage units. The semipermanent storage employs a large twister memory invented at Bell Laboratories. The temporary storage employs a ferrite sheet memory.

(a)

Figure 20.5 (a) An aluminum sheet with 2816 magnetic spots, which forms part of the high-reliability semipermanent memory of ESS No. 1. Each spot contains one bit of information, a "0" if the spot is magnetized and a "1" if it is not. The spots are read with the twister wire. There are 128 of these cards in a module, and 16 modules in a memory unit. This is a "read-only" memory. Its contents can be written by a separate unit but not by the main computer. (b) The principle of the twister memory. (Courtesy Bell Telephone Laboratories.)

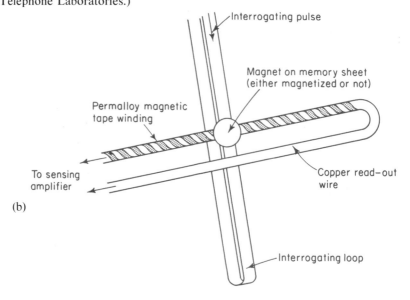

(b)

**THE TWISTER
MEMORY**

The twister memory is used to store the programs and data of semipermanent nature. Its contents can be read by the computer but not written. Only the operator can change its contents. The capacity of one memory unit is 5.8 million bits. Data about lines and trunks which give the telephone number, class of service, and information about features of the line need between 1 and 14 million bits, depending on the office size. Most of the rest of the storage is occupied by programs. Each memory unit is organized into 131,072 44-bit words, and one word is read with a cycle time of 5.5 microseconds.

The data are stored in the form of small magnets attached to aluminum sheets. Such a sheet is approximately $11 \times 6\frac{1}{2}$ inches, and it is shown in Fig. 20.5. The spots in this photograph are the magnets, each containing 1 bit of data. The data bit is either one or zero, depending on whether the magnetic spot is magnetized or demagnetized. Each sheet contains 64 words of 44 bits, one pair for each bit position in the words. One of these is shown diagrammatically in Fig. 20.5. At right angles to these there are 64 interrogating loops, one for each word. To read one word, a pulse is sent through the appropriate interrogating loop. At the intersection of the interrogating loop and the readout wires lies the magnet (the spots in Fig. 20.5).

One of the pair of readout wires is wound around with a spiral of permalloy magnet tape. This is the key to the operation of this type of memory. If at the intersections of the interrogating loop there is no magnet, then the pulse in the interrogating loop will cause the magnetization of the permalloy tape to switch direction and then switch back. This will produce a current pulse in the copper pair which will be detected by a sensing amplifier. If, however, there is a magnet at that intersection, this will not happen. The force of the magnet will be sufficiently strong to hold the magnetization of the permalloy tape in its present state and prevent the interrogating pulse from reversing it. If there is a magnet, then the sensing amplifier will receive no pulse. If there is not a magnet, it will receive a pulse. In other words, if the spot on the aluminum sheet is magnetized, this is a zero condition; otherwise it is a one condition.

The program, or the data about lines, can be modified simply by magnetizing or demagnetizing the spots. To do this the aluminum sheets are removed and a special device used. Such a device can be operated by the computer itself, the operator keying the relevant information into the teletypewriter. Figure 20.6 shows the sheets being loaded into the computer memory units.

**THE FERRITE
SHEET MEMORY**

The ferrite sheet memory is a read/write memory, also with a cycle time of 5.5 microseconds, and again designed with reliability as the main criteria. It is used to store any data which must be changed by the computer, and particularly those data which must be stored moment by moment as a call is being handled. It also stores data used in administration, maintenance, billing the subscribers, and so on.

The amount of temporary storage needed will vary with the number of lines the central office handles. With the range envisaged for No. 1 ESS it will vary from 100,000 to 4 million bits. These stores are built in units of four modules, each unit holding 196,608 bits and thus being organized into 8192 words of 24 bits each.

Figure 20.6 Memory cards such as that in Fig. 20.5(a) being inserted into the ESS No. 1 semipermanent store. (Courtesy AT&T.)

Figure 20.7 This one-inch square ferrite sheet, perforated with 256 tiny holes, is part of the high-reliability temporary memory of ESS No. 1. The ferrite around each hole acts as a "core" and stores one bit of information. Three wires threaded through the holes and a conductor plated onto the sheet are used to write information into and read it out of memory. Sheets are stacked in a module and four modules make up a Call Store, each holding 196,608 bits of erasable information. (Courtesy Bell Telephone Laboratories.)

Each module consists of 768 ferrite sheets. Each of these magnetic sheets is about 1-inch square and contains 256 small holes. It is shown in Fig. 20.7. Each hole can store 1 bit. One set of wires needed for this is plated on the ferrite sheets, thus lessening the difficulty of the expensive core-threading operation.

PROGRAMMING   The programs for controlling the central office operation total a few hundred thousand instructions. These are stored permanently in the twister memory. The programs are written so as to be the same for all sizes of central office, with the differences all in a parameter table.

The programs fall into a variety of different priority categories. At the top of the priority hierarchy for normal processing would be programs for sampling the dial tones or pulses. Near the bottom would be programs for gathering statistics about the number of calls. The machine is equipped with 10 levels of interrupt, of which only the bottom 3 are used during normal processing. Most of the programs which are not involved with error conditions are in the lowest of these. Work is allocated to them by the scheduling loop of a supervisory program, illustrated in Fig. 20.8. This supervisory program also transfers control to subroutines when necessary. When each program is finished, control is given back to the scheduling routine, which decides what to do next.

The 10 interrupt levels are labeled A to K (I is omitted), as shown in Fig. 20.8. Levels J and H, if needed, are given control every 5 milliseconds to ensure that none of the essential tasks are bypassed. An error detected by the various error-detection circuits causes an immediate interrupt to one of the levels A to G. Any program can be interrupted by a higher-level interrupt condition.

On any real-time system where reliability is important, significant parts of the programs relate to error detection, switchover to alternative components, and fallback to a different or degraded mode of operation. It is easy to design a system which will switch to alternative configurations when failures occur but much more difficult to design the programs that will control this switching and cause the minimum disruption to processing.

When an error occurs, an "interrupt sequence" circuit transfers control to the programs in one of the interrupt levels A to G in Fig. 20.8. For example, if an error is detected in the semipermanent memory, an interrupt to level E programs will occur. If the computers disagree in the execution of an instruction, an interrupt to level C will occur—providing, of course, that the machine is not already in a high mode of operation. The purpose of these programs is to establish as quickly as possible a system configuration that will still work correctly. It will then operate the appropriate switches to set up this configuration and return control to the processing programs.

It is possible that this still might not be good enough because the computer on which the interrupt occurred is itself in trouble. A working processing unit is

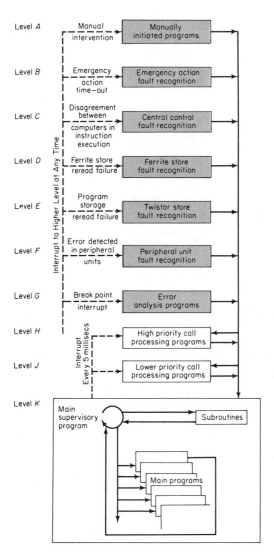

Figure 20.8 The hierarchy of the programs.

needed for system recovery, and a faulty one might be trying to carry out the recovery. This situation is overcome by an ingenious device called the emergency action facility. When an interrupt to a recovery program occurs, an emergency action timer is started. This timer counts down 40 milliseconds regardless of what else is happening. If a correct recovery occurs, it will be stopped. If not, it starts an emergency action circuit, which again operates regardless of processing unit.

The emergency action circuit will connect together various possible system configurations. It will do this without using programming. When it has established a configuration, it will attempt to transfer control to the programs of interrupt level *B* (Fig. 20.8). The purpose of these programs is to discover whether that configuration is working or not. When the transfer is made a clock called the *sanity* timer is started. The level *B* programs continually reset this if they are working correctly. If not, after 128 machine cycles of not being reset, the sanity timer instructs the emergency action circuit to try a new configuration. This process continues, the system "searching" for a configuration which works. The emergency action devices and sanity timer are completely independent of the health of the rest of the system. When a working configuration is found, the level *B* program then reestablishes the call-processing ability of the system.

As with any real-time program of this size its preparation was a major task. Many support programs were needed to accomplish it. These were mostly run on off-line conventional data-processing equipment. ESS is programmed in an assembler language which is compiled on a conventional computer. Much of the testing of the programming also takes place on an off-line computer in a simulated fashion. The computer simulates a stream of input calls and lists how the programs handle them, along with the times that would be taken. The programming of telephone exchanges will be faster and more error-free when higher-level languages are used with a powerful assortment of application macro-instructions.

ESS NO. 2   ESS No. 2, designed to handle 100 to 10,000 lines, is basically similar [2]. Its intended use is in nonmetropolitan areas, where it will stand largely unattended. For this reason, maintenance tests, traffic and plant measurement, translation changes, and other factors were designed so that they could be handled remotely. The engineers visit the office only when repairs, replacements, and reconnections are needed and for occasional preventative maintenance.

Economies are gained in ESS No. 2 by not giving it a high call-handling capacity. In general, the processing has been kept as simple and economical as possible. Some features are needed in rural areas that are usually not found in a metropolitan area — for example, the ability to handle party lines with many parties.

THE PROCESSING   Let us now describe the processing of a typical call in
OF A TYPICAL CALL   No. 1 or No. 2 ESS to show how computerized exchanges work.

1. When a subscriber picks up his telephone, the electronic exchange must detect that service is needed. The line in question will be inspected by the scanner every 100 milliseconds, using a ferrod sensor. When the customer lifts his telephone, the flow of current is detected. The computer, comparing the 24-bit word it receives from the scanner with that it received 100 milliseconds ago and stored in the ferrite store, will detect a change. The computer will then examine the change more closely and will see that 1 bit has changed, relating to the line in question.

2. The computer next connects this line to a digit receiver by instructing the switching con-

trols to set up an interconnection path. The ferrod sensor is disconnected from the line until the completion of the call, and a dial tone is connected to the line.

3. On hearing the dial tone, the customer will dial. If he has a Touchtone telephone, this type of digit receiver will be used; otherwise a rotary dial pulse receiver. The computer will read the digit receiver for the line every 10 milliseconds; the clocks interrupts every 5 milliseconds thus ensuring that none of these readings is missed.

4. If the telephone number dialed is to a telephone connected to this central office, the computer will look at the condition of that line, as recorded on the last scan. If it is busy, the computer will connect a "busy" tone to the calling line. If not, it will switch a ringing tone to the called line, again by instructing the switching controls to make an interconnection to a ringing circuit.

5. At this time, both the calling and the called line must be supervised. If the caller hangs up, this act must be detected and the call abandoned. Such detection is not now done by the ferrods and scanner but by other circuits associated with the digit receiver and ringing-tone circuit.

6. When the called party answers, the ringing tone is removed from the line, and the computer instructs the switching controls to complete the necessary interconnection.

7. While the parties talk, the calling and called lines are supervised so that the termination of the call can be detected. On a trunk call, this step would be done by the ferrods on the trunk circuit.

8. When the call is terminated, the computer disconnects the circuit and restores the lines to their former statuses. The 100-millisecond scanning takes place as initially.

9. The computer then completes a record for accounting purposes. To do so, it will have timed the trunk call.

In all cases throughout this operation, signals on the lines and changes in line state do not of themselves cause system actions to take place, as was the case on electromechanical exchanges. Such conditions are detected by the computer, by scanning, and the computer decides what to do. In this mode of operating lies the potential for building a more "intelligent" communication network in the future, capable of many operations not possible with today's telephone exchanges.

REFERENCES

1. W. Keister, R. W. Ketchledge, and H. E. Vaughan, "No. 1 ESS: System Organization and Objectives," *Bell System Technical Journal*, 45 (Sept. 1964), pp. 1831–1844.

2. A complete issue of the *Bell System Technical Journal* on the subject of the No. 2 ESS was published in October 1969.

3. "Stored Program Controlled Switching, System AKE," manufacturer's booklet from L. M. Ericsson Telephone Company, Stockholm 32, Sweden, 1975.

4. "Metaconta Telephone Switching System Type 11A," a brochure published by the International Telephone and Telegraph Corporation, New York.

5. Toyoda, Kazuo, et al. "Electronic Switching System for Field Trial DEX-2," *Fujitsu Scientific and Technical Journal,* 6 (June 1970), pp. 1–43.

6. "Electronic Switching Systems 1 TXE2 Electronic Exchanges," a booklet published by the GPO, London.

7. R. W. Downing, J. S. Novak, and L. S. Tuomenoksa, "No. 1 ESS Maintenance Plan," *Bell System Technical Journal,* 45 (Sept. 1964), pp. 1961–2020.

8. J. A. Harr, F. F. Taylor, and W. Ulrich, "Organization of No. 1 ESS Central Processor," *Bell System Technical Journal,* 45 (Sept. 1964), pp. 1923–1960.

21 CONTROL SIGNALING

In addition to transmitting *information,* telephone and indeed all telecommunications networks must also transmit *control signals.* These are signals between the network control devices, which enable them to set up, disconnect, and control the transmissions. Many such supervisory functions are needed to set up a call through a complex network and maintain it in operation.

The telephone network is rather like a gigantic nervous system, with the control signals carrying sensing and control information throughout the system. When the telephone network is used by data-processing machines *another* set of supervisory signals are used between the terminals and data-handling devices. This second level of signals is treated by the telephone network as though the signals were information signals. If a network is specifically constructed for data transmission, then the control signals it uses may be entirely different from those used on the telephone network (in Chapters 23 and 24 we will discuss some such networks). In this chapter we will discuss the control signaling used on the Bell System and other telephone systems.

CONTROL SIGNALING   Coded control signals on telephone networks have the following functions:

1. Audible communication with the subscriber: a dial tone to tell him he may dial, a busy signal to indicate that his call cannot be completed, a ringing tone to indicate that the number he called is ringing, a noise to tell him he has done something wrong, and so on.

2. Transmission of the number dialed, to switching offices that will attempt to connect it.

3. Transmission of information between offices saying that a call cannot be completed.

4. Transmission of information between offices saying that a call has ended and that the path can be disconnected.

5. A signal to make a telephone ring or make subscriber equipment "alert."

6. Transmission of information used for billing purposes.

7. Transmission of information giving the status of equipment or trunks in the network. This information may permit calls to be routed by paths which avoid trouble spots or congested routes.

8. Transmission of information used in diagnosing and isolating system failures.

9. Control of special equipment such as echo suppressors, TASI equipment, and satellite channel equipment.

Some control signals convey digital information—the number a subscriber has dialed is passed from office to office. Some signals tell the *status* of certain equipment. For example, it is necessary to know whether a called customer has answered or not. On a long-distance call, when the called customer picks up his telephone the originating office will start charging. It must therefore know when this status changes from "on-hook" to "off-hook." Similarly, it is necessary to know the availability of registers in offices the call may be switched through. These and other status signals can all be sent with two-state signalling.

Other signals are necessary to *control* certain equipment. For example, a line must be seized for a call. Later, after the connection is made a signal is sent instructing that the path be disconnected. Again these actions require two-state signals. On some telephone systems the signals are continuous. In other words, a voltage or other two-state indication is permanently present. On others, the control signal is a short, specially coded message. These are referred to, respectively, as "continuous" and "spurt" signaling.

There is a variety of other signals which fit into one of these three categories. For example, signals for returning a coin in a coin box, other signals for coin-box control, signals for alerting operators, signals for "ringing forward" or sending a ringing signal to the originating end, auxiliary charging functions, and signals for indicating that a call does not terminate in a country but is merely passing through and language digits to indicate the language to be used if an operator is called in for assistance on the call.

Network signaling is basically a problem of transmitting data through the network. Most of the techniques employed, however, are much slower and cruder than the techniques of data transmission. They were designed to be inexpensive and reliable on analog voice circuits.

Many of the control signals are transmitted *before* the transmission of information, for example, dialing, routing, and ringing signals needed to set up a call. Some, for example, detecting when a telephone handset is replaced, must be sent at the same time as the information transmission.

Telephone signaling systems, invented as they were for voice communication, have usually not taken the advent of data transmission into consideration. As we shall see there are certain ways in which data could interfere with telephone networks' supervisory signals. Computer equipment used on a public network must both make use of the signaling arrangments of that network (otherwise it cannot establish and maintain a call) and at the same time tailor its transmission so as to avoid interfering with the signaling and accidentally triggering any circuit that it should not. This can be quite complex, especially in countries where some old plant is still in use and could conceivably be reached by the computer transmission. Much of the old telephone equipment was built to last, and it has. Often data transmission devices designed for one country have met with trouble when exported because they are not compatible with the signaling arrangements of the other country. Some American equipment has failed to find a market in Europe for this reason.

On the other hand, computer systems would benefit from better signaling techniques than those used in the telephone networks—much faster connection time, better diagnostics, and fast reaction to failure or unavailable equipment.

INTERNATIONAL STANDARDIZATION   It is desirable that one telephone system be able to communicate with another, and so standardization of signaling procedures is necessary. Standardization exists *within* countries; for example, the numerous American telephone systems conform (with some minor deviations) to Bell standards. However, technology is changing, and old signaling systems will have to coexist with newer ones. National telephone systems grew up to some extent independently of one another, and so many different incompatible signaling systems exist in different countries.

CCITT defines several different standard signaling systems in the hope that signaling practice will become more standardized throughout the world. Some major national systems, however, do not conform to CCITT recommendations, and the recommendations themselves give a large diversity of options in order to accommodate various existing systems.

Table 21.1 lists the CCITT signaling systems. The words in the table are explained below. Full details of these are published in the reference material at the end of this chapter.

TWO AREAS OF SIGNALING   Control signaling, like the transmission, can be thought of as being in two parts which operate differently: signaling between the subscriber and his local telephone exchange, and signaling between switching offices.

The methods in use between the telephone and central office are to a large extent determined by considerations of the telephone itself and its human user. The signals between automatic offices, on the other hand, are entirely machine-to-machine signals. They are related to the multiplexing and carrier methods that are in use. With modern central offices the office often acts as a buffer between the subscriber and the trunks and between the trunks and their own independent, and more efficient, signaling. Between the subscriber and the central office, fewer sig-

Table 21.1 CCITT signaling systems

	Described in Reference	In-Band Signaling	Out-Band Signaling	Common-Channel Signaling	Analog Two-Frequency Signaling	Analog Multifrequency Signaling	Digital	Suitable for Operation over Satellites	Suitable for Operation with TASI Circuits (Chapter 15)	Suitable for Operation with Time-Division Exchanges (Chapter 26)
CCITT Signaling System No. 3 (obsolete)	1	✔						No	No	No
CCITT Signaling System No. 4	2	✔			✔			No	No	No
CCITT Signaling System No. 5	3	✔				✔		Yes	Yes	No
CCITT Signaling System No. 5 (bis)	4	✔				✔		Yes	Yes	No
CCITT Signaling System No. 6	5	✔				✔		Yes	Yes	No
CCITT Signaling System No. R1	6		✔				✔	Yes	No	Yes
CCITT Signaling System No. R2										
Analog version	7	✔				✔		No	No	No†
PCM version	8		✔				✔	No	No	No†

†May be used on links connecting time-division and circuit-switching exchanges.

naling requirements exist. The need for cheap and reliable devices in the telephone set is the dominating factor.

DIALING

The introduction of the telephone dial led to automatic switching. When a subscriber lifts his handset the dc voltage on the local loop changes and informs the central office that the subscriber wants to dial a call. The central office equipment puts a *dial tone* on the line, telling the subscriber that he may dial. His dialing generates the train dc pulses shown in Fig. 19.8, and these pulses cause the local switching equipment to operate as described in the previous two chapters.

The simple make-and-break pulses are produced inside the telephone by a cam. The pulse rate on most systems is nominally 10 per second and varies between 7.5 and 12. The same pulses can be generated by tapping the receiver rest to make and break the circuit. With telephone coin boxes of an earlier design it was possible to obtain a connection without putting money in the box, by rapidly tapping the telephone rest. In this way the appropriate signal could be sent to the central office without using the dial. Today more ingenious methods are needed.

When a data-processing machine uses the public network and "dials up" remote terminals or computers an automatic calling unit under control of the computer must generate pulses like those in Fig. 19.8.

A new and faster form of subscriber "dialing" is spreading rapidly, especially in North America. In North America it is to be referred to as Touchtone
dialing. The telephone no longer has a dial. In its place is a group of 12 (originally
10) small square keys labeled 1 to 9, 0, *, and #. Each key sends a different audible beep composed of two frequencies down the line. By touching seven or
more of them in sequence a number can be quickly dialed. The train of tones is
received by special equipment in the central office which interprets them and
stores the digits in a register with the same functions as that in which ordinary
telephone dial pulses are stored. Thus the major switching equipment of the exchange is no different when Touchtone dialing is used; only that part for interpreting the signal is different. Many central offices in the United States are now
equipped for Touchtone signaling. Subscribers can have a Touchtone telephone
installed for a slight additional cost.

Each note from the Touchtone keys consists of two frequencies, one from
the group 697, 770, 852 and 941, and the other from the group 1209, 1336, 1477,
and 1633. Thus a self-checking code is produced with $4 \times 4 = 16$ combinations.
Only 12 of these combinations are used on the conventional, 12-key telephone. A
16-button dial is reserved for military communications, for security purposes, and
for multiple priority preemption. The 16 frequencies are agreed upon as an international CCITT standard, shown in Fig. 21.1. If a receiving office detects more or
less than two of these frequencies, this is an error condition.

The signaling between the central office and telephone thus consists of the
following:

1. A means of dialing: Touchtone or rotary dial.

2. A two-state dc signal to indicate whether the telephone handpiece is on the rest or not.

3. Various audible tones intended for the ear of the user, including a dial tone and a busy
   signal.

4. A signal which makes the telephone ring—the ringing tone.

The ringing tone on a subscriber loops is a 20-Hz signal (subaudible) of 75 to
150 volts. Where the signal must travel over a carrier channel, a frequency of
1000 Hz modulated by a frequency of 20 Hz is used.

**OFFICE-TO-OFFICE
SIGNALING**
The signaling between offices which are not step by
step is normally independent of that between the subscriber and the central office. Furthermore, on many
systems—for example, the Bell System—when a telephone call passes over
several links, as in Fig. 13.7, the signaling over the separate links is independent.
Different signaling "languages" can be used over differently engineered links if
necessary.

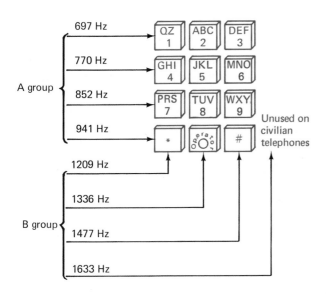

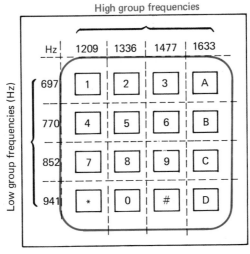

CCITT recommendation Q23 for the allocation of frequencies on the push-button set.

Figure 21.1 The frequencies used on key telephones such as the AT&T *Touchtone®* telephone. Each key pressed generates two frequencies, one from the A group and one from the B group. These frequencies are an internationally accepted CCITT standard.

Signaling which is independent on each link is referred to as *point-to-point signaling*. Each point relays the signal to the next point. Not all systems have this independence. In Europe a somewhat different system is common, using what is called *end-to-end signaling*. Here when a call is routed via toll offices *A, B, C,* and *D, A* sends a signal to *B* saying that this is not a call terminating at *B* but a transit call. *B* allocates a transit register (only) to *A* and tells *A* to go ahead. *A* sends the digits giving the address of *D*. *B* acknowledges these to *A* and sends a signal to *C*, again saying that this is a transit call. The logic circuitry of *B* now takes no further part. *C*'s go-ahead message travels to *A* over the path that *B* has set up. *A* sends *D*'s address again. *C* acknowledges this and sets up a path to *D*. *C*, as with *B* before, then releases its transit register and drops out of the act. *D* now tells *A* to transmit, and *A* sends the rest of the telephone number to *D*.

When the telephone call is over, messages must again be sent to *B* and *C* telling them to disconnect the circuit.

With point-to-point signaling, *A* would send all the information about the routing to *B*. The register and logic in *A* which handles this would then be dismissed. *B* would send the information to *C*, and then *B*'s register and logic would be dismissed. *C* would select the final trunk to *D* and send to *D* the number of the called party (often four digits), droping the other routing digits. This is somewhat quicker than end-to-end signaling and permits a flexible use of different interoffice signaling techniques.

IN-BAND AND OUT-BAND SIGNALING   On most signaling systems, signals between offices travel inside the voice bandwidth. This is referred to as *in-band* signaling. An alternative, used less commonly, is for a channel to carry both the speech frequency band and a separate narrow signaling band, usually employing a single frequency. The CCITT Signaling System No. R2 (Reference 7), for example, employs a signaling frequency of 3825 Hz (± 4 Hz). This is called *out-band* signaling.

In-band signaling has several advantages. The control signals can go anywhere the voice signals go. In-band signaling can therefore be used over any type of line plant. If PCM links are included in the circuit (Chapter 14) the control signals can be converted to PCM form along with the speech. The application of out-band signaling is limited to appropriately designed carrier systems. With in-band signaling a call on a faulty line segment can be easily switched to any other speech channel. It is not possible to set up a call with in-band signaling on a faulty speech channel, though this could happen with out-band signaling. In-band signaling permits several different frequencies to be used, whereas a signal frequency is used with out-band signaling.

In-band signals are not transmitted while conversation is taking place. The detection circuit must, however, "listen" during the conversation so that it is ready to respond to the signals when the conversation ends. There is a danger that it might be fooled and pick up a sound which it thinks is a signal during the conversation. The signals are chosen so that this will not normally happen during speech.

However, when data are sent, the data might, unless precautions are taken, accidently trigger the equipment which listens for control signals.

Out-band signals run no danger of interference from speech. They are unaffected by compandors and echo suppressors. Also signaling can take place during the telephone conversation. However, an out-band signal needs extra bandwidth and extra electronics to handle the signaling band. Signaling rates are slower because the signal has been confined to a narrow bandwidth, usually a single frequency.

DC OFFICE SIGNALING   There are several types of signaling between offices. If the offices are close (or use old-fashioned equipment) dc pulsing like that in Fig. 19.8 is employed.

There are several variations of dc office signaling. A system called *revertive pulse signaling* is used which is faster than that in Fig. 19.8. Here a start pulse from the transmitting office causes a pulse generator in the receiving office to start, and this sends its pulse back to the transmitting office. A counter at the transmitting end counts these and sends a stop signal when the required digit to a register, and the counter is freed for the next digit. Another system, called *panel call indicator signaling,* sends one of five possible signal levels. Two positive voltages, an open line, and two negative voltages are used. The larger positive voltage is used to mean *end-of-pulsing,* and the other four are used to code digits.

MULTIFREQUENCY   The most common form of signaling between toll of-
OFFICE SIGNALING   fices uses multifrequency tones (MF). *Multifrequency signaling* uses six frequencies placed in that part of the voice spectrum where different channels have the smallest deviation in loss. On the Bell System the frequencies used are 700, 900, 1100, 1300, 1500, and 1700 Hz. Digits are coded as two out of the first five of these frequencies and are sent between *start-of-digit-transmission* and *end-of-digit-transmission* codes. Table 21.2, shows the combinations of frequencies used in North America and on CCITT Signaling System No. 5 (Reference 3). If the receiving office detects an invalid signal, it can send a signal requesting retransmission.

Telephone systems in other countries use different frequencies. Britain uses different frequencies for each direction of transmission as specified in CCITT Signaling Systems No. R2, [7]; thus:

Forward Direction	Reverse Direction
1380	540
1500	660
1620	780
1740	900
1860	1020
1980	1140

Table 21.2 Frequencies used in North America for inter-office signaling†

	Signal	Pair of Frequencies	Frequency Tolerance
KP1	(start-of-digit transmission for a national call)	1100 + 1700	
KP2	[start-of-digit transmission for an international call from an intermediate (transit) exchange]	1300 + 1700	
Digits:	1	700 + 900	
	2	700 + 1100	$\leqslant \pm 6$ Hz
	3	900 + 1100	
	4	700 + 1300	
	5	900 + 1300	
	6	1100 + 1300	
	7	700 + 1500	
	8	900 + 1500	
	9	1100 + 1500	
	0	1300 + 1500	
ST (end-of-digit transmission)		1500 + 1700	

†These are the frequencies of the CCITT Signaling System No. 5 (although North American signaling does not conform in every respect to CCITT No. 5).

The MF signals are sent over the normal voice channels and are transmitted like speech. They may be sent either by a switchboard operator or, more rapidly, by automatic equipment. If the switchboard operator accidentally presses two keys at once, her error is detected. The reader may possibly have heard these interoffice signals. On some systems the operator's signaling is occasionally audible, and sometimes the automatic signaling can be faintly heard due to crosstalk. The quiet listener may hear a faraway flurry of diminutive discordant notes.

A TYPICAL TELEPHONE CALL

Figure 21.2 shows the signaling that occurs in setting up a typical toll telephone call in North America. The frequency 2600 Hz is transmitted continuously on all voice channels between toll offices when the channel is free. This frequency also acts as a disconnect signal, indicating that the voice channel should return to its unused status.

Figure 21.2 shows one voice channel between toll offices in what may be a trunk of many channels. When the subscriber dials the number it reaches his local central office and possibly toll office by dc pulsing as in Fig. 19.8. The toll office selects a free voice channel in an appropriate trunk and stops the 2600-Hz tone. The office at the end of that trunk detects the break in the 2600-Hz signal and is alerted to receive a toll telephone number. The number is sent in the MF code of

Table 21.2. One toll office passes the number to another until the called central office is reached. The central office rings the called telephone.

When either party replaces his receiver the call is disconnected and the toll offices tell each other this by transmitting the 2600-Hz tone again.

Unfortunately it is possible to interfere with the telephone trunking mechanism by transmitting the 2600-Hz tone from the subscriber's telephone. An AT&T story has it that a New York shirt manufacturer once broke his front tooth in such a way that he transmitted a brief 2600-Hz whistle every time he said the word "shirt" on the telephone. An Eastern Airline office in Atlanta was plagued by telephone disconnects for seven years and then discovered that they were caused by the shrieks of exotic birds in the hotel lobby cocktail lounge. Worse, Captain Crunch breakfast cereal packets were once delivered with a toy whistle which produced a pure 2600-Hz tone. A brief 2600-Hz tone received by a toll office causes it to free the voice channel in question and place a 2600-Hz tone on the channel to the next toll office, as in the last diagram of Fig. 21.2.

In the early 1970s a generation of students discovered that amusing games could be played with Bell System toll signaling. They were referred to by the press as "Phone Phreaks." Several phone phreaks ended up in prison. The best of the phone phreaks acquired amazing capabilities in manipulating the world's telephone networks.

SIGNALING AND   With some data transmission machines in-band signal-
DATA TRANSMISSION   ing has caused a problem.

Data must be sent over the communications line in such a manner that they do not interfere with the signaling system. The communication system recognizes an in-band control signal by the fact that it has substantial energy at the signaling frequency and little at other frequencies. This would not be true of the human voice, which is smeared across a range of frequencies different from that of control signals. To transmit data we must either ensure that they avoid the signaling frequencies completely, which would limit the usable bandwidth, or else that they are smeared across the band so that when there is energy at the signaling frequency there will always be sufficient energy at other frequencies to prevent the data from being mistaken for a control signal. The latter will make the best use of channel capacity and will be accomplished in the modem design. Figure 21.3 shows the Bell System signaling frequencies. In Britain the Post Office uses a single-frequency signaling band at 2280-Hz, but also there is some equipment with two-frequency signaling at 600 and 750 Hz. This is shown in Fig. 21.4. On the European international service the frequencies used are again different—a two-frequency system operating at 2040 and 2400 Hz.

The telecommunication companies impose restrictions on users, or manufacturers, of data transmission equipment to ensure that they do not interfere with the signaling. Figure 21.5 shows the relative signaling levels permitted for data by Britain's Post Office. The two troughs correspond to the signaling frequencies in

When there is no call on the system, the toll offices whistle at each other continuously at 2600 hertz. A break in this tone indicates that a signal setting up a call is about to be sent.

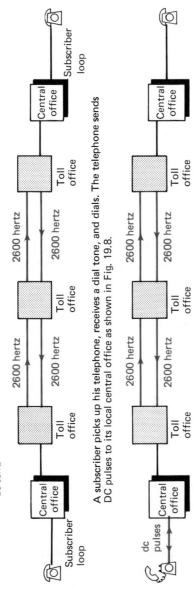

A subscriber picks up his telephone, receives a dial tone, and dials. The telephone sends DC pulses to its local central office as shown in Fig. 19.8.

The central office records the date, time, originating number and dialed number, for billing purposes. It relays the area code and number to its local toll office.

The toll office selects the best free interoffice trunk for the call ("free" means that the trunk has a 2600 hertz tone on it). A sender circuit in the toll office sends the area code and number down that trunk using multifrequency signaling (MF) with the signals shown in Table 21.2.

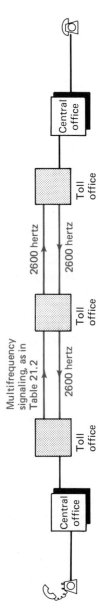

Multifrequency signaling, as in Table 21.2

The next toll office on the route relays the multifrequency signals

416

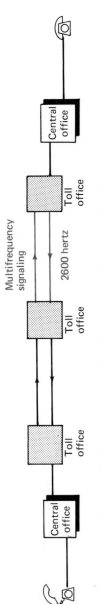

Multifrequency signaling

2600 hertz

The last toll office on the route gives the number to the central office. The central office switches a 20 hertz ringing tone to the requisite subscriber loop, and to the trunk to the caller.

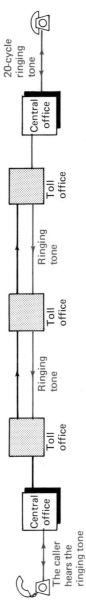

20-cycle ringing tone

Ringing tone

Ringing tone

The caller hears the ringing tone

The called party picks up his telephone, and the call proceeds.

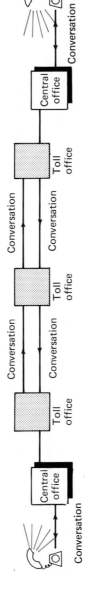

Conversation

Conversation

Conversation

Conversation

Conversation

Either party terminates the call by replacing his handset. His central office detects the absence of current and notifies its toll office. The toll office places the 2600 hertz signal back on the intertoll trunk. The central office which originated the call records its time of termination, for billing purposes.

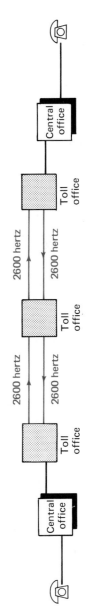

2600 hertz

2600 hertz

2600 hertz

2600 hertz

The intertoll trunk is now ready to be allocated to another call.

Figure 21.2   The signaling involved in a typical toll telephone call in North America.

417

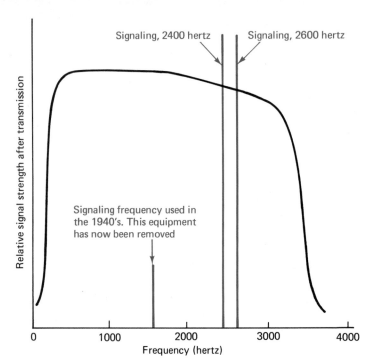

Figure 21.3 In-band signaling frequencies on Bell System Telephone Plant.

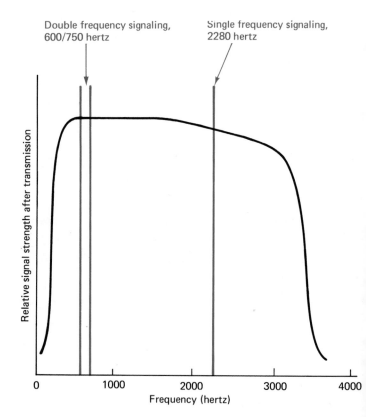

Figure 21.4 In-band signaling frequencies on British telephone plant.

use. Data transmission machines or modems must be designed so that the relative signal levels cannot exceed those indicated in these troughs or else so that sufficient energy is sent elsewhere to prevent the automatic devices from assuming that a signal is being sent. A guard tone in the area *ABCD* of Fig. 21.5, for example, could enable signals to be sent in the signaling zone from 2000 to 2430. The use of a guard tone, however, is less efficient than selecting a modulation method which guarantees that the signal is spread across the band. Care also has to be taken that intermodulation products from other channels cannot have strengths greater than those in Fig. 21.5.

The restrictions on modems and data-processing equipment design imposed by in-band signaling do not in general apply to leased communication lines because these do not carry the signals needed for switching purposes.

COMMON-CHANNEL
SIGNALING

All the above techniques make the signals travel with the voice channel. This is called *channel-associated signaling.* An alternative is to carry the signals over paths completely independent of the traffic circuits. One independent signaling channel can carry the signals for many traffic channels. This approach is called *common-channel signaling.*

Common-channel signaling is more complex than sending signals with the channel, but as computer circuits drop in cost it becomes increasingly attractive.

Figure 21.5   Power level restrictions for data transmission over public switched connections in Britain. Some *private* circuits over carrier channels also need restrictions of this form. (Redrawn from Reference 9.)

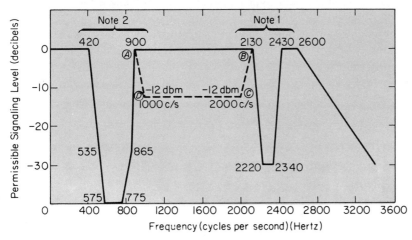

Notes:
1. Signal components up to 0 dbm may be permitted within this area if *always* accompanied by signals in area *ABCD*
2. Signals having components within this area may be permitted if characteristics preclude false operation of trunk—signaling equipment

Table 21.3 Typical signaling speeds of the four types of interoffice signaling

Type of Signaling	Signaling Speed (digits/sec)
DC pulses for step-by-step operation (Fig. 19.8)	1.2
MF (multifrequency tones) (Table 21.2)	6–10
Common channel signaling	25–160
PCM signaling	1000

Furthermore, many switching offices are now becoming computerized, and common-channel signaling can be controlled by these computers. In effect common-channel signaling is a distributed computer network carrying very short messages.

Common-channel signaling gives much faster signaling speeds (see Table 21.3) and a greatly enhanced signaling capacity. Such signaling could be used for new types of network management functions such as diagnostics and the varying of call routing as traffic patterns vary. The network could be made immune from casual "phone phreaking."

CCITT Signaling System No. 6 is an international specification for common-channel signaling. AT&T's version of CCITT Signaling System No. 6 is called *CCIS, Common-Channel Interoffice Signaling*. It is designed for signaling between computerized switching offices and with these offices will enable telephone calls to be set up in much less time than today.

The signaling data are built up of units of 28 bits, of which 8 are error-detections bits. These data are sent over a speech channel using a four-phase modem operating at 2400 bits, as shown in Fig. 21.6.

Common-channel signaling, however desirable, will not take over quickly because so many offices exist using channel-associated signaling, and these offices were designed to last.

Figure 21.6 Common-channel signaling as in AT&T's CCIS and the international CCITT signaling system No. 6.

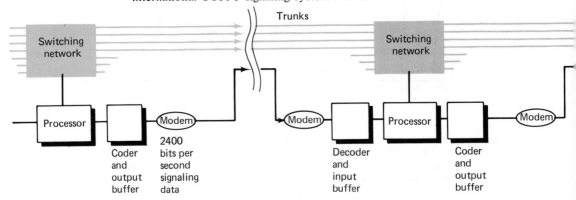

AN ORDER WIRE   An order wire was a separate channel used on old tele-graph and telephone systems for control purposes. Some order wire channels transmitted signals from control mechanisms, but most were used for engineer's voice messages.

The term "order wire" is appearing again in current designs, but now it can be a computer-controlled time-division-multiplexed, or other digital, channel. It is used on some satellite systems in which the satellite channels are assigned on a demand basis.

PCM SIGNALING   The T1 carrier as seen in Chapter 14 has a bit stream of 8000 bps which travels with each voice channel. Other PCM signals are also well endowed with time-division-multiplexed data streams that can be used for signaling purposes.

In-band signaling can be transmitted over PCM links with no problem. The signals are simply digitized along with the voice. It would, however, be advantageous to implement digital signaling systems for PCM links which employ the spare bits in the voice channel which were left available for signaling. A special version of the CCITT Signaling System No. R2 is designed to do this [8], and PCM versions of signaling system No. 6 are being considered. Figure 21.7 shows the CCITT recommendations for common channel signaling and channel-associated signaling with 1.544 Mbps PCM transmission [10].

It is clear that PCM signaling and common-channel signaling will eventually change the nature of telecommunication signaling as the switching and control functions become increasingly computerized or at least controlled by digital logic.

Telecommunication networks are evolving toward being vast computer-controlled systems with digital PCM transmission and signaling. However, dc pulse and MF signaling are likely to be widely used for the foreseeable future, and there will be incompatibilities between the signaling on different systems.

REFERENCES

1. *CCITT Signaling System No. 3,* CCITT recommendations Q.76 to Q.79 in *The CCITT Red Book,* published by the International Telecommunications Union, Geneva, 1961.

   Note: References 2 to 8 describe in detail the specifications of current CCITT signaling systems. They are all taken from *The CCITT Green Book,* Vol. VI: Telephone Signaling and Switching, published by the International Telecommunications Union, Geneva, 1973.

2. *CCITT Signaling System No. 4,* CCITT recommendations Q.120 to Q.139.

1. With common-channel signaling:

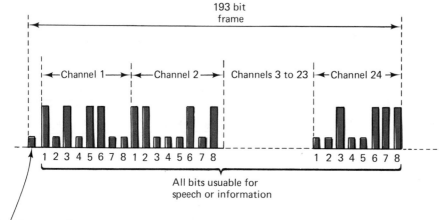

All bits usuable for
speech or information

Odd frames: Frame alignment signal 101010. . . repeated
Even frames: Common channel signaling (4000 bits per second)
              This is called the 5-bit.

2. With channel-associated signaling:

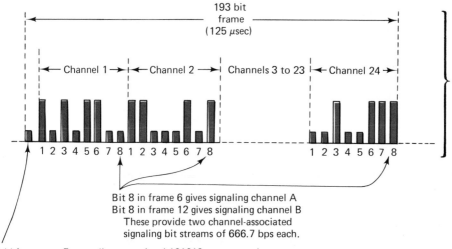

12 such frames
numbered 1 to 12
form one multiframe

Bit 8 in frame 6 gives signaling channel A
Bit 8 in frame 12 gives signaling channel B
These provide two channel-associated
signaling bit streams of 666.7 bps each.

Odd frames: Frame alignment signal 101010. . . repeated
Even frames: Multiframe alignment signal 001110. . . repeated. This is called the S-bit
              If the S-bit in frame 12 is modified from 0 to 1, this is an alarm signal
              indicating the frame alignment has slipped.

Figure 21.7 Signaling with PCM systems operating at 1.544 bps employing CCITT Recommendation No. G.733 [10].

3. *CCITT Signaling System No. 5,* CCITT recommendations Q.140 to Q.164.

4. *CCITT Signaling System No. 5 bis,* CCITT recommendations Q.200 to Q.218.

5. *CCITT Signaling System No. 6,* CCITT recommendations Q.251 to Q.295.

6. *CCITT Signaling System No. R1,* CCITT recommendations Q.310 to Q.332.

7. *CCITT Signaling System No. R2,* CCITT recommendations Q.350 to Q.368.

8. The digital version of the latter is in recommendations Q.357 to Q.359.

9. M. B. Williams, "Characteristics of Telephone Circuits in Relations to Data Transmission," *Post Office Electrical Engineers' Journal,* London (Oct. 1966).

10. *PCM Transmission at 1544 kb/s,* CCITT Recommendation No. G.733, *The CCITT Green Book*, Vol. III: Line Transmission, International Telecommunications Union, Geneva, 1973.

22 PRIVATE BRANCH EXCHANGES

When a corporation or building has many telephones it may have its own switching facility which interconnects these telephones with one another and also with the public network or other lines leaving the premises. Whereas central offices and toll offices are part of the *public* switching facilities, internal telephone exchanges are privately owned or rented. They are referred to as private branch exchanges (PBXs). A PBX can be manual, with all the connections being made by operators. This is referred to as a PMBX (private manual branch exchanges). Many PBXs are completely automatic. An automatic PBX is called a PABX (private automatic branch exchange). The term PBX is sometimes used to imply a PABX. In areas such as North America where most private branch exchanges are automatic PBX is used almost exclusively (Fig. 22.1). Figure 22.2 shows a manual PBX.

PBXs have developed along lines similar to public exchanges. First they were entirely manual, then step-by-step Strowger switch operation was used (Fig. 19.9), then common control was introduced (Fig. 19.11), and finally computerized PABXs were installed (Fig. 20.2). PBXs handle a much smaller number of lines than public exchanges, so the more expensive techniques were less likely to be economically viable. About 90% of PBXs serve less than 200 telephone extensions and consequently many step-by-step PABXs are still being installed.

Computerized PABXs are particularly interesting because the capability of the computer allows the PABX to perform many new types of functions, as we will see. Today computer control can be economical only for large or medium-sized PABXs, but the cost of small computers is dropping rapidly, and some of the new functions they make possible can help to lower a corporation's telephone bill.

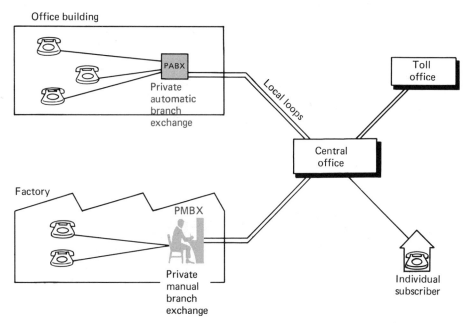

Office building

Local loops

PABX
Private automatic branch exchange

Toll office

Central office

Factory

PMBX

Private manual branch exchange

Individual subscriber

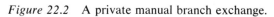

Figure 22.1   Private branch exchanges.

Figure 22.2   A private manual branch exchange.

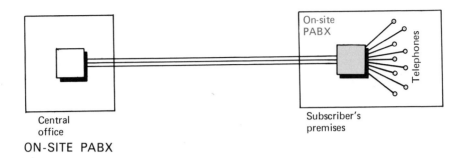

ON-SITE PABX

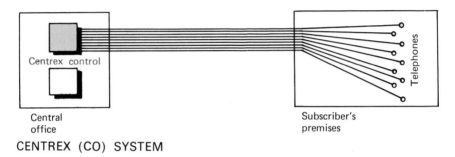

CENTREX (CO) SYSTEM

Figure 22.3 Comparison of an on-site PABX with a Centrex system which does the switching at the central office location. See Figs. 22.3 and 22.4.

CENTREX SERVICE   While PBX equipment has traditionally been on the customer's premises, the equipment used for *Centrex* service may not be. Centrex service gives each station (each telephone) its own number, which can be dialed directly on the public network. The switching equipment *may* be located on the customer's premises (Centrex CU), at a location from which several customers can be served, or at a central office (CO). Most commonly it is at a central office.

Figure 22.3 compares a central office Centrex system with an on-site PABX. It will be seen that more wire pairs are needed with the Centrex system. Substantial *concentration* is used with the PABX, the number of wire pairs to the building being smaller than the number of telephones served. The subscriber location, on the other hand, does not need to provide space for switching equipment; it is all at the central office where maintenance is quicker and easier.

Figure 22.4 compares the telephone switching actions of a Centrex system and a typical PABX. With the Centrex system, incoming calls reach their destination without intervention by the attendant. The attendant is not involved in any of the five switching operations shown in Fig. 22.4(b). With the PABX in Fig.

22.3(a) the attendant is needed for incoming calls, for setting up three-party calls (conference calls), and when a user wishes to transfer a call to another station.

Another advantage of Centrex service is that the central office equipment can record details of all telephone calls made by each station. With an on-site PBX it merely records the calls made by the location. A bill is provided listing the time of day, date, duration, telephone number, and cost of each nonlocal call made by each station. This listing can be used to control a location's costs.

PBX PACKAGES   The approach of some telephone companies to marketing PBXs is to sell a package of services rather than a particular piece of hardware. With this approach the hardware can change, but the package of services remains the same.

Box 22.1 lists possible PABX features. Many of them do not exist on electromechanical PABXs, but all of them exist on *some* PABX or else have been actively discussed as a useful PABX feature.

AT&T offers two Centrex packages. CENTREX I and CENTREX II, and three electromechanical PBX packages, 100, 200, and 300. It is the 200 package that is illustrated in Fig. 22.3. The features of the 100, 200, and 300 packages are noted on Box 22.1. A computerized PBX such as AT&T's Dimension system offers many more features.

Note: AT&T uses the term PBX exclusively, not PMBX or PABX.

COMPUTERIZED PABXs   There is much to be gained from computerizing private branch exchanges as well as the public offices. Initially the cost was too high. Systems such as No. 1 and No. 2 ESS are economical only in large sizes. No. 2 can be justified when it can handle more than 1000 lines. The cost of small computers, however, has dropped dramatically and now small computerized PABXs are economical with under 100 lines. There is a large marketplace for such machines; the U.S. alone has about 190,000 PBXs of 100 lines or more.

There have been two types of approaches. Both give important pointers to the directions in which this technology is likely to progress. The first is a *time-sharing* approach. The computer is still large, like ESS No. 1, but takes over the operations of what would have been many PABXs. The switching is controlled in the local central office rather than on the subscribers' premises. The Bell System No. 101 ESS operates in this way. Using technology similar to No. 1 ESS, it controls up to 32 time-division switching units in separate subscriber locations. The central office approach has the advantage that each extension can have its own external telephone number.

The second approach is to use a stand-alone system which is programmed to perform new types of functions which make it economically viable. Many new functions can give a more appealing telephone service, but two are of particular

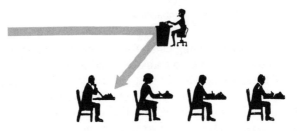

Incoming calls reach an attendant who switches them to the desired station.

Outgoing calls can be dialed directly.

Internal stations can be dialed directly.

A station can have a call transferred by flashing the attendant.

428   Internal conference calls can be set up by the attendant.

Figure 22.4 (a) Switching facilities available with AT&T Series 200 PBXs.

555-2368

555-2369   555-2370   555-2371   555-2372

Each station nas its own external telephone number. Incoming calls can reach stations directly without the intervention of the attendant.

555-2369   555-2370   555-2371   555-2372

Outgoing calls can be dialed directly.

555-2369   555-2370   555-2371   555-2372

Internal calls can be dialed directly.

555-2369   555-2370   555-2371   555-2372

A station can transfer a call automatically.

555-2369   555-2370   555-2371   555-2372

A station can add a third party automatically to a call originating from outside.   **429**

Figure 22.4 (b) Switching facilities available with AT&T Centrex II systems.

BOX 22.1 Features of PABXs
(conventional and advanced)

Features of some conventional PABXs	Features of Bell System PBX Packages		
	PBX Series 100	PBX Series 200	PBX Series 300
• *Attendant's consoles.* Operators at one or more consoles assist with the switching of calls. Varying numbers of switching functions are automatic. (See Fig. 22.3 and 22.4.)	X	X	X
• *Push-button station selection.* The attendant has a status light and button for each station she controls.		X	X
• *Station-to-station dialing.* A station can dial any other station attached to the PABX without assistance from the attendant.	X	X	X
• *Direct outward dialing.* A station can dial a call on the public network without assistance from the attendant.	X	X	X
• *Station hunting.* There are several lines to the PABX location. When the first line dialed is busy, the others are tried.	X	X	X
• *Call transfer by attendant.* A station receiving a call can signal the attendant and request that the call be transferred to another station.	X	X	X
• *Call stacking ("camp-on") by attendant.* When one or more calls are received for a station which is already busy, the attendant can put the calls in a waiting state until the station is free.		X	X
• *Call waiting ("camp-on") signal.* A person receiving a call can be signaled to inform him that another call is waiting.		X	X
• *Conference calls set up by attendant.* Calls between more than two telephones (sometimes not more than three) can be connected by the attendant.		X	X
• *Night service to a few designated stations.* The external lines serving the PABX can each be switched through to a certain station when there is no attendant.	X	X	X

BOX 22.1 Continued

	Features of Bell System PBX Packages		
Features of some conventional PABXs (cont.)	PBX Series 100	PBX Series 200	PBX Series 300
• *Power failure transfer.* Certain prearranged stations can continue operation when there is a commercial power failure.	X	X	X
• *Station restriction from outgoing calls.* Designated stations are prevented from making outgoing calls.	X	X	X
• *Secretary consoles.* A secretary can have a console for handling the calls of one or more employees.	X	X	X
• *Key telephone stations linked to secretaries.* An employee can signal his secretary and vice versa. Secretaries can intercept calls.	X	X	X

Additional features of the Bell System Centrex II Package

- *Each station has its own external telephone number.*
  As illustrated in Fig. 22.3

- *Direct inward dialing.*
  Stations can be dialed from the outside without intervention by the attendant.

• *Call transfer by individual stations.* A station can transfer a call from the outside to another station without assistance of the attendant.	X
• *Consultation hold by individual stations.* A user can place a call on "hold" while he dials another station; he can then return to the interrupted call.	X
• *Third-party add-on by individual stations.* A user receiving a call can dial another station, thereby setting up a three-way conference call.	X
• *Night answering service.* When the attendant console is unoccupied any station may answer incoming calls by dialing a code.	X

- *Listing of charges incurred by stations.*
  A listing is provided for each station of the nonlocal calls dialed, giving the number, time, duration, and cost of each call.

BOX 22.1 Continued

Other desirable PABX features offered on some computerized systems

- *Automatic call forwarding.*
  A person may go to a different location and inform the PABX of its extension number; the PABX will forward his calls to that number.

- *Automatic call stacking.*
  When calls arrive for a station which is busy they will be automatically queried, possibly with a spoken "wait" message being played to the caller.

- *Automatic call distribution.*
  When calls may be answered by any of a group of stations the calls are automatically distributed to the first free station.

- *Automatic call back.*
  When a user places a call and the number is busy, the user may instruct the PABX to call him back when it is free.

- *External number repetition.*
  When a user dials a long number and it is busy, the user may instruct the PABX to remember the number so that it can repeat the dialing.

- *External conference calls dialed by individual stations.*
  A user can set up an external conference call with the help of the attendant.

- *Users who move can retain their numbers.*
  The number can apply to a new station without rewiring, merely by changing the tables used by the PABX.

- *Distinctive ringing.*
  Different ringing tones are used for different categories of calls, for example internal, external, secretary, a call from a given extension, a call from head office. The user is thus given some indication where the call is from before he picks up his phone.

- *Pushbutton to dial pulse conversion.*
  Signaling conversion is performed so that pushbutton telephones can be used even when the local central office accepts only rotary-dial pulsing.

- *Timed operator reminders.*
  If an operator rings an extension which does not answer, the operator will be alerted after 30 seconds; the caller is not kept waiting indefinitely.

- *Abbreviated dialing.*
  Commonly used lengthy numbers are replaced by two-digit numbers.

- *Intrusion signal.*
  A signal is automatically sent to a user if a third party comes on the line (for privacy protection).

- *Automatic call transfer.*
  Incoming calls to a busy station are automatically transferred to another designated station.

- *Alarm-clock calling.*
  Users can register in the PABX a time at which they wish to be called.

BOX 22.1 Continued

- *Paging by attendant.*
When a subscriber cannot be located the attendant can page him (possibly by a radio pager). He dials his own number and is connected to the party trying to contact him.

- *Paging dialed by stations.*
A paging operation can be initiated automatically by any user.

- *Do-not-disturb facility.*
A user may dial a code requesting that no telephone calls be sent to him.

- *Call chaining.*
When a user makes many successive calls to a remote location he is not disconnected at the end of each call so that he has to redial but is automatically transferred to the remote operator or PABX.

- *Traffic monitoring and measurement.*
Traffic is continuously monitored and a manager can obtain a traffic report at any time.

- *Corporate network connections.*
The PABX makes connections to a corporate network including tie lines, CCSA, foreign exchange lines, WATS lines, and specialized common carrier facilities.

- *Priority access to corporate network facilities.*
A priority structure is used so that certain subscribers are given priority access through corporate facilities and do not normally receive network busy signals.

- *Facilities for interconnecting computers and terminals.*
The PABX is designed to handle data traffic, possibly from a terminal without modems. The PABX may be directly coupled to a data-processing system.

- *Data-collection facility.*
The PABX automatically scans data-collection devices and assembles the data for retransmission.

- *Contact monitoring and operation.*
The PABX automatically monitors contacts, e.g., for fire or burglar protection or for process-control applications, and may operate certain contacts.

- *Security features.*
Security provisions are provided to help ensure privacy.

Additional features with sound-recording devices

- *Prerecorded messages.*
Spoken messages can be automatically played to callers kept waiting and for other conditions.

- *Automatic telephone answering.*
When a user does not answer the PABX can play a prerecorded message from him, possibly requesting that a message be left.

433

BOX 22.1 Continued

- *Automatic message reception.*
  Messages for a user who does not answer can be spoken to the PABX, which records them for that user.

- *Remote reading of messages.*
  A user who is traveling can dial his PABX and instruct it (using an appropriate security code) to play back the messages that have been left for him.

- *Dictation services.*
  Users may dictate memos to the system, to be typed by a typing pool.

- *Reminder messages.*
  Users may speak reminder messages to the system, to be played back to a given extension at a given time.

- *Music or advertising*
  Prerecorded sound is played to users who are kept waiting.

Additional features which can lower telephone costs

- *Automatic restrictions on station usage.*
  Each station has restrictions specified concerning what calls it may and may not make, by day and by night.

- *Facility for dialing personal calls.*
  Users may dial personal calls by prefacing them with a code; the users are then billed for their personal calls.

- *Automatic minimum-cost routing on corporate networks.*
  A call is routed to a number accessible via a corporate network by whatever is the cheapest route, e.g., first choice: tie line; second choice: WATS line; third choice: direct distance dialing (applicable only to calls of certain priorities).

- *Remote access for corporate network facilities.*
  The corporate network may be accessed from certain telephones outside the corporation to obtain lower-cost long-distance calls.

- *Automatic monitoring of charges incurred by users.*
  Charges incurred by all users are continuously monitored and may be inspected by a manager at any time. A complete listing of charges incurred can do much to reduce telephone usage.

- *External access to corporate network.*
  Public telephones can use corporate tie-lines, WATS, and FX lines via the PABX.

economic value. First, the machine can be programmed to control and to lower overall telecommunication costs. Second, in addition to connecting telephones it can carry out data-processing functions; the IBM 3750 PABX, for example, connects terminals to a data-processing system and can scan data-collection equipment and contact-monitoring points. Both of the latter functions are, perhaps, more likely to be provided by independent vendors than by the established common carriers. The first of these functions is designed to take revenue away from the common carriers, and the second is more in the province of the data-processing industry than the telephone industry.

Figure 22.5 The connection of subscribers to a No. 101 ESS computer that operates in a time-shared fashion to provide the functions of up to 32 private branch exchanges.

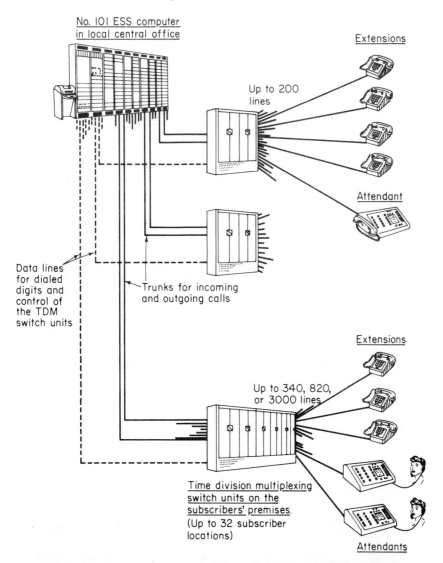

ESS NO. 101   Figure 22.5 shows the connection of subscribers to a
                          No. 101 ESS at the central office location (8). The sys-
tem can handle up to 32 separate subscriber locations. Most subscribers have a
unit that permits the connections of up to 200 telephone extensions. Other large
units can handle 340, 820, or 3000 extensions.

The switching is done in the small cabinet shown in Fig. 22.6. This uses
time-division switching, which is described in Chapter 25. The organization of the
switch unit is shown in Fig. 25.5. Fifty time-division paths through the switch unit
are available to the 200 telephone users, a ratio that is usually more than adequate.
If the average subscriber uses his telephone one tenth of the time, then the probability of failing to obtain a circuit will be extremely low.

The digits dialed or keyed by a telephone user go to the ESS 101 computer
on one of two data lines connecting the computer and the switch unit. The other
data line is used by the computer to control the switching—in other words, to assign the time-division paths to the appropriate lines. In this way, the switching is
under control of the computer program. There are voice trunks from the switch
unit to the computer, for incoming and outgoing calls, and these trunks are
switched in the same way.

SOLID-STATE   ESS 101 is interesting because of its time-shared com-
SWITCHING   puter control of remote switches. Some of the stand-
                         alone computerized PABXs are interesting because
they use no moving parts in their switching matrix. The switching is done entirely

Figure 22.6 The time-division switching unit on the subscriber's
premises, when an AT&T ESS 101 is used to replace private branch
exchanges. This is the unit shown connected to 200 telephone handsets
in Fig. 22.5. (Courtesy of Bell Telephone Laboratories.)

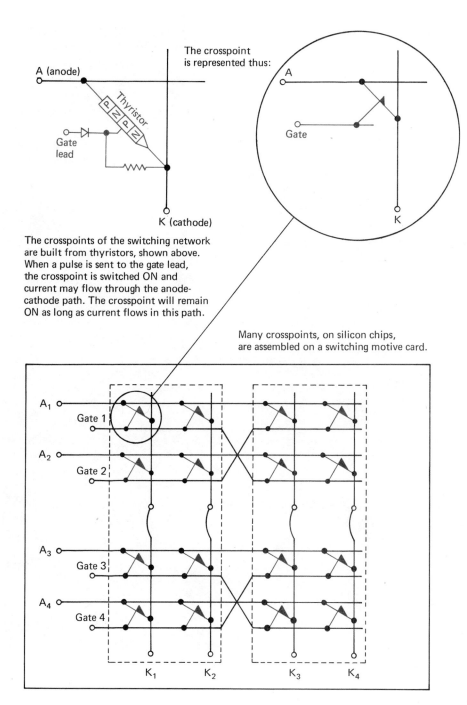

The crosspoint
is represented thus:

A (anode)

Thyristor

Gate lead

K (cathode)

The crosspoints of the switching network
are built from thyristors, shown above.
When a pulse is sent to the gate lead,
the crosspoint is switched ON and
current may flow through the anode-
cathode path. The crosspoint will remain
ON as long as current flows in this path.

Many crosspoints, on silicon chips,
are assembled on a switching motive card.

Figure 22.7 Solid-state switching: fast and reliable.

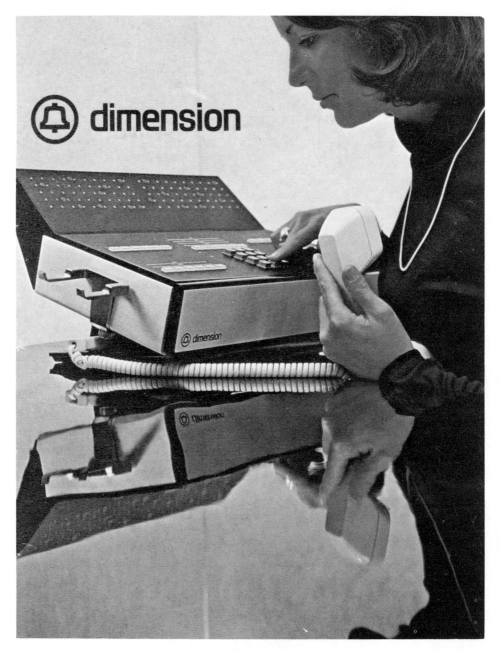

Operator console

Figure 22.8   The AT&T Dimension PBX.

The AT&T Dimension PBX

Some of the many programmed features of the Dimension system:

Outgoing Trunk Queuing helps you when outside lines are busy. Ordinarily, you would have to keep on trying, or have the console attendant keep trying for you. But now a Dimension system will "remember" your extension and call you back as soon as a line is free and your extension is available.

Three-Way Conference Transfer lets you put one person on hold, call another person and consult with him in private, then include him in a conference call with the original person. Or you can excuse yourself and let the two of them confer.

Automatic Route Selection lets you take full economic advantage of your WATS lines, FX lines or the Bell System DDD network by automatically choosing the most economical way to make that long distance call at that time.

Distinctive Ringing means that without picking up your phone, you can tell if the call coming in is from another extension or from the outside or sometimes even from a specific person inside your company. It's all done through a series of distinctive rings. A Dimension system can even tell you if someone is trying to reach you while you're on the phone—through a series of unique tone signals.

Automatic Callback — Calling is a feature that works hard for you when you're trying to get through to someone in your company and the line is busy. If you dial a code into your Dimension system, the PBX will keep trying that person's phone and when his line and your line are free, it will put the call through by ringing your phone first and then the other person's phone.

Timed Reminders — If an operator puts your call through to an extension which does not answer, the operator is alerted after 30 seconds; you are not left hanging.

Computer

in solid-state circuitry and hence is extremely fast and not prone to mechanical problems (Fig. 22.7). The maintenance concerns associated with electromechanical switching disappear.

EXAMPLES OF The AT&T Dimension PBX is a small computer with
COMPUTERIZED PABXs a solid-state switching network through which PAM
(pulse amplitude modulation) pulses flow, like those in
Fig. 14.2. Telephone calls reaching the PBX are sampled 8000 times per second and thus converted to streams of PAM pulses. After these have been switched they are filtered to convert them back to analog signals.

The left-hand side of Fig. 22.8 shows the operator console of the Dimension system, and the right-hand side shows the cabinet housing the computer and switching network. One such cabinet can handle up to 120 extensions, and the system capacity can be expanded up to 400 extensions.

The frequently used programs are held in the main memory of the computer. Infrequently used programs, including maintenance diagnostics, are held on a small magnetic tape. The tape also contains a copy of the main programs for backup in case of power failure.

The Dimension system has most (but not all) of the features listed in Box 22.1. New features can be added easily to the system by programming. A tape unit is used to load new feature programs into the main memory.

The IBM 3750, a solid-state PABX installed outside North America, is designed to handle data as well as voice. Any terminals or data-processing machines may be interconnected by dialing from a telephone, the telephone being free for voice transmission once the connection is made. For inexpensive data collection, push-button telephones may be used, and the 3750 assembles the messages; checks each character; adds control information, such as date, time, and terminal identification; and directs the data to a computer, tape punch, or other device. Voice answerback may be used for controlling data collection or other terminal operations. The machine can read the setting of switch contacts—those on the customer's premises or those remotely connected. It can similarly operate contacts. Such contacts may be used for many different purposes—for example, fire or burglar detection, process control functions, or remote meter reading.

Some PABXs can be equipped with a paging unit, which consists of a radio transmitter. Wandering users carry pocket radio receivers. Any extension user can page a person directly by dialing. That person hears "beep" signals from his pocket device, and when he responds he is automatically connected to the caller.

For reliability, two computers are used in some PABXs. At any one moment one computer is "active" and the other is "standby." The network is under control of the active computer, and the standby computer is always ready to take over. The standby computer continuously checks its own circuitry and monitors the active computer. If the active computer fails, an automatic switchover takes place, and the standby machine becomes "active." As solid state circuitry becomes more reliable so PABXs with a single computer become more appealing.

Because of the flexibility of stored-program control in such machines, a variety of different facilities and restrictions can be associated with each individual extension. Among the facilities that may or may not be assigned to any extension in the IBM 3750 are

1. Outgoing calls permitted via the attendant.
2. Direct outward dialing of *local* calls by *day*.
3. Direct outward dialing of *local* calls by *night*.
4. Direct outward dialing of *national* calls by *day*.
5. Direct outward dialing of *national* calls by *night*.
6. Direct outward dialing of *international* calls by *day*.
7. Direct outward dialing of *international* calls by *night*.
8. *Intrusion allowed.* The extension may be interrupted for a pending call.
9. *Interrupt.* The extension user may signal a busy extension that he wished to interrupt.
10. *Add-on third party.* The extension is allowed to initiate a three-way call.
11. *Paging.* The extension user is allowed to dial a paging signal.
12. *Contact monitoring.* Allows the extension to sense and operate a contact.
13. *Numeric data collection.* Allows numeric data to be collected through an extension.
14. *Alphanumeric data collection.* Allows alphanumeric data to be collected through an extension.

A recorded voice announcement attachment is available; it could respond to external callers in certain circumstances or could send messages to extension users under computer control.

NUMBERING SCHEME   To carry out functions other than the straightforward dialing of telephone numbers a user must have some means of informing his PABX what he wants to accomplish. He usually has only the telephone dial (or Touchtone keys) to use for this purpose. Special meaning must therefore be associated with the dialing of certain digits. An advanced PABX will normally have an instructional brochure telling its users how to employ the special dialing codes. An alternative is to provide many extra keys on the telephone handset for carrying out special functions, as is done, for example, on the key telephones of Executone, Inc.

A common example is the dialing of a digit (often 9) to obtain an outside line. Sometimes an 8 is dialed preceding the dialing of any number on the corporate tie-line network. A 0 is dialed to reach the attendant. On some systems a 1 is dialed for manager-to-secretary or secretary-to-manager calls. Dialing 20 before an outside number indicates, on some systems, that it is a personal call and will be billed to the individual in question.

Some dialing codes such as the above are used before a call is placed. On advanced PABXs another set of dialing codes is designed to be used in the middle of an established call. For example, on a typical IBM 3750 system a 2 is dialed followed by an extension number to transfer a call to another extension. A 2 can be dialed to place an existing call on "hold" while its recipient dials and talks to another subscriber. Dialing 4 returns to the "held" call, and dialing 3 would set up a conference call between the three parties. Dialing 0 in the middle of a call connects its recipient to the operator while putting the caller on "hold."

Some dialing codes carry out special operations. For example, extensions so authorized may dial 22 to bar all incoming calls—a much appreciated do-not-disturb signal. Incoming calls may be unbarred again by dialing 23. Extensions so authorized may interrupt busy extensions by dialing 6. This code places a soft "camp-on" signal on the busy extension, requesting its user to put his call on "hold" for a moment or to terminate it.

If a user dials a lengthy number and receives a "busy" signal, he may instruct the PABX to store that number by dialing 7. When he tries to reach the number again he dials 27, and the PABX dials the stored number. If he dials an extension and the person is not there, he may dial 5 to initiate a paging operation. If the person is wearing a small radio paging unit, he will hear a paging signal. (Alternatively, loudspeaker paging could be used.) The paged person goes to the nearest telephone and dials 25 followed by his own extension number and is automatically connected to the extension which paged him.

Many other dialing codes are used on some systems. The use of programmed control gives a PABX planner the type of flexibility that data-processing designers have in the choice of codes.

ADD-ON FACILITIES   Box 22.1 lists some features not incorporated in the conventional PABXs obtainable from the common carriers. Many users, however, retain their common carrier PABX; in some countries they do not have a choice. Because of this, in some countries many different devices are coming on to the market which operate in conjunction with an established PABX to provide additional features. Such features include traffic-monitoring, automatic call distribution, paging, automatic telephone answering, alarm-clock features, scanning data-collection terminals, and other teleprocessing features. In some installations one finds an electromechanical PABX connected to a small computer for providing add-on functions.

Possibly the most salable of the add-on functions are those which can reduce overall telephone costs. For example, monitoring equipment can show exactly how the telephone facilities are being used and hence enable management to control unnecessary use or the use of corporate facilities for personal calls. An add-on device manufactured by the North Electric Company, called a *communications extender,* gives dialed access from external telephones to a corporate PABX and hence enables remote locations to take advantage of it. A user dials the telephone number of the device and then dials a security code (to prevent unauthorized use). The communications extender is connected directly to a PABX and permits the calling telephone to do whatever could be done from a PABX extension. In particular it may be connected to corporate network facilities so that the user is given direct access to corporate tie lines, WATS lines, or foreign exchanges lines. An executive at home in Los Angeles may thus call an employee at home in New York over the corporate network facilities (at lower cost than Direct Distance Dialing).

23 MESSAGE SWITCHING

The previous three chapters have discussed the switching of telecommunication *lines*. The switch forms an electrical circuit between the lines, forming what the industry sometimes describes as a "copper" path between users. The switching of physical circuits is referred to as *space-division switching*. There is a different type of switching which is applicable mainly to data or message traffic in which the switch does not make and break electrical circuits. Instead each line is connected to a facility which can store the messages received and then retransmit them down another line. As with circuit switching this store-and-forward facility can be manual, automatic, or semiautomatic. Today it is likely to be done by computer.

With circuit switching the subscribers or machines at each end of a connection can converse with one another as rapidly as they please. They have a two-way connection for the duration of the call. Message switching is designed for the one-way delivery of messages and is not suitable for real-time conversational interactions with computers. (The next chapter discusses packet switching, a fast type of store-and-forward operation which *is* used for interactive computing.)

ADVANTAGES OF MESSAGE SWITCHING   Message switching of data traffic has a number of advantages over circuit switching:

1. The communication network can be designed to operate in an efficient manner with higher line utilization than a circuit-switched network. To phrase that another way, a lower-cost network can be designed for the same traffic volumes. Traffic peaks are smoothed out; they do not need to be handled in real-time. Many terminals can be attached to one line.

2. The simultaneous availability of calling party and called party is not necessary. The calling party sends a message to the switching computer, and the computer delivers it when the called party can accept it. The calling party need never receive "busy" signals.

444

3. "Blocking" occurs on a circuit-switched network when all paths to a destination are occupied. There is no blocking on a message-switched network. The message is stored until a path becomes free. When the traffic is heavy on a circuit-switching network some calls will not be connected. When the traffic is heavy on a message-switched network the delay before delivery increases. This property makes a message-switched network more flexible in adapting to peaks of traffic.

4. A message-switching network can send one message to many destinations. The messages may be addressed to a stored "mailing list."

5. Message priorities can be established so that high-priority traffic is delivered rapidly.

6. Error control and recovery procedures can be built into a message-switched network. All messages may be numbered, stamped with transmission time and date, and filed for later retrieval.

7. A message-switching network can carry out code conversion, thereby permitting incompatible machines to communicate. For example, conversion between Baudot code (used for telex), ASCII code (used for TWX and computer data), and EBCDIC code is performed. The Graphnet corporation performs conversions so that ASCII-coded data can be delivered on conventional analog facsimile machines.

8. Messages sent to inoperative terminals may be intercepted and either stored or rerouted to other terminals.

For many years in conventional telegraphy it has been considered vital to saturate the capacity of the costly long-distance lines insofar as possible. This was achieved by using message-switching centers to relay messages. Lines from these centers connected multiple terminal locations, as shown in Fig. 23.1.

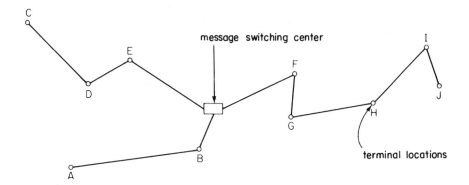

Figure 23.1

PUBLIC NETWORKS   Message switching is used in both public and private networks. The telex network uses circuit switching, and operators at telex machines can carry on a slow conversation with one another. Most other telegraph systems use message switching. When the transmission distances are long the economics tend to favor message switching. Consequently, telex-type circuit switching was emphasized in Europe with its short distances and high population density, and message switching was more common in North America.

The U.S. international carriers use message switching to give efficient utilization of costly transoceanic circuits. Bell Canada, Western Union, and now smaller competing companies offer a variety of message-switched services. Western Union's services include the following:

1. PMS, Public Message Service, for the delivery of telegrams.

2. Mailgram, in which telegrams, including those broadcast to many receivers, are transmitted to a city close to their destination and delivered by mail.

3. TCS, Telex Communications Service, which interconnects telex, TWX, and PMS services.

4. SICOM, Securities Industry Communication, which gives stock market information to its subscribers and provides special services for brokers.

5. Bank Wire Service, which passes credit and other information between banks.

6. INFOCOM, which offers subscribers what appears to them to be a private message-switching system to which only authorized users have access and which serves a specified set of locations. In practice many such private systems are derived from a common network which incorporates both shared and private message-switched channels and telex circuit-switched channels.

PRIVATE NETWORKS   Many organizations have their own private message-switching systems. Some of these are small, serving less than 100 terminals. Some are gigantic international systems serving thousands of locations. The military has pioneered some of the most advanced systems. Some organizations combine to share a private network. For example, some bank networks interconnect many banks, who share the cost. Most of the world's airlines are interconnected by message-switching networks.

TORN-TAPE   The early message-switching centers were manually
SWITCHING CENTERS   operated. They are referred to as *torn-tape* switching centers because the operators tear off transactions punched in paper tape and retransmit them. Such systems were common up to the early 1970s but are now being replaced by more automatic equipment.

Figure 23.2 shows a torn-tape center. Messages received are preceded by the address or addresses to which they are to be sent. They are punched into paper tape. Operators examine the tape, read the address or addresses, tear off individual messages, and carry them to a rack where they await retransmission. Queues of messages waiting to be sent build up as on the right-hand side of Fig.

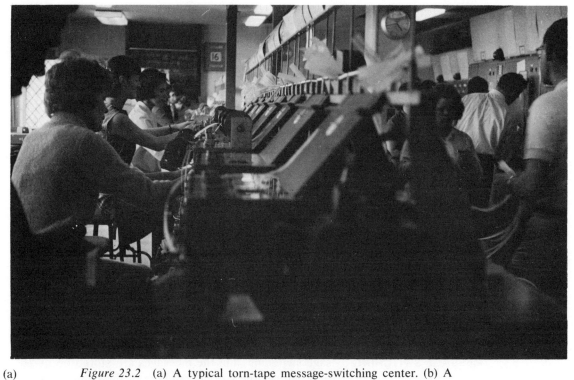

(a)

Figure 23.2 (a) A typical torn-tape message-switching center. (b) A queue of messages on a rack awaiting retransmission in the above center.

(b)

447

23.2. A torn-tape center may have many paper-tape reperforators receiving messages and many paper-tape readers transmitting them, with operators scurrying between these machines. An operator transmitting a message selects the appropriate station on a multistation line by pressing a button on the transmitter console.

Large centers of this type have employed as many as 60 operators.

**SEMIAUTOMATIC
SWITCHING CENTERS**
Figure 23.3 shows a somewhat more mechanized type of switching center. Messages are still stored in paper tape, but now it is not torn by operators. Continuous tape is punched and waits until an operator can attend to it. By reading the address in the message and pushing the correct buttons on the console, the operator can set up a connection to another reader, this time one associated with an outgoing line. The message is then read and transmitted from a second paper-tape loop as shown.

In some systems the job of the console operator in Fig. 23.3 is mechanized, thus giving a fully automatic switching center, still using paper-tape storage.

**COMPUTERS FOR
MESSAGE SWITCHING**
Today computer systems are in common use for message switching, and the messages are stored on disks or whatever the storage medium of the computer is.

One of the largest commercial message-switching centers is the Collins Radio System at Cedar Rapids, Iowa. This uses 10 computers, 24 hours of every day, to switch airline messages. Its volume totals more than $\frac{1}{4}$ million messages per day.

A computerized switching center can be considered to be an automation of the manual and semiautomatic systems in Figs. 23.2 and 23.3. Its advantages over these are

1. Messages are delivered more rapidly (much faster than the torn-tape system).
2. Operating errors are eliminated.
3. Priority indicators in the messages can be easily recognized and acted upon.
4. Transmission errors can be handled with automatic retransmission.
5. Messages can be analyzed and various automatic actions taken if required.
6. Routing a message to several different destinations can be done easily. The computer can book address lists or send broadcast messages.
7. Messages can be stored for later retrieval if desired.
8. Processing of certain message types can be carried out.
9. For message volumes above a certain limit it becomes the cheapest system.

Figure 23.3 A semi-automatic message-switching center using continuous paper tape loops.

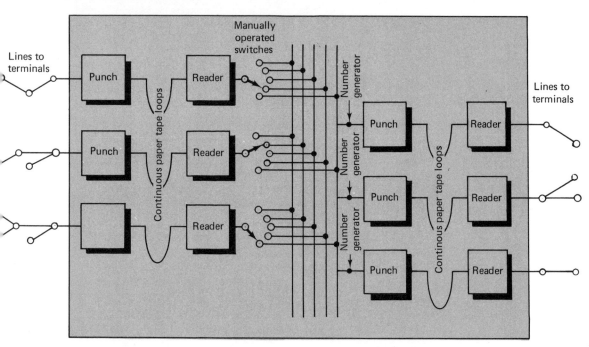

The single computer may have one disadvantage over the above methods. If a computer failure occurs, the entire exchange is out of action until it is repaired. Most of the likely failures in the other methods put only one line or, at worst, a few lines out of action. The mean time to repair the computer when it fails may be of the order of 2 hours. On some networks a break of 2 hours is tolerable. On others it is not. If it is not, two computers must be used, giving a duplexed system, so that if one fails. the other takes over. The standby machine may be doing billing or other data-processing functions.

Some message-switching systems have special no-break power supplies to maintain operation when mains power fails. A motor-driven alternator may be used, driven off a battery that is kept fully charged by being floated across the mains. Such a system can provide at least 2 hours of electricity supply when the

Figure 23.4   Configurations for computer message switching.

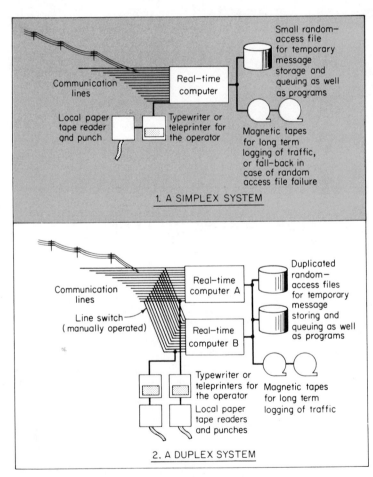

mains fail. This gives ample time to start up a secondary diesel electric set to keep the battery charged.

The facilities desirable on a message-switching computer are a small random-access file for the temporary storing of messages being routed and a serial-access file, often magnetic tape, for logging messages if they are to be kept for a long period of time. It must, of course, have the facilities for handling all the various types of communication lines used. The memory size is often 64,000 to 250,000 bytes or the equivalent in binary words. Figure 23.4 sketches typical configurations.

In a data-processing system, message switching may be performed along with the other work and regarded as one element of the complete system. In this case it may be done by a separate line-control computer that feeds messages also to the main computer, or one computer may handle message routing and data processing at the same time, in parallel. When message switching was first performed by computer, it was usually done by stand-alone machines dedicated to this function. However, as multiprogramming techniques and software improve more systems combine message switching with some data processing. If a system handles, say, 2000 messages per day, even a small computer is left with much idle time that can be absorbed in other work. On the other hand minicomputers are rapidly dropping in cost.

FUNCTIONS OF
THE SYSTEM

Box 23.1 lists the functions that a message-switching computer performs.

The program to carry out these functions resides mainly in main memory, although exception routines may be on a drum or disk to be called in when wanted, for example, when emergencies or line outages are being handled. Also in main memory is a set of tables giving the addresses of the terminals and the lines they are on. This enables the computer to find the correct terminal for a message with a certain destination code. Other tables indicate the status of each line and whether there are messages waiting to be transmitted down it.

The organization of the files of messages is a major design question in this type of system. The queues of messages that are kept on the files vary from one period to another. But messages must be written on the files and retrieved from them in the minimum time. On a disk file in which the read/write heads are physically moved to seek a record, the data must be placed so that the seek times— relatively long in terms of computer speeds—are not too long.

A common method of organizing such files is to allocate areas for each output terminal. As messages are received, they are written sequentially in these areas. These are rather like the pigeonholes for letters in a club or college common room. When the attendant receives letters, he places them in the pigeonholes of the persons who are to receive them. If a pigeonhole becomes full because many letters are received for one person, an overflow area is available. The computer

BOX 23.1   Functions performed by a message-switching computer

1. The system accepts messages from distant terminals. The terminals are often teleprinters and paper-tape readers, but other devices may be used, such as card readers and special input keyboards. The system may also accept messages from other computers.

2. On receipt of a message it analyzes the message's header to determine the destination or destinations to which the message must be sent.

3. The system may analyze the header for a priority indication. This will tell the program that certain messages are urgent. They must jump any queues of messages and be sent to their destination immediately.

4. It may analyze the header for an indication that some processing of the message is necessary; for example, statistical information from the message may be gathered by the system.

5. The system detects any errors in transmission of the incoming message and requests a retransmission of faulty messages. This retransmission may be automatic.

6. It detects format errors in incoming messages insofar as possible. Types of format errors that may be picked up include the following:

   (a) *Address invalid.* The address to which the message is to be sent is not included in the computer's directory.
   (b) *Excessive addresses.* There are more than the given maximum number of addresses allowed.
   (c) *Incorrect format.* An invalid character appears in the message in an incorrect location.
   (d) A priority indicator is invalid.
   (e) *Originator code error.* The address of the originator is not included in the computer's list.
   (f) Incorrect character counts.

7. The system stores all the messages arriving and protects them from possible subsequent damage.

8. It takes messages from the store and transmits them to the desired addresses. One message may be sent to many different addresses. In doing this, it does not destroy the message held in the store. The store is thus a queuing area for messages received and messages waiting to be sent as well as a file in which messages are retained.

9. The system redirects messages from the store and sends them to the terminals requesting them. It may, for example, be asked to resend a message with a specified serial number or to send all messages from a given serial number.

10. Systems in use store messages in this manner for several hours or, on some systems, several days. Any message in the store is immediately accessible for this period of time.

BOX 23.1 Continued

11. The system may also maintain a permanent log of messages received. This will probably be done on a relatively inexpensive medium, such as magnetic tape, and not on a random-access file.

12. If messages are sent to a destination at which the terminal is temporarily inoperative, the system intercepts these messages. It may automatically reroute them to alternative terminals that are operative. On the other hand, it may store them until the inoperative terminal is working again.

13. It may intercept messages for other reasons. For example, the system may be programmed to send a message to the location of an important person, although he may be moving from one place to another. The person in question indicates his current location to the computer, and the computer diverts messages for him to that location. The system may handle messages on a priority basis. There may be one urgent priority level so that these messages are sent before any others. Some systems have more than one level of priority, priority level 1 being transmitted before priority level 2, priority level 2 being transmitted before priority level 3, and so on. The system may notify the operator in the event that any priority queue becomes too great. A simple system may have no priority scheme, messages being handled on a first-in, first-out basis.

14. The system maintains an awareness of the status of lines and terminals. It is programmed to detect faulty operation on terminals where possible, to make a log of excessive noise on lines, and to notify its operator when a line goes out. The system maintains records of any faults it detects.

15. On a well-planned system the messages should be given serial numbers by the operator sending them. The computer checks the serial numbers and places new serial numbers on the outgoing messages. When serial numbers are used, the system can be designed to avoid the loss of any message. This is especially important in the event of a computer failure or of a switchover in a duplex system.

16. At given intervals, perhaps once an hour, the system may send a message to each terminal, quoting the serial number of the last message it received from that terminal. The terminal's operator then knows that the switching system is still on the air.

17. The system may conduct a statistical analysis of the traffic that it is handling.

18. It may be programed to bill the users for the messages sent. It may, for example, make a small charge per character sent from each terminal and bill the terminal location appropriately.

19. It produces periodic reports of its operation for its operator. These may include reports on the status of all facilities, error statistics, reports giving the number of messages in each queue, message counts, and so on.

sends messages from these areas to where they are required. If the retrieval of any message is demanded, the machine can search the appropriate "pigeonhole" for the message.

The active area of the file at any one instant is relatively small. As the day proceeds and messages are sent, the active area may move across the file in such a way that the seek times within that area are always small.

PRIORITY MESSAGES   When computerized message switching is used the delivery time can be much shorter than it is with non-computerized switching. Of particular value is the possibility of appending a priority code to certain messages to ensure that the computer makes them jump the queues. When an executive requires immediate attention to a message he may give it a high priority and perhaps telephone its recipient while the message is being sent to discuss its contents. As long as there is a means of ensuring that only a small proportion of messages have that priority it is possible to guarantee fast delivery.

The SITA (Société Internationale de Telecommunications Aeronautique) network for exchanging messages between airlines in different countries has four levels of priority, which seem to be typical of the needs of other systems also:

Top priority. Delivery time: 3 seconds (for the transfer of single-address messages between computer systems).

Priority 2. Delivery time: up to 2 minutes (for example, for passing urgent messages between people).

Priority 3. Delivery time: up to 30 minutes.

Priority 4. Delivery time: up to 12 hours (or overnight delivery).

LARGE NETWORKS   Many corporations have a private message-switching system with one computer center. Terminals are connected on multipoint lines to that computer center. When the network is very large or the traffic volume very high, it become economical to have more than one switching location.

Figure 23.5 shows a large network. At the center is a computer which carries out all the functions listed in Box 23.1. The peripheral machines *B, C, D,* and *E,* may carry out all these functions, or a subset of them, the purpose being to lower the overall system cost. The peripheral machines could merely be multiplexors or concentrators which funnel the traffic into the higher-speed lines. On the other hand, the peripheral machine could be a switch which forwards a message from, say, terminal *F* to terminal *L* without it going via computer *A* but nevertheless which does not file the message. It may send it on a one-way trip to computer *A* for filing if necessary and may send billing information to computer *A*.

If the traffic volume is high, it may be economical for the peripheral machines in Fig. 23.5 to be full-fledged message-switching computers.

In some organizations a machine which performs on-line data processing has message switching as one of its lower-priority functions. In this case the machines *A*, *B*, *C*, *D*, and *E* of Fig. 23.5 would all be data-processing computers, but now they relay messages between any of the terminals that are connected to them. Most airline reservation computers, for example, perform this message-switching operation at the same time as their reservation processing.

Various other computer network configurations have message switching as one of their functions.

Figure 23.5   A large network with very long distances may use more than one computer for message routing.

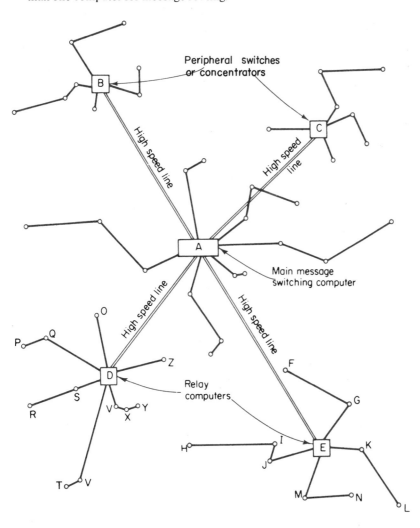

TWO-LEVEL   As with circuit switching or any other form of switch-
NETWORKS   ing, more than one level of switching is needed when a
                      very large number of points are to be interconnected.
With message switching, usually two levels are sufficient in a hierarchical structure. The separate message-switching computers are interconnected with a network of higher-speed lines. This high-level network is analogous to the trunks of
the telephone network. The lower-level network from the terminals to their nearest switch is analogous to the local loops of the telephone network.

The high-speed connections between machines *A, B, C, D,* and *E* in Fig.
23.5 could be regarded as a high-level network, and the multipoint lines from these
machines to the terminals could be regarded as low-level local networks. The
protocol on the high-level lines could be different from that on the low-level lines.
The high-level network in Fig. 23.5 uses the shortest and least expensive line layout. Sometimes additional, redundant, lines interconnect the machines to give
enhanced network reliability. The traffic is shared between the redundant paths,
but if a line or computer fails, it may be possible to bypass it. Figure 23.6 shows
the growing network used by SITA for airline traffic. The solid lines in this figure
represent the high-level network. The dotted lines represent the local distribution
networks.

Figure 23.6 The SITA message-switching network carrying reservations and other messages between airlines. The solid lines are a high-
level "trunk" network between the separate message-switching computers, each of which files and switches traffic for its local area.

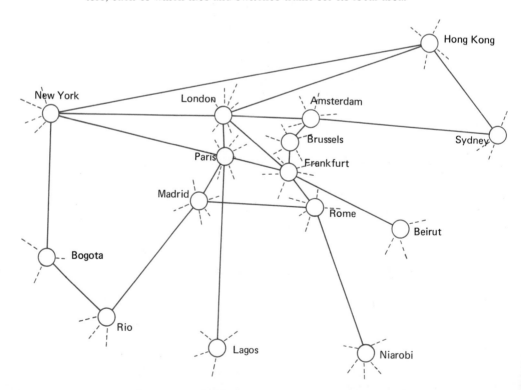

24 PACKET SWITCHING

CCITT Definitions:

packet: A group of binary digits including data and call control signals which is switched as a composite whole. The data, call control signals, and possibly error-control information are arranged in a specified format [1].

packet switching: The transmission of data by means of addressed packets whereby a transmission channel is occupied for the duration of transmission of the packet only. The channel is then available for use by packets being transferred between different data terminal equipment [1].

Note: The data may be formatted into a packet or divided and then formatted into a number of packets for transmission and multiplexing purposes.

Store-and-forward switching has existed for decades in telegraphy where it is called *message switching*. At the end of the 1960s a new type of store-and-forward switching came into experimental use, called *packet switching*. Whereas message switching is intended primarily for non-real-time people-to-people traffic, packet switching is intended primarily for real-time machine-to-machine traffic, including terminal-to-computer connections, and is employed to build computer networks. These differences in purpose are such that there are major differences in operation between message-switching and packet-switching networks. The most important difference is in the speed of the network. A packet-switching network may be expected to deliver its packet in a fraction of a second, whereas a message-switching system typically delivers its message in a fraction of an hour.

Public packet-switching networks are being constructed in many countries. The technique should, perhaps, be regarded as being competitive with *circuit switching* of data rather than message switching, because either circuit switching or packet-switching could be used for fast computer-to-computer connections. Packet switching, as we shall see, has some very important advantages over switching as it is practiced on today's telephone and telegraph networks.

ARPANET   The first operational packet-switching network was
AND TELENET   ARPANET, shown in Fig. 4.9. ARPANET was
                       designed to interconnect university computer centers
and other centers where ARPA-funded projects were in progress. (ARPA is the Advanced Research Projects Agency of the U.S. Department of Defense.) The packet-switching techniques developed for ARPANET are now being used on a public network operated by Telenet Communications Corp. The Telenet network is shown in Fig. 24.1.

The centers connected in Fig. 24.1 are linked with a network of wideband lines operating at 50,000 and 56,000 bps. Each of these lines terminates in a small network computer. The network computer has two main functions. First, it acts as a link between the network and the data-processing equipment which uses the network. Second, it carries out the switching operation, determining the route by which the data will be sent and transmitting them.

The customer's computers which the network serves are called *host* computers. When one host computer sends data to another, it passes the data with a destination address to its local network computer. The network computer formats the data into one or more blocks, called packets. Each packet contains the control information needed to transmit the data correctly. The packets are transmitted from one network computer to another until they reach their destination. The final network computer strips the transmission-control information from the packets, assembles the data, and passes them to the requisite host computer.

A network computer receiving a packet places it in a queue to await attention. When it reaches the head of the queue, the computer examines its destination address, selects the next network computer on the route, and places the packet in an output queue for that destination. A packet-switched network is usually designed so that each network computer has a choice of routing. If the first-choice routing is poor because of equipment failure or congestion, it selects another routing. The time taken for one network computer to relay the packet to the next on the route is a very small fraction of a second. The name *hot-potato network* was once used for packet switching because each network computer passes a packet on to the next as fast as it can, like a hot potato.

The packet thus zips through the network, finding the best way to go at each node of the network and avoiding congested or faulty portions of the network.

We might compare a telecommunications network with a railroad network. With circuit switchings there is an initial switch-setting operation. It is like send-

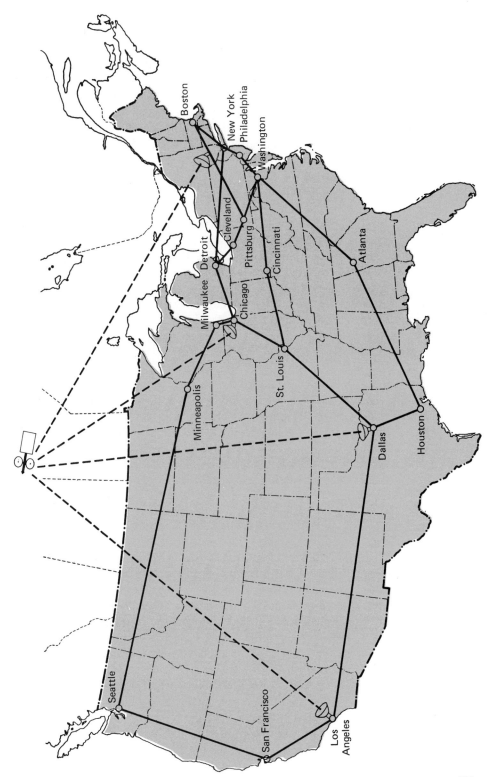

Figure 24.1 The initial 18-city line layout for the United States' first public packet-switching network, operated by the Telenet Communications Corp.

ing a vehicle down the track to set all the switches into the desired position; the switches remain set, and the entire train travels to its destination. With packet switching the cars of the train are each sent separately, with no initial setup operation. When each car arrives at a switch, the decision is made where next to send it. If the network is lightly loaded. the cars will travel to their destination by a route which is close to the optimum. If the network is heavily loaded, they may bounce around or take lengthy or zig zag paths, possibly arriving in a sequence different from that in which they departed.

PACKETS   The packets might be thought of as envelopes into which data are placed. The envelope contains the destination address and various control information. The transmission network computers should not interfere in any way with the data inside the envelopes. Indeed the system should be designed with security safeguards so that network computers cannot pry into the contents of the envelopes.

There is a maximum envelope size, and so long messages have to be cut up into slices and sent in several envelopes. After transmission the slices must be joined together in the right sequence.

Figure 24.2 shows the structure of the ARPANET, or Telenet, packet. It has a maximum length of 1024 bits. The text (data being sent) is preceded by a start-of-message indicator and a 64-bit header. It is followed by an end-of-message indicator and 24 error-detection bits. The header contains the destination address, the source address, the link number, and the packet number which is used to ensure that no packets are lost and that packets in error are transmitted correctly, a message number with an indication of whether there is more of the message following in another packet, and some special-purpose control bits.

NETWORK AND   An error-detecting code is used on each packet trans-
ERROR CONTROL   mitted. This code is checked by every network computer receiving the packet. If the packet is received correctly, an acknowledgment message is sent to the network computer which transmitted the packet. Each network computer retains each packet until it receives the acknowledgment of successful transmission. If it does not receive it by a certain time after transmission, then it retransmits the packet. If acknowledgment fails to be received several times, a different route will be tried. A switching computer deletes the packet from its storage once it receives an acknowledgment of correct receipt.

In addition the destination network computer which finally assembles and delivers the data transmitted sends an acknowledgment to the sender to confirm that the message was delivered correctly.

The error control which takes place between the host computers or terminals and the network is independent of that which takes place *within* the network,

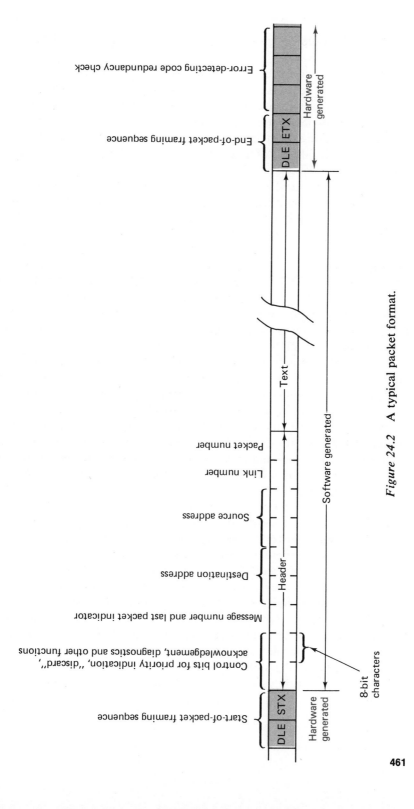

Error-detecting code redundancy check

End-of-packet framing sequence

DLE | ETX

Hardware generated

Text

Packet number

Link number

Source address

Software generated

Header

Destination address

Message number and last packet indicator

Control bits for priority indication, "discard", acknowledgement, diagnostics and other functions

8-bit characters

Start-of-packet framing sequence

DLE | STX

Hardware generated

Figure 24.2 A typical packet format.

461

and depends on the design of the terminal or host computer line-control procedures.

Messages used for controlling the network have the same basic format as data packets and are transmitted in the same way. Their first few bits indicate that they are a control message. Control messages have functions such as telling a neighboring network computer not to send any more messages because the input queue is full, telling the neighboring computer that it can start again, diagnosing problems, sending details of network delays or breakdowns, and sending details of traffic volumes.

The program used in the network computers can be loaded remotely by transmitting it through the network. To accomplish the loading a small program resides permanently in a protected portion in each computer.

THE NETWORK INTERFACE   The ARPA network uses two types of network computer, called an IMP (interface message processor) and TIP (terminal interface processor). The IMP connects one or more data-processing computers ("host" computers) to the network. The TIP connects up to 64 terminals (with no host computer) to the network. Figure 24.3 illustrates TIPs connected to a Telenet central office.

One of the objectives of packet switching is to permit communication between machines which are incompatible in that they could not be connected directly. An IMP or TIP may be programmed to communicate with several such incompatible machines. It uses one protocol for communicating with the host computer or terminal, and another protocol for communicating over the network to another IMP or TIP. The host computers or terminals may use different coding and control procedures and may transmit at different speeds. There is one common speed and protocol for the packet transmission, and the network computer does the conversion.

The ARPA network is connected to a wide variety of different and incompatible computers and terminals from many different manufacturers.

The Telenet network computers are essentially similar to the ARPANET IMPs and TIPs. Telenet refers to its network computer as a *central office*. Figure 24.3 illustrates the types of device connections permitted to a Telenet central office. A user can gain access to the network either via the public network or via a leased line, possibly using its own multiplexors or concentrators.

The customer computer requires some software routines for communicating with the Telenet system. These are referred to as *Network Control Programs* (NCP). They handle the signaling and error-control protocol between the customer computer and its local Telenet central office.

To handle a wide variety of incompatible terminals the system defines one standard terminal, called a *network virtual terminal* (NVT), which has its own

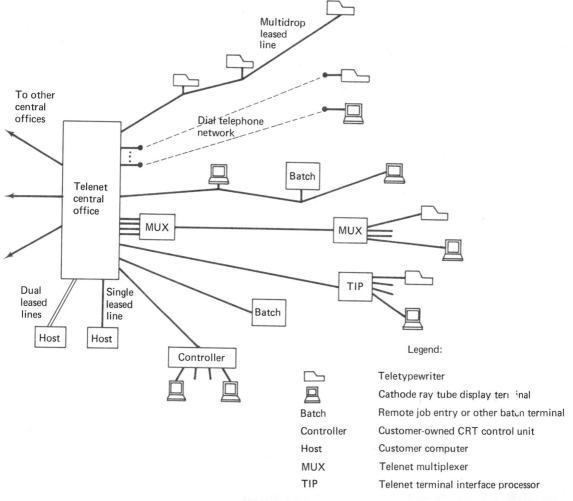

Legend:

⌐	Teletypewriter
💻	Cathode ray tube display terminal
Batch	Remote job entry or other batch terminal
Controller	Customer-owned CRT control unit
Host	Customer computer
MUX	Telenet multiplexer
TIP	Telenet terminal interface processor

Figure 24.3 Proposed local distribution arrangements from the Telenet central office.

character set and control procedure. The TIPs of ARPANET, and the central offices of Telenet, convert terminal transmissions to those of the network virtual terminal. This enables a customer computer to interact with a large variety of terminal types without special software having to be written for each.

ROUTING ALGORITHMS When a packet-switching computer receives a packet addressed to another location, it must determine which of the neighboring nodes of the network to send it to. The computer will have a programmed procedure for routing the packet. A variety of different routing strategies are possible:

1. Predetermined Routing

The route may be determined before the packet starts on its journey. The packet then carries routing information which tells the network computers where to send it. The determination of the route may be done by the originating location, or it may be done by a "master" station controlling the entire network.

An argument posed for predetermined routing is that most of the traffic consists, not of lone wandering packets, but of packets involved in "sessions" in which many packets pass back and forth between the same two locations. Typical sessions are *dialogues* between a terminal user and a computer in which many messages and responses are sent, *data entry operations* in which an operator keys in many transactions, *file transfers* in which many packets of data are sent, and *monitoring operations* in which a remote computer monitors or controls a process. With predetermined routing, the routing decision can be made once for the entire session. Any packet which is part of that session contains the same routing instructions. If a line or network computer fails during the session, a recovery procedure will be necessary in which a new route is established.

The alternative to predetermined routing is that each network computer makes its own routing decision for each packet. ARPANET and most proposed packet-switching systems employ this *nonpredetermined* routing. Wiht nonpredetermined routing the network computer has more processing to do but the packet envelope may be shorter because it contains a destination address, not an entire route.

2. Calculated Routing

The addresses of the destination nodes in a network may be chosen in such a way that it is possible for any interim node to determine which way to send a packet by performing a simple calculation on its address. If a node has received information about a failure in that direction, it may calculate a second-best routing.

Calculated routing is simple but in general too inflexible and hence unlikely to be used in practice.

3. Static Directory Routing

With directory routing, each node has a table telling it where to send a packet of a given destination. Figure 24.4 shows a possible form of such a table. The table shown gives a first-choice and a second-choice path. If the first-choice path is blocked or inoperative, a node will use the second-choice path. (There is no need for the table in the illustration to give a third choice because no node has more than three lines going from it.) As a packet travels through the network, each node does a fast table look-up and sends it on its way.

4. Dynamic Directory Routing

The previous method uses a fixed table. A more versatile method is to use a table which can be automatically changed as conditions of the network change.

There are several possible criteria that could be used in selecting the entries for a table such as that in Fig. 24.5. They include

(a) Choosing a route with the minimum number of nodes.

(b) Choosing routes which tend to spread the traffic to avoid uneven loading.

(c) Choosing a route giving a minimum delay under *current* network conditions.

The last of these conditions implies that the table will be constantly modified to reflect the current delays on the network caused by congestion or failure. Figure 24.5 shows figures proportional to the delays on the network at one time and a routing table for node C which takes these delays in consideration. As part of the time delay at each node is caused by queuing, the delays and the optimum routing table will change as the traffic patterns and volumes change.

The question now arises: How should a node be informed of what the network delays are? Several methods have been suggested. Paul Baran suggested that the delay on any route $x \rightarrow y$ is similar to the delay traveling in the opposite direction $y \rightarrow x$. Consequently, a node could obtain approximate information about network delays by finding out how long it had taken each packet reaching it to travel from its originating node. The time of departure would be recorded in each packet so that each node could determine this transit time. The method is rather like asking travelers on a rush-hour traffic system, "What is it like where you came from?"

Another system is for each node to send a service message at intervals to each of its neighbors. The message will contain the time it was originated and also that node's knowledge of whether the delays have changed. The recipients will record how long the message took to reach them, and if this is substantially differ-

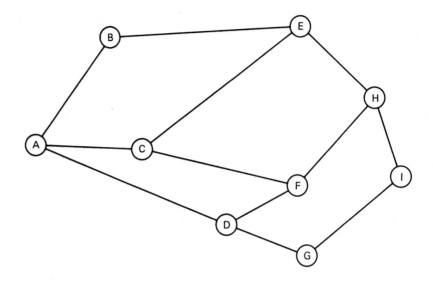

Packet destination	First choice node	Second choice node
A	A	F
B	A	E
D	A	F
E	E	A
F	F	A
G	F	A
H	E	F
I	E	F

Figure 24.4 A static routing table for node C in the above network, giving the node to which C should route a packet for a stated destination.

The numbers are proportional to the delays occurring on the network:

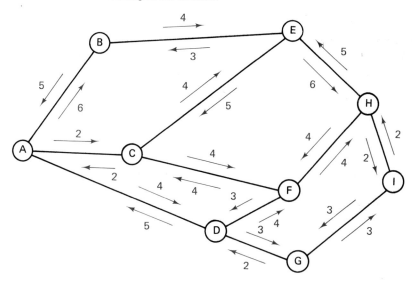

Packet destination	Next node
A	A
B	E
D	A
E	E
F	F
G	A
H	F
I	F

Figure 24.5 A dynamic routing table for node C intended to minimize transit delays under current network conditions.

ent from the previously recorded delay, they will include this information in the next service message they send. Knowledge of changes in transit times will thus be disseminated throughout the network.

ADAPTIVE ROUTING   A scheme in which the routes selected vary with the conditions of the network is called *adaptive routing*. The ARPA and Telenet networks use adaptive routing, each node sending a service message every half second. Most proposed public networks in other countries also use adaptive routing.

Adaptive routing sometimes results in oscillatory behavior, the routing pattern oscillating rapidly backwards and forwards under peak conditions. Minor changes in the routing algorithm can affect the routing behavior under heavy loading in ways which are difficult to predict without simulation of the network. Detailed study of routing algorithms is needed because they can substantially affect the peak network throughput.

It is possible that packets in such a network could fail to reach their destination because of temporary equipment failure or a data error in the address. Such packets might be passed indefinitely from one node to another if something did not stop them. To prevent this occurrence, a count field is used in each packet, and the number of nodes that have relayed that packet is recorded in it. When the count exceeds a certain number, the packet is returned to its point of origin. This process protects the network from becoming clogged with roving, undeliverable messages.

PRIORITY   It is possible to assign different priorities to different packets. It is of value to do so because some messages travel under much tighter time constraints than others. A packet-switching network commonly handles a mixture of real-time traffic and non-real-time traffic, for example. The real-time packets could be given priority over the non-real-time ones, which means that a switching computer would handle them first. In conditions of extreme congestion non-real-time traffic could be stored by a network until the congestion abates.

In the ARPA network short character streams (under about 100 characters) are handled as high-priority messages. They occupy single packets. It can be shown with queuing theory that the *average* time in a queue is shorter if priority is given to those items which can be served quickly. In giving priority to single-packet messages ARPA lowers the overall *average* response time. Many of the messages which compose man-computer dialogues at teletype machines or typewriterlike terminals are less than 100 characters, so the network transit time appears fast to these users. The scheme has the merit of simplicity but is not necessarily the best priority scheme because occasionally longer messages may need high priority, for example, in a graphics or visual-display-unit man-computer dialogue. ARPANET does not have the facility to store non-real-time traffic during periods of congestion.

NETWORK
CONTROL CENTER

One of the data-processing computers attached to the ARPA and Telenet networks is called the *Network Control Center* (NCC). One or more such centers monitor the status of the network, detecting and diagnosing failures and problems. To accomplish this function each network computer sends a status or diagnostic message to the Network Control Center at regular intervals. Any repair work or corrective actions will be initiated from the Network Control Center.

The Network Control Center computer maintains statistics about the network traffic and performance. It does the customer billing and can provide a customer with statistics of his operations. In addition, a file is maintained of all hardward and software connected to the network and is available for customer usage. Periodic bulletins are sent to subscribers, and subscribers can make on-line inquiries about the procedures they must use to employ network facilities.

GEOGRAPHIC LAYOUT

The geographic layout of a packet-switching network using adaptive routing can be changed easily. It will probably be typical of such networks, as with ARPANET, that the traffic patterns will change often, sometimes in a major and unpredictable fashion, and so the network itself must be changed to accommodate the new traffic. Figure 24.6 shows how the layout of ARPANET changed in the first 5 years of its life.

Given a particular traffic pattern the task of designing an efficient network is complex enough to need computer methods. A variety of computer algorithms have been developed for this purpose. There are two conflicting objectives in choosing the network layout. First, the cheapest configuration is required that gives the requisite response times. Second, a configuration is desirable which gives high availability — in other words, when a network computer or transmission link fails there is a high probability that alternative routes will be available. The designer must adjust the balance between cost and availability.

TWO SEPARATE
FUNCTIONS

There are two separate functions that must be carried out by nodes of a packet-switching network: the interface with the users and the switching of the packets. In the ARPA network both functions are carried out by the same machines — the IMPs or TIPs. Other proposals, including proposals for public switched networks in several countries, have employed separate machines for the two functions.

If the functions are separated, the switch can be a relatively small computer. The steps to be executed in switching one packet are relatively few, and so a high data throughput is possible with a fast machine. It has been estimated that more

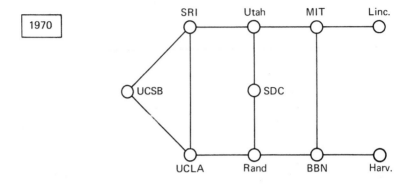

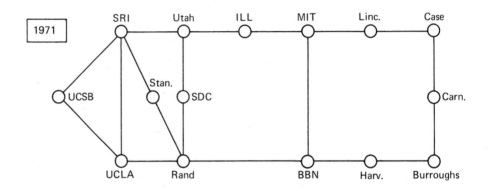

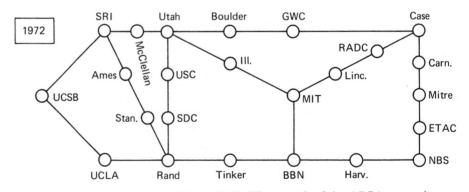

Figure 24.6   The growth of the ARPA network.

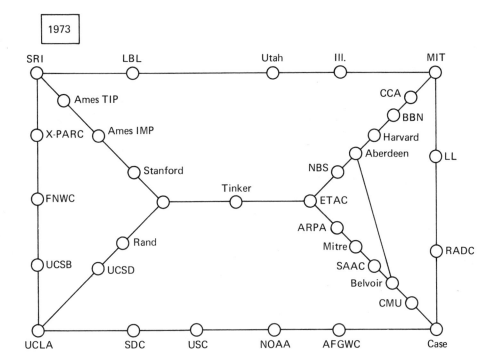

1973

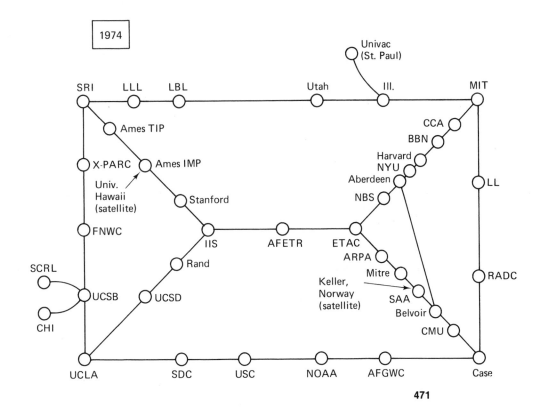

1974

than 1 million bps could be switched using today's technology, and so the switch might be geared to PCM transmission channels of 1.544 or 2.048 million bps. Unlike a message-switching computer, the packet-switching computer does not have any accounting or filing functions to perform. It needs no peripheral storage. It merely receives packets, queues them, determines their routing, and retransmits them. During idle moments it will send diagnostic packets and update its routing table.

The subsystem which interfaces with the user has somewhat more complex functions than the switch. They may include

1. *Creating the packets.* Data from a terminal or host are placed into one or more envelopes, and control information needed for transmission is written.

2. *Reassembly of data.* After transmission data, often sent in more than one packet, are reassembled. For some uses the entire message may be assembled prior to delivery; for others it may be better to deliver the data a block at a time as they arrive, provided that the blocks are in sequence.

3. *Host-network protocol.* The interface computer will observe a protocol for communicating with the host computers to ensure that the interchange functions correctly and that no data can be lost. Different protocols may be needed for different types of host.

4. *Terminal protocol.* A protocol for communicating with user terminals will be observed. Some of the user terminals may be far away, connected to the interface computer by links incorporating multiplexors, concentrators, polling, public network dialing, or other procedures.

5. *Special-function protocols.* Special protocols may be used for functions such as the transfer of files, the use of graphics, "conference calls" in which more that one user participates in one dialogue, mailbox services in which the network passes messages between users and holds them until the users see them, facsimile transmission, transmission of exceptionally high security, and others.

6. *Session control.* Many transmissions, as discussed earlier, are part of a "session" in which multiple messages are sent, as in a human telephone conversation. In this case the interface computer may control the session. It will store an envelope header for the session so that it does not have to be recreated for each message. It may use a session-oriented protocol with the host computer or terminal. It may allocate a high priority to the packets of certain sessions, possibly after establishing during the setup procedure for the session whether this priority is sustainable with the current network load.

TWO-LEVEL NETWORK Various authorities, including D. W. Davies and his coworkers at the British National Physical Laboratory [2] and the designers of the PCI (Packet Communications Incorporated) network, which received FCC approval but was not built; have advocated that a packet-switched network should be constructed having two levels. The lower level would be the local area network, corresponding broadly to the central office and local loops of the telephone network. This would carry out the function of interfacing

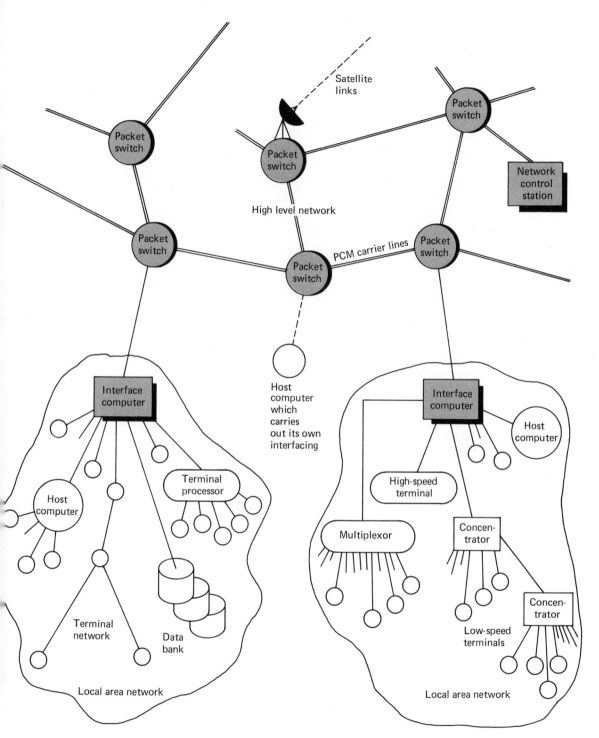

Figure 24.7 Various proposals for public packet-switched networks have separated the switching function and the user interface function, to form a two-level network as here.

with the host computers and terminals. The higher level would be the long-distance packet transmission and switching network, corresponding broadly to the trunking network of the telephone system.

Figure 24.7 shows such a two-level network. The shaded machines on the diagram are components of the public network, and the others in the local area networks are the user machines. User machines are normally connected to the network via the network interface computers. In locations where a network interface computer is not yet installed, however, a host computer may do its own interfacing, creating its own packets and transmitting them to the nearest switch over any convenient communication line.

A wide variety of different terminals can be attached to the interface computer. Indeed, one of the chief advantages of this type of network is that entirely different types of terminals can intercommunicate. They can have widely different speeds and use different codes. They can be synchronous or start-stop. They can use polling or contention, or can be alone on a line. They can use different types of error control. A telex machine could communicate with a visual display unit. If desired, they can be connected via multiplexers or concentrators. Above all, they can be inexpensive, for most of the costly terminal features such as buffering and elaborate line control are not really needed. The interface computer maintains a list of the characteristics of all the terminals attached to it, their control mechanisms, speeds, and transmission codes. If the code differs from the network transmission code, the interface computer converts it as the packet is being assembled. When the packet is received by the destination line-control computer, it is converted, if necessary, to the code of the receiving terminal. It is estimated that one interface computer could handle more than 1000 terminals in this way. The software complexity could, however, become severe.

An additional function of the interface computer is to collect the information necessary for logging and billing subscribers.

The high-level network can use links of different speeds for moving the packets. As satellite technology develops the high-level network of such a system would probably incorporate satellite transmission. Where the high-level network is installed by the same authority that operates the telephone network, the links can be the PCM links that carry telephone voice traffic. Packet transmission and voice transmission can thus share facilities.

SATELLITE CHANNELS   Satellite channels are a natural and valuable addition to a packet-switched terrestrial network. ARPANET was the first such network to use satellite channels. Telenet plans satellite usage, illustrated in Fig. 24.1, employing 1.544-million bps channels.

Unlike land lines the satellite channels operate in a broadcast mode. A signal which goes up to the satellite comes down to all stations listening to that channel. This has several advantages:

1. The cost of the satellite channel is independent of distance.

2. Isolated locations can be quickly linked to the network. The network may be extended to foreign locations or interconnect with Asian, European, or South American networks.

3. The channel can be shared between far-distant locations depending on their instantaneous needs. One moment there may be a traffic peak in California and later one in New York. The same channel can be assigned to these locations on a demand basis.

4. The satellite is shared between different time zones across the earth, at one time carrying mostly East Coast traffic and later mostly West Coast traffic.

5. The high bit rate of the channel permits a high throughput from any station connected to it for a brief period if needed. If 10 stations share the 1.5 Mbps channel, their average throughput may be about 0.15 Mbps but one station may demand all 1.5 Mbps for half a second or so.

6. Satellite channels spanning the network improve the overall network reliability, having the ability to bypass failed parts of the terrestrial network.

The main disadvantage of using a satellite is the quarter-second delay in one-way transmission. It is desirable that packets not normally be routed over multiple satellite links. The choice of optimum routing procedures is complicated by using a satellite. The protocol needed for controlling the *demand assignment* of the broadcast channel is also an extra complexity. A special satellite IMP (SIMP) is used for the satellite links.

TELEPHONE SWITCHING VERSUS PACKET SWITCHING
The switching and multiplexing facilities of the telephone network have evolved over the years to meet the needs of *people* talking to *people*. The needs of *machines* talking to *machines* are different in at least four critical respects.

First, the transmission rate needed by machines varies from very low rates such as those used in telegraphy to very high rates. Very high-rate teleprocessing is not used much today because appropriate transmission links are not available at a reasonable cost. If they had been available for the last decade or so, the computer industry would have developed very differently. When a computer sends, say, 100,000 bits of data, there are some circumstances in which it is satisfactory for it to travel over low-speed lines taking many minutes. There are other circumstances in which it would be desirable to obtain it in a second or so, if possible, because it is needed as part of a real-time operation. Most of the physical links of the telephone network are of sufficient bandwidth to transmit 100,000 bits in a fraction of a second, as we saw in Part II of this book, but the switching does not have the versatility to provide high-speed transmission for a short time when needed.

A second disadvantage of telephone switching is the long time that it takes to connect a call. The time taken from the start of dialing to when the call is connected is often 20 seconds or more. In many potential computer applications, when a terminal or computer needs data it needs them fast, and the prospect of a 20 second wait each time rules out the use of telephone switching.

Third, in many uses of computers the transmission of data is sporadic. A response is sent to a terminal. The terminal user thinks about it for a while and then sends a brief message to the computer. A dialogue proceeds in which the messages and responses are interspersed with substantial pauses. If the dialogue is taking place over a switched public telephone line transmitting at 4800 bps, the gaps may be a 100 times longer than the times when data are being transmitted. If the voice line is a PCM link with a potential throughput of 56,000 bps, the situation appears worse. In 1 minute of connection time 3 million bits could be sent, but less then 3000 are actually sent in many typical dialogues. If the connection could be switched and disconnected in a small fraction of a second, the lines could still be used efficiently, but such is not the case on the telephone network. Because of this inefficiency, many organizations resort to leased lines rather than switched lines, with private concentrators and multiplexors [3].

A fourth problem of telephone switching is that blocking sometimes occurs on the network, and a call cannot be switched as required because all the registers in an exchange or all the circuits on a given route are occupied. The person attempting to place the call receives a network "busy" signal.

An appropriately designed packet-switching network could improve upon these four snags of the telephone network in that

1. Low-speed or high-speed burst data transmissions can be accommodated.

2. There is no dialing, and hence effectively no call connect time. The only delay is the packet transit delay of a fraction of a second.

3. When the transmission is sporadic, packets are sent only when needed and other users are not excluded from the trunk during its idle periods.

4. When all trunks are occupied the caller is not rejected with a "busy" signal. Instead his packet waits in a queue until a trunk becomes free — normally a wait of a fraction of a second. It is not true to say that there is *never* blocking on a packet-switched network, because when the network is flooded the nodes must eventually run out of queuing space. Their buffer capacity is exceeded. However, the normal reaction to peak traffic will be an increase in transit time due to queuing rather than the rejection of calls.

ALLOCATION OF CAPACITY   Perhaps the most important of these four differences relates to the way the available information-carrying capacity is shared between its many users. The physical channels — microwave links, coaxial cables, and even wire-pairs — have a high usable bandwidth. The telephone network divides up the channels *by frequency* so that each user has a small bandwidth for as long as he wants it. All users are switched to channels of the same bandwidth. If a user wants to play his trumpet over the telephone, he will not be switched to a hi-fi trunk. A packet-switched network, on the other hand, divides up the capacity *by time* so that users are given a high bandwidth for a brief moment of time. A burst of data therefore travels to its destination quickly, leaving the capacity free for the next user. Figure 24.8 illustrates this difference.

The readers should note that packet switching is not the only way to allocate the information-carrying capacity flexibly. Any form of fast computer-controlled switching could be designed to assign different bandwidths for different periods of

Figure 24.8   The circuit-switched telephone network slices up the available information-carrying capacity in a different way to a packet-switched network. The shaded blocks show the allocation to one user.

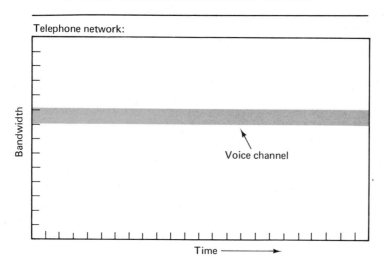

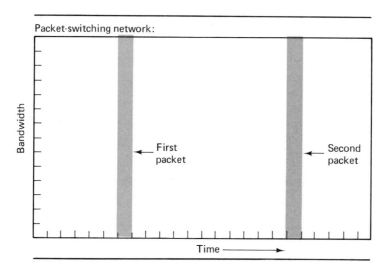

BOX 24.1 Comparison of circuit switching, message switching, and packet switching

Main Characteristics of Circuit-Switching Systems	Main Characteristics of Message-Switching Systems	Main Characteristics of Packet-Switching Systems
The equivalent of a wire circuit connects the communicating parties.	No direct electrical connection	No direct electrical connection.
Real-time or conversational interaction between the parties is possible.	Too slow for real-time or conversational interaction	Fast enough for real-time or conversational interaction.
Messages are not stored.	Messages are filed for later retrieval.	Messages are stored until delivered but not filed.
The switched path is established for the entire conversation.	The route is established for each individual message.	The route is established dynamically for each packet.
There is a time delay in setting up the call and then negligible transmission delay.	Substantial delay in message delivery.	Negligible delay in setting up the call; delay of usually less than 1 second in packet delivery.
Busy signal if called party is occupied.	No busy signal if called party is occupied.	Packet returned to sender if undeliverable.
Effect of overload: Increased probability of blocking causing a network busy signal. No effect on transmission once the connection is made.	Effect of overload: Increased delivery delay.	Effect of overload: Increased delivery delay (but delivery time is still short); blocking when saturation is reached.

BOX 24.1 Continued

Electromechanical or computerized switching offices are used.	Fairly complex message-switch center is needed, with filing facilities.	Small switching computers are used with no filing facilities.
Protection against loss of messages is the responsibility of the end users.	Elaborate procedures are employed to prevent loss of messages; the responsibility of the network for the message is emphasized.	Some protection against loss of packets; end user protocols can be employed in message protection because of the conversational interaction.
Any length of transmission is permitted.	Lengthy messages can be transmitted directly.	Lengthy transmissions are chopped into short packets.
Economical with low traffic volumes if the public telephone network is employed.	Economical with moderate traffic volumes.	High traffic volumes needed for economic justification.
The network cannot perform speed or code conversion.	The network can perform speed or code conversion.	The network can perform speed or code conversion.
Does not permit delayed delivery.	Delayed delivery if the recipient is not available.	Does not permit delayed delivery (without a special network facility).
Point-to-point transmission.	Permits broadcast and multiaddress messages.	Does not permit broadcast and multiaddress messages (without a special network facility).
Fixed bandwidth transmission.		Users effectively employ small or large bandwidth according to need.

time according to the user demands. Particularly appropriate is *time-division switching,* which will be discussed in the following chapter. A computer-controlled time-division network could overcome all the above objections to telephone switching, except for the avoidance of blocking by the use of storage in the network for queuing.

ADVANTAGES OF PACKET SWITCHING   The advantages of packet-switching networks derive from two different aspects. It is worthwhile to differentiate between them. The first aspect is the switching and routing technique; the second is the storage and message manipulation offered by the interface computers. Either aspect could be retained without the other. Switching computers could interconnect compatible hosts or terminals; interface computers could be used with a different form of switching network.

The main advantages of the switching and routing technique are as follows:

1. Fast response times.
2. High availability because of distributed routing.
3. High-speed data bursts can be handled as well as low-speed requirements.
4. No blocking except when the network storage is completely flooded.

The main advantages of the interface computers are as follows:

1. Machines using incompatible codes and control procedures can be made to communicate.
2. Terminals operating at different speeds can be connected because of the buffering.
3. Data for a terminal that is busy can be held until the terminal becomes free, rather than a "busy" signal being given.
4. End-to-end protection is possible against transmission errors or message loss.
5. The sending of one message to a named list of destinations is possible.
6. The network techniques can be made transparent to the host computer or terminal.

Box 24.1 compares the main characteristics of circuit switching, message switching, and packet switching.

COST   The primary advantages of packet switching—fast response time, ability to handle large bursts quickly, and network reliability—derive from network features which are expensive. High-speed lines are needed. The higher their speed, the better able the network is to adapt to varying demands for burst transmission and fast responses. Distributed networks are needed to meet the reliability criteria.

If there are many users for a packet-switching network, so that the average line utilization is high, then the cost per 1000 bits to the users becomes lower than the rates for the switched telephone or telegraph networks. However, the economics of packet switching is volume-sensitive. It begins to look attractive economically only when there is a high volume of user traffic.

Telenet's charge is $.60 for 1000 packets plus a charge for an access port at the Telenet central office ranging from $.90 to $1.40 per hour (prorated to the nearest second) or a fixed monthly charge ranging from $100 to $400 per month. This is a fraction of public long-distance telephone or telegraph rates. It is interesting to note that the charge is independent of the distance of transmission. This independence may have an effect on the centralization of the user's computer facilities.

Packet switching, then, is an appealing way to build a public network which can attract a very high volume of users. It is not economically attractive today for a medium-sized corporation building its own private network. It may be economically justifiable for very large corporations but having a network which integrates voice and data transmission would usually be better. Certain new inter-organization networks could use packet switching with profit, for example, networks for the electronic transfer of money to and from banks, hotel and travel reservation networks, intercorporate billing and mail networks, government networks combining traffic from many departments, and the value-added common carriers.

REFERENCES

1. "CCITT Fifth Plenary Assembly," *The CCITT Green Book,* Vol. VII: Telegraph Technique, International Telecommunications Union, Geneva, 1973.

2. D. W. Davies, "The Principles of a Data Communication Network for Computers and Remote Peripherals," presented at IFIP Congress, Edinburgh, 1968.

3. Discussed in detail in the author's *Systems Analysis for Data Transmission,* Prentice-Hall, Englewood Cliffs, N.J., 1972.

25 TIME-DIVISION SWITCHING

For a century telecommunication authorities have divided up the capacity of a channel by giving all users the same type of subchannel, as illustrated in Fig. 25.1. Where switching offices existed they were designed to switch this type of subchannel. Thus the telephone network provided switched telephone channels. The telex network provided switched telex channels. Message-switching networks provided low-speed message channels. Television networks provided switchable television channels.

It cannot be stressed too strongly that we have now entered a different era in telecommunications when the capacity of a network needs to be switched in a much more flexible manner. The channels should be switchable between different types of users in a time-varying manner. One time the user at a given location may want a voice channel, at another a high-speed data channel, and at another an interactive data channel with a high peak-to-average transmission ratio; in the future he may want a video channel, and so on. If the transmission network is to serve the users as well as it indeed can, the transmission capacity should be dynamically switchable between these bandwidths according to the users' instantaneous requirements.

There are several reasons switching should swing from the allocation of fixed-bandwidth subchannels to the time-varying assignment of differing bandwidths:

1. The bandwidths available today are much higher than in previous decades and promise to become much higher still. It is desirable to use these high bandwidths to their full potential.

2. PCM transmission techniques permit all manner of signals, e.g., voice, video, facsimile, and data, to be freely intermixed.

3. High-speed control equipment can be built to interleave the different signals economi-

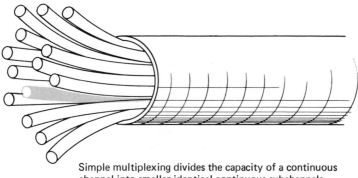

Simple multiplexing divides the capacity of a continuous channel into smaller identical continuous subchannels. Conventional switching offices switch these subchannels. A more elaborate form of switching is needed which dynamically allocates subchannels of <u>widely varying</u> capacity to users according to their instantaneous needs.

Figure 25.1   Simple multiplexing divides the capacity of a continuous channel into smaller identical continuous subchannels. Conventional switching offices switch these subchannels. A more elaborate form of switching is needed which dynamically allocates subchannels of *widely varying* capacity to users according to their instantaneous needs.

cally in a time-varying manner. To have done so in earlier decades would have been prohibitively expensive.

4. The needs of machines talking to machines are fundamentally different from those of people talking to people. To force machine-to-machine transmission into the type of network needed for people-to-people transmission is to cripple the potential capabilities of today's machines.

5. In today's networks real-time and non-real-time signals can be intermixed, and this is a key to network efficiency.

6. Satellite channels are not, in essence, point-to-point channels but can interconnect users dispersed over much of the earth's surface. To take advantage of such channels highly flexible allocation of their capacity among different users is needed.

MORE THAN PACKET SWITCHING   The designers of packet-switching networks recognized the need to build switched *data* networks in which the capacity is allocated in a rapidly time-varying fashion. One second a user may be receiving a high-speed burst of data; the next he is not receiving or transmitting anything. At one time he uses his link for fast-response man-computer dialogue, and at another for slow batch transmission.

    Today's packet-switching networks assume that all their users transmit *data*. The majority of the bandwidth of a corporation's telecommunication facilities, however, is employed for transmitting voice signals. For the foreseeable future it is likely that voice, video, and facsimile will outpace computer data in the demand for bandwidth, both within corporations and in public telecommunications. To switch bandwidth in a time-varying manner among voice, data, and other uses, in a way which employs the bandwidth efficiently, requires a switching technique more flexible than those discussed in the previous chapters.

Time-division switching may provide the answer.

TIME-DIVISION   When time-division *multiplexing* is used, either of
TRANSMISSION   PCM signals or data, as described in Chapter 14, the
AND SWITCHING   type of switching called *time-division switching*
becomes economical.

For decades, telecommunications organizations have completely separated the design of switching equipment from that of transmission equipment. With PCM transmission and time-division switching the two are intimately associated so that they cannot be designed in isolation from one another.

Figure 25.2 illustrates the principle. Here balls are shown rolling down a chute and being sorted by gates that open and close at exactly the right times. There are four types of balls, *A, B, C,* and *D,* and they must be switched to four outgoing paths.

Figure 25.2   Synchronous time-division switching of a time-multiplexed stream.

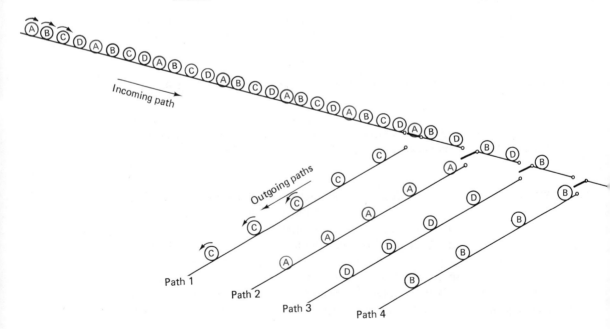

The balls may be thought of as being samples of four signals traveling together in a time-division-multiplexed fashion. Signal *C* is to be switched to path 1, signal *A* to path 2, and so on. The gate to path 1 opens at exactly the right moments to make the *C* samples travel down path 1. At other times it may be desirable to make signal *C* travel down path 4. This is accomplished by changing the timings with which the gates open.

The balls of Fig. 25.2 could represent *bits* arriving at a switch; they could represent *8-bit bytes;* or they could represent larger *blocks* of data. The same principle applies: The separation of the time-multiplexed elements is a switching process. They may be arriving at the rate of several million bits per second. There may be hundreds or thousands of paths, and the gates open and close at electronic speeds like the logic circuits in a computer.

In a switch designed for telephone calls only, the balls rolling down the chute could represent the PAM (pulse amplitude modulation) samples of Fig. 14.2. Telephone speech can be encoded as 8000 PAM pulses per second. The switches have to open and close far less often with PAM pulses than with PCM bits (56,000 bits per second) and hence the switching cost can be lower with PAM pulses than with digital bits. This is how the relatively inexpensive AT&T *Dimension* PBX operates. The Dimension system samples the incoming speech, forming PAM samples, interleaves those on a bus like that in Fig. 25.2, and then switches them by opening and closing the gates under the control a computer which has the switching instructions.

VARYING-SPEED STREAMS   In Fig. 25.2 the signals being switched are of equal speed. One of the most attractive features of time-division switching for future networks is that it can handle streams of different speeds. The capacity of a high-speed channel and switch can be dynamically allocated between users needing high, medium, and low transmission rates.

Figure 25.3 illustrates the principle of a transmission line and switch handling channels of different speeds. A high data rate channel is switched to path 2. Low data rate channels are switched to paths 1 and 3.

When a user requires a channel in a time-division network employing different channel speeds, he will indicate the destination of the call, the channel speed he wants, and possibly the duration of the call. A network controller will then attempt to make a reservation for these requirements through the transmission channels and switches. If the operation is under computer control, the reservation could be for a very short block of time. One user device might request a 200-bps channel for 3 minutes, while another might request a 256,000-bps channel for 1 second. Both of these may be derived from a digital carrier operating at say 2.048 million bps. With time division of the capacity under computer control, subchannels of large or small capacity may be allocated to users for very brief periods of time.

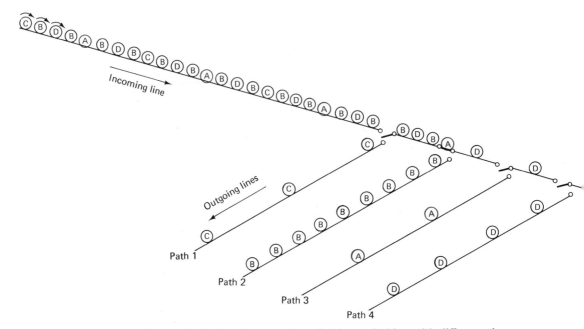

Figure 25.3 Synchronous time-division switching with different channel speeds.

While some users specify the duration of their call when the reservation is made, others may request a channel for the foreseeable future and at a later time request the disconnection of that channel. A telephone caller (or the equipment serving him) may, for example, request a channel of 64,000-bps capacity for indefinite duration. When the call ends the capacity is to be returned to a pool of transmission capacity available for allocation to other callers.

TIME-DIVISION
TELEPHONE
SWITCHING
Most of the time-division switching equipment employed by telephone companies switches fixed-capacity channels, as in Fig. 25.2, but nevertheless promises major advantages over conventional telephone switching.

The incoming path of Fig. 25.2 may be a PCM or digital transmission line. There are clear economic benefits in this association of time-multiplexing switching with time-multiplexed transmission. Time-multiplexed switches are also used to switch channels which enter and leave the switch in a nonmultiplexed form. Figure 25.4 shows a number of lines entering a simple switch. The signals are time-multiplexed by the switch on to a fast bus, and the resulting stream is demultiplexed by accurately timed gates in accordance with the user's requirements.

When a switch such as that in Fig. 25.4 switches telephone voice signals,

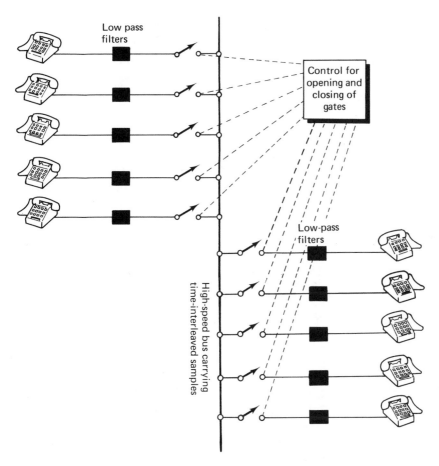

Figure 25.4   A simple time-division telephone switch.

and not data, there is no need to completely digitize the voice. Instead it can be left in the form of pulses which represent the instantaneous value of the analog signal – the PAM (pulse amplitude modulation) pulses of Fig. 14.2. There are one eighth as many PAM pulses as there would be bits if the signal was completely digitized; therefore, a higher throughput can be handled by the gates.

It will be seen that time-division switching is quite different from conventional switching, in which a physical path is permanently connected (called space-division switching). The number of switch points is much lower. If there are N lines, then N switch points are needed. If, on the other hand, N lines are to be interconnected physically, then N^2 switch points are needed. The latter figure can be reduced by multistage switching with a limited number of simultaneous interconnections; if no more than one tenth of the lines are permitted to be interconnected at one time, $0.21\ N^2$ switch points could be used. Thus for a switch interconnecting 100 lines, time-division switching needs 100 switch points, whereas open space-division switching is likely to need at least 2100. Time-division switching is lower in cost today than space-division switching for many applications. As the cost of fast logic circuitry drops lower, time-division switching will become increasingly economical.

ESS NO. 101   An example of time-division switching is found in the
SWITCH UNITS   way subscribers are connected to a Bell No. 101 ESS (discussed in Chapter 22). A subscriber location might have, say, 200 telephone extensions. Any extension must be able to dial numbers on the public network. In the No. 101 ESS, the dialed digits are interpreted in a computer at the local central office, which then controls the switching. All that is required on the subscriber's premises is a small cabinet of electronics, which samples the signals on the voice lines and switches them in a time-division fashion.

    The organization of the equipment in this cabinet is shown in Fig. 25.5. There are two buses onto which the samples of signals on the voice lines are switched. The speech is sampled 12,500 times per second. Thus there are 80 mi-

Figure 25.5 The organization of the time-division switching units that are connected to the Bell System No. 101 electronic switching system (as shown in Figs. 22.5 and 22.6).

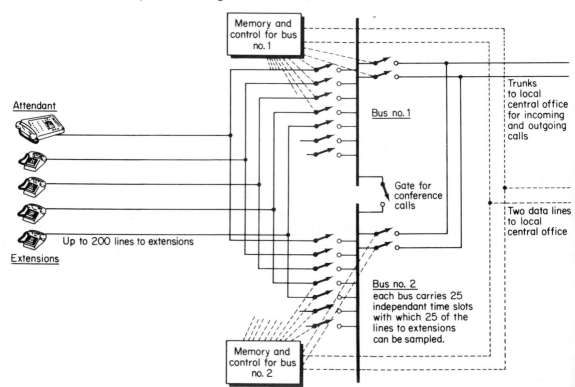

croseconds between sampling times. The duration of each sampling is approximately 2 microseconds, with a guard interval of 1.2 microseconds between samples. Therefore there can be 25 independent sets of samples, or 50 for both buses, giving a total of 50 time slots to be divided among 200 speech lines. No more than a quarter of the extensions can be in operation at one time—a higher traffic-handling capacity than normally encountered on private branch exchanges of this size.

The buses thus carry $12,500 \times 25 = 312,500$ PAM samples per second each. Samples are gated to the appropriate speech lines under the control of electronic circuitry, which is directed by the computer at the local central office. For this purpose, two data lines go from the time-division switch unit to the central office computer. One carries the dialed or Touchtone digits; the other carries the switching instructions to the switch unit.

The use of two buses doubles the number of time slots available to the customer, but perhaps more important, it greatly enhances reliability. If one bus, or its controls, fails, the other can handle any of the extensions. In this way, 25 time slots instead of 50 are then available. The gate shown between two buses also gives a convenient means for establishing conference calls, Figure 25.6 shows the bus arrangement of a larger switch unit. Here there are up to four buses. Each carries 60 independent time slots, giving a total of 240, which means 3 million samples per second. Subgroups of 32 speech lines are connected through the time-division switch to subgroup buses. The subgroup buses in turn are connected through more time-division switches to the 60-time-slot buses.

The upper limit of this technology today is several million PAM samples per second.

**CONNECTED
WITH TRUNKS**　　　　PAM samples are not used for long-distance transmission purposes because, being susceptible to noise and distortion, they do not have the advantages of binary encoding (PCM, as described in Chapter 14). In trunk switching, therefore, PCM samples can flow through a time-division switch.

When digital transmissions are switched, as with PCM trunk switching, it is common to employ a buffer storage. Figure 25.7 shows such a buffer with 120 cells each of which can hold 8 bits (one PCM sample). 120 telephone calls are time-division-multiplexed. A time slot from incoming call 1 is read into cell 1, a time slot from incoming call 2 is read into cell 2, and so on. The contents of the cells are read out in a different sequence, depending on the call-switching instructions that have been set up, and are time-multiplexed onto the output channel. In this way, time slots are interchanged between the incoming and outgoing lines. The storage must adjust to the time differences of the incoming samples, and because it does so, it is sometimes referred to as an *elastic store*.

To handle 120 calls of 8000 samples per second each, a switching time of $1/(120 \times 8000)$ seconds, or approximately 1 microsecond, is needed. The switch-

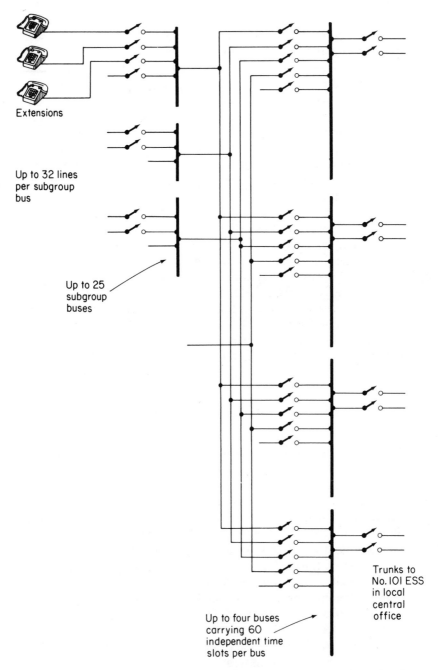

Extensions

Up to 32 lines
per subgroup
bus

Up to 25
subgroup
buses

Trunks to
No. 101 ESS
in local
central
office

Up to four buses
carrying 60
independent time
slots per bus

Figure 25.6 A time-division switching unit organization like that in
Fig. 25.5 but capable of handling 800 telephone extensions.

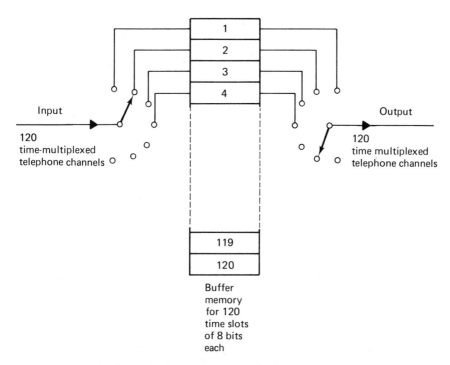

Figure 25.7   Use of a buffer memory for switching 120 telephone channels. 120 telephone channels require a switching time of about 1 microsecond. If a larger number of telephone channels are to be switched then
time-division stages are intermixed with space-division switching stages
as in Fig. 25.8.

ing speed of reasonable-cost circuitry sets a limit to the number of calls that can be
handled by a single time-slot interchange unit. To switch a large number of calls,
many such units are employed, and calls going between them must themselves be
switched. Fast circuit-switching units are employed for this purpose (referred to
as *space-division switching*). A larger electronic switch thus consists of a combination of time-division and space-division units.

ESS NO. 4   This is the manner of operation of the Bell System
                              trunk switch, ESS No. 4, illustrated in Fig. 25.8.
In ESS No. 4 the telephone voice-sampling interval of 1/8000 second is further subdivided into 128 time slots of 0.977 microsecond each. This is the basic
cycle time of the switch. In each 0.977 microsecond time slot an 8-bit PCM
sample is transferred from an incoming trunk to an outgoing trunk, and another
sample is transferred in the opposite direction to give two-way conversation. One
time-slot interchange unit, like that in Fig. 25.7, handles 120 voice channels. The

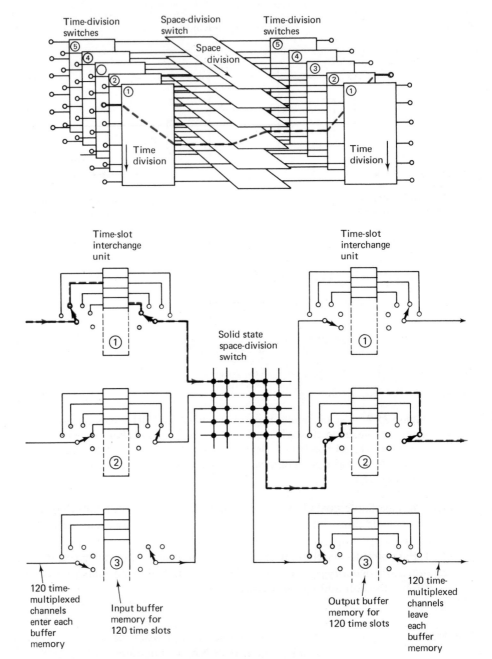

Time-division switches
Space-division switch
Time-division switches

Space division

Time division

Time division

Time-slot interchange unit

Time-slot interchange unit

Solid state space-division switch

① ② ③

① ② ③

120 time-multiplexed channels enter each buffer memory

Input buffer memory for 120 time slots

Output buffer memory for 120 time slots

120 time-multiplexed channels leave each buffer memory

The ESS No. 4 trunk switch can switch calls between more than 100,000 trunks by a set of time-division switches interconnected by a space-division switch. The dotted line shows the path of a typical call through the exchange.

Figure 25.8 The ESS No. 4 trunk switch can switch calls between more than 100,000 trunks by a set of time-division switches interconnected by a space-division switch. The dotted line shows the path of a typical call through the exchange.

8-bit samples from these 120 voice channels are read into the buffer storage and then read out onto the appropriate outgoing lines.

Many of these time-slot interchange units are employed in parallel and are interconnected by a space division switch, as shown in Fig. 25.8. Other designs of time-division switches have employed different combinations of space division and time division.

When a call is received by ESS No. 4 a *trunk-hunt* program is used first, to search for a free trunk to the required destination. Then a *path-hunt* program searches for a free path out of the many possible paths (in space and time) that could interconnect the two trunks. The commands for each switching stage are then written in a cyclically repeating memory and read out 8000 times per second to control the time-slot interchanges.

ESS No. 4 can interconnect more than 100,000 trunks and can handle more than three times the traffic of the largest electromechanical trunk switch. Many of the incoming trunks are analog, not digital. Before these can be switched their calls must be converted to PCM form, and then after switching converted back again, if necessary. Even with these conversion steps ESS No. 4 is more economical than electromechanical toll switches. With digital trunks, however, it is much more economical.

Fig. 25.9 shows the console of ESS No. 4.

Figure 25.9   The console of ESS No. 4, a large digital time-division trunk switch.

**RESERVED AND
NONRESERVED
CHANNELS**

In the time-division schemes discussed so far, the subchannels are *reserved* before the transmission takes place. A request for a subchannel is made and the necessary time slots are allocated to it. The allocated stream of time slots cannot then be used for any other purpose until the subchannel is "disconnected." With burst data transmission the reservation may be for a very short period of time, but nevertheless the transmission cannot commence until the reservation is made.

Another form of time division is that in which no reservation is made. This form of operating is suitable when the transmission does not need to be *continuous* but can be broken up in time in a nonsynchronous fashion. Telephone and television, for example, require continuous transmission. Data transmission does not usually need to be continuous but can procede in bursts providing that the transmitting and receiving machines have buffers in which the stream of data can be assembled and disassembled. When a burst travels through a network and a reservation has not been made for it at the switching points, it may be delayed. It is therefore necessary to be able to store the bursts at the switching points until time slots are free which the bursts can use.

Time division with reserved channels may be referred to as *synchronous* time division, because the fragments of the calls fit into regular preassigned time slots. Nonreserved time division may be called *asynchronous,* because no regular timing pattern is observed.

With synchronous time division the time slots which a call occupies determine where that call will be switched. The switch has to be given a routing instruction only once, at the start of the call. With asynchronous time division each fragment of a call must carry the address of its destination, and the switch is given a routing instruction for each fragment.

Packet switching is a form of asynchronous time-division switching. In today's packet-switching networks the entire leased channel carries packets, and nothing else. The control signals are themselves packets. Another possibility is to intermix reserved time slots and nonreserved time slots such as packets. If this is done, then the advantages of packet switching, enumerated in the previous chapter, can be combined with the capability of synchronous time division to carry *continuous* transmissions such as voice.

**COMBINED
SYNCHRONOUS
AND ASYNCHRONOUS
TIME DIVISION**

Most data transmission is subject to peak rate requirements which are much higher than the average rate; as indicated in Table 5.1, the ratio of peak rate to average rate on interactive systems often exceeds 1000. Hence an asynchronous operation such as packet switching is much more efficient than any scheme which allocates continuous fixed-capacity subchannels to each user. On the other hand, voice and some other types of transmission require continuous channels, and so, if a switching system is to handle both voice and data, a form of switching which combines synchronous and asynchronous time division is desirable.

Synchronous and asynchronous time division can be combined in either a fixed or dynamic fashion. To combine them in a fixed fashion one or more channels in a synchronous time-division system can be reserved for data transmission or other nonsynchronous operation. Certain channels on a time-division system could be reserved for packet switching, for example.

To combine synchronous and asynchronous operation in a dynamic fashion would require specially built switching equipment. This equipment would give priority to reserved channels. Time slots are first allocated to the reserved channels, and any time slots left over are available for nonreserved transmission. Figure 25.10 shows a combination of reserved channels and time slots which are available for other use because they are not reserved. In addition to having enough storage for the maximum number of reserved time slots, the switch would have to have storage to queue the nonreserved traffic when time slots are not immediately available for it.

Each unreserved time slot in Fig. 25.10 must carry addressing and control information with it. The time slots must be long enough for this to be economical — probably several hundred bits in duration. To maximize the utilization of the channel, three priorities might be observed:

> priority 1:   continuous channels (e.g., voice)
>
> priority 2:   interactive or real-time data
>
> priority 3:   batch or non-time-dependent data

Also to maximize channel utilization, time slots of different sizes may be employed. It is uneconomic to divide batch data up into small time slots each with addressing and control information, and it is uneconomic to put brief interactive responses into large time slots.

There are many possible variations in the details of the scheme expressed in principle in Fig. 25.10.

As the merging of computer technology and telecommunications pursues its inevitable course we can expect to find more uses of flexible demand-assigned time-division transmission and switching.

Figure 25.10   Combination of reserved channels for continuous transmission and nonreserved time-slots available for non-continuous transmission.

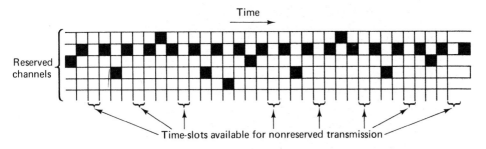

26 CONCENTRATION AND DEMAND ASSIGNMENT

When a fixed number of facilities are available for use they can be assigned to users in a fixed or a variable manner.

Suppose that a sales office has 100 employees each of whom need a desk when they work at the office. Most of the employees, however, work at the office only occasionally. The rest of the time they are out traveling or with customers. The office, therefore, has only 20 desks, and when an employee comes into the office he takes any free desk. This is a *variable* allocation of desks, and it clearly needs fewer desks than fixed allocation.

In today's telephone systems the local loops are normally assigned on a fixed basis, and the trunks are assigned on a variable basis, like the above office desks. The trunks are therefore used much more efficiently than the local loops.

CONCENTRATORS Suppose that there are 100 subscribers in a locality who use their telephones no higher proportion of the time than the above salesmen use their desks. There is, in theory, no need for 100 channels to connect them to their local switching office. Twenty channels could be used with some means of allocating a channel to a subscriber when he needs it. This technique is called *concentration*. There are various ways in which it can be done, and hence there is a variety of devices called concentrators.

Note that there is a fundamental difference between *concentration* and *multiplexing*. With multiplexing all subscribers can have a channel simultaneously if they want one. With concentration they cannot. In the above example, if all 20 channels are in use and a twenty-first subscriber requests a channel, he will be unlucky. He either receives a *busy* signal or must wait until a channel becomes free. Such is the nature of concentration. It takes advantage of the fact that not all the users are active all the time. Probability or queuing calculation is needed in the design of concentrator facilities [1].

A concentrator mechanism could be used in an apartment building and in a town street, if it were economical, to connect a large number of subscribers to a smaller number of channels to the local central office. It is sometimes used in rural areas, where subscribers are a long distance from their central office, to avoid having many lengthy wire-pair connections to the central office.

The design of a concentrator depends on the type of signal it is to concentrate. Concentrators are frequently used in data networks where their design is adjusted to the type of data traffic [2]. They read data into storage from lines with low utilization and then retransmit it over one or more lines with high utilization. A concentrator for telephone lines may be an electromechanical device which scans a bundle of lines searching for a free one. It may be a solid-state circuit which concentrates PCM traffic to travel over a digital trunk.

An economically attractive configuration for future networks is to have subscribers connected to time-division concentrators that are linked to the nearest exchange by PCM lines (Fig. 26.1). These lines would probably carry 24 channels, like the Bell T1 carrier, or 30 channels, like the CCITT 2.048 Mb/s lines.

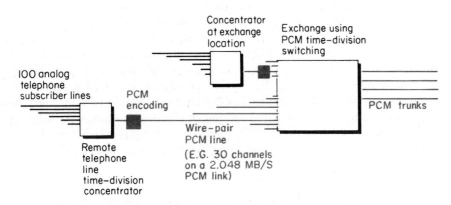

Figure 26.1   A remote telephone concentrator designed for a PCM line.

The exchange would use time-division switching. Queuing calculations for such a system, based on today's telephone traffic, suggest that between 100 and 200 subscribers making calls at random could be connected to one such concentrator and a good grade of service would be given. That is, there would be a very low probability of a subscriber obtaining a busy signal because the concentrator and PCM line did not have a free channel. A smaller number of subscribers than this might be connected when they do not make calls at random. Nevertheless, such a scheme would give much higher utilization of the wire pairs into the exchange than today's methods.

Any such arrangement would have to be linked into today's telephone network. Most of the present telephone exchanges are here to stay, for a long time. However, new capacity has to be added to most of the exchanges and local networks, and one way to do so is to operate a **PCM** system alongside the existing plant. With traffic doubling every six or seven years and with most of today's plant being near saturation, we are likely to have a new network superimposed on the old.

DEMAND ASSIGNMENT
Suppose that the sales office described above has four separate departments staffed, respectively, by salesmen, maintenance men, design engineers, and the administrative staff. Each department could have its own desks, which it may or may not assign dynamically. The demand for desks, however, will vary from time to time; sometimes the salesmen will want many desks and sometimes not. A high demand for salesmen desks may occur when the demand for design engineer desks is low. If the desks could be quickly reassigned from one department to another on a basis of the demand at that instant, then the organization could use fewer desks than if they were permanently assigned to the four departments. This is referred to as *demand assignment*.

The toll telephone network uses demand assignment. Its trunk-switching offices assign different groups to different destinations on a demand basis.

A particularly important form of demand assignment occurs on satellite channels. It has one fascinating difference from the conventional dynamic assignment of channels. Although there is one satellite and hence a single cluster of channels to assign, the users of this cluster are widely scattered geographically. To give efficient utilization of the facility the channels should be constantly reassigned between one earth station and another. It is rather as though the sales organization we discussed had many offices and the capability existed to magically switch desks from one office to another as the demand for desks varied.

DEMAND-ASSIGNED MULTIPLE ACCESS
The capability to switch channels between multiple access points on a demand basis is referred to as *demand-assigned multiple access*. When traffic fluctuates widely, fixed assignment of satellite channels to separate geographical locations will lead to inefficient utilization of the satellite capacity. The satellite is sufficiently costly that it is economic to use elaborate control equipment to achieve demand-assigned multiple access.

FDMA AND TDMA
Like simple multiplexing, demand assignment can be achieved by frequency division, time division, or space division. Space division is of limited use because even if a satellite has multiple directional spot-beams, each spot-beam covers a large portion of the earth. Many

of today's satellites requiring demand assignment have only one main antenna beam (e.g., the WESTAR and ANIK satellites). Hence *frequency-division multiple access* or *time-division multiple access* is used. They are referred to as FDMA and TDMA, respectively. Either can be employed with any existing satellite transponder.

An FDMA system makes available a pool of frequencies and assigns them, on demand, to users. A TDMA system makes available a stream of time slots and assigns them, on demand, to users.

One channel, derived either by frequency division or time division, is reserved to function as the control channel. The signals on this channel convey each station's requests for capacity, and inform each station about the channel assignments.

CONTROL OF DEMAND ASSIGNMENT   The control equipment must maintain a single set of data showing which channels are assigned to which locations. When a location needs a new channel assignment it requests this via a control channel.

In some systems a central location keeps track of the instantaneous channel allocations. In theory, the satellite itself could be the controlling location. The reader might imagine a demand assignment multiple-access system operating with a little old lady sitting in the satellite talking to the users like a PBX operator and pushing plugs into a panel to connect them. In reality, if the satellite *were* the control point, an on-board computer would be used. It is desirable, however, not to complicate the satellite too much because it must have extremely high reliability. Furthermore, demand assignment is needed with *today's* satellites. Therefore, if a controlling location is used, it must be on earth, and a transmission to it and back via satellite is needed when a station requests a new channel. In Fig. 26.2 the centralized control is performed by a computer at earth station 2.

Some satellite systems use a *central* controlling location. Others have *decentralized control* in which any station requiring space can make a bid on a common control signaling channel at random, somewhat like an auction. This is referred to as a *contention* form of control.

Using contention a single-signaling channel is allocated which all stations listen to. Each station maintains a record of which channels are free. If station *A* wants to set up a channel to station *C*, it selects one of the free channels and makes a request to use that channel to transmit to station *C*. Station *C* hears the request and responds. Transmission then begins on the selected channel, and all stations update their records to show that the channel is no longer free.

Two stations might, be chance, select and request the same free channel at the same instant. When this happens neither receive it, and each makes a new request, selecting a channel at random.

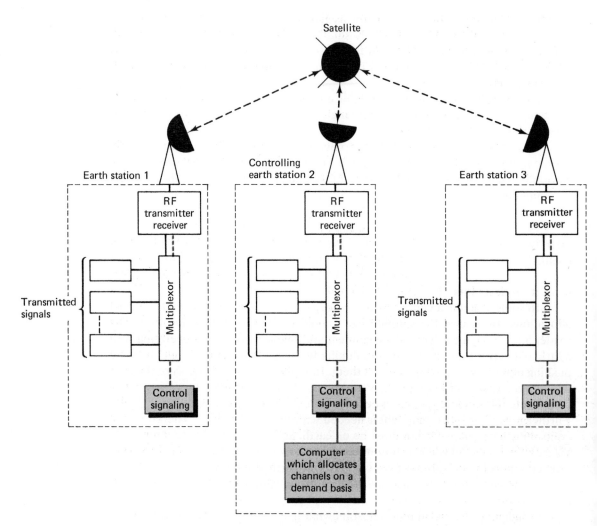

Figure 26.2 Centralized control of demand-assignment performed by a computer at earth station 2.

SPADE   An FDMA system in common use is the SPADE system [3, 4]; SPADE, devised by Comsat, stands for
single-channel-per carrier, PCM, multiple-access, demand-assigned equipment.
The goals of the SPADE project were stated as follows [4]:

a. to provide efficient service to light traffic links,

b. to handle overflow traffic from medium-capacity preassigned links,

c. to allow establishment of a communications link from any earth station to any other earth station within the same zone on demand,

d. to utilize satellite capacity efficiently by assigning circuits individually, and

e. to make optimum use of existing earth station equipment.

SPADE increased the traffic over existing INTELSAT IV links and made the establishment of new links more economical. By permitting variability in channel assignment and routing, SPADE enabled INTELSAT transponders to be used more efficiently. The INTELSAT system is characterized by large earth stations separated by large distances. For a domestic system characterized by large numbers of small earth stations, a technique to enable these stations to share the satellite flexibility is a key to economic viability.

Figure 26.3 is a diagram of the SPADE system. At the top of the figure is a transponder of 36-MHz bandwidth, in a satellite with many such transponders. Each earth station using that transponder transmits and receives the entire bandwidth but then extracts from it only those channels which are allocated to it. The equipment used with INTELSAT IV derives 800 channels with 45-kHz spacing from the bandwidth; 794 of these can be working voice channels, and 6 are housekeeping and control channels.

There is no central control location. Instead a control-signaling channel of high quality (low error rate) is employed, and each station monitors this channel continuously. The control channel carries a bit rate of 128,000 bps, and the control signals from each earth station are time-multiplexed onto the channel. The stations make bids for voice-channel allocation on a *contention* basis. Each station has a time slot allocated to it once every 50 milliseconds. When a station makes a bid for a channel it will have to wait at least 540 milliseconds before it receives a response because of the propagation delay into space and back. One station may have several requests pending at once.

The transponder at the top of Fig. 26.3 is, in effect, a pool of 794 allocatable channels, and a free channel can be seized by any earth station when it needs it. Earth station C in the diagram is using channels f_1 and f_3. It ignores all other channel frequencies. Channel f_1 is connected at this instant to earth station B, and channel f_3 is connected to earth station D.

When earth station C finishes with channel f_3 it will send a signal that that channel is free again. Every earth station receives the signal and assigns channel f_3 to its pool of available channels. Earth station A may then request that it use channel f_3 to communicate with earth station B. Earth station B acknowledges the request, and A and B switch a voice channel to their modems using frequency f_3.

The earth station may be connected to a telephone exchange with many thousands of lines coming into it. The SPADE equipment thus performs a concentration and a demand-assignment function.

Not all transponders in a satellite will be connected to SPADE equipment. Other transponders may use other techniques for allocating bandwidth.

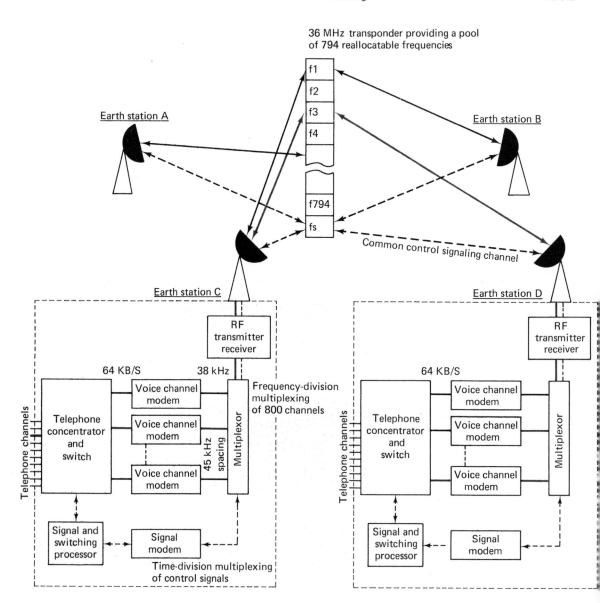

Figure 26.3 The operation of the SPADE frequency-division demand-assignment system.

**OTHER FORM OF
DEMAND ASSIGNMENT**

Although the above technique is part of the glamorous world of satellites, it still looks like "plain old telephone service" in that it allocates only voice channels. We can repeat the comments of the previous chapters that these days there are other types of telecommunication users who, ideally, do not want to be allocated *continuous channels of fixed bandwidth.* We would therefore like to have satellite demand assignment of video channels, continuous data channels of widely varying speeds, noncontinuous data channels, and short bursts of data suitable for inter-active systems.

Analog signals sent by satellite can be sent more efficiently if they are digit-ized before transmission. SPADE equipment, in fact, digitizes the voice channels before putting them through the modems shown in Fig. 26.3 and then frequency-multiplexing them. If satellites transmit digital signals, it would seem that the right way to perform demand assignment must eventually be time-division rather than frequency-division multiple access. The earth station equipment will be high-speed computers. Time-division multiple access has a further advantage. It modu-lates all of the signals on to *one carrier,* whereas SPADE uses 800 carriers. The 800 channels of a SPADE system need substantial guard bands between them. The channels, each carrying one voice signal, are spaced 45 kHz apart. An all-digital TDMA system can be made to carry more voice or data channels than an FDMA system, but needs high-speed computers at the earth stations.

Demand-assignment systems exist which are designed purely for data trans-mission. Ingenious forms of packet switching have been used on satellite links and work very effectively [5, 6]. Fortunately the diverse types of equipment can all use the satellite at the same time, each employing a different transponder. The transponders operate entirely independently.

Future demand-assignment equipment may, however, combine the needs of the diverse users and thereby give a further saving of transponder and equipment capacity.

REFERENCES

1. These calculations are discussed in the author's *Systems Analysis for Data Transmission,* Prentice-Hall, Englewood Cliffs, N.J., 1972.

2. Concentrators are discussed in Reference 1 above.

3. E. R. Cacciamani, "The SPADE System as Applied to Data Communications and Small Earth Station Operation," *COMSAT Technical Review,* 1, No. 1 (Fall 1971).

4. B. I. Edelson and A. M. Werth, "SPADE System, Progress and Application", *COMSAT Technical Review,* 2, No. 1 (Spring 1972).

5. N. Abramson, "Packet Switching with Satellites", *AFIPS Conference Proceedings, National Computer Conference,* 42 (1973).

6. L. Kleinrock and S. Lam, "Packet Switching in a Slotted Satellite Channel," *AFIPS Conference Proceedings, National Computer Conference,* 42 (1973).

PART IV IMPERFECTIONS

Few of today's communication links have been specifically designed for data transmission. From the computer systems analyst's point of view the vast networks designed for telephone and television have certain imperfections but nevertheless are immensely valuable in carrying the power of computers to users everywhere. By employing certain compromises we can engineer around the imperfections. First, measurements must be made of the behavior (or misbehavior) of the vast organisms which comprise today's telecommunications.

27 NOISE AND DISTORTION

On any telecommunication link there will be noise and distortion of some degree. A variety of techniques can be used in engineering the link to reduce this to an acceptable minimum.

Unfortunately, the acceptable criteria for data transmission differ from those for other uses of the link, such as voice. There can be a considerable degree of impulse noise and distortion on a voice line without the speech becoming unintelligible, annoying, or too unnatural. Similarly, teletype lines can be noisy; if telegram arrives with a few incorrect letters, it is still basically readable and understandable to a human being.

Voice transmission differs from data transmission in two fundamental ways. First, with voice we have an intelligent agency at each end of the line. If a burst of noise or other failure prevents the listener from hearing a word, he can either guess what the word should have been or else ask the speaker to repeat it. If he cannot hear, he will ask the speaker to speak louder or clearer. This highly flexible intelligence does not exist when machine talks to machine, so rigid procedures are devised which are as fail-safe as possible. Second, the information conveyed by the human voice is at a very much slower rate than that at which we want the machines to "talk." The normal rate of speaking is equivalent to something like 40 bps of written words, and this is "coded" with a very high degree of redundancy. We can usually follow the meaning of what is being said if we hear only about half of the words. On the other hand, data transmission over voice channels can take place at 4800 bps and higher. It is because of the low rate of information in speech and because of the adaptability of the human ear that distortions and noise levels damaging to data transmission have been quite acceptable in the engineering of the world's telephone lines.

In this chapter we will outline the types of noise and distortion that are common and the effects they have. Some types can be overcome or minimized by suitable design of the terminal equipment or by slower rates of transmission. Others cannot be prevented but must be faced up to in the design of systems, by means of error detection and correction, and retransmission, techniques. Tight controls on the accuracy of data transmitted must be built into the computer programs.

SYSTEMATIC AND
FORTUITOUS
DISTORTION

We may classify the disturbances into two types: *systematic* and *fortuitous*.

Systematic distortion is that which occurs every time we transmit a given signal over a given channel. Knowing the channel we can predict what is going to occur. The pulses may *always* be narrower or *always* be distorted in a certain way. Given frequencies will *always* have a certain minimum phase delay. *Fortuitous distortion* is something which occurs at random, and so it is not predictable, except in terms of probability. Examples of fortuitous distortion are white noise, impulse noise, chatter from the switchgear, crosstalk, atmospheric noise, sudden changes in signal phase, and brief losses of signal amplitude. Fortuitous distortion refers to *transient* impairments rather than continuing conditions on the line.

Systematic distortion is then something which might possibly be compensated for electronically so that its effects are eliminated. Fortuitous distortion is more difficult to compensate for, though steps can be taken to minimize its effects and repair the damage it does. It may be possible to correct systematic distortion so that it never actually damages data. Fortuitous distortion, on the other hand, is occasionally going to produce an extra large noise burst or impulse which destroys or creates one or more bits at random.

Box 27.1 summarizes the main types of transmission impairments of concern. We will first discuss the various types of fortuitous distortion.

WHITE NOISE

White noise was described earlier and illustrated in Fig. 16.2. This is the random hiss that forms a background to all electronic signaling. It cannot be removed, and so it sets a theoretical maximum on the performance of any communication link and on the various modulation methods. The amplitude of the signal after attenuation must be kept sufficiently far above the white noise background to prevent an excess of hiss on radio or telephone circuits or an excess of errors in data transmission. On the majority of lines the signal-to-white-noise ratio is better than 30 decibels, although occasionally a dialed call will encounter a much worse ratio than this.

Occasionally there will be spikes of white noise higher than the majority and peaks of other noise types such as crosstalk which may add to the white noise. The error rates in the various types of equipment due to white noise alone can be calculated theoretically.

If white noise were all we had to worry about, the design of data transmis-

BOX 27.1 Types of impairments which afflict
communication circuits

Systematic Distortion (Static Impairments)	Fortuitous Distortion (Transient Impairments)
Loss	White noise
Attenuation distortion	Impulse noise
Delay distortion	Crosstalk
Harmonic distortion	Intermodulation noise
Frequency offset	Echoes
Bias distortion	Changes in amplitude
Characteristic distortion	Radio
	Line outages
	Radio fading
	Changes in phase
	Phase jitter
	Dropouts

sion systems would be more straightforward. Unfortunately, there are other types of noise and distortion which are far less predictable and more disastrous in their effects.

IMPULSE NOISE   Unlike white noise and the various types of systematic distortion described later, *impulse noise* can have peaks of great amplitude which saturate the channel and blot out data. Impulse noise is the main source of errors in data. The duration of the impulses can be quite long relative to the speed of data transmission—sometimes as long as 0.01 second, for example. This would be heard merely as a sharp click or crack to a human listener and would not destroy any verbal intelligence, but if data were being transmitted 75 bps, one bit might be lost. For speeds of 4800 bps a group of 50 or so bits would be lost.

Often a noise impulse removes or adds two or more adjacent data bits, and this means that odd/even *parity checking may not detect the error.* A more sophisticated form of error-detection code is needed, as discussed in a later chapter.

Figure 27.1 gives an illustration of the effects of a burst of noise. In this example the signal-to-noise ratio is low, as is indicated by the amplitude of the white noise relative to that of the signal. In this diagram the sampling is shown taking place at one instant. Some systems take a series of samples throughout the intended duration of the pulse and thus lessen the probability of short noise spikes giving incorrect results.

There are many causes of impulse noise, some of which can be controlled but most of which cannot without a complete reengineering of the telecommunication facilities. Some impulse noise is audible during telephone conversations and some goes unnoticed. Stray clicks and crackles are all too familiar. Impulse noise comes from a variety of different sources. It may come from within the communication channel itself or from a source external to the channel.

External noise is picked up by induction or capacitance effects. Sharp voltage changes in adjacent wires or equipment induce noise spikes in the communication channel. Many of the audible clicks which are of high amplitude and which damage one or several adjacent bits come from switchgear and telephone exchanges. Sometimes one can hear the rapid sequence of clicks generated by another person dialing. All relay operation is a potential source of noise if the shielding or suppression is not adequate. Any switches or relays which make or break circuits carrying current cause a sharp voltage change and so can induce an equivalent voltage change in nearby sensitive circuits. Sometimes the power sup-

Figure 27.1   Transmission in the presence of bad noise.

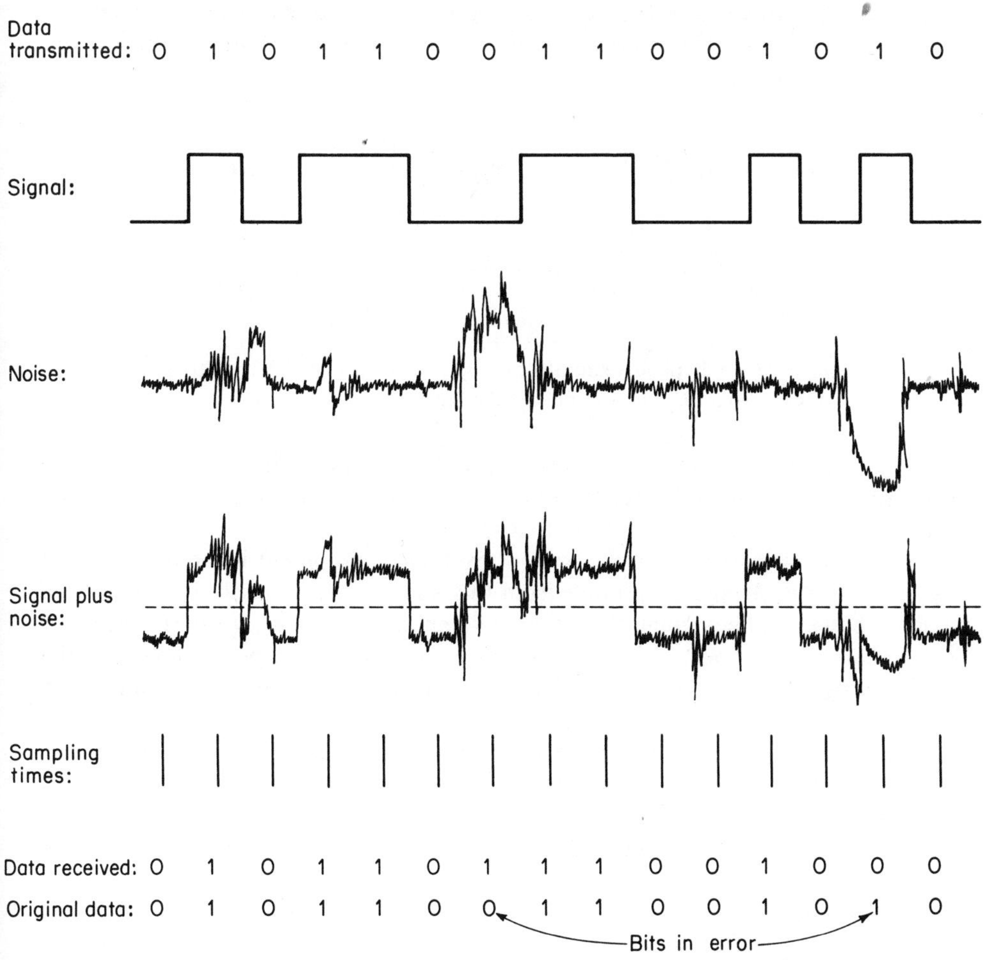

| Data transmitted: | 0 | 1 | 0 | 1 | 1 | 0 | 0 | 1 | 1 | 0 | 0 | 1 | 0 | 1 | 0 |

Signal:

Noise:

Signal plus noise:

Sampling times:

| Data received: | 0 | 1 | 0 | 1 | 1 | 0 | 1 | 1 | 1 | 0 | 0 | 1 | 0 | 0 | 0 |
| Original data: | 0 | 1 | 0 | 1 | 1 | 0 | 0 | 1 | 1 | 0 | 0 | 1 | 0 | 1 | 0 |

Bits in error

ply may induce hum or higher-frequency components into the communication channel.

The inductive or captive coupling through which noise is induced may be in the exchange. It may be coupling between adjacent cable pairs which are physically close. The noise generated by relays and switches may travel down wires a long way before reaching the low-level transmission signals, and may come from plant in separate buildings.

In some manual exchanges, particularly private branch exchanges, trouble has been experienced due to operators plugging into the data connection and listening to see whether the connection is free, as they might do with voice. Many large exchanges in cities, especially outside the United States, contain much old plant and this can cause more noise than the newer exchanges. Exchanges with step-by-step switches (Chapter 19) are worse than those with crossbar switches. Often these switches are the most important noise source on a line. Figure 27.2 shows characteristics of a typical noise burst caused by step-by-step selection switches. Amplitudes of 100 millivolts are common in such disturbances. They can occur at points where the signal strength is low and may last up to several hundred milliseconds. The periodicity is caused by the step-to-step motion of the selector (see Fig. 19.7).

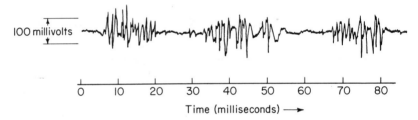

Figure 27.2   Typical example of impulse noise caused by Strowger switches (Fig. 19.7) in a public exchange.

Noise of the type in Fig. 27.2 usually causes errors in data to occur in a cluster rather than singly. Chapter 29 gives some statistics on the quantities of data errors and the degree of clustering that is found.

In years to come, when the present exchange equipment is largely replaced by electronic exchanges, the noise from this source will probably be much less. Today, however, lines going through public switching offices are more prone to noise than private lines. Other sources of impulse noise on public lines are the various ringing tones that are used, and the devices used during maintenance such as breakdown test sets and buzzers.

External noise can also come from atmospheric sources. Open-wire pairs hanging between telegraph poles can pick up atmospheric static. They are affected by distant lightning flashes and sometimes by contacts with trees or other foreign objects. Sometimes power lines or radar transmitters can cause interference. Radar interference has caused trouble with computer systems installed at airports, sometimes with the transmission of data and sometimes with the computer. Electric trains or electrical machinery sometimes cause periods of severe noise.

Impulse noise also originates from *within* the communication channel itself. This may be caused by circuit faults such as poor-quality soldering and dirty relay contacts and jacks. Nonsoldered twisted joints may cause noise due to changes in temperature and slight movements of the joint. In all these cases a variation in the contact resistance causes a fluctuation in voltage.

Many circuits carry more than one channel, as explained in Chapter 14. In such systems there is usually a small amount of crosstalk between one channel and another. The parameters of the multiplexing scheme and voice transmission equipment are chosen so that the effect of this is very small, and it is unlikely to affect data. However, an extra strong signal or impulse on one channel will exceed an overload point in the amplifiers and other devices, and various effects of this will be felt in the other channels. The signal which causes the overload may be a noise impulse, or it may result merely from the fact that all the signals being transmitted by the same multiplex device happen to be at a peak at that instant. The sum of the peaks extends the capacity of the channel for distortion-free transmission. This fortuitous adding together of peaks occurs in most multiplex systems, especially during busy periods. Harmonics and modulation products spill into other channels.

Large impulses in any communication system tend to overload the amplifiers, repeaters, and other electronic equipment they pass through on their journey. If an impulse is very large, it will momentarily render each amplifier or repeater inoperative, and this will tend to prolong the disturbance. It will tend to generate resonant frequencies characteristic of each amplifier and filter in its path. These frequencies are usually close to the maximum frequency the system is designed to handle and so again will tend to add to the disturbance. Amplifiers tend to convert severe amplitude disturbances into frequency disturbances so that they interfere with frequency modulation as well as amplitude modulation.

There are many different means of reducing the effects of impulse noise. First, good screening can be used and careful planning of the circuit paths to minimize induction, especially from switching equipment and relays. Second, multiplex systems are designed so that crosstalk and peak overloads are minimized. Third, the amplifiers, filters, repeaters, equalizers, and other equipment on the line are designed to lessen the noise effects sufficiently for voice. For data transmission, further equalization can be employed in the form of conditioning. Fourth, the choice of modulation method has an effect on the accuracy of transmission.

The computer systems designer cannot change the properties of the line he is given. Choice of the most suitable modem, however, is sometimes in his hands. He can lease a line with "conditioning" to minimize distortion, and he can ensure that the cabling and equipment on the user's premises are located and designed so that noise is not picked up there (as has often been the case).

CROSSTALK   "Crosstalk" refers to one channel picking up some of
                             the signal that is traveling on another channel. Occa-
sionally you hear faint fragments of somebody else's conversation on the telephone. It occurs between cable pairs carrying separate signals. It occurs in multiplex links in which several channels are transmitted over the same facility
(Chapter 14). It occurs in microwave links where another antenna picks up a
minute reflected portion of the signal for another antenna on the same tower. In
both of the latter cases the level of crosstalk noise is very small, because the system is designed with strict criteria for the maximum allowable crosstalk.

    Often the strongest source of crosstalk is induction between separate wire
circuits. Any long telephone circuits running parallel to each other will have crosstalk coupling unless they are perfectly balanced, which is not the case in practice.
Crosstalk between wire circuits will increase with increased length of circuit, increased proximity, increased signal strength, or increased signal frequency.

    Crosstalk may originate in exchanges or switching centers where large num-
bers of wires run parallel to each other around the exchange. It may originate in a
subscriber's building. It can be caused by capacitive as well as inductive coupling.

    However, on most systems it is barely greater, and often less, than the level
of white noise, and so, like white noise, it does not normally interfere with data
and is not annoying during speech. Occasionally, sometimes due to faults in the
exchange, crosstalk becomes louder, and it it possible to hear another person's
voice on the telephone. There will also be momentary peaks in crosstalk which do
interfere with data. With these exceptions, however, the level of crosstalk is
known or measurable and can be treated like white noise for the purposes of selecting transmission parameters so that no interference with data will normally
occur.

INTERMODULATION   There are certain undesirable types of data signals
NOISE   which can *cause* bad crosstalk. On a multiplexed chan-
                             nel, many different signals are amplified together, and
very slight departures from linearity in the equipment cause "intermodulation"
noise. The signals from two independent channels intermodulate each other to
form a product which falls into a separate band of frequencies, just as two sound
waves may "beat" to form a sound oscillation of a different frequency. The result
of this may fall into a band of frequencies reserved for another signal. Such products arising from large numbers of pairs of channels combine to form low-amplitude babble, which adds to the background noise in other channels. However, if
one signal were a *single frequency,* then when it modulates a voice signal on another channel, this voice might become clearly audible in a third independent
channel. One telephone user in this case would hear the conversation of one other.
Privacy is important on the telephone, and so the telephone company attempts to
restrict the power of any single-frequency signal to a suitably low level.

    The guilty single frequency could arise from data transmission in one of two ways. First, a repetitive code in a data signal could cause it unless the modem were specifically designed to prevent this. Many data-processing machines send repetitive codes to each other as part of their "line-control" procedures, for example to keep machines in synchronization while data are not being transmitted. Second, the modem itself, if not designed to avoid doing so, might transmit a single frequency when not transmitting data. Often one can hear this when dialing a computer. When the connection is established the apparatus at the other end will send a single-frequency "data tone" down the line to tell you that the connection is established. On some modems this frequency will always be present when there are no data. Such techniques could cause intermodulation crosstalk unless the signal strength is restricted. If it is restricted, then we are effectively reducing the capacity of the channel, as seen from Shannon's equation

$$C = W \log_2 (1 + \frac{S}{N})$$

This problem can be overcome by good modem design. When there are no data, the modem must not transmit a fixed-frequency tone, unless it is of very low amplitude. Further, the data must be randomized so that repetitive data patterns are not converted into a signal in which a single frequency dominates. Modems with these properties are now available and place no code restriction on the user. When they are used they can be permitted to send a high-strength signal over the carrier network and so not cause undesirable intermodulation crosstalk.

This is especially important for transmission over bandwidths wider than a single telephone channel. The power of the signal and the power of the noise are both proportional to the bandwidth. The limitation imposed upon a single-frequnecy transmission is, however, independent of bandwidth. Therefore, with this restriction, the higher the bandwidth the lower the signal-to-noise ratio will be for such signals.

ECHOES   Echoes on transmission lines are similar to crosstalk in their effects on data transmission. Where there is a change in impedance on the transmission line a signal will be reflected so that it travels back down the line at reduced amplitude, thus forming an echo. The signal-to-echo power ratio can occasionally become less than 15 decibels, though it rarely falls below 10 decibels. It can, however, be greater than white noise or crosstalk.

Echo suppressors are used on long lines, as was discussed in Chapter 10. They are not generally of value, however, in the problem of echoes in data transmission. Their action is triggered by the detection of human voice signals, and they are normally disabled when the circuit is used for data. In voice telephony, echoes of the speaker's voice become annoying when he hears them with a time delay measured in tens of milliseconds. In data transmission delays of a fraction of a millisecond are significant, and it is the listener rather than the talker who is affected by them. Multiple reflections down a two-wire path, or echoes formed at the junction of a two-wire and four-wire circuit, reach the receiving machine or modem. If they are of sufficient volume, they can cause errors in data.

SUDDEN CHANGES IN AMPLITUDE

Sometimes also the amplitude of the signal changes suddenly. This may be due to faults in amplifiers, unclean contacts with variable resistance, added load or new circuits switched in, maintenance work in progress, or the switching to a different transmission path.

These sudden changes can have an effect on certain data transmission systems which could result in the loss or addition of a bit. The effect they have depends on the type of modem in use.

LINE OUTAGES

Occasionally a communication circuit fails to be operational for a brief period of time. These "outages" may be caused by faulty exchange equipment, storms, temporary loss of carrier on a multiplex system, or other reasons, giving a brief period of open or short circuit. Often maintenance work on the lines, repeaters, and exchanges is the cause of brief interruptions. This gives rise to two concerns. First, the data may be damaged by signal losses of a few milliseconds. Second, signal losses of 10 seconds or more may cause a serious break in system availability. Line outages are of serious concern to the designers of some types of computer systems and are discussed in the next chapter.

RADIO FADING

Radio links are subject to fading. Many long-distance telephone circuits travel over microwave paths, and fading sometimes occurs. Small fades are compensated for by the radio automatic gain control. Large fades, however, may cause a serious degradation of the signal-to-noise ratio. Heavy rain and snow may cause fading. Violent gales may cause slight movement of the microwave dish. On rare occasions objects such as helicopters or flocks of birds may come into the transmission path. Sometimes microwave paths have been interfered with by new buildings which may cause microwave reflections. In the next chapter we will discuss atmospheric fading of microwave paths.

Other radio transmission is not very often used for data links to computers (at the time of writing), although there are certain applications for which it is invaluable and certain locations where it is the most economic form, or even the only possible form, of transmission. High-frequency radio is subject to many variations and sudden changes which are not found in wire transmission or microwave, such as deep and variable fading.

On local walkie-talkie radio transmission, the signal passes directly from the receiver to the transmitter. Long-distance radio, including the normal AM radio in the home, relies on reflection by the ionosphere. In the latter case, the path variations are very much greater. There are many propagation paths, and these suffer large daily and seasonal variations. The different propagation paths interfere with each other and so cause severe amplitude and phase variations. Different frequencies will suffer these effects in different amounts, and the effects constantly change so that they cannot easily be compensated for. Long-distance radio with ionospheric reflection is normally not regarded as good enough for data transmission, although where it is the only facility available, it is used at a low transmission rate with special modems.

CHANGES IN PHASE   The phase of the transmitted signal sometimes changes. Some impulse noise causes both attenuation and phase transients. Transients which affect only the phase are also common, especially on long lines. Figure 27.3 shows a brief change in phase. Sometimes the phase slips and returns as in Fig. 27.3, and sometimes it slips without returning. Typically the phase change occurs in less than 1 millisecond, but sometimes a gradual rotation of phase occurs.

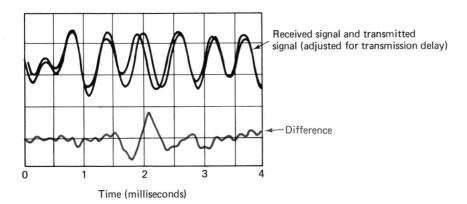

Time (milliseconds)

Figure 27.3   A typical transient change in phase. Brief phase changes are imperceptible in speech but sometimes cause data errors, especially when phase modulation is employed.

SYSTEMATIC   The types of noise and distortion discussed above
DISTORTION   were all "fortuitous." Those below are "systematic"
                    and so can be compensated for partially or completely in the design of the electronics.

In general, as the speed at which data are to be transmitted increases, so the need for uniformity in the transmission characteristics becomes greater. On private leased lines, steps can be taken to ensure that this uniformity measures up to certain standards. When dialing for a connection, however, it is not certain which path the call will take, and there are likely to be certain connections in the network on which the distortion will be high. Loaded cables, for example, which are not properly terminated or which have discontinuities give very undesirable transmission characteristics. It is possible that a short section of a nonloaded cable can be switched into the link, again increasing its distortion level. Some of the loading coils may be faulty or missing. The proportion of faulty or poor-quality sections in a public network differs from one area to another. Some networks still have a high proportion of old telephone plant in them. Some of the early cables used very heavy loading; this gives bad delay distortion, which is discussed below. In most of the major industrial countries this old plant is being replaced with modern equipment having more suitable characteristics for data transmission, but in some areas the replacement rate is slow.

When dialing for a connection on the public network, then, there is small probability that a line with several nonuniform characteristics might be obtained. This would give a higher proportion of errors than normal, and especially so if a high transmission speed was being used. Redialing might establish a different route between the two points and avoid the bad segment. Often, however, the error rate is not high enough to indicate to the user at the time that he has a bad line. He may never know that he is experiencing 2 or 3 times, or perhaps even ten times, the normal error rate.

LOSS   The loss of signal strength on a circuit is typically about 16 decibels. It may vary because of aging equipment, amplifier drift, temperature changes, and other causes. Changes are adjusted for during routine maintenance.

ATTENUATION   The attenuation of the transmitted signal is not equal
DISTORTION   for all frequencies, as ideally one would like it to be. This was shown in Chapter 10 in the discussion of bandwidth. The increase in attenuation at the edges of the band is indeed what demarcates the bandwidth, as shown in Fig. 10.3.

The attenuation of a typical cable pair, *within* the voice band, is approximately proportional to the square root of the frequency. To compensate for this variation in amplitude and to reduce attenuation as discussed before, the cable may be loaded by adding inductance at intervals. Similarly, on multiplex systems carrying many voice channels, filters are designed to yield a flat amplitude-frequency curve. However, some variation in amplitude across the band remains, and this is referred to as amplitude-frequency distortion. The effect of this is to distort the receivable signal slightly.

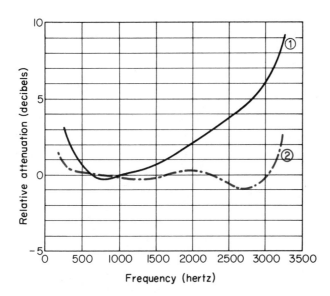

① Is a typical attenuation–frequency curve for a leased
telephone line (without equalization). With today's modems
this line would typically be used for transmission at 1200
or 2400 bits per second

② Is for the same line with equalizers. This line could now be
used for transmission at speeds up to 9600 bits per
second. See also Fig 27.8

Figure 27.4   Variation of attenuation with frequency on telephone lines.

The solid line in Fig. 27.4 shows the attenuation-frequency curve of a typical telephone line. Attenuation relative to that at 1000 Hz is shown. It is common that, as in this case, the attenuation level does not vary more than 10 decibels between 400 and 3000 Hz. This line would be satisfactory for transmission at 1200 bps. If, however, it were desirable to transmit data on the line at speeds of, say, 4800 bps, or greater, a further flattening of the attenuation-frequency curve would probably be needed, and to do this an *equalizer* might be used at each end cf the line. The equalizer would give a somewhat greater overall attenuation but a flatter frequency-attenuation curve. The dotted line in Fig. 27.4 shows the effect of the equalizer. Here again attenuation is shown relative to that at 1000 Hz.

Much worse results than these, however, are sometimes obtained on old telephone plant or equipment that is not functioning correctly. In some areas this is encountered on the switched public network.

Figure 27.5 shows results of measurements made on the AT&T switched public network in the United States [1]. These measurements were made on over 1100 test calls. About 25% of these were local calls not involved with the long-

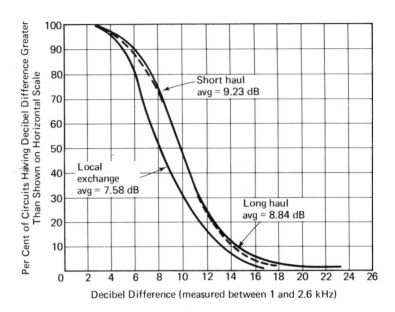

Figure 27.5 Variation of attenuation with frequency on AT&T switched public network. The curves relate to the difference in attenuation between 1000 and 2600 hertz, and show the percentage of lines having different values. About 30% of the local exchange lines, for example, have a difference greater than 10 decibels [1].

distance switching plan. About 25% were short-haul long-distance calls with distances up to 500 airline miles. The remaining 50% were long-haul, 400 to 3000 miles long. Figure 27.5 shows the difference in attenuation between frequencies of 1000 and 2600 Hz. It indicates that about half of the local exchange calls have a variation of 8 decibels and the other half about 10 decibels.

HARMONIC   Harmonic distortion refers to distortion in which the
DISTORTION   signal attenuation varies with amplitude. For example
                      a 1-volt signal may be attenuated by one half while a
5-volt signal is attenuated by two thirds. If a sine wave is transmitted on such a channel, it is flattened at the peaks. Pulse shapes are therefore not reproduced faithfully at the receiver. If harmonic distortion is considerable the demodulation operation is effected.

The flattening of the sine wave is equivalent to adding harmonics of low amplitude to the transmitted signal. If a sine wave of frequency f is transmitted, harmonic distortion would be equivalent to also transmitting a low amplitude sine wave of frequency 2f and a lower amplitude sine wave of frequency 3f. Hence it is

called harmonic distortion, and is measured in terms of the relative power of these
2f and 3f harmonics. Thus a signal with harmonic distortion might have a signal-
to-second-harmonic ratio of 25 decibels and a signal-to-third-harmonic ratio of 35
decibels. Keeping these ratios low is important for high-speed data transmission.
Telephone companies attempt to control the harmonic distortion of their circuits.

DELAY DISTORTION   The phase of the signal likewise is not transmitted line-
arly. The signal is delayed more at some frequencies
than at others. This is referred to as phase-frequency distortion or *delay distortion*. Some frequencies reach the receiver ahead of others. Figure 12.7 illustrated how signals on wire pairs are propagated with different speeds at different
frequencies.

It is a serious form of distortion in data transmission. It has only been corrected to a limited extent on the voice channels that we may wish to send data
over, because the human understanding of speech is not greatly affected by it. The
ear is a relatively slow-acting organism. It is normally necessary for a sound to
exist for 0.2 second in order to be recognized. If we have delay distortion of 0.05
second, the speech is still intelligible, though normally the distortion is not nearly
that great.

If there were no delay distortion, the curve of phase of the received signal
plotted against frequency would be a straight line, as in the top left-hand diagram
of Fig. 27.6. In reality it is a curve like that in the right-hand diagram of Fig. 27.6.
Envelope delay is defined as the slope of such curves and may be measured in
microseconds. In effect, envelope delay at a given frequency is the delay that

Figure 27.6   If there were no delay distortion, the curve of phase of
the received signal plotted against frequency would be a straight line.
In reality it is a curve like that in the right-hand diagram. "Envelope delay" is defined as the slope of this curve and may be measured in microseconds.

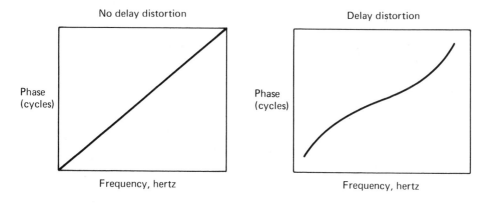

No delay distortion

Delay distortion

Phase
(cycles)

Phase
(cycles)

Frequency, hertz

Frequency, hertz

would be suffered by a very narrow bandwidth signal transmitted at that frequency. Because there is delay distortion, it varies with frequency, as shown in Fig. 27.7. It is sometimes simply referred to as the *delay* at a given frequency.

Figure 27.7 shows the effect of delay distortion on the transmission of a square-edged pulse. As was noted in Chapter 10, a square-edged pulse train is, in effect, composed of many frequencies, and so the edges of the pulses begin to distort as the wave travels to its destination. Again, if parallel transmission were used on the line, as is described in Chapter 17, in such a way that 1 bit of a character were transmitted at 800 Hz, the next at 1000, and so on, then, because of delay distortion, some of these bits would reach the receiver before others. The main effect of delay distortion, however, is on the more elaborate forms of modulation. Delay distortion must be kept below a certain level in order that the fastest and most efficient types of modem work correctly.

Figure 27.8 shows the delay distortion encountered on a typical telephone

Figure 27.7   The effect of delay distortion on the transmission of square-edged pulses over a line of limited bandwidth.

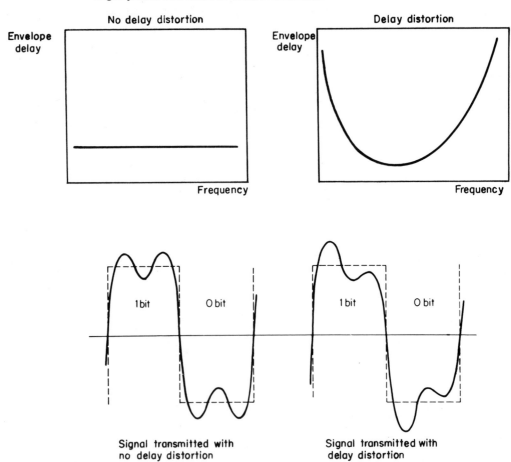

No delay distortion   Delay distortion

Envelope delay   Envelope delay

Frequency   Frequency

1 bit   0 bit   1 bit   0 bit

Signal transmitted with no delay distortion   Signal transmitted with delay distortion

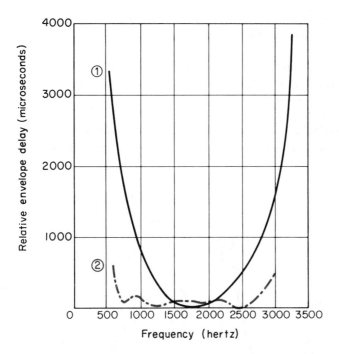

① Is a typical envelope delay curve for a leased telephone
line (without equalization). With today's modems this line
would typically be used for transmission at 1200 or 2400
bits per second.

② Is for the same line with phase equalization. This line
could now be used for transmission at speeds up to
9600 bits per second. See also Fig. 27.4

Figure 27.8 Variation in envelope delay with frequency on telephone
lines.

line. The solid curve shows the delay of different frequencies in microseconds,
relative to that at about 1800 Hz (which arrives first in this case). Just as *equalizers* are used to compensate for attenuation distortion, so a form of equalizer can
be employed to compensate for delay distortion. The signal at 1800 Hz is slowed
down to the speed of those at the outer frequencies. A typical result of this compensation is shown by the dotted line in Fig. 27.8.

As with the curves for attenuation in Fig. 27.4, an unequalized circuit with
the characteristics of the solid line would be likely to be used for transmission at
1200 or 2400 bps. If a leased telephone line is to be used for transmission at 4800
or higher bps, it may be equalized beforehand. To do this, the line has attached to
it circuits containing inductance and capacitance adjusted to give similar attenuation and delay at all frequencies.

**STANDARD PROFILES
FOR LINE DISTORTION**   CCITT makes recommendations that certain types of lines should have delay and attenuation distortion characteristics lying within given profiles. To a large extent, national telecommunication organizations accept these recommendations and attempt to engineer their lines to the specifications laid down. In addition, national bodies lay down distortion specifications for certain tariffs. A data transmission user with a leased line can, if he wishes, check that his line does lie within the specified profiles. Where a voice line is to be used for speeds higher than 2400 bps, this has been done, because a low bit error rate would be achieved only if the line did in fact meet its obligations for low-distortion operation.

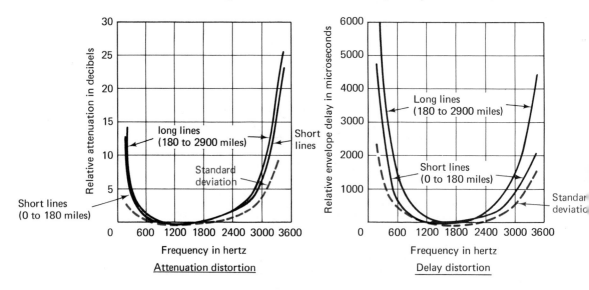

Figure 27.9   Attenuation distortion and delay distortion of Bell System toll connections. Conditioning can be applied to leased circuits, for a fixed monthly charge, to lessen both forms of distortion as shown in Fig. 27.12. (Redrawn from Reference [3].)

Figure 27.9 shows the attenuation and delay distortion on a Bell System toll connection. The curves were drawn from an AT&T survey which measured attenuation and delay on large numbers of circuits [2]. The standard deviation in Fig. 27.9 indicates the variations found between circuits. Some old telephone plant has characteristics much worse for data transmission than those of Fig. 27.9. Figure 27.10 shows measurements made on a line in England. Both the attenuation and the delay rise rapidly at frequencies about 1800 Hz. This is caused mainly by lengthy sections of wire-pair circuits with heavy loading — in this case toll-connecting trunks (junction circuits). In newer plant frequency-division-multiplexed groups would be used giving voice circuits with distortion characteristics more similar to those in Fig. 27.9.

Some old telephone circuits have characteristics even worse than those in

Fig. 27.10. Long wire-pair circuits were usually heavily loaded to correct attenuation distortion which affected speech, but delay distortion was largely ne-
glected. On the other hand, very little delay distortion is found in the higher-frequency media such as coaxial cables and microwave. As more and more calls are
transmitted over these newer media, the incidence of bad delay distortion
diminishes.

Modern circuits of widely differing physical nature are engineered to have
characteristics close to those of Fig. 27.9. For example, Fig. 27.11 shows characteristics of a tropospheric scatter circuit in Alaska. The attenuation and delay differ from day to day, but the distortion remains similar to Fig. 27.9.

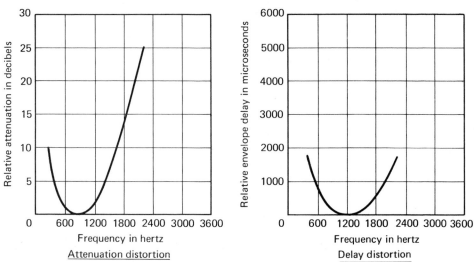

Figure 27.10   Some older telephone lines have much worse attenuation
and delay distortion than those in Fig. 27.9. These curves are for a short
telephone connection in England. The right-hand side of each curve is
worse than those in Fig. 27.9, and this is mainly caused by long sections
of wire-pair circuits with heavy loading. The use of modern carrier equipment on the wire pairs would give curves more like those of Fig. 27.9.
(Redrawn from Reference [4].)

FREQUENCY OFFSET   Signals transmitted over some channels suffer a fre-
                            quency change. That is, if 1000 Hz per second are
sent, 999 or 1001 might be received. This is sometimes caused by the use of multiplex systems for carrying each voice band at a different frequency. The oscillators
used for generating the carrier supplies for modulation and demodulation are not
precisely at the same frequency. When demodulation occurs the entire band suffers a frequency change because the carrier used for demodulating does not have
exactly the same frequency as that which was used for modulating. Frequency
shift is sometimes overcome by transmitting the carrier with the signal. A change

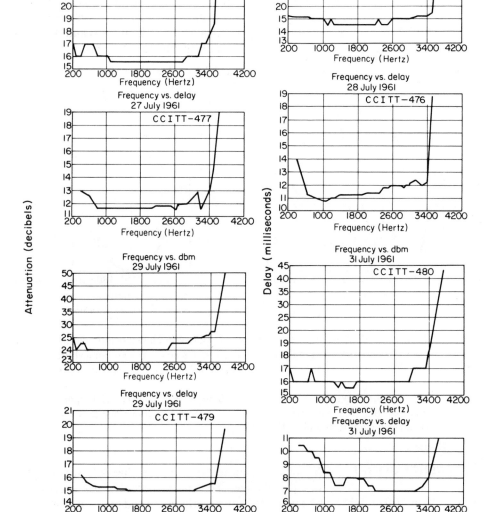

Figure 27.11 Modern circuits of widely differing physical nature are engineered to have characteristics close to those of Fig. 27.8. Here, for example, are curves showing the attenuation and phase distortion on a tropospheric scatter circuit in Alaska. (Redrawn from Reference [5].)

of about 20 Hz could be permitted without causing unpleasant distortion of the
human voice. However, the CCITT recommendation is that frequency offset
should be limited to ±2 Hz per link, and most circuits conform to that. A circuit
with five links in tandem might then have a frequency shift up to ±10 Hz. Under
faulty conditions, this value may be exceeded for short periods, but rarely exceeded by enough to interfere with data transmission. On the Bell System, frequency shift on the long-haul carrier system is generally held to less than ±1 Hz
per line section. The overall private line data channel objective is ±10 Hz, because there are several line sections in tandem.

**BIAS AND
CHARACTERISTIC
DISTORTION**
Repeaters are sometimes used to reconstruct pulses,
producing a new, clean, square-edged pulse out of
what by that time had become a distorted pulse. Similarly, the output of modulation systems are "sliced" to
give square-edged pulses. A form of systematic distortion that occurs can result in
all the pulses being lengthened or shortened. This was illustrated in Fig. 7.7 and is
referred to as bias distortion. If all the 1-bit pulses are lengthened, it may be called
"positive bias," or "marking bias," and if they are shortened, it may be called

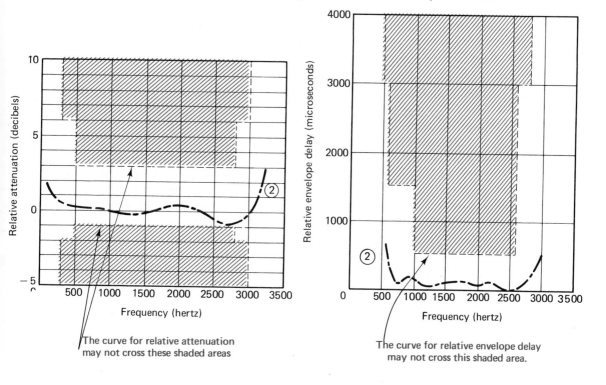

The curve for relative attenuation
may not cross these shaded areas

The curve for relative envelope delay
may not cross this shaded area.

Figure 27.12 Type C2 conditioning. Curves 2 from Figs. 27.4 and 27.8
are redrawn above. It will be seen that they both meet the specification
for C2 conditioning. Without the attenuation equalization and delay equalization shown in Figs. 27.4 and 27.8, they would not meet the specifications. Box 27.2, shows the specifications for other types of conditioning.

"negative bias," or "spacing bias." Bias distortion changes sign when the one and zero bits are interchanged, or, in other words, does the opposite to a 1 to what it does to a 0. Bias distortion may be caused by a decision threshold for the pulse regeneration being set at the wrong value and thus can usually be adjusted.

A similar type of systematic distortion is called "characteristic" distortion. Here the effect is not reversed when the 1 and the 0 are interchanged but is similar for either of them. For example, a single 1 or 0 may be shortened in transmission, whereas a long mark or space representing adjacent 1s or 0s may lengthen. This may be caused by some nonlinear characteristic of the transmission, possibly caused by bandwidth restriction or intersymbol interference.

CONDITIONING
Many of the impairments which effect circuits can be controlled so that they do not exceed certain limits. Table 27.1 indicates which of the impairments are controlled. Some are controlled by telephone company internal practices, which may limit the impairments to levels in the CCITT recommendations. An additional degree of control is sometimes applied to leased voice-grade circuits to improve their properties for data transmission. This, mentioned earlier, is referred to as *conditioning*. A monthly charge is made for conditioning in addition to the cost of the circuit.

There are two types of conditioning used in North America which control different impairments. *C conditioning* applies attenuation equalization and delay equalization to the line so that attenuation and delay distortion lie within certain limits. *D conditioning* controls harmonic distortion and the signal-to-noise ratio. A system designer may employ either form of conditioning, or both together. Conditioning is described in the FCC tariff No. 260.

There are five grades of **C** conditioning, as specified in Box 27.2. The maximum allowable distortion levels for each grade are specified in the tariff. Figure 27.12 plots the allowable distortion with C2 conditioning. The curves for the variation in attenuation and delay must not cross the shaded sections of the diagrams. C2 conditioning is the most commonly used grade for data transmission.

Modern high-speed modems, for example the Paradyne M-96, automatically equalize the circuit they operate on, effectively doing the same as **C** conditioning. With such modems **C** conditioning does not help performance. Less expensive modems can benefit from **C** conditioning.

The effects of attenuation distortion and delay distortion are thus reversible. The effects of noise and harmonic distortion cannot be reversed by the modem. Consequently AT&T introduced a second type of conditioning, **D** conditioning, which specifies limits to the permissible noise-to-signal ratio and harmonic distortion. **D** conditioning is intended for use with modems which transmit 9600 bits per second over voice-grade lines.

D-conditioned lines will meet the following specifications:

Signal to C-notched noise: 28 db
(C-notched noise is a standard measurement relating to the energy levels of a typical telephone signal.)

Signal to second-harmonic distortion: 35 db
Signal to third-harmonic distortion: 40 db
D1 conditioning relates to point-to-point channels.
D2 conditioning relates to 2- or 3-point channels.

To achieve **D** conditioning, the routing of the leased circuits will be chosen so as to avoid noisy or old-fashioned equipment, and to avoid step-by-step switching offices. In some countries it would be impossible to do this because there is much old equipment and most of the switching offices use Strowger switches.

Table 27.1   Circuit impairments which can be controlled either
by *conditioning* or by telephone company practices.

ATTENUATION DISTORTION	CONTROLLED BY C CONDITIONING
DELAY DISTORTION	
SIGNAL TO NOISE RATIO	CONTROLLED BY D CONDITIONING
HARMONIC DISTORTION	CONTROLLED BY TELEPHONE
IMPULSE NOISE	COMPANY INTERNAL PRACTICES
FREQUENCY OFFSET	
ECHOES	
PHASE JITTER	
RADIO FADING	
CHANGES IN PHASE	
ATMOSPHERIC GAIN HITS	NOT CONTROLLED
DROPOUTS	

SUMMARY   Data transmission is afflicted by both static impairments of the circuits (systematic distortion) and transient phenomena such as impulse noise and brief variations in phase (fortuitous distortion). A modem well designed for a given line with appropriate conditioning if specified can withstand substantial amounts of the static impairments such as attenuation and delay distortion without producing errors. If static impairments exist, however, the data transmission will be *more vulnerable* to errors from the transient disturbances.

AT&T conducted an analysis of the cause of errors on the Bell System, and Table 27.2 shows the results [5]. Transmission of 2000 bps was used, with an AT&T 201 data set. The transient disturbances causing the errors are subdivided arbitrarily into *short* transients lasting less than 4 milliseconds and long transients lasting more than 4 milliseconds. The most common signal disturbances were short transients. To avoid biasing the percentages in the table, calls giving a very high error rate—more than 1 bit error in 10,000—were not included; 3.5% of the calls have such an error rate, and it was caused mainly by impulse noise.

In Chapter 29 we will discuss how many data errors can be expected.

BOX 27.2 Five grades of channel conditioning which may be employed on Bell System leased lines, for fixed monthly charges

---

GENERAL PURPOSE — CHANNEL CONDITIONING
F.C.C. TARIFF 260, ISM SEC. 21

Channel conditioning in available on Series 2000 and Series 3000 channels, and on channels for similar purposes furnished under Series 5000 and 8000, as follows:

Type C1 — *For a two point or multipoint channel*
- the envelope delay distortion shall not exceed:
  between 1000 and 2400 Hertz, a maximum difference of 1000 microseconds
- the loss deviation with frequency (from 1000 Hertz reference) shall not exceed:
  between 1000 and 2400 Hertz, −1db to +3db
  between 300 and 2700 Hertz, −2db to +6db
    (+ means more loss)

Type C2 — *For a two point or multipoint channel*
- the envelope delay distortion shall not exceed:
  between 1000 and 2600 Hertz, a maximum difference of 500 microseconds
  between 600 and 2600 Hertz, a maximum difference of 1500 microseconds
  between 500 and 2800 Hertz, a maximum difference of 3000 microseconds
- the loss deviation with frequency (from 1000 Hertz reference) shall not exceed:
  between 500 and 2800 Hertz, −1db to +3db
  between 300 and 3000 Hertz, −2db to +6db
    (+ means more loss)
Note: On a multipoint channel arranged for switching, conditioning in accordance with the above specifications is applicable only when in the unswitched mode.

Type C3 — *For access lines and trunks associated with a Switched Circuit Automatic Network or Common Control Switching Arrangement*

BOX 27.2   Continued

Access Lines
- the envelope delay distortion shall not exceed:
  between 1000 and 2600 Hertz, a maximum difference of 110 microseconds
  between 600 and 2600 Hertz, a maximum difference of 300 microseconds
  between 500 and 2800 Hertz, a maximum difference of 650 microseconds
- the loss deviation with frequency (from 1000 Hertz reference) shall not exceed:
  between 500 and 2800 Hertz, −0.5db to +1.5db
  between 300 and 3000 Hertz, −0.8db to +3db
    (+ means more loss)

- the envelope delay distortion shall not exceed:
  between 1000 and 2600 Hertz, a maximum difference of 80 microseconds
  between 600 and 1600 Hertz, a maximum difference of 260 microseconds
  between 500 and 2800 Hertz, a maximum difference of 500 microseconds
- the loss deviation with frequency (from 1000 Hertz reference) shall not exceed:
  between 500 and 2800 Hertz, −0.5db to +1db
  between 300 and 3000 Hertz, −0.8db to +2db
    (+ means more loss)
Note: Conditioning in accordance with the above specifications is limited to:
  Each Interexchange of Local Access Line between the customer's station and switching center.
  Each Trunk — between switching centers.

Type C4 — *For a two point, three point or four point channel*
  the envelope delay distortion shall not exceed:
  between 1000 and 2600 Hertz, a maximum difference of 300 microseconds
  between 800 and 2800 Hertz, a maximum difference of 500 microseconds
  between 600 and 3000 Hertz, a maximum difference of 1500 microseconds
  between 500 and 3000 Hertz, a maximum difference of 3000 microseconds

BOX 27.2 Continued

the loss deviation with frequency (from 1000 Hertz reference) shall not exceed:

between 500 and 3000 Hertz, −2db to +3db
between 300 and 3200 Hertz, −2db to +6db
 (+ means more loss)

Note: On a three point or four point channel, conditioning in accordance with above specifications is applicable only between one exchange (that is designated by the customer as the control point) and each of the other two or three exchanges.

the envelope delay distortion shall not exceed:
between 1000 and 2600 Hertz, a maximum difference of 100 microseconds
between 600 and 2600 Hertz, a maximum difference of 300 microseconds
between 500 and 2800 Hertz, a maximum difference of 600 microseconds
the loss deviation with frequency (from 1000 Hertz reference) shall not exceed:
between 300 and 3000 Hertz, −1.0db to +3.0db
between 500 and 2800 Hertz, −0.5db to +1.5db
 (+ means more loss)

REFERENCES

1. From AT&T contribution A.T. 43, No. 13 in *The CCITT Red Book,* International Telecommunications Union, Geneva, 1961.

2. F. P. Duffy and T. W. Thatcher, Jr., "Analog Transmission Performance on the Switched Telecommunications Network," *Bell System Technical Journal,* 50 (April 1971), pp. 1311–1348.

3. M. B. Williams, "The Characteristics of Telephone Circuits in Relation to Data Transmission," *The Post Office Electrical Engineer's Journal,* London (Oct. 1966).

4. M. I. T., Lincoln Laboratory, *Transmission Testing on Alaska Forward-Propagation Tropospheric Scatter Circuits,* International Telecommunications Union, Geneva, (June 1963).

5. M. D. Balkovic, H. W. Klancer, S. W. Klare, and W. G. McGruther, "High-Speed Voiceband Data Transmission Performance on the Switched Telecommunications Network," *Bell System Technical Journal*, 50 (April 1971), pp. 1349–1384.

Table 27.2 An analysis of the causes of data errors on the Bell System when transmitting at 2000 bps with an AT&T 201 data set [5].

	Percent of Errors†	Percent of Line Signal Disturbances	Average Number of Errors Per Disturbance
Additive impulses lasting less than 4 msec	45	79	3
Other short transients lasting less than 4 msec, e.g., signal attenuation and phase transients	18	79	3
Additive signals lasting more than 4 msec, e.g., crosstalk or an increase in noise level	25	6	17
Amplitude changes lasting more than 4 msec, e.g., dropouts	5	4	8
Phase changes: the received phase differed from the original phase for more than 4 msec	7	11	2

†3.5 percent of the lines tested had an error rate greater than 1 bit in 10,000, and these were excluded from this table. On these lines the dominant cause of the errors was additive impulse noise, and they accounted for 66% of all the data errors.

Figure 27.13 Terminal test board of MCI, the specialized common carrier. From this board in New York City, all customer circuits are checked for transmission quality. (Photo by Harry Newton.)

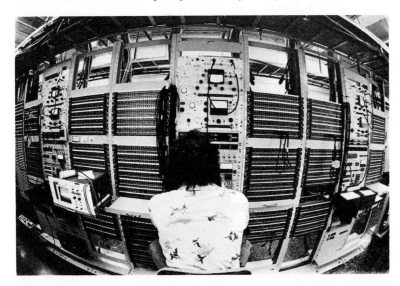

28 LINE FAILURES

Communication links throughout the world occasionally fail. For a certain period of time the signal cannot be transmitted. The computer system designer is concerned with the probability or frequency of failure and how long the failures last.

When telecommunications are used for batch processing, line failures are a nuisance but not a disaster. To an increasing extent, however, telecommunications are being used for real-time systems and interactive terminals. A high level of outages can be extremely frustrating to the terminal users and on some systems intolerable. The failure rate that can be tolerated depends on the functions of the system. The computer industry is learning slowly, and sometimes painfully, that terminals used for vital functions in industry must have very high dependability. The technology for designing on-line systems which are inoperative for only a few minutes a year is becoming understood, and such an objective is likely to be achieved by the end of the 1970s. However, most telecommunication links which interconnect the components are likely to be inoperative for several *hours* per year, sometimes many hours.

Thus, whereas in the 1960s data errors were regarded as the most serious problem of line performance, in the future, outages will probably by the most serious problem.

The duration of line failures ranges from a fraction of a second to several days. Failures of several days are generally rare. Failures of several seconds are not uncommon. When the outage time exceeds a certain duration it will normally be reported to the telephone company as a line failure. In general, the longer the outage time, the less likely it is to occur. In a sense, noise on communication lines causing damage to a few bits or characters is a transmission failure of very short duration, and this is more common than longer failures.

Outages ranging from a fraction of a second to many seconds are referred to as *dropouts*. Longer outages are referred to as *line failures*. We will discuss the data errors caused by noise in the next chapter. In this chapter we will discuss dropouts and line failures.

UNAVAILABILITY   The outage rate of communication lines is quoted in terms of their *availability*. The availability, measured over the period of time when transmission is required, is

$$\frac{\text{total time that line is transmitting}}{\text{total time that transmission is required}}$$

An availability of 0.98 would thus mean that the line was operating correctly 98% of the time.

The term *unavailability* is also used:

$$\text{unavailability} = 1 - \text{availability}$$

There are many factors which can contribute to a data transmission link being inoperative other than line failures. These include modem failures, operator errors, software errors, failures of power supply to the transmitting or receiving machine, and failures of terminals, multiplexors, concentrators, line control units, or other equipment constituting the data transmission path. In this chapter, *availability* (or *unavailability*) refers only to the communication line. The other types of failure make the overall data transmission availability worse than the availability figures quoted.

If several components of a link are in series (in tandem) as in Fig. 28.1, and have availabilities $a_1, a_2, \ldots, a_n$, then the overall availability of the link is $a_1 a_2 \ldots a_n$ if the elements of the link fail independently. Thus if the terminal and the equipment which connects it to the line have an availability of 0.995, the line has an availability of 0.98; and if the computer and the equipment which connects it to the line have an availability of 0.975, then the overall availability as seen by the terminal operator will be $0.995 \times 0.98 \times 0.975 = 0.951$. The terminal will be inoperative for about 10 hours per month, which is far too much for many applications.

Figure 28.1   Availabilities of independent elements in series are multiplied to give the overall availability.

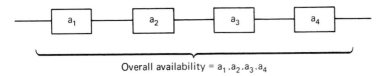

Overall availability = $a_1 . a_2 . a_3 . a_4$

STANDARD DEVIATION   The situation is made much worse by the fact that the failures are not uniformly distributed. Most failure statistics have a high standard deviation. For numbers of failures per month, the standard deviation is typically about equal to the mean. This indicates that in some months the above system would be inoperative for less than 10 hours, while in others it would be much worse. If the number of failures per month follows an exponential distribution (an assumption which makes a reasonable fit to the observed statistics), then in 0.14 of the months the failure rate will be twice the average, and in 0.05 of the months it will be three times the average. Once every 4 years, it will be four times the average—a situation that would often be regarded as totally inacceptable.

SELF-CORRECTING FAILURE?   Line outages fall into two groups: those caused by conditions which are self-correcting, i.e., repair work is not generally needed, and those which are equipment failures. Brief temporary outages are caused by noise or fortuitous distortion as described in the previous chapter. Other self-correcting outages are caused by workmen on the line, by microwave fades, by the switching to alternative equipment, and occasionally by sticking switch contacts or intermittent equipment failure.

Automatic switching to alternative channels in the same cable or path, or to alternative equipment, occurs fairly frequently and causes the transmission link to be inoperative for a fraction of a second. By far the most common communication line outage is a dropout lasting for a small fraction of a second. Although it is the most common outage type, it is not a major contributor to the overall unavailability, nor is it too serious because of its brief duration.

MICROWAVE FADING   The majority of today's nonlocal telephone calls in North America travel over microwave radio, and microwave is subject to radio fading. Small fades are compensated for by the radio automatic gain control. Worse fades cause a degradation of the signal-to-noise ratio and hence cause data errors. Large fades attenuate the signal to the point where a communication outage occurs.

There are two main types of microwave fades, as illustrated in Fig. 28.2. The most common is called *selective fading* and typically lasts for a few seconds to a few hundred seconds. It is frequency-selective; that is, it affects one microwave channel but not the channels at nearby frequencies. It is caused by atmospheric distortion bending one microwave path more than another so that signals reach the receiving antenna by more than one path, as shown in the second drawing of Fig. 28.2. Signals from the different paths interfere with one another, tending to cancel each other out at certain frequencies in certain places, just as two waves from stones dropped close to each other in a pond cancel each other at certain

points on the pond surface. Not far from the cancellation point on the pond surface the waves are strong—and similarly with microwave fading. If a second receiving antenna is used near the first, they can be arranged so that one receives the
signal strongly at times when the other does not. The use of compensating antennas is referred to as *space diversity* in microwave links.

The interference is also highly dependent on frequency. The location at
which the signals cancel will be different for different frequencies. The same phe-
nomenon is observed with interference of *light* waves and is the cause of the rainbow coloring of light interference patterns. A selective fade affects one microwave
channel but not the channels at nearby frequencies. Switching to alternative channels of different frequency on the same path can therefore protect the link. The
provision of spare channels in the frequency band is referred to as *frequency diversity*. Some microwave systems reserve 4 channels out of 20 on a fully loaded
route for this form of protection. Spare channels are also needed to protect against
outages due to equipment failure and maintenance activity. Both *frequency diver-*

Figure 28.2   Two types of microwave fading.

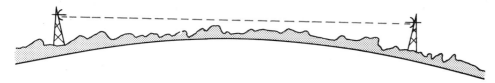

1   Line of sight transmission.

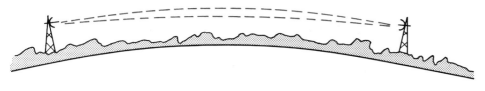

2   Selective fading. Atmospheric distortion causes
multiple path transmission. The separate transmission paths interfere with each other at certain
frequencies, causing signal loss of durations
ranging from a few to a few hundred seconds.

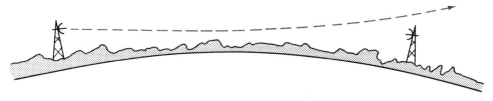

3   Flat fading. Under certain rare atmospheric
conditions the signal is refracted so that
signal loss may occur for several hours.

sity and *space diversity* increase the cost of the link. There is thus a trade-off between channel availability and channel cost.

Near to the point on a pond surface at which the waves cancel out are larger areas where the waves are of smaller size than if there were no interference. Similarly, with microwave links, the deep fades which eliminate the signal (40-decibel fades) usually last only a few seconds, whereas smaller fades (say 10 decibels) last up to a few hundred seconds. This variation in the depth of fades is illustrated in Fig. 28.3.

A less common type of microwave fading which is more harmful when it occurs is called *flat fading*. These fades are illustrated at the bottom of Fig. 28.2 and occur only under unusual atmospheric conditions. Ground fog causes a gradient in the dielectric constant of the atmosphere and so refracts the radio beam so that the main part of its energy does not reach the receiving antenna. This occasionally occurs in areas of high moisture content with stable calm air conditions at night or early morning. Widely spaced frequencies are affected, so neither frequency diversity nor space diversity give protection from flat fading. The best protection is an appropriate choice of microwave path with sufficiently short radio hops. Flat fading occasionally causes microwave outages of as long as several hours.

Neither selective nor flat microwave fades occur much during the prime working hours. They occur mainly at night and tend to be confined to certain months of the year. During winter months when humidity is low signal variations are usually small. On clear summer nights with little or no wind, nonuniform distributions of temperature and humidity can create major variations in the dielectric constant of the lower atmosphere, and microwave performance suffers.

The AT&T objective for reliability on a long-haul microwave system (4,000 miles, two-way) is that the outage of a channel should not exceed 0.0002, or $1\frac{3}{4}$ hours per year [2].

Figure 28.4 shows typical outages for such a communication path, with varying numbers of protection channels.

The automatic switching to protection channels will itself cause a transmission outage of less than a second.

EQUIPMENT FAILURE   The main cause of long outages is equipment failure. The duration of the outages, and hence the line availability, depends on how quickly the failure is reported and how quickly the telephone company mobilizes the repairmen. Most equipment failures are corrected in an hour or several hours either by replacing equipment or by substituting an alternative line. If the failure occurs in an office with maintenance men on site, it can usually be corrected in less than an hour. If it occurs at a remote site, it may take 2 to 4 hours. A few failures are more catastrophic in nature and take a long time to repair.

Figure 28.5 shows an analysis of the duration and causes of interruptions on

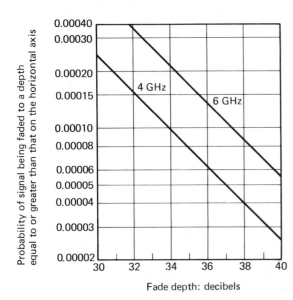

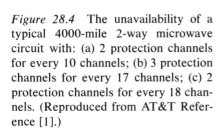

Figure 28.3 The variation of fade probability with fade depth on a typical microwave link. (Reproduced from Reference [1].)

Fade depth: decibels

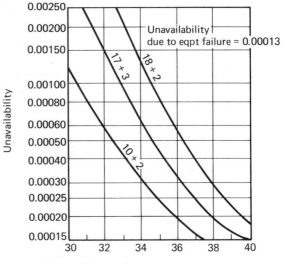

Figure 28.4 The unavailability of a typical 4000-mile 2-way microwave circuit with: (a) 2 protection channels for every 10 channels; (b) 3 protection channels for every 17 channels; (c) 2 protection channels for every 18 channels. (Reproduced from AT&T Reference [1].)

Fade margin: decibels

a data circuit routed 892 kilometers via cable (no microwave) between Copenhagen and Rotterdam [3]. It shows typical causes of outages over a 2-month period. All the equipment failures in this illustration were corrected in less than 1 hour.

TIME OF DAY   Line availability varies somewhat with the time of day. Microwave fades tend to occur at night. Most other failures occur most frequently during the day and usually when the traffic is heaviest. More switching to alternative circuits occurs during heavy traffic, and failures caused by repair work on the line tend to occur at busy periods. The unavailability

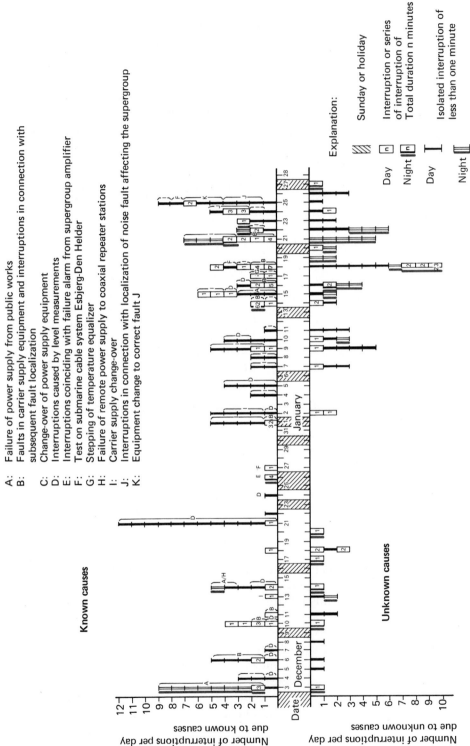

Figure 28.5 An analysis of the duration and causes of interruptions on a data circuit routed via cable (no microwave) between Copenhagen and Rotterdam (892 kilometers) [3].

538

at 9 A.M. to 5 P.M. is often two to four times that at night. On the other hand, catastrophic failures may not be repaired quickly at night.

MEAN TIME TO REPAIR   The unavailabililty due to equipment failures depends on how quickly the failures are reported and repaired. It is clear from statistics kept by data transmission customers that this time is highly variable. If the process is slow and inefficient, it can typically take 4 hours or so. If it is fast and efficient, it may take less than 1 hour.

The telephone company customer can influence the availability of his leased lines by maintaining efficient error detection and reporting procedures and by maintaining close relations with the telephone company. Fowling [4] reported that on the leased lines used by British European Airways the mean repair time dropped *by a factor of 5,* from 4 hours to less than 1 hour, as the personnel learned how to define a fault and how to help the telephone company expedite its repair. Other companies have also achieved a major improvement in line availability by tightening the human organization and in some cases tackling the telephone company at a high management level.

In one detailed survey, a group of communication-based systems acknowledged to be well-managed were compared with a general cross section of communication-based systems. It was found that average systems had five times the unavailability of the well-managed systems. Much of the reason had to do with customer-operator inefficiencies as well as relations with the common carrier. In the poorly managed organizations arguments about who was responsible for the failures would delay their isolation and repair.

A telephone company learns about most of the failures when subscribers complain. On the newer data lines such as AT&T's Dataphone Digital Service (DDS) there is a high degree of performance monitoring of the lines so that failures can be detected quickly. Most analog telephone lines are not monitored, though some faults in switching equipment are detected automatically.

When a customer complaint is received, a trouble ticket is issued indicating the time the complaint is received, the nature of the complaint, the actions taken, and the time testing is completed. AT&T refers to such a ticket as a C1. Elaborate statistics of the C1s are kept. In a busy section of New York at the time of writing, leased links serving data-processing equipment had an average about 0.3 C1s per month, giving a line availability of the order of 0.998. 0.3 C1s per month seems to be a typical figure for the North American leased lines in general.

The designer of data-processing systems, however, needs to qualify this statistic with two important facts. First, telephone company C1 statistics relate only to the problems that are reported to them. Comparison of typical data transmission users' trouble logs with the telephone company C1 listings revealed that less than half of the troubles were in fact reported so that they became C1s. Many of the troubles are in fact of relatively short duration and self-correcting. They are caused, for example, by excessive noise, operator errors, or microwave fades. The telephone company statistics on line outages include only those cases in which the line was still inoperative when the telephone company tested it.

GOOD AND BAD LINES

A still more important statistic for the data-processing designer is that there is likely to be an order of magnitude difference between the *average* outage rate and that of *troublesome* lines. Circuit trouble distributions tend to be highly bimodal. Some lines, perhaps routed via old or marginal equipment, are persistently troublesome. The following breakdown of a number of reported line outages is typical. It relates to a user with 18 long-distance leased voice-grade lines. Five of the 18 lines were exceptionally bad [5], as shown.

	Reported Line Outages Per Month
LINE 1	2
LINE 7	4
LINE 9	3
LINE 10	13
LINE 13	3
Average of all 18 lines	0.32

When a system is likely to be critically dependent on a line, the system designer should appreciate that while the mean line availability is quite good, the availability of a critical line may be as low as 0.98, and occasionally it is lower. Occasionally the statistics show lines with availabilities as low as 0.90. If the line in question serves process control equipment, stock broker terminals, supermarket cash registers, or other devices on which outages are harmful, the devices should not be *solely* dependent on the line. The system should be designed so that the devices in question have their own local minicomputer or other stand-alone capability with which the system can survive when the line is out.

The above argument is especially important because the duration of the failures of the troublesome lines is often several hours. An examination of the telephone company C1 logs revealed that approximately 30% of all C1s in a large sample of data links took longer than 2 hours to repair. The average time to restore a channel after a microwave equipment failure on the Bell System is quoted by AT&T to be $2\frac{1}{4}$ hours [6]. These times are measured from when the failure report is received by the telephone company, which is usually some time after the failure actually occurs.

Table 28.1 shows 4 months in the life of a somewhat troublesome, leased, multidrop. voice-grade line. It is taken from the telephone company trouble tickets and shows the causes of the various troubles [5].

If a data-processing user has a leased line which is substantially more troublesome than the average, he should ask the telephone company to reroute it via

Table 28.1 Summary of AT&T trouble tickets for 4 months
in the life of a somewhat troublesome leased,
multidrop, voice-grade line [5]

Line 107	Time Received	Time Up	Stations Affected	Comments on Trouble Ticket
Jan. 10	1245	1335	All of Chicago	Noisy carrier
Jan. 19	1140	1210	All of Kansas City	Noisy carrier
Jan. 25	1125	1245	St. Louis	Intermittent response to poll
Jan. 25	1125	1250	All of Kansas City	Intermittent response to poll, customer provided equipment
Jan. 26	1900	1915	All of Kansas City	No response, customer provided equipment
Jan. 26	1015	1130	All of Kansas City	No response, customer provided equipment
Feb. 02	1300	1820	All of Kansas City	Open loop
Feb. 03	1010	1315	All of Chicago	Noisy carrier
Feb. 04	0920	1005	All of Chicago	Noisy carrier − no response
Feb. 09	1400	1440	All of Kansas City	No response to poll
Feb. 10	1645	1730	All of Kansas City	No response to poll
Feb. 11	0840	0910	All of Kansas City	Customer action
Feb. 12	0945	1600	Kansas City	Open loop, low levels
Feb. 15	1018	1100	Kansas City	Low levels, customer action
Feb. 15	1212	1430	St. Louis	No loop back
Feb. 15	1408	1425	Chicago	No response, customer provided equipment
Feb. 16	0930	1400	St. Louis	No response, customer provided equipment
Feb. 18	0940	1010	Chicago	No response, customer provided equipment
Mar. 10	1618	1805	All of Chicago	Open loop
Mar. 11	0905	0911	Kansas City	Customer provided equipment
Mar. 20	1345	1430	Kansas City	No response
Mar. 21	1012	1330	All of Chicago	Heavy noise
Apr. 10	0930	1015	Chicago	Low levels
Apr. 11	1010	1040	Bloomington	No loop back

different equipment. If this is done with all troublesome lines, the overall availability for the system can be substantially improved. Even so we suggest that a systems analyst designing a system in which availability is vital should assume that the critical line may be inoperative 2% of the time.

CATASTROPHIC OUTAGES

On rare occasions severe damage occurs to transmission lines, which cannot be repaired quickly. A bulldozer tears up a local loop, a vehicle crash knocks down a telephone pole, or line termination equipment is destroyed by fire. Outages of a day or more are rare occurences in North America but seem to be more common in some other countries. In Europe and South America long outages are sometimes caused by administrative problems such as a replacement for a critical item not being available. Occasionally strikes result in long outages.

In large cities the probability of catastrophic telephone outages appears to be increasing, probably because the wires under the streets are becoming old and corroded. New York telephone users have been plagued with the effects of floods and fires and broken underground cables. Good quality maintenance of local loops is becoming more expensive. We may face an era of worsening local loop availability.

TYPICAL
OUTAGE RATES

Figure 28.6 summarizes typical circuit availabilities. A line at 45° to the axes on this chart is a line of constant availability. The position of such a line is different for different causes of failures. On the left of the chart is the availability for very brief outages caused by noise which shows up as data errors. In the next chapter we will discuss these. To the right is a line with availability 0.0002, which is AT&T's objective for microwave links and which corresponds closely with measurements of outages of from 1 to 100 seconds.

AT&T's availability objective for DDS lines is somewhat better: 0.0004, shown by the dotted line [9].

For longer failures, the availability line moves to the right again. For outages due to equipment failure a typical repair time is from 1 to 3 hours, and a typical availability is 0.002. While 0.002 may be typical, a substantial proportion of lines are much worse for failures longer than 1 hour, as shown by the rightmost line on the chart.

Figure 28.6 was drawn using statistics from voice-grade lines in North America. Statistics on outages from other industrial countries appear broadly similar. Table 28.2 shows an analysis of outage durations on a leased voice line from London to Rome [7]. The data in Table 28.2 is plotted as crosses in Fig. 28.6. Note that while the overall unavailability of the line in Table 28.2 seems high, 0.0575, most of this is contributed by one outage of longer than 1 day. It is generally true that one such long outage can degrade the overall availability figure.

WIDEBAND LINES

Less statistical data exist on the outage rate of wideband lines. What little there are suggest that their outage rate is roughly comparable to leased voice lines. Table 28.3 shows a 10-month summary from the ARPA network [8].

PROTECTIVE ACTION

To protect itself from line failures it is desirable that a system be able to take one of two actions:

1. Set up automatically and rapidly an alternative transmission path.

2. Take an alternative data-processing action which temporarily circumvents the need for transmission.

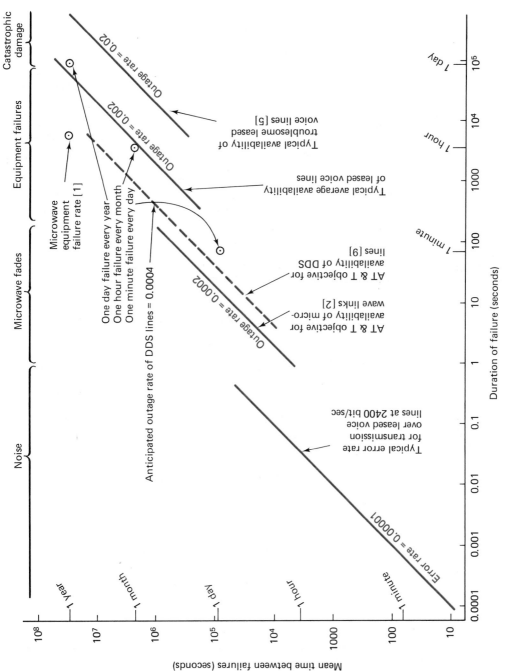

Figure 28.6 Typical outage rates.

Table 28.2 An analysis of outage durations on a leased
voice circuit between London and Rome [7]

Range of Outage Duration (min)	Number of Outages	Relative Frequency (%)	Time Lost (min)	Percentage of Total Time Lost	Fractional Unavailabilities
0.0001 to 0.001	685	69.68	0.50	0.017	0.0000101
0.001 to 0.01	83	8.44	0.38	0.013	0.00000770
0.01 to 0.1	105	10.68	2.96	0.104	0.0000600
0.1 to 1	68	6.92	29.59	1.042	0.000600
1 to 10	23	2.34	62.68	2.208	0.00127
10 to 100	14	1.42	382.89	13.489	0.00776
100 to 1000	4	0.41	737.40	25.979	0.0149
1000 and up	1	0.10	1622.00	57.143	0.0329
Total	983		2838.40		0.0575†

†The total unavailability is seriously affected by one outage of 27 hours and 2 minutes. The total
unavailability not counting this outage would be 0.0246.

Table 28.3 Availability of leased
50,000-bps lines on the
ARPA network [8]

Month	Average Line Unavailability
Sept. '71	0.0059
Oct.	0.0166
Nov.	0.0165
Dec.	0.0321
Jan. '72	0.0102
Feb.	0.0123
Mar.	0.0136
Apr.	0.0088
May	0.0111
June	0.0041
Mean	0.0131
Standard deviation	0.0078

The latter is a complex subject and highly application-dependent. Small
inexpensive computers and storage units are employed at the terminal locations.
The use to which they are put differs from one type of system to another. The
peripheral computers generally have many functions other than those related
solely to availability.

The automatic setting up of an alternative transmission path may be a good
solution to line outage problems. In some systems, *distributed* networks are used

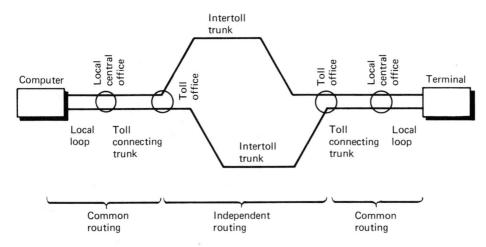

Figure 28.7 Attempts to achieve automatic alternate routing do not give complete protection because the alternate circuits are routed over the same local facilities.

in which more than one line is routed by different paths between critical locations, as, for example, in the ARPA network. In other systems a line-control unit can detect leased-line failures and automatically dial an alternative path over the public network.

Unfortunately, while such alternative routing appears to be the perfect solution, in practice part of the alternative route normally uses the same transmission facilities as the route it replaces. Consequently both the alternative circuit and the one it replaces often fail simultaneously. In particular both circuits are likely to travel over the same local loops to the computer or terminal locations.

Figure 28.7 shows two transmission circuits between a terminal and computer. Either of them may be a leased voice-grade line, public dial line, or wideband line. The two circuits are in the same wire-pair cable connecting the computer to its local central office and again in the loop connecting the terminal to its local central office. Both central offices are connected to toll offices on toll-connecting trunks, and in these also the circuits probably share common physical facilities. They sometimes share the same multiplexed group. Between the toll offices the circuits may be genuinely separate and independent. If the terminal is connected to the same central office or toll office as the computer, the circuits may share common facilities all the way.

LOCAL LOOP OUTAGES There are few statistics available which shed light on what proportion of outages would affect two differently routed circuits simultaneously. However, a few general conclusions can be drawn. Most microwave links are intertoll trunks, and so microwave fades would usually affect only one of the differently routed circuits. Most dropouts due to automatic channel switching similarly affect the intertoll trunks. On the other hand, most catastrophic failures or outages of longer than 3 hours are local problems and would be likely to affect both circuits. Local failures due to construction work, wet cables caused by flooding, ice storms, tree branches falling on overhead lines, vehicle crashes, urban sabotage, and so on, damage both circuits and would seriously affect unavailability because of the long repair times.

In one study of system outages in Canada local loops contributed more downtime than trunks. Fowling [4] comments that on the British European Airways system most communication unavailability is due to local loops. In some comparisons of unavailability for different line lengths, long lines are not much worse than very short ones, suggesting again that a high proportion of the problems are in the subscriber loop or toll-connecting trunk. Figure 28.8 shows the causes of 4800 bps circuit outages during the first year of operation of the Datran system [10] and it will be seen that the subscriber loop outages are dominant.

Can a subscriber obtain two separately routed loops? Sometimes, especially for subscribers in city blocks, this is possible at reasonable cost. Often, however, separately routed cables do not already exist, and the cost of installing them is too great. It is very great in rural locations. In some cases the telephone company will not specify whether or not a standby dial-up line uses the same cable as the leased line it would replace. Many solutions to the problem have been suggested, in-

Figure 28.8   Causes of 4800 bps circuit outages during the first year of operation of the Datran system.

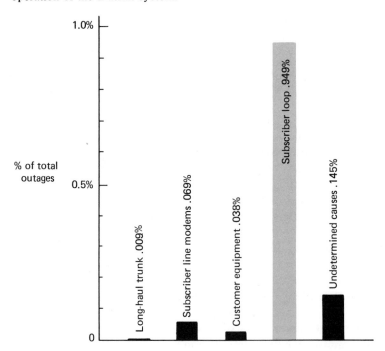

Figure 28.9   Catastrophic outages, of abnormal length, are often those involving local loops. The outage cannot be bypassed because all circuits, dial and leased, go through the same cable.

cluding the use of microwave or millimeterwave radio, and infrared or optical transmission. Western Union has a program for installing "survivable" local distribution facilities in business communities, with each user being wired to two Western Union locations. Another backup procedure might be to have a loop from more than one common carrier—say, AT&T and Western Union—but in most cases both would be provided by the local telephone company and routed in the same cable. Most installations outside major cities will probably not be able to obtain alternative loops for years to come.

    The local loop and intertoll trunks may remain the Achilles' heel which prevent the computer industry from obtaining dependability of transmission facilities and force it to seek *system* dependability by the use of "distributed intelligence."

REFERENCES

1. W. Y. S. Chen, "Estimated Outage in Long-Haul Radio Relay Systems with Protection Switching," *Bell System Technical Journal,* 50, No. 4 (April 1971).

2. *Ibid.,* p. 1471.

3. *Tests by the Danish Telecommunications Administration, Third Plenary Assembly of the CCITT, Geneva 1964,* International Telecommunications Union, Geneva, 1964.

4. Fowling, "Developing an International Dedicated Computer Network," *INFOTECH State-of-the-Art Report No. 6 — Computer Networks,* 1971, pp. 157–176, INFOTECH, London.

5. DMW Corporation, from a study by Dr. Dixon Doll, a consultant in the design of data transmission systems.

6. Chen, *op. cit.,* p. 1461.

7. CCITT contribution COM Sp. A/No. 64, "Data Transmission Test on a Multipoint Telephone Network in Europe," *The CCITT Blue Book,* International Telecommunications Union, Geneva, 1964.

8. A. A. McKenzie, B. P. Cosell, J. M. McQuillan, and M. J. Thorpe, "The Network Control Center for the ARPA Network," First International Conference on Computer Communication, 1972.

9. AT&T specifications for Dataphone Digital Service, 1974.

10. E. V. Farinholt, *Domestic Digital Transmission Services Signal,* Armed Forces Communications and Electronics Association, April 1975.

29 DATA ERRORS

Because of the noise on communication lines, especially impulse noise, there will be a number of errors in data transmitted. We cannot prevent all the errors occurring; all we can hope to do is to detect them and somehow correct them. In the next chapter we will discuss the various ways of dealing with errors. In this chapter we will summarize the quantity and nature of errors that can be expected.

Many measurements of transmission errors have been made on communication lines. Telecommunication companies throughout the world have statistics about the error rates and patterns that can be expected on different types of lines. In studying this mass of documentation, a pattern emerges. There is generally a similarity between a good-quality line in one location and a similar type of line elsewhere. This is not surprising as there has been much international standardization of the design parameters. However, some lines do deviate from this pattern. Just as a few lines have exceptionally poor availability, so some have exceptionally high error rates.

The systems designer or analyst needs to have some knowledge of the error rates and patterns his system will encounter. On some systems this has been obtained by taking measurements on the network that will be used. Typically, a tape of data is transmitted many times over the lines in question. The data received are then compared with the original, and a computer program produces statistics about the errors that occurred. This is how most of the figures in this chapter were obtained.

Often, however, such machines are not available or it is not practical for the user to test the lines he wants to use. Systems analysts constantly complain that they do not know what error rates to expect and cannot obtain any figures for this. To persons in this position it is suggested that they use the figures in this chapter as being typical.

1. TELEGRAPH AND TELEX CIRCUITS

ERROR RATES   The CCITT Study Group on Data Transmission has
                         made and analyzed tests on the world's 50-baud leased
telegraph channels and on the international telex circuits of many countries [1].
The results are as follows:

Most Probable Error Rate

1. In point-to-point service:

    On elements: one- to two-bit errors in 100,000 transmitted.
    On characters: one- to eight-character errors in 100,000 transmitted.

2. On switched telex circuits:

    On elements: one- to two-bit errors in 100,000 transmitted.
    On characters: four- to five-character errors in 100,000 transmitted.

The character error rate of 1 in 100,000 corresponds to one error in approximately 4 hours, and 8 in 100,000 to one error in approximately ½ hour. The most probable time interval without error was found to be approximately 1 hour.

BURSTS   A considerable proportion of the errors were found to
                         come in *bursts*. *Bursts of errors* are defined differently
by different authorities. Here a *burst of errors* was defined as elements in error
separated by less than 10 nonerroneous elements. The proportions of such bursts
were found to be as follows:

isolated errors on elements:	50 – 60%
bursts with two errors:	10 – 20%
bursts with three errors:	3 – 10%
bursts with four errors:	2 – 6%

These figures are important for predicting the effectiveness of different
error-detecting codes.

Time periods in which the "start" condition remains on the line for more
than 300 milliseconds are referred to as "dropouts," the line being regarded as
temporarily out of service. These are not included in the preceding figures.

The CCITT recommends that the above performance be taken as that of
standard 50-baud telegraph and telex circuits.

VARIATION WITH   The number of line segments that comprise the tele-
NUMBER OF   graph or telex connection has an effect on the error
LINE SECTIONS   rate. The curves in Fig. 29.1 are taken from a study of
                         the telex network in Germany, which reveals this [2].
In this study, test messages were transmitted from 300 different subscriber
sets throughout Germany to an evaluation center. The messages were sent from

Bit error rate

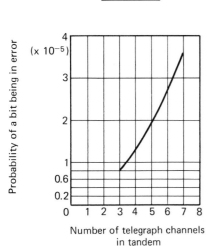

Character error rate

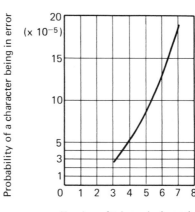

Block error rate

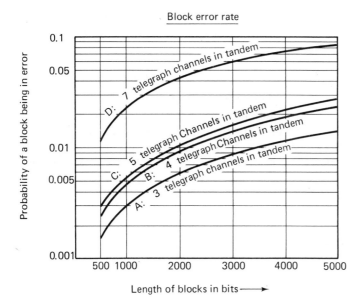

Figure 29.1   Mean bit, character, and block error rates on Telex circuits – from a survey of Telex connections in Germany [3].

plastic tapes containing 1000 characters repeated several times. The connections used were subdivided into categories depending on the number of voice-frequency telegraph sections they comprised. They were selected in such a way that the ratio of these connection types is equivalent to the distribution of actual connections in Germany, as follows:

Category	Number and Kind of Voice Frequency Telegraph Systems		Number of Connections
A (3 sections)	3 AM		1355
B (4 sections)	3 AM	1 FM	804
C (5 sections)	5 AM		745
D (7 sections)	5 AM	2 FM	103

Category A: Local exchange – central exchange – central exchange – local exchange.

Category B: Subexchange – local exchange – central exchange – central exchange – local exchange.

Category C: Subexchange – local exchange – central exchange – central exchange – local exchange – subexchange.

Category D: Subscriber for toll service – subexchange – local exchange – central exchange – central exchange – local exchange – subexchange – subscriber for toll service.

It will be seen that the error rate is dependent on the number of line sections and that lines of six or more sections are worse than the CCITT performance figures quoted above. Only a small number of connections in Germany exceed six sections, however.

BLOCK ERROR RATE   Where data are to be sent in blocks, it is desirable to know how the error rate varies with the block length so that an estimate can be made of the number of blocks that have to be retransmitted and so that the systems analyst can select an optimum block length. The bottom chart of Fig. 29.1 plots the probability of a block error against the block lengths. Again the variation in error rate with number of line sections is clear. These results can be regarded as typical of 50-baud telex networks.

2. SUB-VOICE-GRADE CIRCUITS

200-BAUD CIRCUITS   The character error rate over the world's 200-baud circuits is generally somewhat better than over telex circuits. Here the voice channel is typically divided into 480-Hz spacing (360 Hz in some countries), as opposed to 120 Hz for 50-baud transmission. The slightly better performance is attributed to the fact that the error-free intervals are about the same and that more characters are transmitted during these periods.

Table 29.1 shows the results of transmitting data over 200-baud connections in Germany [3].

Table 29.1   Results of measurements of errors in data transmitted on the German 200-baud telegraph network

Number of VF Telegraph Sections	Number of Transmitted Characters	Number of Erroneous Characters	Average Error Rate	Number of Error Sequences Following Each						With x Erroneous Characters Other Immediately						Number of Disconnections Resulting in the Nonprinting of Characters			Number of False Clearings
				$x=1$	$x=2$	$x=3$	$x=4$	$x=5$	$x=6$	$x=7$	$x=8$	$x=9$	$x=10$	$x=11$	$x=12$	<40 ms	<80 ms	<120 ms	
1	2	3	4	5	6	7	8	9	10	11	12	13	14	15	16	17	18	19	20
2	8,046,367	201	2.50×10^{-5}	144	2	1	1	2	—	—	1	1	1	—	—	1	1	2	6
3	3,936,849	115	2.925×10^{-5}	61	2	2	1	3	—	—	—	—	1	—	1	—	—	1	6

3. START-STOP TRANSMISSION OVER TELEPHONE CIRCUITS

TELETYPE TRAFFIC   Much teletype transmission today takes place over dial-up telephone circuits. In 1969–1970 AT&T conducted a detailed survey of the errors occurring in start-stop transmission using 10-bit ASCII characters over the toll telephone system [4]. Twenty-one million characters were transmitted at 15 characters per second over 534 different telephone connections. The results are shown in Figs. 29.2, 29.3, and 29.4.

In start-stop transmissions the character error rate is of interest rather than the bit error rate. Certain bit errors can cause more than one character to be in error. The start of a new character is determined by a transition from a stop interval (mark) to a start bit (space) as shown in Fig. 3.5. A disturbance which produces a false transition during a stop interval, or makes a start bit unrecognizable, will cause the receiving terminal to assume that a character is starting at the wrong instant. This loss of character synchronization may persist for several characters, causing a string of character errors.

The overall average character error rate was 1.46×10^{-4}. In other words, one character in 6850 was incorrect. Interestingly, the number of lost characters is substantially greater than number of erroneous characters. The overall average lost character rate was 6.81×10^{-4}; i.e., one character in 1467 was lost. This corresponds to 2.7 lost characters per page of average single-spaced typed text and 0.6 characters in error. The loss of characters on dialed telephone circuits is caused by *dropouts,* as discussed in the previous chapter. *Many of the data error statistics that are published do not include loss caused by dropouts.* The German study above, and some other studies published by the CCITT, exclude the effects of line outages longer than 300 milliseconds, for example.

The *average* lost character rate is a misleading statistic because most of the loss occurs in a relatively small number of calls. One percent of the calls contain about 90% of the lost characters.

The *average* error rate is less misleading, though the distribution of errors is also skewed. Eighty percent of the calls contain fewer errors than the average, but the probability that a character following an error is also in error is as high as 0.45. Errors, like lost characters, tend to be grouped together.

Figure 29.2 shows cumulative distributions of error rates and lost character rates. The connections are categorized by distance into short, medium, and long circuits as follows:

>   short: 0–180 miles
> medium: 180–725 miles
>   long: 725–2900 miles

Figure 29.3 shows the tendency of errors to occur in bursts. The horizontal line on the top chart shows the overall average error rate. The other curves show the error rates for a distance of L characters after an error has occurred. For 1000 characters after an error has occurred the error expectancy will be about 20 times greater than the average. For 100 characters after an error has occurred the error expectancy will be more than 100 times greater than normal. For 5 characters after an error the error expectancy is about 0.3.

This clustering of errors is beneficial when automatic retransmission of messages or blocks in error is used. If the errors were randomly distributed, a larger number of messages would contain an error. In practice, some messages have many characters in error (or lost), but consequently there are longer intervals between errors occurring, so fewer messages have to be retransmitted.

The systems analyst is concerned with how many messages or blocks have to be retransmitted because they contain either an error or a lost character. He needs to select a block message length which will give a reasonable compromise between high overhead caused by many messages and high retransmission rates caused by errors.

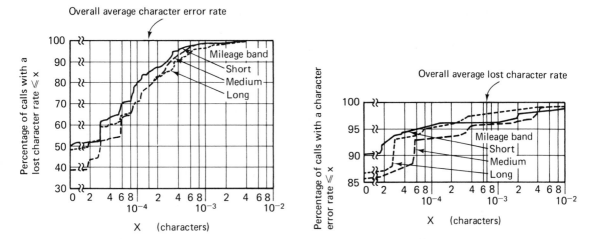

Figure 29.2 Cumulative distribution of average character error rate and average lost character rate for ASCII 10-bit characters transmitted at 15 characters per second over Bell System toll telephone circuits [4].

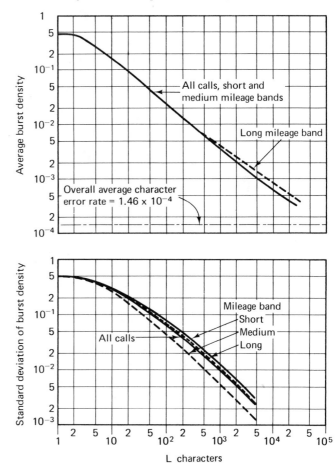

Figure 29.3 Mean and standard deviation of burst density during the transmission of ASCII 10-bit characters at 15 characters per second over Bell System toll telephone circuits [4]. Burst density is defined as the character error rate during an interval of L characters following a character error.

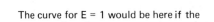

The curve for E = 1 would be here if the
same number of errors occurred at random

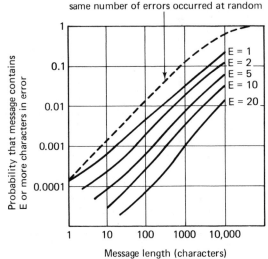

Figure 29.4 Probability that messages contain one or more errors. Transmission of ASCII 10-bit characters at 15 characters per second over Bell System toll telephone circuits [4].

Figure 29.4 shows the probability that a message transmitted contains a number of errors equal to or greater than 1, 2, 5, 10, or 20. The systems analyst is mainly concerned with the probability that it contains 1 or more errors.

If the occurrence of errors were uniformly distributed, the probability of a message containing an error, P_{error}, would be proportional to the message length:

$$P_{error} = 1.46 \, L \times 10^{-4}$$

where L is the number of characters in the message.

In fact the curve for $E = 1$ (one or more error) in Fig. 29.4 is approximately of the form $P_{error} = AL^B$ for values of L between 10 and 10,000, where A and B are constants.

B is a constant < 1. The fact that B is less than 1 indicates that the errors are clustered.

The equation which fits the data from this study is approximately

$$P_{error} = L^{0.85} \times 10^{-4}, \quad 10 \leqslant L \leqslant 10,000 \tag{29.1}$$

To determine the probability that a message will have to be retransmitted the probability that one or more of its characters is *lost* must be added to the probability that it has an *error*. If the lost characters were uniformly distributed, the probability of a message having a character lost would be

$$P_{loss} = 6.81L \times 10^{-4}$$

However, the losses are not uniformly distributed, and the published results of the AT&T study do not give enough information to calculate the probability of loss when clustering is taken into consideration. We therefore assume that the ratio of best characters in a cluster to the average lost character rate is the same as for the characters in error. In other words,

$$\frac{\text{actual } P_\text{loss}}{P_\text{loss} \text{ if losses were uniformly distributed}} = \frac{\text{actual } P_\text{error}}{P_\text{error} \text{ if errors were uniformly distributed}}$$

P_loss may then be estimated as follows:

$$P_\text{loss} = \frac{P_\text{loss} \text{ if losses were uniformly distributed} \times \text{actual } P_\text{error}}{P_\text{error} \text{ if errors were uniformly distributed.}}$$

$$= \frac{6.81L \times 10^{-4} \times L^{0.85} \times 10^{-4}}{1.46L \times 10^{-4}}$$

$$= 4.66L^{0.85} \times 10^{-4} \qquad\qquad (29.2)$$

The probability that a message will have to be retransmitted because it is faulty is

$$P_\text{fault} = P_\text{error} + P_\text{loss}$$

$$= L^{0.85} \times 10^{-4} + 4.66L^{0.85} \times 10^{-4}$$

$$= 5.66L^{0.85} \times 10^{-4} \qquad\qquad (29.3)$$

The curve in Fig. 29.5 is plotted by adding P_loss, i.e., $4.66L^{0.85} \times 10^{-4}$, to

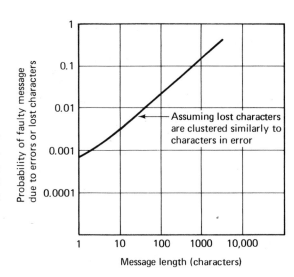

Figure 29.5 Estimated probability that messages will have to be retransmitted (transmission of ASCII 10-bit characters at 15 characters per second over Bell System toll telephone circuits) [4].

the values for $E = 1$ from Fig. 29.4. We suggest that a systems analyst should use Fig. 29.5 or the above equation, in absence of any better information, to estimate the numbers of messages that will be faulty or retransmitted.

4. HIGHER-SPEED USE OF TELEPHONE CIRCUITS

When data are transmitted over telephone circuits the error rate tends to be greater at higher transmission speeds and is dependent on the modem design. The error rate tends to increase rapidly as transmission speeds are raised above 4800 bps, and at such speeds is so variable that reliable statistics are not published. Below 4800 bps a voice-grade line with a good modem typically encounters about 1 bit error in 100,000 bits transmitted. Lines with error rates worse than 1 bit in 100,000 usually have an audible level of noise which can be a nuisance in the reception of speech. The *mean* error rate is worse than this *typical* error rate because a small proportion of lines behave badly.

At speeds of 9600 bps transmitted over voice lines, typical error rates on modems which do not use automatic error correction have been quoted to be as high as 1 bit error in 2000 but much variation is found from one line to another and from one modem to another. The performance at high speeds will probably improve as modem design improves and can be made acceptable by using a combination of error correction and error detection, as will be described in the following chapter.

Again, errors tend to come in bursts. When trouble comes, it comes fast and furious. Bursts of error are the rule rather than the exception, and they sometimes go on for hundreds of bits. This clustering has both a good effect and a bad effect. The number of error-free messages which do not have to be retransmitted is much higher than if the same number of errors occurred at random. On the other hand, the clustering makes it much more difficult to devise safe error-detecting or error-correcting codes.

AT&T conducted major surveys of data error rates on voice-grade lines in 1960 [5] and 1969–1970 [6]. Figures 29.6, 29.8, and 29.9 show results from the more recent survey. Figure 29.7 compares that survey with the previous one.

Figure 29.6 shows the percentage of calls having different error rates. The calls are again categorized as short-distance (0 to 180 miles), medium-distance (180 to 725 miles), and long-distance (greater than 725 miles). With 1200-bps transmission, about 80% of all calls had error rates better than 1 error bit in 100,000. Half of the short calls had error rates better than 1 bit in 5 million. However, more than 1% of the calls had error rates worse than 1 bit in 1000. The error rates become higher as speed is increased. At 4800 bps about half of the calls have fewer than 1 error bit in 100,000, but about 5% of the calls are worse than 1 error bit in 1000.

Users employing leased lines find that usually their lines have an error rate low enough for error detection and retransmission techniques to give a good

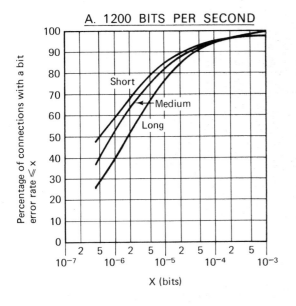

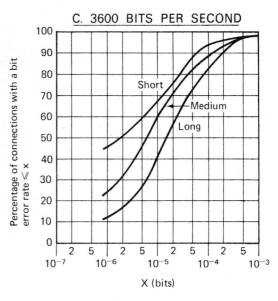

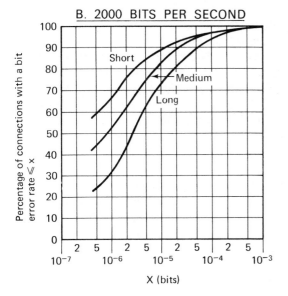

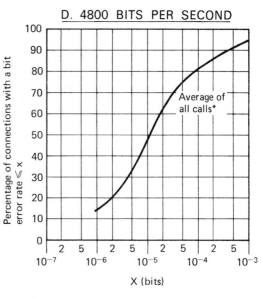

*Not broken
down into
short, medium,
and long
distance calls
because of
inadequate
sample size

Figure 29.6   Error rates on Bell System voice-grade trunks at different
transmission speeds using AT&T modems [6].

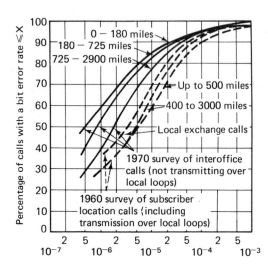

Figure 29.7 Comparison of the results of AT&T's 1960 and 1970 surveys of data transmission over toll telephone lines at 1200 bits per second [5][6]. The 1970 results are better but the transmission excludes local loops on which some bit errors originate. The 1970 results also exclude certain noisy central offices.

throughput. Occasionally, however, they have a bad line which consistently gives a low throughput or poor response times because many messages have errors and have to be retransmitted. Just as with a line plagued by outages, users should complain about lines with consistantly high error rates. Often the common carrier can reroute the leased line via different equipment that gives better performance.

The mean error rate on all the test transmissions was only slightly better than 1 error bit in 10,000. This mean is dragged down by a small proportion of bad calls. Quoted by itself it does not represent fairly the behavior of a typical line.

Unfortunately the lines that data were transmitted over in the 1969–1970 Bell System study did not go from subscriber to subscriber; they went from central office to central office. In other words, local loops were not tested – only interoffice trunks. Many of the errors that occur in practice are generated on local loops and on subscriber premises. A few local loops and associated equipment such as PBXs are exceptionally noisy. Sometimes badly routed wiring on a subscriber's premises consistently generates noise. Furthermore, the Bell study deliberately avoided certain panel and step-by-step central offices which are noisy. They often avoid them in practice in the routing of leased data circuits. The data error rates found in practice may therefore be worse than those in Figs. 29.6, 29.8, and 29.9.

Figure 29.7 compares AT&T's 1969–1970 survey with that in 1960. It will be seen that the 1969–1970 results are better; however, the 1960 survey included transmission over local loops, whereas the 1969–1970 survey did not. The loss and delay of local loops was simulated in the latter study. It is not certain how much of the improvement is due to the more modern telephone plant and how much is due to the exclusion of noise induced by the outside plant. The highest dotted curve in Fig. 29.7, however, shows results for calls transmitted over local loops *only*, in 1960. It will be seen that these calls had a substantial error rate, and this suggests that if local loop transmission had been included in the 1969–1970 survey its curves might have been closer to the 1960 curves.

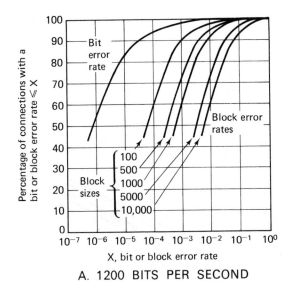

A. 1200 BITS PER SECOND

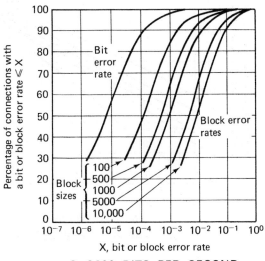

C. 3600 BITS PER SECOND

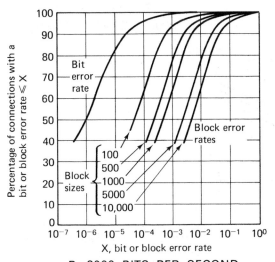

B. 2000 BITS PER SECOND

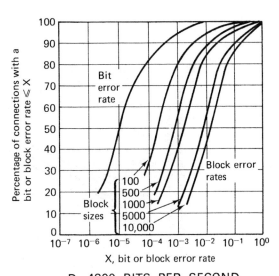

D. 4800 BITS PER SECOND

Figure 29.8 Curves showing how block error rates vary for voice-grade trunks on the Bell System [6].

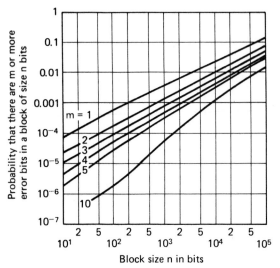

A. 1200 BITS PER SECOND

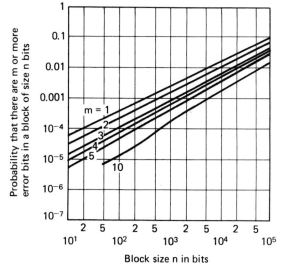

B. 2000 BITS PER SECOND

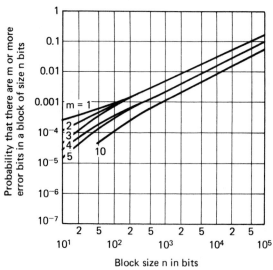

C. 3600 BITS PER SECOND

Figure 29.9 Probabilities of different numbers of errors in a block on Bell System voice-grade trunks [6].

562

Table 29.2 AT&T tests of international circuits at a data rate of 9600 bits per second.

Test Route	Length of test (hours)	Average bit error rate	% of ten-second intervals achieving the following bit error rates:		
			10^{-5}	10^{-4}	10^{-3}
Philippine Is. to Hawaii	11.5	3.2×10^{-4}	16	65	99
	9.4	5.8×10^{-5}	93	98	100
Hawaii to Philippine Is.	12.3	1.6×10^{-4}	24	67	100
	10.2	3.8×10^{-5}	97	99	100
Hawaii to Washington, D.C.	1.0	7.3×10^{-5}	62	97	100
	2.0	9.0×10^{-5}	97	99	100
Washington, D.C. to Hawaii	2.4	4.8×10^{-4}	11	57	96
	1.2	8.3×10^{-5}	95	98.5	99
	0.7	3.0×10^{-3}	39	85	91
Japan to Hawaii	10.2	1.3×10^{-5}	98	99	100
	14.2	1.2×10^{-4}	81	92	99
Hawaii to Japan	10.5	1.4×10^{-4}	87	93	98
	13.6	3.5×10^{-5}	69	96	100
Japan to Philippine Is.	13.0	2.3×10^{-4}	60	81	94
	10.3	3.3×10^{-5}	92	97	99
Philippine Is. to Japan	12.6	4.1×10^{-4}	51	75	93
	11.6	2.4×10^{-5}	95	99	100
Okinawa to Philippine Is.″	0.8	1.2×10^{-2}	0	0	0
	3.9	6.3×10^{-4}	39	40	91
Philippine Is. to Okinawa″	0.9	6.6×10^{-3}	0	0	0
	4.6	9.4×10^{-4}	0	1	72
Okinawa to Japan″	8.4	3.9×10^{-4}	45	71	96
	8.7	1.1×10^{-3}	23	63	86
Japan to Okinawa″	8.5	9.9×10^{-4}	41	65	73
	7.5	4.0×10^{-4}	43	62	87

″Indicates dominant tropospheric scatter link.

Table 29.2 shows the results of some AT&T tests [7] over international circuits at 9600 bps. The tests used a single-sideband AM modem transmitting 4800 4-level symbols per second and employing adaptive equalization. The data was transmitted between telephone company test sites. It is interesting to note that trans-continental links sometimes give better performance than links across a city. They use relatively modern equipment and there is little noise under the ocean.

BURSTS OF ERRORS   As we found with lower-speed transmission the errors are highly clustered. There tend to be lengthy trouble-free intervals interspersed by bursts of bit errors. Consequently, the numbers of blocks or messages that contain an error cannot be deduced from mean bit error rates for a connection. Figure 29.8 shows the block error rates for different block sizes. These curves may be used by a systems analyst in determining what would be an economical block size to use — avoiding too much retransmission.

Figure 29.9 shows the probabilities of having different numbers of errors in a block of a given length. The top curve, the probability that the block contains one or more errors, is approximately of the form

$$P_{\text{error}} = An^B$$

where n is the number of bits in the block and A and B are constants.

The approximate equations for the three transmission speeds are

$$1200 \text{ bps:} \quad P_{\text{error}} = 1.2n^{0.81} \times 10^{-5} \tag{29.4}$$

$$2000 \text{ bps:} \quad P_{\text{error}} = 1.0n^{0.80} \times 10^{-5} \tag{29.5}$$

$$3600 \text{ bps:} \quad P_{\text{error}} = 2.5n^{0.77} \times 10^{-5} \tag{29.6}$$

CARRIER FAILURES   The error statistics represented by Fig. 29.9 and Eqs. (29.4) to (29.6) do not take *carrier failures* into consideration. Many quoted data error statistics do not indicate the carrier failures or dropouts, just as the teletype statistics discussed above for characters in error did not include *lost* characters.

Each of the modems used was equipped with a carrier detector which indicates an ON condition during normal reception of signals. An OFF indication occurs during periods of dropout or when there are severe line disturbances such as high-amplitude impulse noise. The carrier OFF conditions typically last from about 10 to 300 milliseconds. They cause many messages to have to be retransmitted but are not included in the error statistics.

Table 29.3 summarizes the carrier failures that occurred in the Bell study.

Table 29.3 Statistics on carrier failures during data transmission on Bell System toll telephone cir-
cuits [6]

	1200 bps	2000 bps	3600 bps
Speed	1200 bps	2000 bps	3600 bps
Duration of calls	1800 sec	1800 sec	1200 sec
Bits per call	2.16×10^6	3.6×10^6	4.32×10^6
Percentage of calls with one or more carrier failures			
Short-distance	7.0	9.1	14.3
Medium-distance	7.9	14.8	7.2
Long-distance	10.3	21.7	14.6
Average	8.4	15.2	12.03
Number of carrier failures per call (median call)	3	2	1
Number of carrier failures per million bits transmitted (calculated approximately from the above figures)	$\dfrac{3 \times 0.084}{2.16} = 0.12$	$\dfrac{2 \times 0.152}{3.6} = 0.084$	$\dfrac{1 \times 0.1203}{4.32} = 0.028$

If the carrier failures were uniformly distributed, the probability of a message being faulty because of a carrier failure, $P_{failure}$, would be

$$1200 \text{ bps}: \quad P_{failure} = 0.012n \times 10^{-5}$$

$$2000 \text{ bps}: \quad P_{failure} = 0.0084n \times 10^{-5}$$

$$3600 \text{ bps}: \quad P_{failure} = 0.0028n \times 10^{-5}$$

Carrier failures are much less subject to clustering than bit errors. They are, in fact, related to the prolonged periods of disturbance that *cause* bit error clustering. As the figures in the above equations are small, these probabilities of carrier failures may be added to the error probabilities to give a probability of faulty messages, P_{fault}:

$$1200 \text{ bps}: \quad P_{fault} = (1.2n^{0.81} + 0.012n) \times 10^{-5} \qquad (29.7)$$

$$2000 \text{ bps}: \quad P_{fault} = (1.0n^{0.80} + 0.0084n) \times 10^{-5} \qquad (29.8)$$

$$3600 \text{ bps}: \quad P_{fault} = (2.5n^{0.77} + 0.0028n) \times 10^{-5} \qquad (29.9)$$

It is suggested that a systems analyst use these equations (in absence of anything better) to evaluate probabilities of retransmission and hence to determine optimum block sizes.

MULTIPLE ERRORS   A systems analyst is concerned not only with the probability of having to retransmit a message but also, and perhaps more seriously, with the probability that an error will fail to be detected. We will discuss error-detecting codes in the following chapter. Their effectiveness in catching errors depends on the probability of different numbers of errors occurring in the same message, and hence the curves for $E = 2$ and higher in Fig. 29.9 are of value in assessing the capability of such codes.

LINES IN   The performance of telephone lines is fairly similar
DIFFERENT   throughout the world. Most telephone administrations
COUNTRIES   claim that an error rate less than 1 bit in 100,000 is obtained "under normal conditions" over a leased circuit and over most switched circuits. In all countries a few bad circuits have error rates much worse than this objective. The proportion of bad-quality circuits varies from one area to another. A few areas, usually with obsolete telephone plant, are exceptionally poor.

Figure 29.10 compares measurements of the German telephone system with those of the 1960 AT&T survey. Both curves are for local exchange calls with a transmission rate of 1200 bps. The results are remarkably similar.

Figure 29.10   A comparison of error measurements made in America and Germany. Both curves are for 1200 bit/sec transmission over the switched public telephone network using local exchange calls. (*This curve is from CCITT *Red Book,* Vol. VII, published by the International Telecommunication Union, Geneva, March 1961. **This curve is redrawn from CCITT *Blue Book,* Vol. VIII, Supplement No. 29, "Report on Data Transmission Tests on the German Telephone Network," published by the International Telecommunication Union, Geneva, November 1964.)

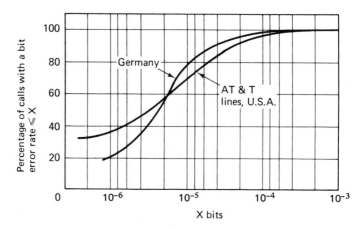

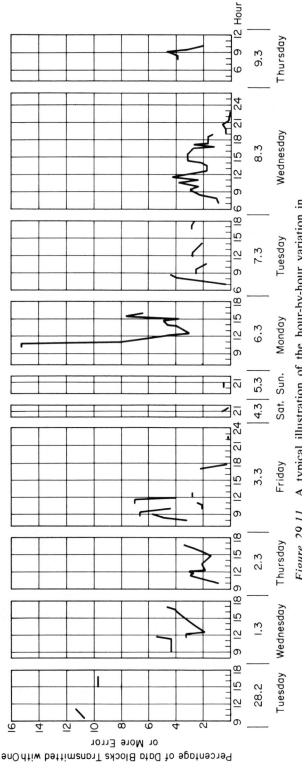

Figure 29.11 A typical illustration of the hour-by-hour variation in error rate for data transmitted over a switched public network. These measurements were made on the Stuttgart Telephone Network.

567

**HOUR-BY-HOUR
VARIATION IN
ERROR RATES**

When transmitting data over the switched public telephone network considerable variation in the error rate is found from one period of time to another. A typical illustration of this is shown in Fig. 29.11. This shows the result of transmitting blocks of data over the local public network of Stuttgart, Germany. The blocks received in error were counted electronically. It will be seen that occasionally there are periods when the error rate is several times higher than the average. The error rate is generally higher during periods of high traffic intensity. Measurements on other telephone networks show similar results. The curves in Fig. 29.12 show variations in error rates on telephone lines in Europe. The error rates are highest when the circuits are most heavily loaded.

While traffic peaks cause a change in error rate which is somewhat predictable, *maintenance work on the lines or in the exchanges can cause sudden and unpredictable peaks*. These frequently cause a greater error rate for a period than the normal daily variations.

Figure 29.12   Hour-by-hour variation of error rates encountered. Both illustrations use transmission at 1000 bits per second over voice-grade lines. ((1) Extract from CCITT contribution COM Sp. A/No. 7, July 1961, by the Chile Telephone Company. (2) Extract from CCITT contribution COM Sp. A/No. 16, September 1961, by N. V. Philips Telecommunicatie Industrie.)

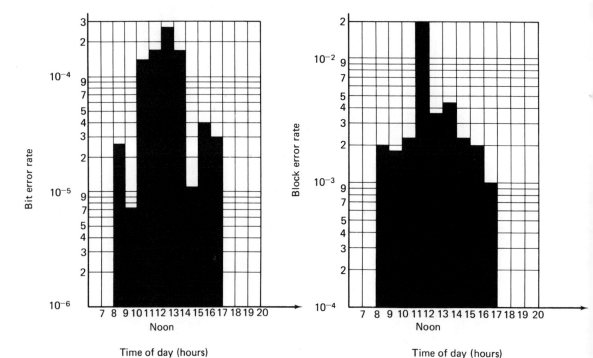

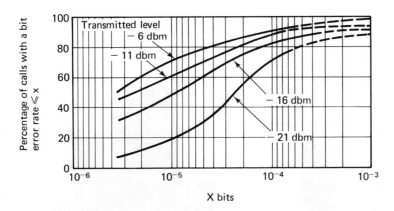

Figure 29.13 The error rate varies widely with the transmission voltage used. Incorrectly adjusted terminal equipment can give rise to error rates higher than those shown in this chapter. These measurements were made on AT&T public switched line. (Curves reproduced from CCITT *Red Book,* Vol. VII, published by the International Telecommunication Union, Geneva, March 1961.)

VARIATION IN ERROR RATE WITH TRANSMISSION VOLTAGE
The number of errors have been found to vary widely with the voltage level at which the signal is transmitted. If the voltage is low, then the noise, which does not change its level, does more damage. Figure 29.13 illustrates how wide this variation is. Perhaps a more significant measure than the transmission voltage would be the lowest level of the signal when it passes through the exchanges, as it is here that much of the impulse noise originates.

The statistics in this chapter have all been collected on lines with the transmission voltage adjusted for near-optimum performance. In practice, if the line termination equipment is not correctly adjusted, then the error rate can be expected to be worse than in these illustrations. Users sometimes have exceptionally bad error performance due to maladjustment of their equipment. If performance is consistently poorer than that in the curves of this chapter, an engineer should check to see that the modems are adjusted correctly. If performance is still poor on a leased line, this should be discussed with a common carrier, as some adjustment on their part might be needed.

WIDEBAND DATA CHANNELS
Tests carried out on wideband data channels have generally indicated a lower bit error than on voice-grade channels. But, as with voice-grade channels, it is clear that some wideband channels perform substantially worse than the average.

ARPA has been operating a network of 50 Kbps lines (Fig. 24.6) since 1970.

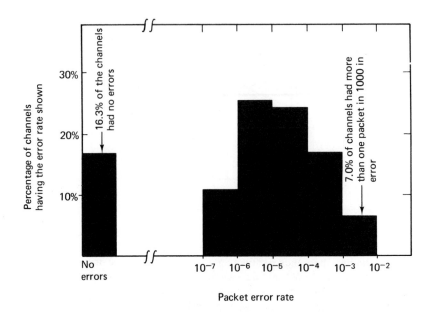

Figure 29.14   Error rates on 50 kbps lines used on ARPANET measured during 7 days of continuous operation (average packet length: 218 bits).

The error performance of the lines has generally been good and only a small proportion of the packets transmitted have to be resent because of errors. Fig. 29.14 shows the result of seven days continuous measurement of the network [11] at a time when 86 wideband lines were in use. The measurements counted the numbers of packets that had to be retransmitted because of errors. The packets during the period in question had an average length of 218 bits. The average error rate was one faulty packet in 12,880 (12,880 packets contained 2.8 million bits). Fourteen of the lines had no detected errors during the seven days. On the other hand 6 lines had error rates worse than 1 faulty packet in 1000. The worst line had 1 faulty packet in 340.

Table 29.4 shows the result of measurements made in Europe using wideband modems on 48 kHz *group* channels of different types. The majority of the tests show an error rate better than 1 bit error in a million bits. However some links have a performance an order of magnitude worse than the average. Fig 29.15 shows the distribution of error rates on the tests done in Britain.

The low result, marked with an asterisk in Table 29.4, was obtained using transmission at night when there was little other traffic. There tended to be a high error concentration in short periods, especially on the Italian tests marked with double asterisks. The transmitted blocks contained 511 bits, and on some tests the average number of error bits in an error block was a quarter of the total. The telephone administration explained these noisy spasms as being due to causes

Table 29.4 Results of tests of data transmission at 40,800 and 48,000 bits per second over a variety of different circuits [12, 13]

	Average Number of Bit Errors in 10^7 Bits	Average Number of Block Errors in 10^5 Blocks	Average Number of Error Bits in in Error Block
			(Blocks of 511 Bits were Used)
Transmission Tests Made in Italy at 40,800 bps			
Local loop	2.1	1.1	˙10
Coax. cable	43	1.4	152**
Coax. cable	0.021*	0.018	6
Coax. cable	150	7.7	104**
Microwave	85	2.5	173**
Coax. cable	72	2.1	177**
Coax + Microwave	11.6	1.8	33
Coax + Microwave	0.84	0.65	6
Coax + Wire pair	15.8	6.9	12
2 systems in tandem	58.4	14.3	21
Ditto + local connections	122	17.8	35
Transmission Tests made in Britain at 48,000 bps			
Microwave	1.9	1.8	5
Microwave	6.9	6.7	5
Cable	0.97	1.2	4
Cable	0.34	0.3	6
Microwave	3.0	3.0	2
Microwave	0.25	0.25	2
Cable	10.8	15.2	4
Cable	0.72	0.24	15
Microwave	1.1	0.75	7
Microwave	34.6	50.6	4
Cable	3.7	2.0	9
Cable	0.20	0.28	4

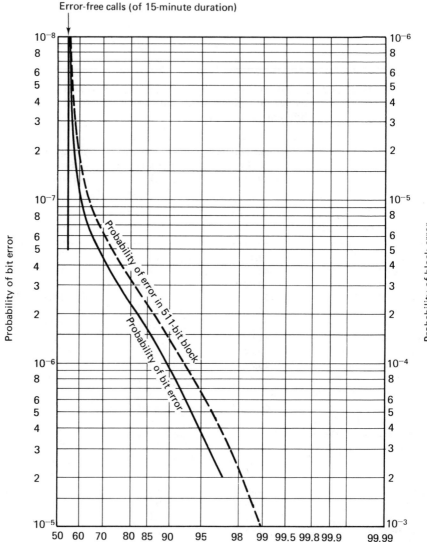

Figure 29.15 Error rates in 48 kbps transmission over 48-kHz *group* channels in Britain [13].

such as "human intervention for reroutings, random failures over the system, recovery automatic switchings, and so on." [12] This seems typical of wideband channels on telephone networks throughout the world. The incidence of noisy spasms is sufficiently low that they would have negligible serious effects on appropriately designed data-processing systems, except in a few very special applications.

REFERENCES

1. CCITT Special Study Group A, *Data Transmission Contribution 92,* Annex XIII (Oct. 18, 1963), p. 131.

2. "Tests Conducted on the German Telex Network with 50-Band Start-Stop Transmission," *The CCITT Blue Book,* Supplement No. 4, International Telecommunications Union, Geneva, Nov. 1964.

3. "Measurements of Errors in Data Transmitted on the German 200-Baud Telegraph Network," *The CCITT Blue Book,* Supplement No. 6, International Telecommunications Union, Geneva, Nov. 1964.

4. H. C. Fleming and R. M. Hutchinson, Jr., "Low-speed Data Transmission Performance on the Switched Telecommunications Network," *Bell System Technical Journal,* 50 (April 1971), pp. 1385–1406.

5. A. A. Alexander, R. M. Gryb, and D. W. Nast, "Capabilities of the Telephone Network for Data Transmission," *Bell System Technical Journal,* 39 (May 1960), pp. 431–476.

6. M. D. Balkovic, H. W. Klancer, S. W. Klare, and W. C. McGruther, "High-Speed Voiceband Data Transmission Performance of the Switched Telecommunications Network," *Bell System Technical Journal,* 50 (April 1971), pp. 1349–1384.

7. "4800 and 9600 Bit per Second Data Transmission over International Circuits," U.S. contribution to *The CCITT Green Book,* Vol. VIII, Data Transmission, International Telecommunications Union, Geneva, 1973.

8. *Extract from CCITT contribution Com Sp. A/No. 16, September 1961, by N. V. Philips Telecommunicatie Industrie.*

9. "2400 bit/second data transmission tests on the Italian Switched Public Network," *The CCITT Green Book,* Vol. VIII, Data Transmission, International Telecommunications Union, Geneva, 1973.

10. *The CCITT Red Book,* Vol. VII, International Telecommunications Union, Geneva, 1961.

11. L. Kleinrock and W. E. Naylor "On Measured Behaviour of the ARPA Network," *AFIPS Conference Proceedings,* Vol. 43, AFIPS Press, Montvale, N.J., 1974.

12. "Wideband Transmission Tests," Italian Administration contribution to *The CCITT Green Book,* Vol. VIII, Data Transmission, International Telecommunications Union, Geneva, 1973.

13. "48 Kbps Transmission Tests," United Kingdom Post Office contribution to *The CCITT Green Book,* Vol. VIII, Data Transmission, International Telecommunications Union, Geneva, 1973.

30 THE TREATMENT OF ERRORS

On many data transmission systems the elimination of errors is vitally important. There are two basic ways to control errors: (1) Detect the error and have the erroneous message retransmitted, and (2) use a scheme for correcting the error without retransmission. There are multiple variations of these two techniques.

In addition to transmission errors, the errors made by operators constitute a significant problem in the design of on-line systems. The number of errors made by the operators of the input/output devices on a large system usually far exceeds the number of errors caused by noise or distortion on the transmission lines. It is usually important that accuracy controls be devised for the human input. A tight network of controls is necessary to stop abuse or embezzlement on many systems. It is also important to ensure that nothing is lost or double-entered when hardware failures occur on the system or when switchover takes place [1]. In this chapter we will confine our attention to errors caused by transmission.

IGNORE THE ERRORS? The easiest approach is to ignore the noise — and this is often done. The majority of telegraph links in operation today, for example, have no error-checking facilities at all. Part of the reason is that they normally transmit English-language text that will be read by human beings. Errors in English language caused by the changing of a bit or of a small group of bits are usually obvious to the human eye, and we correct them in the mind as we read the material. Telegrams that have figures as well as text in them commonly repeat the figures. This inexpensive approach is also taken on computer systems where the transmission handles verbal text. For example, on administrative message-switching systems, it is usually acceptable to have transmission to and from unchecked telegraph machines. If the text turns out to be unintelligible, the user can always ask for a retransmission.

On some systems for telemetry and statistical data transmission, errors are detected and the erroneous values ignored. This can be the case with statistical data where erroneous samples can be discarded without distortion. It is used on systems for reading remote instruments where the readings are changing slowly and an occasional missed reading does not matter. The advantage of a detection-only scheme is that it requires a channel in only one direction — sometimes a great advantage with tracking and telemetry systems.

An error rate of 1 bit in 10^5 is perhaps not quite so bad as it sounds. Suppose that we considered transmitting the text of this book, for example, and coded it in 5-bit Baudot code. If 1 bit in 10^5 was in error, in the entire book there would be about 30 letters that were wrong. The book would certainly still be readable, and the majority of its readers would not notice most of these errors. The human eye has a habit of passing unperturbed over minor errors in text. However, such an error rate would be intolerable in the transmission of computer programs, financial data, and many other types of commercial information.

CALCULATIONS OF THE   Calculations should then be done to estimate the effect
EFFECT OF ERRORS   of the expected error rate on the system as a whole.
                                        On some systems the effect of infrequently occurring errors is cumulative, and it is in such situations that special care is needed in eliminating errors. For example, if messages cause the updating of files and an error in the message causes an error to be recorded on the file, then on some systems it is possible that as the months pass the file will accumulate a greater and greater number of inaccuracies.

Suppose, for example, that teletype transactions with 1 bit in 10^5 in error update a file. Suppose that, on the average, a record is updated 100 times a month and that if any one of 20 5-bit characters is in error in the transmission, then the record will be updated incorrectly. After 6 months 45% of the records will be incorrect. If an error-correction procedure on the telecommunication lines reduces the rate of undetected errors to 1 bit in 10^7, then 0.6% of the records are likely to be incorrect at the end of 6 months. With 1 bit in 10^8, 0.06% records are likely to be wrong. Probability calculations of this type need to be done on various aspects of the system when it is being designed.

DETECTION OF ERRORS   To *detect* communication errors, redundancy is built
                                          into the messages transmitted. In other words, more bits are sent than are needed for the coding of the data alone.

Redundancy can be built into individual characters. This is usually done by using odd-even parity checks. It is also done by using certain character codes, such as a 4-out-of-8 code. A simple parity check (see Box 30.1) is not very effective in data transmission (although often used) because a burst of noise frequently destroys more than 1 bit in a character. The higher the transmission speed, the

BOX 30.1   The ineffectiveness of simple parity checks

A parity bit is sometimes used in each character to detect errors in the character. This technique is of little value in telecommunications (although surprisingly often used) because noise impulses are often long enough to destroy more than 1 bit:

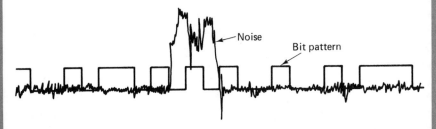

The parity check detects the error when 1 or an odd number of bits are changed. It does not detect it when an even number of bits are changed.

The following curve shows the probability that on a typical telephone line an error bit will have another error bit closely following it [2]:

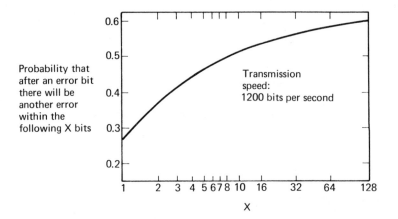

Probability that after an error bit there will be another error within the following X bits

Transmission speed: 1200 bits per second

Consider a parity check used as part of an 8-bit character. If the character is in error, the probability that the first erroneous bit is the first bit of the character is approximately $\frac{1}{8}$. The parity check will fail to detect the error if there is another incorrect bit in the next 7 (ignoring as yet the probability of having 3, 5, or 7 bits in error). The probability of this situation is 0.49. The probability that the first erroneous bit of the

BOX 30.1 Continued

error character is the second bit is also approximately 0.49, and then 6 error-free bits must follow if the parity check is to work. The probability of having an error in the following 6 bits in 0.48. Adding together all the possible 2-bit errors in an 8-bit character, we find, from the above curve, that the probability of the parity check not working is $\frac{1}{8} \times 0.49 + \frac{1}{8} \times 0.48 + \frac{1}{8} \times 0.47 + \frac{1}{8} \times 0.45 + \frac{1}{8} \times 0.41 + \frac{1}{8} \times 0.37 + \frac{1}{8} \times 0.36 = 0.38$. From this we conclude that approximately 30% of the time the simple parity check will fail to detect the errors.

On the other hand, appropriately chosen polynomial checks on messages (with less redundancy than character parity checks) can be devised to catch virtually every error.

more likely this is to be true. A noise impulse 2 milliseconds in duration, for example, may destroy only 1 bit if the transmission rate is 600 bps, but at 1200 bps it may wipe out 2 bits, at 4800 bps 4 or 5 bits may be destroyed, and so on. If an even number of bits are destroyed, the error will not be detected.

The 4-out-of-8 code and other such character codes are a little safer because to be undetectable errors must be compensating; in other words, if a 1 bit is changed into a 0, than a 0 bit must also be changed into a 1. This sometimes happens with the bursts of noise and oscillating effects on communication lines. It is more likely to happen at higher bit rates and with high-speed modulation techniques.

With other than the slowest transmission speeds it is better to forget about character checks and use redundancy within a *block* of characters. A block of many characters will be followed by a cluster of error-checking bits—perhaps in the form of one or more error-checking characters. Some transmission schemes have character checking—for example, a parity bit on each character—in addition to the check to the block. It can be shown, however, that much more efficient error detection results from using all the redundant bits for block checking, not character checking.

ERROR-DETECTING CODES   The probability of a transmission error remaining undetected can be made very low, either by a brute-force method that gives a high degree of redundancy or, better, by an intricate method of coding. Many systems in actual usage do not employ a very secure error-detection scheme; instead they compromise with one that is not too expensive but that will probably let one error in a thousand or so slip through. It is important to understand that a very high level of safety can be achieved, without a high level of redundancy, with good error-detecting codes.

A class of codes referred to as *polynomial codes* are in common use. Many of the codes referred to by other names (Hamming codes, cyclic codes, Bose-Chaudhuri codes, fire codes, the codes of Melas, various interleaved codes, and the simple odd-even parity check) are special cases of polynomial codes. (See the bibliography at the end of the chapter for detailed references to these.)

All these codes can be described in terms of the properties of division of polynomials. If the data contain K bits, they can be represented by a polynomial of order $K - 1$. To protect the message a group of r redundant bits is transmitted with it, and these are represented by a second polynomial of order $r - 1$. The second polynomial is obtained by dividing the first by a third fixed polynomial (see Reference 2). The receiving machine carries out the same division as the transmitting machine and sees whether the results are the same as the r redundant bits. If they are not, then a transmission error has occurred.

The transmission line produces bursts of errors in which any pattern of bits can be changed. Some patterns of bits will occasionally defeat the error-detection operations by changing the error-checking bits as well as the data message in a compensating fashion. The objective is to select a method of generating the check pattern which is as impervious as possible to compensating errors.

As illustrated in Box 30.2, polynomial codes can be made powerful enough to detect virtually all telecommunication errors. Nevertheless, many systems today use codes which allow some errors to slip through.

**DEALING
WITH ERRORS**

Once errors are detected it is generally desirable that the transmission system take automatic action to correct the fault. Some data transmission systems in common use leave the fault to be corrected by human means. On a system in which terminal operators originate each transaction, the operator may be told to reenter a transaction when an error is detected. This is acceptable because most of the errors are caused by the operator miskeying, and only 1 message in 1000 contains a transaction error.

Most systems automatically request the retransmission of data in error.

**HOW MUCH IS
RETRANSMITTED?**

Systems differ in how much they require to be retransmitted when an error is detected. Some retransmit only one character when a character error is found. Others retransmit many characters or even many messages.

There are two possible advantages in retransmitting a *small* quantity of data. First, it saves time. It is quicker to retransmit 5 characters than 500 when an error is found. However, if the error rate is, say, 1 character error in 20,000, the percentage loss in speed does not differ greatly between these two cases. It *would* be significant if a block of 5000 had to be retransmitted.

Second, when a large block is retransmitted, it has to be stored somewhere until the receiving machine has confirmed that the transmission was correct. Often there is no problem. In transmitting from paper tape, for example, the tape reader merely reverses to the beginning of that block. The paper tape is its own message storage. The same is true with transmission from magnetic tape or disk. With transmission from a keyboard, however, an auxiliary storage, or *buffer,* is needed if there is a chance that the message may have to be retransmitted automatically. Several input devices may share a common control unit, and this unit contains the buffer storage.

The *disadvantages* of using small blocks for retransmission are first that the error-detection codes can be more efficient on a large block of data. The proportion of redundant bits is smaller. Second, where blocks of data are sent synchronously, a period of time is taken up between blocks in control characters and line turnaround procedures. The longer the block, the less significant is this wasted time.

The well-designed transmission system achieves the best compromise between these factors, and a systems analyst may determine the optimum block lengths.

Let us consider some examples of different retransmission quantities.

1. *Retransmission of one character.* Characters are individually error-checked, as with a 4-out-of-8 code. As soon as a character error is detected, retransmission of that character is requested. Sometimes referred to as an echo check, such a scheme is used only on short, slow, and usually in-plant links.

2. *Retransmission of one word.* The Western Union EDAC (Error Detection Automatic Correction) System uses four-character blocks in one of its models. Four 5-bit teletype code characters are transmitted synchronously, with a 5-bit error-checking pattern, from a buffer. A 3-bit control signal is returned by the receiver to indicate correct reception. If this signal is mutilated, the transmitting machine repeats the transmission. It does so until it receives confirmation of correct receipt, and then the four characters in storage can be erased.

3. *Retransmission of a message or record.* Many systems retransmit a message or record, which is often of variable length. The record may be retransmitted from backspaced tape, from a variable-length area in computer memory, or from a buffer in a control unit attached to the transmitting device. If a buffer is used, there may be a maximum size for the amount that can be retransmitted. If the message exceeds that size, it is broken into separate messages that are linked together with a control character indicating that a given transmission has not completed the message.

4. *Retransmission of a block of several messages or records.* When the transmission speed is high, it becomes economical on many systems to transmit the data synchronously in large blocks. These blocks may be longer than one "message" or "record." On most systems, when an error occurs in any of the records, the whole block of records is retransmitted. By using machines with good logic capabilities, it would be possible to resend only the faulty record and not all the other records in the block.

BOX 30.2 The power of polynomial
error-detecting codes

It can be shown that an appropriate polynomial error-detecting code using r redundant bits gives the following protection [3]:

single-bit errors:	100% protection
two bits in error (separate or not):	100% protection
an odd number of bits in error:	100% protection
an error burst of length less than $r + 1$ bits:	100% protection
an error burst of exactly $r + 1$ bits in length:	$1 - (\frac{1}{2})^{(r-1)}$ probability of detection
an error burst of length greater than $r + 1$ bits:	$1 - (\frac{1}{2})^{r}$ probability of detection

The last two terms assume an equal probability of any error pattern. In practice some error patterns are more prevalent than others, so some generating polynomials of a given r are better than others.

It will be seen that polynomial codes can provide a high measure of protection. As r becomes larger, so the measure of protection against bursts becomes greater. This is especially so as long bursts are rarer than short bursts on telephone lines. We could, if it were necessary, make r large; in other words, we could use a high-order generating polynomial. In this way we could produce a *very* high measure of protection indeed.

To illustrate the potential power of polynomial codes suppose that when one is transmitting fixed-length blocks of 100 data characters (800 bits) over a certain telephone line, there is a probability of 10^{-3} that a block will be perturbed by an error burst of length greater than 17 bits. It will be seen from Fig. 29.9 that this is a pessimistic assumption. If we make $r = 16$ and use 16 redundant bits for protection, then the probability that the burst greater than 17 bits is undetected is $1 - (\frac{1}{2})^{16} = 1 - 1.5 \times 10^{-5}$. The probability of having an undetected error is, then, theoretically of the order of $10^{-3} \times 10^{-5} = 10^{-8}$.

If we make $r = 80$, then the probability that bursts of length greater than 81 bits will be undetected is $1 - (\frac{1}{2})^{80} = 1 - 0.83 \times 10^{-24}$. Bursts of less than 80 bits will always be detected. So now our probability of undetected error is at least $10^{-3} \times 10^{-24} = 10^{-27}$.

This is a much higher degree of protection than is needed for most practical purposes. If we had transmitted data protected in this way from all of the locations in the world where there is now a telephone,

BOX 30.2 Continued

and if we had been transmitting nonstop since the time of Christ (with no equipment failures), it is unlikely that there would yet have been an undetected error anywhere in the world.

Furthermore, we had to add only 10 redundant characters to a message of 100 data characters. So the transmission efficiency is quite high—higher, indeed, than using the ASCII horizontal and vertical parity checking. The cost of the encoding and decoding equipment would have been higher. However, if it were mass-produced in great quantities in *large-scale-integration* circuitry, it might not be significantly higher.

5. *Retransmission of a batch of separate records.* Sometimes a control is placed on a whole batch of records. Like the controls conventionally used in batch data processing, the computer adds up account numbers and/or certain data fields from each record to produce (otherwise meaningless) *totals*. These totals are accumulated at the sending and the receiving end and are then compared. Such a control is often used to detect not only the errors in transmission but also errors in manual preparation of data. When one computer sends a program to another computer, it is vital that there be no undetected error in the program, so the words or groups of characters are added up into an otherwise meaningless hash total. This hash total is transmitted with the program, and only if the receiving computer obtains the same total in *its* addition is the program accepted.

Some form of batch control of this type is often used, where acceptable, as well as other automatic transmission controls. Its use is entirely in the hands of the systems analyst and can be made as comprehensive and secure as he feels necessary.

6. *Retransmission at a later time.* Some forms of validity check might not be capable of being used until the items are processed. They may, for example, necessitate comparing transactions with a master file. They are, nevertheless, valuable error controls, and an originating computer might keep the data in its files until a receiving computer has confirmed this validation.

ARQ
A system which detects an error in data and has it automatically retransmitted is called an ARQ (automatic repeat request) system.

ARQ systems are of two types: *stop-and-wait ARQ* and *continuous ARQ*.

Stop-and-wait ARQ is the most widely used. After sending a block the transmitting terminal waits for a positive or negative acknowledgment from the receiving terminal. Sometimes the negative acknowledgment used is mere absence of a positive acknowledgment. If the acknowledgment is positive, it sends the next block. If it is negative, it resends the previous block. Data transmission codes such as the ASCII code contain special characters for positive acknowledgment (referred to as an *ACK character*) and negative acknowledgment (a *NAK character*).

With *continuous* ARQ the transmitting terminal does not wait for an acknowledgment after sending a block; it immediately sends the next block. While the blocks are being transmitted the stream of acknowledgments is examined by the transmitting terminal. When a negative acknowledgment is received the terminal must determine which block was incorrect. The blocks must therefore be numbered, for example, with a 3-bit binary number (modulo 8). The negative acknowledgment will continue the number of the transmitted block it refers to so that the transmitting terminal can identify it. If the transmission timing is such that the acknowledgment may be received more than eight blocks after it was transmitted, then more than 3 bits will be needed to number the blocks. A satellite link may use 8 bits (modulo 256).

On receiving a negative acknowledgment the transmitting terminal may back up to the block in error and recommence transmission with that block. This is sometimes referred to as a *pullback* scheme. A more efficient technique is to retransmit only the block with the error and not those blocks which follow it. This single-block retransmission requires more logic and buffering in the transmitting and receiving terminals. It is called *selective repeat ARQ*.

Figure 30.1 illustrates stop-and-wait ARQ and the two types of continuous ARQ.

Continuous ARQ with retransmission of individual blocks is the most efficient and also the most expensive. Whether it is worth the extra cost depends on the number of blocks in error and the ratio of block transmission time to the time that elapses before an acknowledgment can be received.

For teletype transmission the time taken to transmit a block is long compared to the time taken to receive an acknowledgment, so stop-and-wait ARQ is normally used. On satellite circuits the transmission rate is high, and the time taken to receive an acknowledgment is as high as 650 milliseconds (the satellite round trip plus two line-turnaround times). Stop-and-wait ARQ would be highly inefficient, so continuous ARQ is normally used. In general, continuous ARQ is needed when the transmission rate is high and the propagation delay or line turnaround time is long. Many of the line-control procedures of the 1960s were designed for stop-and-wait ARQ. Modern line-control procedures are usually designed to allow either stop-and-wait or continuous ARQ.

Continuous ARQ usually needs a full-duplex line, whereas stop-and-wait ARQ can operate over a half-duplex line. Stop-and-wait ARQ is therefore normally used on the switched public telephone network (which gives half-duplex connections). On some full-duplex circuits with continuous ARQ, data are sent in both directions at once, the acknowledgment signals being interspersed with the data.

1   Stop-and-wait ARQ (half-duplex line)

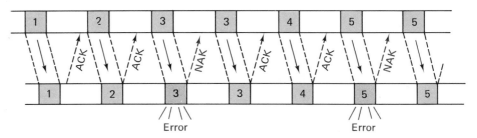

2   Continuous ARQ, with pull-back (full-duplex line)

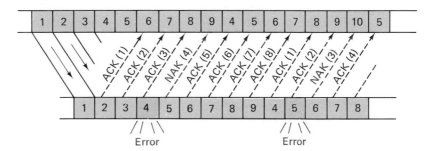

3   Continuous ARQ, with retransmission of individual block (full-duplex line)
       (sometimes called selective repeat ARQ)

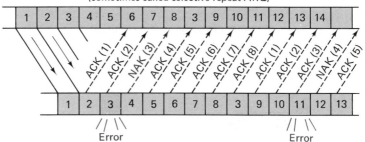

Figure 30.1   Three types of ARQ (automatic repeat request). The continuous ARQ here uses a modulo 8 (3-bit) count.

MESSAGE COUNTS   It is possible that the control characters themselves or end-of-transmission characters could be invalidated by a noise error. If this happens, then there is a danger that complete messages might be lost or two or more messages inadvertently joined together. It is possible that during the automatic transmission process a message could be erroneously sent twice. To prevent these errors, a count may be kept of the records transmitted.

Sometimes, at the start of a block, a control character is sent to indicate whether this is an odd-numbered or even-numbered block. If an odd-numbered block does not follow an even-numbered block, then the block following the last correct block is retransmitted. On some systems, two alternative start-of-transmission characters are used. With other schemes, the ACK characters contain this odd-even check. Two different ACK-type signals may be sent: ACK 0 and ACK 1. On the ASCII code there is only one ACK character, so if this code is used, a two-character acknowledgment sequence may be employed.

A more thorough type of check is to use a count field in each transmitted block. A modulo 8 (3-bit) count is commonly used. The count can be employed in a continuous ARQ scheme, as well as for detecting loss of messages. With fast transmission rates it is possible that more than one consecutive message could be lost in a transmission dropout lasting 100 milliseconds or more. The modulo-8 count would detect up to seven consecutive missing messages.

When a line interconnects several terminals in different locations an addressing (or polling) technique is used to make each device recognize when a message is addressed to it, or to enable a computer to know which device originated a message. It is difficult on multipoint lines to recover from *addressing* errors with certainty by means of answerback schemes with negative acknowledgments. The simplest method is a scheme with positive acknowledgments *only*. The receiver simply ignores incorrect blocks. The transmitter sends a block with a block count and waits for acknowledgment of its correct receipt. If no acknowledgment is received after a specified time, then it resends the block with the same count. One of two things might have gone wrong. First, the block might have been received with an error. In this case, it is retransmitted as would be required. Second, the positive acknowledgment might have been destroyed. In this case, when the receiver receives a second copy, he will send the positive acknowledgment again. Providing that *all* errors are detected by the error-detecting code, this simple scheme will be infallible on point-to-point or multipoint lines.

PERSISTENT NOISE   Sometimes the transmitting machine in an ARQ system will attempt to transmit the same message several times because of persistent noise. The system may be designed so that after a given number of attempts it adjusts the means of transmission in some way. The common way is to lower the transmission speed by adjusting the modulation. The system may have the capability to dial an alternative route.

ERROR CONTROL ON   When high-frequency radio is used for telegraphy, the
HF RADIO CIRCUITS   mutilation of bits is normally much worse than with
                         land-based telegraph circuits. It is subject to severe
fading and distortion, especially in times of high sunspot activity. Because of its
high error rate and general unreliability, its use is avoided insofar as possible for
transmission of computer data. However, it is still used in some more isolated
areas and in ship-to-shore links.

A system of error detection and retransmission in use for radio telegraphy is
the van Duuren continuous ARQ system. This transmits full-duplex, synchronously, the characters being sent in blocks of "words," a 3-out-of-7 code being
used (which permits 35 different combinations, as opposed to 32 with the 5 bits of
normal telegraphy). The START and STOP bits of the Baudot code are stripped
off, and the remaining 5 are recorded into 3-out-of-7 code and transmitted.

If the receiving equipment detects more or less than 3 bits in any character,
transmission of data in the opposite direction is interrupted. An error signal is sent
back to the transmitter of the data now in error. This transmitter then interrupts its
sending, returns to the invalid word, and retransmits it. On a long radio link, one
or more words may have been sent after the message that had the error, depending
on the duration of the transmission path. These words are discarded by the receiver. When the transmitter receives the error indication, it stops what it is transmitting, backtracks to the word in error, and retransmits that and all following
words.

High-frequency radio links can be expected to have an error rate before correction of 1 character in 1000 — sometimes much worse. Most of these errors will
be detected with the 3-out-of-7 code, but there is a certain probability of a double
mutilation that makes a character incorrect while still leaving it with three 1 bits.
The number of *undetected* errors in this case is approximately 1 character in 10
million. The mutilation rate can rise as high as 1 character in 40 or even as high as
1 character in 4 on bad links and at certain bad points of time. If the mutilation
rate is 1 character in 40, the undetected error rate rises to 1 character in 16,000,
and the effective speed of the link would drop to a speed of perhaps 90% of the
nominal speed, depending on the word size and link retransmission time. If the
mutilation rate rises to 1 character in 4, the rate after error detection and retransmission is about 1 character in 160, and the effective speed drops to a fraction of
the nominal speed. This error rate is still usable for human-language messages,
because we can apply our own error-correction thinking.

LOOP CHECK   One method of detecting errors does not use a code at
                        all. Instead, all the bits received are retransmitted back
to their sender, and the sending machine checks that they are still intact. If they
are not, then the item in error is retransmitted. Sometimes referred to as a loop
check or echo check, this scheme is normally used on a full-duplex line or on a

continuous loop line. Again, it uses the channel capacity less efficiently than would be possible with an error-detection code, although often the return path of a full-duplex line is underutilized in a system, for the system does not produce enough data to keep the channel loaded with data in both directions. A loop check is most commonly found on short lines and in-plant lines where the wastage of channel capacity is less costly. It gives a degree of protection that is more certain than most other methods.

ERROR-CORRECTING CODES An error-correcting code attempts to use redundant bits to *correct* errors in messages rather than merely *detect* them. A much higher proportion of redundant bits is needed.

Systems using error-correcting codes alone do not need the return path that an ARQ scheme uses. They are therefore sometimes called *forward error-control* (FEC) schemes.

Unfortunately, whereas the best error-detecting codes can be relied upon to prevent almost all errors slipping through, error-correcting codes are not so trustworthy. Codes may be designed to correct up to a given number of random errors, or alternatively a burst of errors which does not exceed a certain span and which is surrounded by error-free borders of certain minimum width. By suitably interleaving the transmission, error-correcting codes can be made to correct both random errors and burst errors (but not both close together). The nature of the telephone channel is such that it scatters errors in patterns that occasionally defeat the best error-correcting codes. Errors often occur in bursts, and sometimes the bursts go on for hundreds of bits.

Box 30.3 illustrates the nature of bursts of data errors on telephone lines. Figure 30.2 shows the effectiveness of error-correcting codes which correct (1) single random errors and (2) bursts of up to 8 bits with a guard space of 26 bits.

Figure 30.3 shows the sizes of bursts with different transmission speeds. These measurements assume 50 error-free bits between bursts. It will be seen that a substantial proportion of bursts are large. At transmission speeds higher than those in Fig. 30.3 the bursts tend to be much larger.

An efficient burst-correcting code with a number of redundant bits equal to the data bits (halving the transmission speed) can correct bursts of up to 16 bits with a guard space of 50 bits. It will be seen from Fig. 30.3 that 12% of bursts are greater than 16 bits at a transmission rate of 1200 bps. At 2000 bps, 14% of bursts exceed 16 bits. At 3600 bps, *most* bursts exceed 16 bits.

It will thus be seen that whereas ARQ techniques can be made fast and safe, the use of error-correcting codes substantially reduces the transmission rate and is unsafe. In general, error-correcting codes used on half-duplex or full-duplex lines do not give as good value for money, or value for bandwidth, as error-detecting codes coupled with the ability to retransmit automatically data that are found to contain an error.

The curves below [4] relate to bursts of errors of size *B* bits. When an error is detected, *B* bits are counted and then the error-free bits before the next error are counted *(N)*. Suppose, for example, that a stream of bits is represented using E for error bits and C for correct bits:

CCCCCCECCCCCCCCEECECCCCCCECCCEECCCCCCCCCCCCCCECCECCEECCCCC

If we assess this in terms of the 4-bit bursts, the bursts would be as follows, the figures being the number of correct bits between bursts:

bits
between
bursts *(N)*:   4   5   0   9   2

And if it is assessed in terms of 6-bit bursts,

bits
between
bursts *(N)*:   2   3   11   0

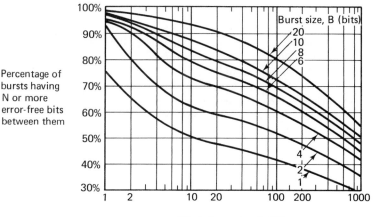

Percentage of bursts having N or more error-free bits between them

Bits between bursts (N)

The curves are from measurements of data transmitted at 1200 bps over Bell System switched public telephone lines. They permit assessment of the effectivenss of burst-correcting codes on such transmission. A burst-correcting code can correct a burst of up to *B* bits provided that it is surrounded by, or followed by, a *guard space* of N error-free bits.

For example, an error-correcting code which could correct 4-bit bursts provided they were followed by zones of 20 error-free bits would correct 70% of such bursts. The effectiveness would be worse when data are transmitted at higher speeds.

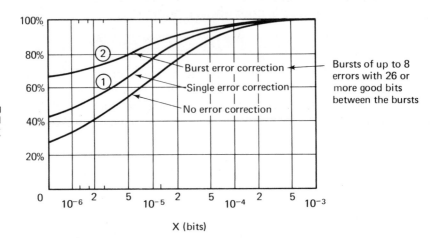

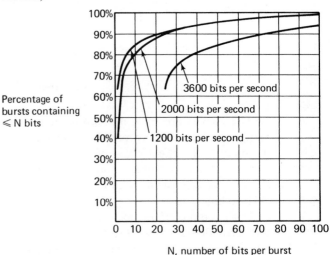

Figure 30.2 Measurements made of the undetected error rate with transmission at 1200 bits per second on Bell System switched public voice lines [4]. The effectiveness is shown of error-correcting codes which (1) correct single random errors and (2) correct bursts of up to 8 errors with guard spaces of 26 good bits. The latter error-correcting code is not highly effective but uses a number of redundant bits approximately equal to the number of data bits—i.e., it effectively halves the transmission rate.

Figure 30.3 Lengths of bursts of errors. A burst is defined as beginning and ending with an error bit, being surrounded by 50-bit error-free zones, and not itself including 50 contiguous error-free bits. The measurements were from an AT&T survey, of its toll telephone connections [5]. It will be seen that some bursts are lengthy, which means that they are not corrected by error-correcting codes of low redundancy. The error bursts tend to be longer when transmission speeds are higher than those shown. (Reproduced courtesy AT&T.)

On channels of characteristics different from today's telephone circuits error-correcting codes may be of great value. They are of value on channels with a high noise-to-signal ratio which tend to give a relatively large number of bit errors randomly distributed. In some circumstances the most economical approach is to engineer a channel so that it gives a high transmission rate at the expense of a high noise-to-signal ratio, and then use error correcting codes to lower the bit error rate. This approach is used on satellite channels. The earth station costs can be much lower if a high noise-to-signal ratio is tolerated giving a bit error rate of, say, 1 error in 1000 bits. Such a channel is acceptable for voice transmission with appropriate codecs, but would be intolerable for data without forward error correction. Error correcting codes reduce the data throughout but improve the error rate from, say 10^{-3} to about 10^{-6}, or 10^{-4} to about 10^{-7}. The degree of improvement depends upon the degree of redundancy. It is still necessary to use error *detection* and retransmission on the channel to catch the errors which are not corrected.

Error-correcting codes are also of great value where, for some reason, it is not possible or convenient to retransmit data, for example, with schemes for one-way transmission to mobile receivers, one-way transmission over cable TV channels, or data broadcasting techniques.

COMBINED
ERROR DETECTION
AND CORRECTION

When the error rate is high and the retransmission time is long, ARQ techniques lower the overall throughput of the channel. Such is the case with the highest-speed modems (9600 bps or higher) used on half-duplex voice lines, or with HF or shortwave radio circuits. In such systems forward error-correction techniques which do not have excessive redundancy may be used in conjunction with ARQ to catch *some* of the errors, and hence lessen the quantity of retransmission needed.

Some modems contain forward error-correction capability. These may be used like ordinary modems within an ARQ system. They introduce a delay in the transmission and turnaround, because the error-correcting code must be applied to a fairly large block of data. After the block is corrected it is passed on to the data-processing machine. It would be better to have a high-speed channel between the modem buffer and the data-processing machine. In general, hybrid schemes tend to be more efficient and less expensive if the error-correction and ARQ techniques are designed in an integrated fashion.

SUMMARY

The bibliography at the end of this chapter gives references containing details of error-detecting and -correcting codes.

Boxes 30.4 and 30.5 summarize the techniques for dealing with errors.

BOX 30.4 Techniques for controlling line errors

Technique	Usage
1. Ignore line errors	Used in some verbal message transmission, e.g., telegrams, message-switched systems. Used when the proportion of operator errors is far greater than line errors.
2. Loop checks	
Characters returned to transmitting device after transmission	Simple. Used on closed loops or short full-duplex circuits.
Characters transmitted twice	Simple. Used on in-plant wiring with capacity to spare.
3. Error detection without automatic retransmission	
Ignore the erroneous item	Monitoring applications where a new reading will be transmitted soon.
Manually initiated retransmission	Used on dialogue systems, e.g., airline reservation systems, where the proportion of operator errors is high and these require manual reentry. Used when the transmitter does not have any form of · storage (e.g., an unbuffered terminal).
4. ARQ (automatic request repeat) (See Box 30.5.)	The safest and most common method. Some form of storage is needed at the transmitting device.
5. Forward error correction	Error-correcting codes have high redundancy and do not correct all error bursts.
Alone	Errors slip through. Should only be used where no reverse channel is available, e.g., mobile receivers, one-way CATV, data broadcasting.
Backed up by ARQ	Used when error rate is so high that this is better than ARQ alone.
Where performed:	
In modem	
In data-processing machine	
6. Retransmit at a later time	
After a balancing run	
After a check against data on files	} Designed by the systems analyst.
After a hash totaling operation	

BOX 30.5 ARQ (automatic request repeat) techniques

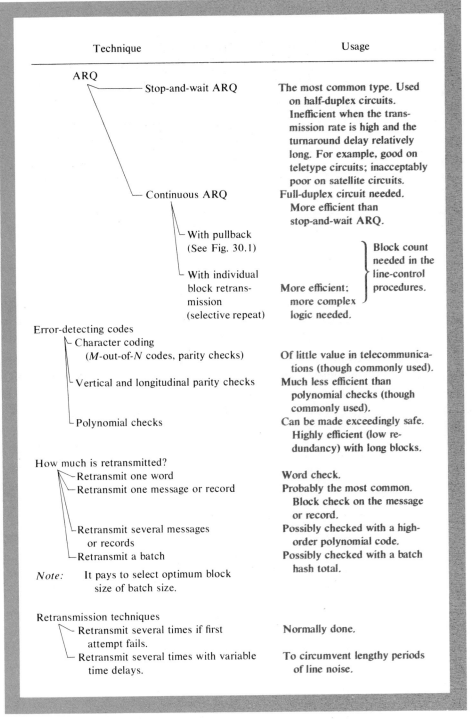

Technique	Usage
ARQ	
Stop-and-wait ARQ	The most common type. Used on half-duplex circuits. Inefficient when the transmission rate is high and the turnaround delay relatively long. For example, good on teletype circuits; inacceptably poor on satellite circuits.
Continuous ARQ	Full-duplex circuit needed. More efficient than stop-and-wait ARQ.
With pullback (See Fig. 30.1)	Block count needed in the line-control procedures.
With individual block retransmission (selective repeat)	More efficient; more complex logic needed.
Error-detecting codes	
Character coding (M-out-of-N codes, parity checks)	Of little value in telecommunications (though commonly used).
Vertical and longitudinal parity checks	Much less efficient than polynomial checks (though commonly used).
Polynomial checks	Can be made exceedingly safe. Highly efficient (low redundancy) with long blocks.
How much is retransmitted?	
Retransmit one word	Word check.
Retransmit one message or record	Probably the most common. Block check on the message or record.
Retransmit several messages or records	Possibly checked with a high-order polynomial code.
Retransmit a batch	Possibly checked with a batch hash total.
Note: It pays to select optimum block size of batch size.	
Retransmission techniques	
Retransmit several times if first attempt fails.	Normally done.
Retransmit several times with variable time delays.	To circumvent lengthy periods of line noise.

BOX 30.5 Continued

Retransmit with modem switched to lower speed.

Can be done automatically on certain terminals with built-in modems. Otherwise the modem speed may be switched manually.

Forward error correction used in addition to ARQ

Sometimes done when transmission is so noisy that throughput would be substantially lowered due to retransmission. (Forward error correction alone is not good enough.)

REFERENCES

1. A discussion of errors other than those caused by transmission lines is in *Security, Accuracy, and Privacy in Computer Systems,* by James Martin, Prentice-Hall, Englewood Cliffs, N.J., 1974.

2. Curve drawn from statistics for transmission at 1200 bps over Bell System switch public telephone lines, *The CCITT Red Book,* Vol. VII, International Telecommunications Union, Geneva, 1961.

3. W. W. Peterson and D. T. Brown, "Cyclic Codes for Error Detection," *Proceedings of the IRE.* (Jan. 1961).

4. Curves redrawn from *The CCITT Red Book,* Vol. VII, International Telecommunications Union, Geneva, 1961.

5. M. D. Balkovic, H. W. Klancer, S. W. Klare, and W. G. McGruther, "High-Speed Voiceband Data Transmission Performance on the Switched Telecommunications Network," *Bell System Technical Journal,* 50 (April 1971).

31 DELAYS AND BLOCKING

There are several possible time delays associated with the use of telecommunication networks: the time to propagate the signals, the time to establish a connection, and, worst, the delay when a "busy" signal is received.

PROPAGATION DELAY   The time delay before a signal reaches its destination over a telecommunication line is of interest to computer systems designers. The propagation delay is a small fraction of a second, because the signals travel at a speed close to the velocity of light. There is little concern about the propagation time of the *message* itself; what is of concern is the time lag in transmitting the control signals which are used for error control or for the control of lines with multiple data terminals.

Figure 31.1 compares the velocity of signals traveling on loaded and unloaded wire pairs with the velocity of radio. Signals at telephone voice frequency on a loaded wire pair travel about 10 miles in a millisecond. By radio, signals travel about 186 miles in the same time.

Telephone systems consist of a mixture of different physical paths, and so the propagation times vary. Figure 31.2 shows the results of measurements of propagation time on the Bell System. The return journey, for example, the time delay before an echo of his voice reaches a talker, would be twice the times shown in Fig. 31.2. The slope of the curve is lower at longer distances because of the increased proportion of the circuits which propagate at high velocities, such as microwave radio or coaxial cable. The mean propagation times in Fig. 31.2 are shorter than those measured a few years earlier because of the higher proportion of microwave and coaxial carriers. A communication circuit is usually longer than the airline mileage between the cities it joins. A conservative estimate for systems analysts calculating propagation delay would be to use a circuit mileage $2\frac{1}{2}$ times the airline mileage.

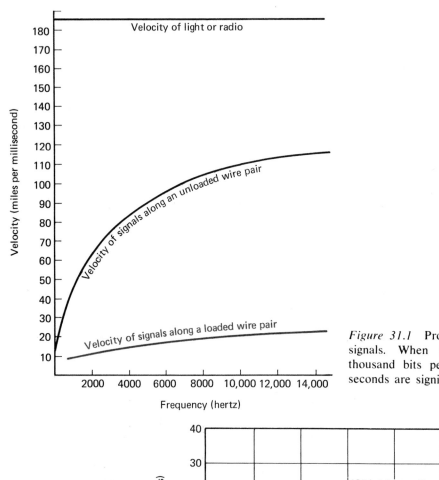

Figure 31.1 Propagation velocity of signals. When transmitting several thousand bits per second, the milliseconds are significant.

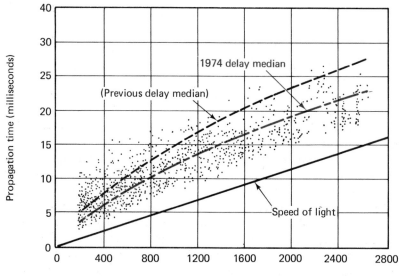

Figure 31.2 The propagation time on Bell System telephone connections. (Redrawn from Bell Laboratories annual report 1973.)

At low transmission rates, for example, 15 characters per second, a propagation time of a few milliseconds is of no concern. At high transmission rates, propagation time affects block-length considerations in error-control schemes. It affects buffer sizes and synchronization techniques in multipoint or distributed networks. A useful way of regarding propagation delay, therefore, is to quote the number of bits that would be transmitted in a time equal to the propagation time. Some examples of equivalent bit volumes are as follows:

Transmission Rate	Line	Equivalent Bits
150 bps	20 miles of a loaded wire pair	0.15
4800 bps	20 miles of a loaded wire pair	5
T1 carrier speed	20 miles of an unloaded wire pair	260
4800 bps	3000-mile circuit in U.S.	120
50,000 bps	3000-mile circuit in U.S.	1,250
9600 bps	Point-to-point satellite link	2,600
50,000 bps	Double-hop satellite link	27,000
T1 carrier speed	Single-hop satellite link	417,000
T2 carrier speed	Double-hop satellite link	3,310,000

At high transmission rates the bit volumes equivalent to the propagation delay are high enough to have a substantial effect on line-control procedures. On a line connecting many locations control signals may have to go backwards and forwards many times (polling) before a particular location can transmit.

**DELAY ON
SATELLITE
CIRCUITS**
Propagation delay is of serious concern with satellite circuits. As mentioned in Chapter 15, each time a signal is relayed via the satellite its propagation time in space is 270 milliseconds or thereabouts depending upon the relative positions of the satellite and earth stations. This time would have a serious effect on most of the teleprocessing control techniques used with land lines. It is therefore necessary either to avoid sending some of the control signals via the satellite or else to devise specialized control procedures for satellite circuits. In most cases a computer should not poll over a satellite channel, for example. New types of control procedures are being employed for satellites.

On circuits with a low transmission rate the transmitting device sends a block of data and waits for confirmation that it has been received correctly. When positive acknowledgment is received the next block is sent. If the transmission rate is high or propagation time is long, the transmitting device may not wait before it sends the next block. It keeps sending continuously and continuously receives a stream of acknowledgments. When a negative acknowledgment is received (or absence of a positive acknowledgment), meaning that a block was received incorrectly, the transmitter must back up and retransmit the block in question. A sufficient number of blocks must be stored by the transmitting device for backup to be possible, and the control signals must be able to indicate which is the block in question. With the long delay of a satellite channel, the error-recovery mechanism will require the transmitter to back up over many bits—often over many messages.

CONNECTION TIME   A much longer delay is the time associated with establishing a connection on a dial network. Electromechanical exchanges take much longer to set up a circuit than electronic exchanges. Step-by-step exchanges take longer than crossbar exchanges.

A trunk call passes through several switching offices. Each is likely to take several seconds to switch the circuits, and time is taken to relay the signaling information (giving the telephone number) between offices. Table 31.1 shows typical times to switch trunk calls in North America. It gives the time that elapses

Table 31.1   Measurements of the Bell System connection times: time from end of dialing to start of audible ringing [2]

Connection Length (airline miles)	Time to audible ringing (sec)			
	1970 Survey		1966 Survey	
	Mean	*Standard Deviation*	*Mean*	*Standard Deviation*
0–180	10.8 ± 1.5	3.7	11.1 ± 0.9	4.6
180–725	14.7 ± 3.2	5.0	15.6 ± 1.0	5.0
725–2900	13.9 ± 2.6	4.5	17.6 ± 2.1	6.6
All connections	11.7 ± 1.8	4.3		

between the subscriber finishing his dialing and his hearing the called party's telephone begin to ring. The table gives the mean and standard deviation of the times. The times follow approximately a normal distribution, which a systems analyst may use to estimate the number of calls that take longer than a certain time. For example, 10% of the values in a normal distribution exceed the mean + 1.28 standard deviations (a figure obtained from a table of the normal distribution). Therefore 10% of the calls between 180 and 725 miles have a connection time exceeding $14.7 + 1.28 \times 5.0 = 21.2$ seconds (the figures are from the 1970 AT&T survey of connection times [1]).

It will be observed that connection times have dropped slightly between the 1966 and the 1970 measurements. This is due to the introduction of more modern switching equipment and more long-distance trunks. The medium-distance calls in the 1970 survey had the longest mean connection times. With the introduction of electronic trunk switching—especially ESS No. 4 offices—trunk connection times will drop further in the future. Connection times are somewhat longer in the countries with less modern equipment than North America, especially those with step-by-step trunk switching, for example, Britain.

Local calls have shorter connection times than trunk calls, typically from 1 to 5 seconds depending on the type of switching office.

There are two types of systems in which public dial-up are used. The most common type is where the operator originates the calls and dials the computer (sometimes dials another terminal). The second is where the computer originates the call and an automatic dialing unit establishes the connection. In the first case, if the computer has no out-dialing facility, it cannot send unsolicited messages to a terminal until the operator of that terminal dials in. In the second case, the system may be designed so that the operator can dial in addition to the computer dialing out. In some systems, however, this situation does not exist, and the computer originates all calls. The operator may load cards or tapes at the terminal, and the terminal awaits a dial-up signal from the computer. This practice occurs, for example, in some message-switching systems. The computer is programmed to dial the terminals one by one to see if they have anything to send—a scanning operation that is slow because of the time taken to dial.

When the computer originates the calls, WATS lines are sometimes used, for they allow an unlimited number of calls to a given area. When the operator originates the calls, the network may be designed with a variety of different line types—telex, TWX, INWATS, and private branch exchanges with leased lines. Often telephone lines are used because of the ubiquitous convenience of the public network.

TIMING CALCULATIONS   Figure 31.3 shows the elements of time that occur when an operator dials a computer on the public telephone network.

1. The operator picks up the telephone and receives a dial tone. On some data sets, she will press the TALK button at this time.

2. She dials the number of the computer. The time taken to do so varies with the number of digits dialed and with operator dexterity. A conservative time to use in the calculations might be 2 seconds per digit for a device with a rotary dial. Timing myself on my office telephone, I find that I take 1.5 seconds per digit dialing random numbers, 2.1 seconds per digit dialing all 9s, and 0.75 second per digit dialing all 1s. With a Touchtone telephone, a reasonable time to allow is 0.5 second per digit.

3. After dialing is completed, the equipment in the switching centers connects the call, taking the time discussed above.

4. On many data sets the operator hears a "data tone" when the connection is completed;

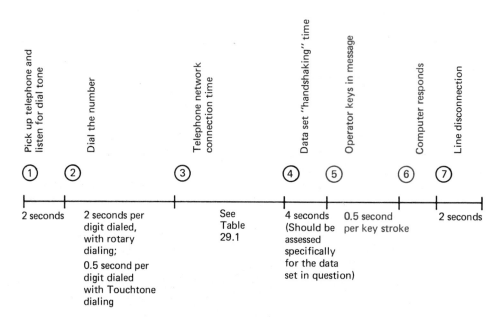

Figure 31.3   Typical times involved when an operator dials a computer.

she presses the DATA key and hangs up. A time of 4 seconds might be used in the calculations for this action.

5. The operator keys in the message, and it is transmitted.

6. The computer responds.

7. The time for disconnecting the line. Until this time element is complete, the computer cannot accept another call on the same line.

The response time may now be defined to include time elements 1 through 6 above. The total time the line into the computer is occupied includes elements 4 through 7. During this time, the line appears busy to another incoming call. This is the time period that would be used in calculations of the number of lines needed. Knowing this time period, the systems analyst can calculate the probability of succeeding in reaching a computer when it is dialed [2].

**COMPUTER
DIALING OUT**

The computer may be equipped with an automatic dialing unit that will enable it to send unsolicited messages to terminals over the public network and to scan terminals on WATS or other suitably tariffed lines to see whether they have data ready to send.

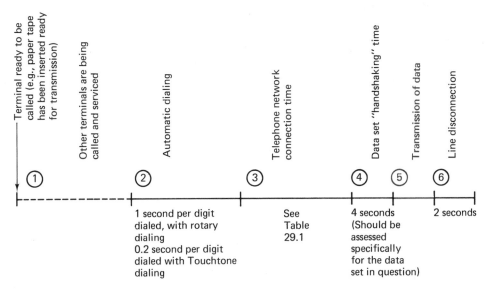

Figure 31.4 Typical times involved when a computer automatically dials terminals.

Figure 31.4 shows typical times involved when the computer dials out:

1. At the beginning of time element 1, the terminal is ready to transmit. The computer, or its transmission control unit, however, is busy dialing other terminals. Eventually it dials the terminal in question.

2. The dialing time, as in the previous case, varies with the number and value of digits dialed. It is at least twice as fast as manual dialing. If the systems analyst cannot obtain exact figures for the equipment in question, he might assume 1 second per digit for rotary dialing and 0.2 second per digit for Touchtone.

3. The same comments as before apply to the telephone network connection time.

4. When the connection is established, the terminal must respond to the "ringing," and a data set handshaking operation takes place. The time for thus establishing the data path will vary from one type of equipment to another. A typical figure of 4 seconds is given in Fig. 31.4.

5. The time for the transmission of data must include the necessary control characters and time-out intervals, if any.

6. Finally, there is the disconnection time, from the end of transmission of data to the time the next call can be initiated – typically about 2 seconds.

The line-control equipment at the computer is tied up for time elements 2 through 6 above, and this time may be used for calculating the rate at which terminals can be dialed.

GRADE OF SERVICE   A more serious form of delay occurs when a telephone number is dialed, possibly the number of a computer, and a busy signal is received. The busy signal may occur in one of two cases. First, the dialed number is genuinely "busy" — a person is using his telephone, or a computer has all its ports in use. Second, the dialed number is not busy, but there are no free circuit facilities which will reach the number. The latter situation is referred to as a *network busy* condition. In North America the telephone user hears a faster busy signal for a network busy condition than for a subscriber busy.

A network busy condition is referred to as *blocking*. Blocking is the inability to interconnect two idle lines because all possible paths between them are in use.

The telephone system is designed so that the probability of blocking is suitably low. The probability of receiving a *network busy signal* is referred to as the *grade of service*. (This term is also less commonly employed for other measures of whether users receive satisfactory performance. When used without further explanation it refers to the probability of blocking.)

Grade of service is expressed as a probability, e.g., 0.02, meaning that 98% of the calls will reach a called instrument if it is free. Commonly the same number is quoted following a letter P. Thus P.02, or simply P02, refers to a network busy probability of 0.02.

Some telephone companies have as their objective a P001 grade of service. On a small system, however, achieving a grade of service of P001 when requests for service occur at random would require that some of the facilities stand idle much of the time. Many small systems are therefore designed to give a poorer grade of service. Some corporate tie-line networks are designed for between P02 and P05. Some PBXs are designed for P01. The Datran System has a design objective of P01. Although some telephone companies have a P001 objective, it is often not met today in congested areas such as major cities.

A systems designer will often be required to select a number of trunks, a number of paths through a PBX, or a number of ports into a computer that give a specified grade of service. Equations and tables are available for accomplishing this [3]. In general the larger the number of such facilities in the group to be used or the larger the population to be served, the greater can be the efficiency of their utilization for a given grade of service. In other words, there are economies of scale. Figure 31.5 illustrates this by plotting the mean number of facilities left idle when providing a given grade of service for different traffic volumes.

Suppose that the traffic volume is such that it would fill 10 trunks, ports, or other facilities. Individual items to be transmitted arrive at random (as with telephone calls). If the grade of service is to be P.1, 13 of the facilities will be needed [3]. If there were less than 13, then more than 10% of the calls would receive busy signals. On average, 3 of the facilities will be idle. If, however, the grade of service is to be P.01, 18 of the facilities will be needed, 8 of them being idle on average. If the grade of service is to be P.001, 21 of the facilities will be needed, 11 of them being idle on average.

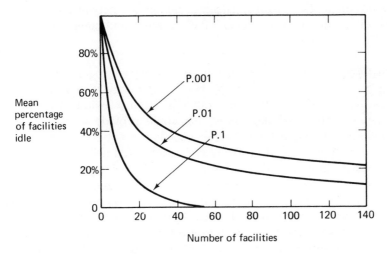

Figure 31.5 The larger the number of facilities and their users, the better the grade of service than can be achieved economically [3].

In summary,

grade of service:	P.1	P.01	P.001
number of facilities required:	13	18	21
average number of facilities idle:	3	8	11
percentage of facilities idle:	23%	44%	52%

If the traffic volume is higher, so that it would fill 75 facilities, then the percentage of facilities idle would be lower, as follows:

grade of service:	P.1	P.01	P.001
number of facilities required:	76	89	100
average number of facilities idle:	1	14	25
percentage of facilities idle:	1.3%	16%	25%

To save money, a corporate tie-line network may be designed with a poor grade of service on its trunk routes. A time-sharing computer may be designed with a small number of ports. A low-cost PBX configuration may be installed. The danger of such attempted economies is that users may become frustrated, may fail to take advantage of the computer, or may take expensive alternative actions such as dialing on the public network instead of the tie-line network, using a different computer, or leaving the line between their terminal and computer open.

FAST-CONNECT SWITCHING A switched network designed specifically for data traffic with the latest technology would have characteristics very different from a switched telephone network. First, modern solid-state switching could provide a switching time of several milliseconds rather than several seconds.

Second, a circuit would probably not be allocated to a man-machine "conversation" for its entire duration as with a telephone voice conversation. A man-machine conversation over a line operating at, say, digitized voice speed — 56,000 bps — would usually consist of bursts of data transmission lasting milliseconds, interspersed by periods of human thinking lasting seconds. It would be grossly uneconomical to keep the line idle during the human "think" periods. The network should therefore take one of two actions. Either the lines should be disconnected at the end of each transmission rather than at the end of the entire conversation, or else some means should be used for interleaving the transmissions of many users.

Third, there is no need for such a network to give a user network busy signals. The network can include some means of storing the data when no path is free. Because an individual transmission does not tie up the facilities for a long time (as would a complete conversation), the data will normally not have to wait long before they are transmitted. The designer will be concerned with delays in the network but not with grade of service. The only time the network will be unavailable to the user will be when an essential facility has failed.

The future will see a variety of different designs for fast, efficient, data networks.

REFERENCES

1. F. D. Duffy and T. W. Thatcher, Jr., "Analog Transmission Performance on the Switched Telecommunications Network," *Bell System Technical Journal,* 50 (April 1971), pp. 1311–1348.

2. James Martin, *Systems Analysis for Data Transmission,* Prentice-Hall, Englewood Cliffs, N.J., 1972, Chap. 41.

3. *Ibid.,* Chap. 33. The calculations are done using Table 11 in that book.

BIBLIOGRAPHY

1. W. W. Peterson and E. J. Weldon, Jr., *Error-Correcting Codes,* 2nd ed., M.I.T. Press, Cambridge, Mass., 1972.

2. F. F. Sellers, Jr., M-Y Hsiao, and L. W. Bearnson, *Error Detecting Logic for Digital Computers,* McGraw-Hill, New York, 1968.

3. D. T. Tang and R. T. Chien, "Coding for Error Control," *IBM Systems Journal,* 8, No. 1 (1969).

4. R. W. Hamming, "Error Detecting and Error Correcting Codes," *Bell System Technical Journal,* (April 1950). (Hamming's original classic paper.)

5. P. Fire, "A Class of Multiple-Error-Correcting Binary Codes for Non-Independent Errors," *Technical Report No. 55,* Stanford Electronics Laboratories (April 24, 1959).

6. N. Abramson, "A Class of Systematic Codes for Non-Independent Errors," *IRE Transactions on Information Theory,* IT-5 (1959), 150.

7. L. H. Zetterberg, "Cyclic Codes from Irreducible Polynomials for Correction of Multiple Errors," *IRE Transactions on Information Theory,* IT-8 (1962), 13.

8. S. H. Rieger, "Codes for the Correction of 'Clustered' Errors," *IRE Transactions on Information Theory,* IT-6 (1960), 16.

9. M. Melas, "A New Group of Codes for Correction of Dependent Errors in Data Transmission," *IBM Journal,* 4 (1960), 58.

10. R. Bose and D. Ray-Chaudhuri, "A Class of Error-Correcting Binary Group Codes," *Information and Control,* 3 (March 1960).

11. D. Hagelbarger, "Error Detection Using Recurrent Codes," presented at the AIEE Winter General Meeting (Feb. 1960).

12. W. Peterson and D. Brown, "Cyclic Codes for Error Detection," *Proceedings of the IRE* (Jan. 1961).

13. S. Lin, *An Introduction to Error Correcting Codes,* Prentice-Hall, Englewood Cliffs, N.J., 1970.

14. M. Muntner and J. K. Wolf, "Predicted Performance of Error-Control Techniques over Real Channels," *IEEE Transactions on Information Theory,* IT-14 (Sept. 1968), 640–650.

15. G. D. Forney, Jr., "Burst Correcting Codes for the Classic Bursty Channel," *IEEE Transactions on Communications Technology,* COM-19 (Oct. 1971), 772–781.

16. S. Y. Tong, "Performance of Burst-Trapping Codes," *Bell System Technical Journal,* 49, No. 4 (Apr. 1970), 477–492.

17. W. K. Pehlert, Jr., "Analysis of a Burst-Trapping Error Control Procedure," *Bell System Technical Journal,* 49, No. 4 (Apr. 1970), 493–520.

18. H. O. Burton, D. D. Sullivan, and S. Y. Tong, "Generalized Burst-Trapping Codes," *IEEE Transactions on Information Theory,* IT-17 (Nov. 1971), 736–742.

19. W. K. Pehlert, Jr., "Design and Evaluation of Generalized Burst-Trapping Error Control System," *IEEE Transactions on Communications Technology,* COM-19 (Oct. 1971), 863–868.

20. E. Y. Rocher and R. L. Pickholtz, "An Analysis of the Effectiveness of Hybrid Transmission Schemes," *IBM Journal of Research and Development* (July 1970), 426–433.

GLOSSARY

Note: An asterisk after a definition indicates that it is a CCITT, q.v., definition, published by the International Telecommunications Union, Geneva.

Ac-dc ringing. A type of telephone ringing which makes use of both ac and dc components—alternating current to operate a ringer and direct current to aid the relay action which stops the ringing when the called telephone is answered.

ACK. A code character indicating a positive acknowledgement, i.e. that a message has been received correctly.

ACU. Automatic Calling Unit, q.v.

A/D. Analog to digital conversion. The process of converting between an analog and a digital signal.

Adaptive equalization. Equalization which is adjusted while signals are being transmitted in order to adapt to changing line characteristics. See also automatic equalization, q.v.

Adaptive routing. Routing which is automatically adjusted to react to network changes such as changes of traffic pattern or failures. The 'experience' used for adaption comes from the data traffic being carried. The term 'alternate routing' is sometimes used to cover failure situations.

Address. (1) A coded representation of the destination of data, or of their originating terminal. Multiple terminals on one communication line, for example, must each have a unique address. Telegraph messages reaching a switching center carry an address before their text to indicate the destination of the message. (2) (Sometimes referred to as "called number".) That group of digits which makes up a telephone number. For example, an address may consist of area code, central office, and line number.

AGC. Automatic Gain Control.

AIS. Automatic Intercept System.

Alphabet. An agreed set of characters used to store or transmit data. A table of correspondence between an agreed set of characters and the signals which represent them.*

Alternating routing. An alternative communications path used if the normal one is not available. There may be one or more possible alternative paths. (*See* Adaptive routing.)

AM. Amplitude modulation, q.v.

AMA. Automatic message accounting, q.v.

Amplitude. The maximum departure of the value of a wave or alternating current from its average value.

Amplitude-change signaling. A telegraph signaling method in which the modulating conditions of the telegraph code are represented by currents (ac or dc) of different amplitude.

Amplitude modulation. One of three ways of modifying a sine wave signal in order to make it "carry" information. The sine wave, or "carrier," has its amplitude modified in accordance with the information to be transmitted.

Analog signal. A signal in the form of a continuously varying physical quantity such as voltage, which reflects variations in some quantity, such as loudness of the human voice.

Analog transmission. Transmission of a continuously variable signal as opposed to a discretely variable signal. Physical quantities such as temperature are continuously variable and so are described as "analog." Data characters, on the other hand, are coded in discrete separate pulses or signal levels, and are referred to as "digital." The normal way of transmitting a telephone, or voice, signal has been analog; but now digital encoding (using PCM) is coming into use over trunks.

Angstrom, A. A unit of length, a ten millionth of a millimeter, used to express the wavelengths of very high frequency signals such as light.

ANI. Automatic number identification.

Anisochronous signal. A signal which is not related to any clock, and in which signal transitions could occur at any time.

Answer signal. A supervisory signal (usually in the form of a closed loop) from the called telephone to the central office, and back to the calling telephone (usually in the form of reverse battery) when the called number answers.

Application program. The working programs in a system may be classed as "applications programs" and "supervisory programs." The application programs are the main data-processing programs. They contain no input-output coding except in the form of macroinstructions that transfer control to the supervisory programs. They are usually unique to one type of application whereas the supervisory programs could be used for a variety of different application types. A number of different terms are used for these two classes of program.

ARPA. Advanced Research Projects Agency of the U. S. Department of Defense, which supports the ARPA resource-sharing computer network.

ARQ. Automatic request for repetition. A system employing an error-detecting code and so conceived that any false signal initiates a repetition of the transmission of the character incorrectly received.*

ASA. American Standards Association.

ASCII. American Standard Code for Information Interchange. Usually pronounced "ask'-ee." An eight-level code for data transfer adopted by the American Standards Association to achieve compatibility between data devices.

ASR. Automatic send-receive teletypewriter machine.

Asynchronous transmission. Transmission in which each information character, or sometimes each word or small block, is individually synchronized, usually by the use of start and stop elements. The gap between each character (or word) is not of a necessarily fixed length. (*Compare with* Synchronous transmission.) Asynchronous transmission is also called start-stop transmission.

Attended operation. In data set applications, individuals are required at both stations to establish the connection and transfer the data sets from talk (voice) mode to data mode. (*Compare with* Unattended operations.)

Attended station. A repeater station normally staffed by maintenance personnel.

Attenuation. Decrease in magnitude of current, voltage, or power of a signal in transmission between points. May be expressed in decibels.

Attenuation equalizer. (*See* Equalizer.)

Audible ringing tone. That tone received by the calling telephone indicating that the called telephone is being rung (formerly called ringback tone).

Audio frequencies. Frequencies that can be heard by the human ear (usually 30 to 20,000 hertz).

Automatic calling unit (ACU). A device which permits a business machine to dial calls automatically.

Automatic dialing unit (ADU). A device capable of automatically generating dialing digits. (*Compare with* Automatic calling unit.)

Automatic equalization. Equalization of a transmission channel which is adjusted while sending special signals. (*See also* Adaptive equalization.)

Automatic message accounting. An automatic recording system which documents all the necessary billing data of subscriber-dialed long distance calls.

Automatic switching system. A system in which the switching operations are performed by electrically controlled devices without the intervention of operators. (*See* Direct distance dialing.)

Automatic volume expandor. (U.K.) A means for increasing the variations in signal amplitude in a transmission system according to a specified law, normally used with signal which was previously compressed. (*See* Compandor.)

Balance transmission line. A line having conductors of the same type, equal conductor resistances per unit length and equal impedances and admittances from each conductor to earth and to other electrical circuits.

Band. A portion of the electromagnetic frequency spectrum.

Bandpass filter. A circuit designed to allow a single band of frequencies to pass, neither of the cut-off frequencies being zero or infinite.

Bandwidth. The difference in cycles per second, or hertz, between the high and low frequencies of a band. The range of frequencies present in any given signal; also the range of frequencies that can be passed by a transmission medium (or a piece of electronic apparatus) without undue distortion.

Baseband or basic signal. The original signal from which a transmission waveform may be produced by modulation. In telephony it is the speech waveform. In data transmission many forms are used, being in effect a succession of bits.

Baseband signaling. Transmission of a signal at its original frequencies, i.e., a signal in its original form, not changed by modulation. It can be an analog or digital signal.

Basic numbering plan. (U.S.) A plan whereby every subscriber station is identified for nationwide dialing by a routing code and a directory number. (Called national numbering scheme in U.K.)

Baud. Unit of signaling speed. The speed in bauds is the number of discrete conditions or signal elements per second. (This is applied only to the actual signals on a communication line.) If each signal event represents only one bit condition, "bauds" is the same as "bits per second." When each signal event represents other than one bit (e.g., *see* Dibit), "bauds" does not equal "bits per second."

Baudot. (Code.) 5-bit, 32 character alphanumeric code used in transmission of information by telex.

BCD. Binary coded decimal—6-bit alphanumeric code.

BEL. Transmission units giving the ratio of two powers. The number of bels is equal to the logarithm to the base ten of the power ratio. (10 decibels, q.v.)

Bell System. The telephone operating companies controlled by American Telephone and Telegraph Corporation.

BEX. Broadband exchange, q.v.

Bias distortion. In teletypewriter applications, the uniform shifting of the beginning of all marking pulses from their proper positions in relation to the beginning of the start pulse.

Bias distortion, asymmetrical distortion. Distortion affecting a two-condition (or binary) modulation (or restitution) in which all the significant conditions have longer or shorter durations than the corresponding theoretical durations.

Binary code. An electrical representation of quantities expressed in the base 2 number system.

Bipolar coding. A method of transmitting a binary stream in which binary 0 is sent as no pulse and binary 1 is sent as a pulse which alternates in sign for each 1 that is sent. The signal is therefore ternary.

Bipolar violation. In a waveform which is mainly bipolar, there may be violations of the bipolar rule, i.e. a "1" pulse that has the same sign as the preceding "1" pulse. A violation may be used deliberately to carry information outside the binary stream.

Bit. Contraction of "binary digit," the smallest unit of information in a binary system. A bit represents the choice between a mark or space (one or zero) condition.

Bit rate. The speed at which bits are transmitted, usually expressed in bits per second. (*Compare with* Baud.)

Bit stuffing. Insertion into a binary stream of an occasional "dummy" bit so that the mean data rate is slightly less than the bit signalling rate of the channel. The position of the stuffed bits must be signalled on a supplementary channel. Also called pulse stuffing.

Blocking. The inability to interconnect two idle lines in a network because all possible paths between them are already in use.

Bridge. To connect a load across a circuit.

Bridged ringing. Any system where ringers on a line are connected across that line. To avoid shunting of the dc component, a capacitor is placed in series with each ringer.

Bridge lifter. A device which removes, either electrically or physically, bridged telephone pairs. Relays, saturable inductors, and semi-conductors are used as bridge lifters.

Broadband. Communication channel having a bandwidth greater than a voice grade channel, and therefore capable of higher-speed data transmission.

Broadband exchange (BEX). Public switched communication system of Western Union, featuring various bandwidth FDX connections.

Broadcasting service. A radiocommunication service of transmission to be received directly by the general public. It may comprise sound, vision, facsimile, or other one-way transmission.

Broadcast repeater. A repeater connecting one incoming channel to several outgoing channels.

Buffer. A storage device used to compensate for a difference in rate of data flow, or time of occurrence of events, when transmitting data from one device to another.

Bulk redundancy. A method of coding an anisochronous channel on a synchronous stream of bits in which a 1 state is represented by a string of 1s while it lasts, and an 0 state by a string of 0s. It is a redundant method because the strings must be long ones to reduce telegraph distortion when the anisochronous signal is reconstructed.

Burst isochronous. A burst isochronous signal consists of bursts of digits synchronized to a clock, interspersed by "silent" periods when no bits are presented. To indicate the bursts and silence a special clock may be provided which operates only when bits are present. This is called a "stuttering clock."

Busy flash signal. In semiautomatic or automatic working, a signal sent back to the outgoing exchange to indicate that the route or the called subscriber is busy (that is, not available). Also referred to as engaged signal.

Busy hour. The continuous one-hour period which has the maximum average traffic intensity.

Busy signal. (1) Audible and/or flashing signal, often 60 impulses per minute (IPM), which indicates that the called number is unavailable, (2) a signal transmitted at 120 IPM which indicates that all voice paths are temporarily unavailable. Also called busy tone in U.K.

Busy test. A test made to find out whether or not certain facilities which may be desired, such as a subscriber's line or trunk, are available for use. Also called engaged test in U.K.

Byte. A small group of bits of data that is handled as a unit. In most cases it is an 8-bit byte and for this unit an alternative word is octet.

Byte multiplexing. In this form of time-division multiplexing, the whole of a byte from one sub-channel is sent as a unit, and bytes from different sub-channels follow in successive time slots.

Byte stuffing. Insertion into a byte stream of some "dummy" bytes so that the mean data rate is less than the rate of the channel. A qualifying bit, if used, can distinguish the dummy bytes, which then appear as a species of control signal.

Cable. Assembly of one or more conductors within an enveloping protective sheath, so constructed as to permit the use of conductors separately or in groups. (1).

CAI. Computer assisted instruction or, sometimes, computer aided instruction. An educational use of a computer in which the student and computer maintain a dialogue in order to instruct the student.

Call. Any demand to set up a connection. Also used as a unit of telephone traffic.

Call duration. The interval of time between the moment when the connection is established between the calling and called stations and the moment when the calling station gives the clearing signal (or the moment when the connection is broken down by the operator).

Called subscriber (or Called party). The subscriber required by the calling subscriber.

Call finder. A switch which finds a calling line among a group of lines and connects it to another device. Also called line finder.

Calling relay. A relay which is controlled over a subscriber's line or trunk line. Also called a line relay.

Calling subscriber (or Calling party). The subscriber who originates a call.

Call restriction. A PABX feature that prevents selected extension stations from dialing toll calls or reaching a toll operator except through an attendant.

Call transfer. A feature that allows the called customer to instruct the switching equipment or operator to transfer incoming calls to another station.

CAMA. Centralized automatic message accounting, q.v.

Camp-on. A subscriber calling a busy number is placed in a waiting condition; both phones ring automatically when the called party hangs up.

Carrier. A continuous frequency capable of being modulated, or impressed, with a second (information carrying) signal.

Carrier, communications common. (*See* Common carrier.)

Carrier frequency. The frequency of the wave (carrier) which is modulated to transmit signals.

Carrier signaling. Any of the signaling techniques used in multi-channel carrier transmission. The most commonly used techniques are in-band signaling, out-of-band signaling, and separate channel signaling.

Carrier system. A means of obtaining a number of channels over a single path by modulating each channel on a different carrier frequency and demodulating at the receiving point to restore the signals to their original form.

Carrier telegraphy, carrier current telegraphy. A method of transmission in which the signals from a telegraph transmitter modulate an alternating current.*

CARS. Community antenna relay service, providing microwave links for carrying CATV signals to the cable head.

CATV. Originally, Community Antenna Television. Now also, Cable Television. It refers to the use of a coaxial cable loops to deliver television or other signals to subscribers.

C-band. 3.9 to 6.2 GHz (in the U.S.). (3900-6200 Hz.)

CCIR. International Radio Consultative Committee (part of the International Telecommunication Union's permanent organization).

CCIS. Common-channel interoffice signaling, q.v.

CCITT. The International Consultative Committee for Telephones and Telegraphs, part of the International Telecommunications Union (ITU) which is an organ of the U.N. CCITT is the forum for international agreement on recommendations for international communication systems, including data.

CCSA. Common-channel signaling arrangement, q.v.

CCSS. Common-channel signaling system, q.v.

C conditioning. A North American term for a type of conditioning, q.v., which controls attenuation, distortion and delay distortion so that they lie within specified limits. (*See also* D conditioning.)

CCTV. Closed Circuit Television. Unlike the over-the-air signals used in broadcast television, CCTV signals are confined to coaxial cables, microwave links, or other transmission media with controlled access.

CENTO system. Central European Treaty Organization—an international microwave system.

Central battery system. A system in which the whole of the energy for signaling and speaking is drawn from a power installation at the exchange. Also called common battery system.

Central office. The place where communications common carriers terminate customer lines and locate the switching equipment which interconnects those lines. Also referred to as an exchange, end office, and local central office.

Centralized automatic message accounting (CAMA). An automatic message accounting system which is located at a central office, but which serves various adjacent central offices. Calls not processed by ANI (automatic number identification), must be routed through an operator who dials the calling number into the equipment.

Centrex. A provision which allows every subscriber to be directly dialed from the outside. Commonly, a Centrex switching equipment is not on the premises, but at the central office.

CEPT. Conference of European Postal and Telecommunications Administrations. A European body which makes recommendations for Telecommunications practice within Europe.

Chad. The material removed when forming a hole or notch in a storage medium such as punched tape or punched cards.

Chadless tape. Perforated tape with the chad partially attached like a hinged flap to facilitate interpretive printing on the tape.

Chaining of i/o commands. The linking together (in a chain) of the commands which initiate Input-Output operations. When one command is finished the next one in the chain begins operation.

Channel. (1) (CCITT and ASA standard.) A means of one-way transmission. (*Compare with* Circuit.) (2) (Tariff and common usage.) As used in the tariffs, a path for electrical transmission between two or more points without common-carrier-provided terminal equipment. Also called circuit, line, link, path, or facility. *Note:* In some usage *channel* implies one-way transmission whereas circuit implies two-way transmission.

Channel, voice-grade. A channel suitable for transmission of speech, digital or analog data, or facsimile, generally with a frequency range of about 300 to 3000 Hz.

Channel group. The assembly of 12 telephone channels, in a carrier system, occupying adjacent bands in the spectrum, frequency-division multiplexed together. (One of the basic building blocks of carrier telephony—also called Group.)

Character. Letter, figure, number, punctuation or other sign contained in a message. Besides such characters, there may be characters for special symbols and some control functions.

Characteristic distortion. Distortion caused by transients which, as a result of the modulation, are present in the transmission channel and depend on its transmission qualities.

Charge period. Periods of the day during which certain rates of charges are applied.

Chip. The substrate upon which LSI circuits are fabricated; sometimes used to refer to the circuits on the chip themselves.

Circuit. (1) A means of both-way communication between two or more points. (2) A group of components connected together to form a specific function.

Circuit, four-wire. A communication path in which four wires (two for each direction of transmission) are presented to the station equipment.

Circuit, two-wire. A circuit formed by two conductors insulated from each other. It is possible to use the two conductors as either a one-way transmission path, a half-duplex path, or a duplex path.

Circuit cord. A device including one or more plugs, with their associated flexible conducting cords, for the purpose of interconnecting circuits on a manual switchboard.

Circuit switching. The temporary direct electrical connection of two or more channels between two or more points in order to provide the user with exclusive use of an open channel with which to exchange information. Also called line switching. Contrasts with message switching and packet switching, q.v.

Circular buffer. A form of queue in which items are placed in successive locations in a store, and are later taken from these locations in the same sequence. Two pointers keep track of the head and tail of the queue. When a pointer reaches the end of the available store it returns to the start. The items in the circular buffer may themselves be pointers to the items in the queue.

Class of office. A ranking assigned to switching points in the telephone network determined by its switching functions, interrelationships with other offices, and transmission requirements. *See* (Office classification.)

Class of service. The categorization of telephone subscribers according to specific type of telephone usage. Telephone service distinctions include, for example, rate differences between individual and party lines, flat rate and message rate, and restricted and extended area service.

Clipping. In a voice-operated telephone circuit, clipping is the loss of initial or final parts of words or syllables due to operation of the voice-actuated devices e.g. TASI equipment, q.v.

Cloax. (Bell System.) Improved-design 9.5-mm diameter coaxial cable.

Clock. A repetitive precisely timed signal used to control a synchronous process such as logic or transmission.

Clock interrupt. A type of interrupt which occurs at regular intervals and is used to initiate processes such as polling, which must happen regularly.

Clock recovery. The extraction, from the signal received on a synchronous channel, of the clock which accompanies the data.

Coax. Coaxial cable, q.v.

Coaxial cable. A cable consisting of an outer conductor concentric to an inner conductor, separated from each other by insulating material. It can carry a much higher bandwidth than a wire pair.

Codec. Coder-decoder (analog-to-digital and digital-to-analog converter). It is used to convert analog signals such as speech, music, or television, to digital form for transmission over a digital medium, and back again to the original analog form.

Code character. The set of conventional elements established by the code to enable the transmission of a written character (letter, figure, punctuation sign, arithmetical sign, etc.) or the control of a particular function (spacing, shift, line-feed, carriage return, phase corrections, etc.); this set of elements being characterized by the variety, the duration, and the relative position of the component elements (or by some of these features).

Code ringing. The alerting of telephone subscribers on multiparty lines by combinations of short and long rings which are different for each subscriber.

Cohesion. The cohesion of a connected network is the minimum number of links which, if they were removed, would divide the network into at least two parts which are not joined by any links.

Coin-collect tone. A low tone which informs the originating toll operator that the change for a call has been collected by the local operator or the coin control circuit.

Coin-denomination tones. The tones produced by gongs in multislot coin telephones, when different coins are deposited. The tones are detected and transmitted to the operator so that the correct amount can be checked.

Coin-return tone. A high tone which informs the originating toll operator that the change for a call has been returned by the local operator or coin control circuit when the connection is not completed.

Collect call. A call in which the caller may specify at the time of booking that the charge should be paid by the called party.

Common battery. A dc power source in the central office that supplies power to all subscriber stations and central office switching equipment.

Common-battery signaling. The method by which supervisory and telephone address information is sent to a central office by opening and closing the circuit at the telephone, i.e., depressing and releasing the switch on the cradle of the handset.

Common carrier. An organization in the business of providing regulated telephone, telegraph, telex and data communications services. This term is applied most often to US and Canadian commercial organizations but is also sometimes used to refer to telecommunications entities (such as government operated PTTs) in other countries. The term used to be used also for transportation companies carrying goods or people.

Common-channel interoffice signaling (CCIS). A technique by which the signaling information for a group of trunks is transmitted between switching offices over a separate voice channel using time-division methods.

Common-channel signaling system (CCSS). A system whereby all signaling for a number of voice paths is carried over one common channel, instead of within each individual channel. Common channel signaling should not be confused with CCIS.

Common control. An automatic switching arrangement in which the control equipment necessary for the establishment of connections is shared, being associated with a given call only during the period required to accomplish the control function. It applies to electronic, crossbar, and some Strowger-switch offices.

Common control switching arrangement CCSA. (U.S.) Switching facilities connected by the telephone company to corporate tie-line networks. Switching of the leased lines in the organization's network is accomplished by common-control central office switching equipment. All stations in the network may then dial one another regardless of distance, and without using exchange or toll facilities. They may also dial outside the network via local, foreign exchange, and/or WATS lines. On many such networks the digit "8" is dialled before the number of an extension on the tie-line network.

Communications satellite. An earth satellite designed to act as a telecommunications radio relay. Most communications satellites (other than Russian ones) are in geosynchronous orbit 22,300 miles above the equator so that they appear from earth to be stationary in space.

Compandor. A compandor is a combination of a compressor at one point in a communication path for reducing the volume range of signals, followed by an expandor at another point for restoring the original volume range. Usually its purpose is to improve the ratio of the signal to the interference entering in the path between the compressor and expandor.

Composite signaling (CX). A dc signaling system which requires a single line conductor for each signaling channel, and which provides full duplex operation. In this system, voice frequencies above 100 Hz are separated from the signaling currents by a filter network known as a composite set. Two composite signaling channels are derived from one pair of wires, and four from a phantom group. Composite signaling channels may also be used for dc telegraph or teletypewriter circuits.

Compressor. Electronic device which compresses the volume range of a signal, used in a compandor, q.v. An "expandor" restores the original volume range after transmission.

Comsat. Communications Satellite Corporation, a private U.S. company established by statute as the exclusive international satellite carrier and representing the U.S. in Intelsat.

Concentrator. A device which connects a number of circuits which are not all used at once to a smaller group of circuits for economical transmission. A telephone concentrator achieves the reduction with a circuit-switching mechanism. A data concentrator buffers incoming data and retransmits it over appropriate output lines.

Conditioning. A procedure to make transmission impairments of a circuit lie within certain limits which are specified in a tariff. Conditioning is used on many telephone lines leased for data transmission to improve the possible transmission speed. The customer pays a monthly charge for conditioning. Two types are used: C and D conditioning, q.v.

Conference call. A call established among three or more stations in such a manner that each of the stations is able to carry on a communication with all the others. Also called conference connection.

Congestion. Any communication network has a limit to the traffic it can carry. Beyond that limit the network must somehow restrict traffic. Congestion means the condition in which traffic is thus restricted.

Contact. Part of a switch, usually noble metal alloy, designed to touch a similar contact to permit current to flow, or designed to break this union to cause a current to cease.

Contention. (1) A "dispute" between several devices for the use of common equipment. Examples are: contention for access to a store, contention by two terminals for the use of a half duplex circuit, and contention by many peripherals for a computer channel. (2) A method of line control in which the terminals request to transmit. If the channel in question is free, transmission goes ahead; if it is not free, the terminal will have to wait until it becomes free. A queue of contention requests may be built up by a computer, and this can either be in a prearranged sequence or in the sequence in which the requests are made.

Control character. A character whose occurrence in a particular context initiates, modifies, or stops, a control operation—e.g., a character to control carriage return.

Control mode. The state that all terminals on a line must be in to allow line control actions, or terminal selection to occur. When all terminals on a line are in the control mode, characters on the line are viewed as control characters performing line discipline, that is, polling or addressing.

Control signals. Signals which pass between one part of communication system and another as part of the mechanism of controlling the system.

CPU. Central processing unit.

Cross-bar switch. A switch having a plurality of vertical paths, a plurality of horizontal paths, and electromagnetically operated mechanical means for interconnecting any one of the vertical paths with any of the horizontal paths.*

Cross-modulation. Interference caused by two or more carriers in a transmission system interacting through nonlinearities in the system.

Crosspoint (semiconductor). A two-state switching device having a low transmission system inpedance in one state and a very high one in the other.

Cross-section. Signal transmission capacity of a transmission system, usually measured in terms of the number of two-way voice channels.

Cross talk. The unwanted transfer of energy from one circuit, called the disturbing circuit, to another circuit, called the disturbed circuit.*

Cross talk, far-end. Cross talk which travels along the disturbed circuit in the same direction as the signals in that circuit. To determine the far-end cross talk between two pairs, 1 and 2, signals are transmitted on pair 1 at station A, and the level of cross talk is measured on pair 2 at station B.

Cross talk, near-end. Cross talk is propagated in a disturbed channel in the direction opposite to the direction of propagation of the current in the disturbing channel. Ordinarily, the terminal of the disturbed channel at which the near-end cross talk is present is near or coincides with the energized terminal of the disturbing channel.

DAA. Data access arrangement, q.v.

Data. Numbers, texts, facts, instructions, etc, which are represented in a formalized manner so that they can be stored, manipulated, and transmitted by machines.

Data access arrangement, DAA. (U.S.) Machines not designed, owned, or authorized, by the telephone company, and without an FCC authorized line interface, can be attached to the telephone network in the U. S. by means of a DAA, a small wall-mounted box containing an isolation transformer, designed to prevent signals harmful to the network being sent down the network lines.

Dataphone. Both a service mark and a trademark of AT&T and the Bell System. As a service mark it indicates the transmission of data over the telephone network. As a trademark it identifies the communications equipment furnished by the Bell System for data communications services.

Dataphone digital system. (See DDS.)

Dataphone 50. An 50 kb/s switched data service offered by the Bell System (in certain cities). Dataphone is a trademark of AT&T to identify the data sets manufactured and supplied by Bell System for use in the transmission of data over the regular telephone network. It is also a service mark (Dataphone Service) of the Bell System.

Data set. Bell System terminology used to describe its various modems (q.v.) which convert the signals of a business machine into signals that are suitable for transmission over communications lines.

Data-signaling rate. It is given by $\sum_{i=1}^{m} \frac{1}{T_i} \log_2 n_i$, where m is the number of parallel channels, T is the minimum interval for the ith channel, expressed in seconds, n is the number of significant conditions of the modulation in the ith channel. Data-signaling rate is expressed in bits per second.*

Data sink. The equipment which accepts data signals after transmission.

Data source. The equipment which supplies data signals to be transmitted.

Dataspeed. An AT&T marketing term for a family of medium-speed paper tape transmitting and receiving units. Similar equipment is also marketed by Western Union.

Data structure. A system of relationships between items of data. To express these relationships when a data structure is stored, lists may be used, or other systems using pointers, etc.

Data terminal equipment. In the context of a data network it is the equipment which is attached to the network to send or receive data, or both.

Datel. Data transmission services offered by European PTTs, using switched public telephone networks.

db. Decibel, q.v.

dbm. Decibel referenced to one milliwatt. Employed in communication work as a measure of absolute power. Zero dbm equals 1 milliwatt.

dBrn. This ratio is expressed in decibels above reference noise. Reference noise is the magnitude of circuit noise that will produce a circuit noise meter reading equal to that produced by 10–12 watt of electric power at 1000 Hz (cycles per second).

D conditioning. A U. S. term for a type of conditioning, q.v., which controls harmonic distortion and signal-to-noise ratio so that they lie within specified limits. It may be used with or without C conditioning, q.v.

DDD. (*See* Direct distance dialing, q.v.)

DDS. Dataphone Digital System. An AT&T network and tariff giving digital (not analog) communication circuits.

Decibel (db). A tenth of a bel. A unit for measuring relative strength of a signal parameter such as power, voltage, etc. The number of decibels is ten times the logarithm (base 10) of the ratio of the power of two signals, or ratio of the power of one signal to a reference level. The reference level must always be indicated, such as 1 milliwatt for power ratio.

Decimonic ringing. A type of party line selective ringing which uses ringing frequencies of 20Hz, 30Hz, 40Hz, 50Hz, and 60Hz.

Dedicated. Used exclusively for a single purpose or by a single subscriber.

Delay distortion. Distortion of the signal which occurs as a result of phase shift being a nonlinear function of frequency.

Delay equalizer. A corrective network which is designed to make the phase delay or envelope delay of a circuit or system substantially constant over a desired frequency range. (*See* Equalizer.)

Delay-pulsing signal (delay dial, stop dial). An off-hook signal from the called end of a trunk, which is sent to the calling end of a trunk, to indicate that it is not ready to be pulsed.

Delay vector. Associated with one node of a packet switching network, the delay vector has as its elements the estimated transit times of packets destined for every other node in the network. Nodes send copies of the delay vector to their neighbors as part of an adaptive routing scheme.

Delta modulation. Method of representing a speech waveform (or other analog signal) in which successive bits represent increments of the waveform. The increment size is not necessarily constant.

Demodulation. The process of retrieving intelligence (data) from a modulated carrier wave; the reverse of modulation.

Diagnostic programs. These are used to check equipment malfunctions and to pinpoint faulty components. They may be used by the computer engineer or may be called in by the supervisory programs automatically.

Diagnostics, system. Rather than checking one individual component, system diagostics utilize the whole system in a manner similar to its operational running. Programs resembling the operational programs will be used rather than systematic programs that run logical patterns. These will normally detect overall system malfunctions but will not isolate faulty components.

Diagnostics, unit. These are used on a conventional computer to detect faults in the various units. Separate unit diagnostics will check such items as arithmetic circuitry, transfer instructions, each input-output unit, and so on.

Dial-normal transmission signal. A secondary dial tone which is returned to an operator to indicate that the rest of a number may be dialed.

Dial off-normal tone (dial key off-normal). The tone that reminds an operator to restore the dial key after a call has been completed into a step-by-step office, and after the called party has answered.

Dial pulse. A current interruption in the dc loop of a calling telephone. It is produced by the breaking and making of the dial pulse contacts of a calling telephone when a digit is dialed. The loop current is interrupted once for each unit of value of the digit.

Dial speed. The number of pulses that a rotary dial can transfer in a given amount of time. The dial speed of a typical rotary dial is 10 pulses per second.

Dial telephone set. (U.S.) A telephone set provided with a dial or similar device. Also called automatic telephone set.

Dial tone. A 90Hz signal (the difference between 350 Hz and 440 Hz) sent to an operator or subscriber indicating that the receiving end is ready to receive dial pulses.

Dial-up. The use of a dial or pushbutton telephone to initiate a station-to-station telephone call, or a simular dial pulsing operation carried out by an automatic machine.

Dibit. A group of two bits. In four-phase modulation, each possible dibit is encoded as one of four unique carrier phase shifts. The four possible states for dibit are 00, 01, 10, 11.

Differential modulation. A type of modulation in which the choice of the significant condition for any signal element is dependent on the choice for the previous signal element.*

Digital data. Information represented by a code consisting of a sequence of discrete elements. (*Compare with* Analog transmission.)

Digital signal. A discrete or discontinuous signal; one whose various states are discrete intervals apart. (*Compare with* Analog transmission.)

Digital speech interpolation (DSI). When speech is digitized, it may be cut into slices such that no bits are transmitted when a person is silent. As soon as he begins to speak, bits flow again. This reduces the number of bits needed to carry a conversation to about 45% of those needed for a continuous full duplex speech channel. It is a faster version of TASI, q.v., designed for digital circuits.

Diplex. Permitting simultaneously and in the same direction the transmission or reception of two signals over a circuit or channel.

Direct distance dialing (DDD). A telephone exchange service which enables the telephone user to call other subscribers outside his local area without operator assistance. In the United Kingdom and some other countries, this is called subscriber trunk dialing (STD).

Direct exchange line. A line serving only the subscriber's main station. Also called exclusive exchange line.

Direct inward dialing. This Centrex service feature allows an incoming call from the exchange network to reach a Centrex station without attendant assistance.

Directory number. The full complement of digits (or letters and figures) associated with the name of a subscriber in the directory.

Directory routing. A message or packet routing system which uses a directory at each node which states, for each destination, the preferred outgoing link. The directory may also show second preferences.

Direct outward dialing. This service feature allows a **PBX** or Centrex station user to gain access to the exchange network without the assistance of the attendant.

Disconnect signal. A signal transmitted from one end of a subscriber line or trunk to indicate at the other end that the established connection should be disconnected.

Distortion. The unwanted change in waveform that occurs between points in a transmission system.

Distortion, amplitude vs. frequency. That distortion in a transmission system caused by the nonuniform attenuation or gain of the system with respect to frequency under specified conditions.

Distortion, bias. (*See* Bias distortion.)

Distortion, delay vs. frequency. That distortion in a transmission system caused by the difference between the maximum transit time and the minimum transit time of frequencies within a certain band under specified conditions. Also called delay distortion and phase distortion.

Distortion, nonlinear. Distortion caused by a deviation from a linear relationship between the input and output of a system or component.

Distributing frame. A structure for terminating permanent wires of telephone central office, private branch exchange, or private exchange, and for permitting the easy change of connections between them by means of cross-connecting wires.

Divided ringing. (*See* Ground-return ringing.)

Double-current transmission, polar direct-current system. A form of binary telegraph transmission in which positive and negative direct currents denote the significant conditions.*

DPCM. Differential pulse code modulation. A form of pulse code modulation in which some redundancy in the signal is removed by the process of subtracting the present strength of a signal from its strength an instant earlier; only this difference is transmitted.

DPSK. Phase shift keyed modulation (q.v.) which encodes the differences between signal elements.

Drop. A connection made available for a terminal unit on a transmission line.

Drop, subscriber's. The line from a telephone cable to a subscriber's building.

Drop out. A brief interruption of a communication signal. (Often caused by deep fades on a microwave path).

Dry circuit. A circuit over which voice signals are transmitted, and which carries no direct current.

DSI. Digital speech interpolation, q.v.

DTMF. Dual tone multifrequency signaling. A method of signaling in which a combination of two frequencies, each from a group of four, are used to transmit numerical address information. The two groups of four frequencies are 697 Hz, 770 Hz, 852 Hz, 941 Hz, and 1209 Hz, 1336 Hz, 1477 Hz, and 1633 Hz. (Touchtone dialing uses these frequencies.)

Duobinary signaling. A method of transmitting a synchronous binary waveform in which neighbouring signal elements influence one another in a controlled manner. The resultant signal is, as a consequence ternary.

Duplex circuit. A circuit used for transmission in both directions at the same time. It may be called "full duplex" to distinguish it from "half duplex," q.v.

Duplexing. The use of duplicate computers, files, circuitry or transmission lines, so that in the event of one component failing an alternative one can enable the system to carry on its work.

Duplex signaling (DX). A signaling system which occupies the same cable pair as the voice path, yet does not require filters. One duplex signaling section is confined to 5000 ohms of loop resistance, though several sections may be used in tandem.

Duplex transmission. Simultaneous two-way independent transmission in both directions. (*Compare with* Half-duplex circuit.) Also called full-duplex transmission).

Dynamic multiplexing. A form of time-division multiplexing in which the allocation of time to constituent channels is made according to the demands of these channels.

E & M signaling. A signaling arrangement characterized by the use of separate paths for the signaling and the voice signals. The M lead (derived from "mouth") transmits ground or battery to the distant end of the circuit, while incoming signals are received as either a grounded or open condition on the E (derived from "ear") lead.

EBCDIC. Extended binary coded decimal interchange code: an 8-bit alphanumeric code.

Echo. An echo is a wave which has been reflected or otherwise returned with sufficient magnitude and delay for it to be perceptible in some manner as a wave distinct from that directly transmitted.

Echo check. A method of checking data transmission accuracy whereby the received data are returned to the sending end for comparison with the original data.

Echo modulation. A method of producing a shaped pulse in which a main pulse with "echoes" is generated, then put through a simple band-limiting filter. The "echoes" are pulses of controlled amplitude occurring both before and after the main pulse.

Echo suppressor. Used to attenuate echos on long telephone connections. The unit, inserted at four-wire points, is voice-actuated, and functions by increasing loss into the transmission path opposite in direction to the one being used. The hang-over time required for the echo suppressor to release is usually in the 15 to 130 millisecond range, depending on the type used, and so may limit the speed with which data transmission can be turned around.

EDP. Electronic data processing.

EHF. Extremely high frequency (30 GHz to 300 GHz.)

EIA.   Electronics Industries Association (U.S.).

EIA interface.   A standardized set of signal characteristics (time duration, voltage and current) specified by the Electronic Industries Association.

e.i.r.p.   Effective isotropic radiated power.

Elastic store (or buffer).   A store or buffer which can hold a variable amount of data. It behaves like a queue, sending the data out in the same sequence as it receives it, but able to vary the rate of output somewhat, subject to the limits of being full and empty. The units of data held might typically be bits or bytes.

End distortion.   End distortion of start-stop teletypewriter signals is the shifting of the end of all marking pulses from their proper positions in relation to the beginning of the start pulse.

End office.   The local central office, q.v., at which subscriber lines and trunks are interconnected. It is designated a class 5 office in the DDD network.

Engaged signal.   British term for "busy signal", q.v.

Envelope.   In an amplitude modulated signal the waveform has maxima and minima at almost exactly the carrier frequency. The location of these maxima and minima can be joined by two smooth curves through the peaks of the waveform, which form the envelope of the waveform.

Envelope delay.   The slope of the phase shift versus frequency at any frequency.

Equalization.   Compensation for the attenuation (signal loss) variation with frequency (attenuation equalization) and propagation time variation with frequency (delay equalization). Its purpose is to adjust channel characteristics so that data can be transmitted at a higher rate or other signals transmitted with less distortion.

Equalizer.   Any combination (usually adjustable) of coils, capacitors, and/or resistors inserted in transmission line or amplifier circuit to improve its frequency response.

Equivalent four-wire system.   A transmission system using frequency division to obtain full-duplex operation over only one pair of wires.

Erlang.   A dimensionless unit of traffic intensity. One Erlang is the intensity at which one traffic path would be continuously occupied, i.e. 1 call-hour per hour, 1 call-minute per minute, etc.

Error burst.   A sudden outbreak of errors in a short amount of time compared to the period of errors immediately before and after the occurrence.

Error control.   An arrangement that will detect the presence of errors. In some systems, refinements are added that will correct the detected errors, either by operations on the received data or by retransmission from the source.

Error-correcting code.   An error-detecting code incorporating sufficient additional signaling elements to enable the nature of some or all of the errors to be indicated and corrected entirely at the receiving end.

Error-detecting and feedback system, request repeat system, ARQ system.   A system employing an error-detecting code and so arranged that a signal detected as being in error automatically initiates a request for retransmission of the signal as being in error.*

Error detecting code. A code in which each telegraph or data signal conforms to specific rules of construction so that departures from this construction in the received signals can be automatically detected. Such codes require more signal elements than are necessary to convey the fundamental information.

Error detecting system. A system employing an error detecting code and so arranged that any signal detected as being in error is either deleted from the data delivered to the data sink, with or without an indication that such deletion has taken place, or delivered to the data sink together with an indication that it has been detected as being in error.

Error rate. The ratio of the number of bits, elements, characters or blocks incorrectly received to the total number of bits, elements, characters, or blocks transmitted.

ESS. Electronic switching system. A computerized telephone exchange. The Bell System ESS 1 and ESS 2 are central offices. ESS 101 gives private branch exchange (PBX) switching controlled from the local central office. ESS 4 is a trunk switch. ESS is now used as an industry term rather than merely a Bell System term.

ETV. Educational television. Similar to Instructional Television (ITV) but generally broadcast and, therefore, available to television receivers in the home.

E-type pulsing signal. In PCI (panel call indicator) pulsing, the end of pulsing signal informs the distant end that all digits have been sent. This signal is required when (PCI) traffic is completed through a crossbar tandem.

Even parity check (odd parity check). This is a check which tests whether the number of digits in a group of binary digits is even (even parity check) or odd (odd parity check).*

EVR. Electronic video recording. The recording of a video signal electronically on magnetic tape, a magnetic disk, or some other material.

Exchange. (1) A telephone switching center. (2) A defined area, served by a communications common carrier, within which the carrier furnishes service at the exchange rate and under the regulations applicable in that area as prescribed in the carrier's filed tariffs.

Exchange, classes of. Class 1 (*see* Regional center); Class 2 (*see* Sectional center); Class 3 (*see* Primary center); Class 4 (*see* Toll center); Class 5 (*see* End office). These form the hierarchy of switching offices used in North America, Class 1 being the highest level trunk switching office, and Class 5 being the local telephone exchange.

Exchange, private automatic (PAX). A dial telephone exchange that provides private telephone service to an organization and that does not allow calls to be transmitted to or from the public telephone network.

Exchange, private automatic branch (PABX). A private automatic telephone exchange that provides for the transmission of calls internally and to and from the public telephone network.

Exchange, private branch (PBX). A private telephone exchange connected to the public telephone network on the user's premises. It may be operated by an attendant supplied by the user. PBX is today commonly used to refer also to an automatic exchange, PABX (q.v.). In North America the initials PBX are commonly used instead of PABX.

Exchange, Trunk. An exchange devoted primarily to interconnecting trunks.

Exchange service. A service permitting interconnection of any two customers' stations through the use of the exchange system.

Expandor. A transducer which for a given amplitude range or input voltages produces a larger range of output voltages. One important type of expandor employs the information from the envelope of speech signals to expand their volume range. (*Compare with* Compandor).

Extended area service. An option whereby the telephone subscriber can pay a higher flat rate in order to obtain wider geographical coverage without additional per-call charges. (In U.S., WATS is such a service).

Extension telephone set. (1) Additional telephone set on the same line but at a different location than the main station. (2) A PBX station.

Facsimile. A system for the transmission of images. The image is scanned at the transmitter, reconstructed at the receiving station, and duplicated on some form of paper.

Fail softly. When a piece of equipment fails, the programs let the system fall back to a degraded mode of operation rather than let it fail catastrophically and give no response to its users.

Fall-back. When the equipment develops a fault the programs operate in such a way as to circumvent this fault. This may or may not give a degraded service. Procedures necessary for fall-back may include those to switch over to an alternative computer or file, to change file addresses, to send output to a typewriter instead of a printer, to use different communication lines or bypass a faulty terminal, etc.

Fall-back, double. Fall-back in which two separate equipment failures have to be contended with.

Fast-connect circuit switching. A form of switching of value for highly sporadic traffic such as interactive data transmission. A circuit is not connected for the duration of a lengthy "session", as would be a telephone connection. Instead it is reswitched rapidly for every message, and disconnected after the message. This technique permits the maximum number of data users (or other sporadic users) to share the transmission and switching facilities.

FAX. Facsimile, q.v.

FCC. Federal Communications Commission, q.v.

FD or FDX. Full duplex. (*See* Duplexing).

FDM. Frequency-division multiplex, q.v.

FDMA. Frequency-division multiple access, q.v.

Federal communications commission (FCC). A board of seven commissioners appointed by the President under the Communication Act of 1934, having the power to regulate all interstate and foreign electrical communication systems originating in the United States.

Feedback. A method of signal regeneration involving a coupling from a high-level point in an amplifier to a lower-level point in the same or a previous stage in such a manner as either to increase or decrease the apparent gain of the amplifier.

Fiber optic waveguides.   Thin filaments of glass or other transparent materials through which a light beam may be transmitted for long distances by means of multiple internal reflections.

Figures shift.   A physical shift is a teletypewriter which enables the printing of numbers, symbols, upper-case characters, etc. (*Compare with* Letters shift).

Filter.   A circuit designed to transmit signals of frequencies within one or more frequency bands and to attenuate signals of other frequencies.

Flashing.   A signal from one operator to another, or from a subscriber held by an operator.

Flat rate.   A fixed payment for service, independent of use, within a defined area, with an additional charge for each call outside the area.

Flooding.   A packet routing method which replicates packets and sends them to all nodes, thus ensuring that actual destination is reached.

Flow control.   In data communication networks which employ storage there is a possibility of congestion if more data flows into a node than flows out of it. To remove this possibility, flow control is needed.

FM.   Frequency modulation, q.v.

Foreign exchange (FX).   Connects a customer's location to a remote exchange. This service provides the equivalent of local service from the distant exchange.

Fortuitous distortion.   Distortion resulting from causes generally subject to random laws (accidental irregularities in the operation of the apparatus and of the moving parts, disturbances affecting the transmission channel, etc.)

Four-wire circuit.   A circuit using two pairs of conductors, one pair for the "go" channel and the other pair for the "return" channel.* A telephone circuit is basically duplex (it carries voice signals both ways). In the local network this is achieved over two wires because the waveforms travelling each direction can be distinguished. In the trunk network, where amplifiers are needed at intervals and multiplexing is common, it is easier to separate the two directions of transmission and use (effectively) a pair of wires for each direction. At this point it is a four-wire circuit.

Four-wire equivalent circuit.   A circuit using the same pair of conductors to give "go" and "return" channels by means of different carrier frequencies for the two channels.*

Four-wire terminating set.   Hybrid arrangement by which four-wire circuits are terminated on a two-wire basis for interconnection with two-wire circuits.

Frame.   One complete cycle of events in time-division multiplexing. The frame usually includes a sequence of time slots for the various sub-channels and extra bits for control, framing, etc.

Frame-grabber.   A device that can seize and record a single frame of video information out of a sequence of many frames.

Framing.   Synchronizing of the equipment at the receiving end of a TDM channel so that it correctly recognizes the frame.

Frequency. The rate at which a current alternates, measured in Hertz, kiloHertz, megaHertz, etc. (Older units of measure: cycles, kilocycles, or megacycles.)

Frequency bands. Frequency bands are defined arbitrarily as follows:

Range-megahertz	*Name*
Below 3	Low frequency (LF)
3 – 30	High frequency (HF)
30 – 300	Very high frequency (VHF)
300 – 3,000	Ultra high frequency (UHF)
3,000 – 30,000	Super high frequency (SHF) (microwave)
30,000 – 300,000	Extremely high frequency (EHF) (millimeterwave)

Frequency-change signaling. A telegraph signaling method in which one or more particular frequencies correspond to each desired signaling condition of a telegraph code. The transition from one set of frequencies to the other may be either a continuous or a discontinuous change in frequency or in phase.

Frequency-derived channel. Any of the channels obtained from multiplexing a channel by frequency division.*

Frequency-division multiple access (FDMA). Communicating devices at different locations share a multipoint or broadcast channel by means of a technique which allocates different frequencies to different users. (*Compare with* Time-division multiple access, TDMA.)

Frequency-division multiplex (FDM). A multiplex system in which the available transmission frequency range is divided into narrower bands, each used for a separate channel.*

Frequency modulation (FM). One of three ways of modifying a sine wave signal to make it "carry" information. The sine wave or "carrier" has its frequency modified in accordance with the information to be transmitted. The frequency function of the modulated wave may be continuous or discontinuous. In the latter case, two or more particular frequencies may correspond each to one significant condition.

Frequency-shift, frequency-shift keying (FSK). Frequency modulation method in which the frequency is made to vary at the significant instants. 1. By smooth transitions: the modulated wave and the change in frequency are continuous at the significant instants. 2. By abrupt transitions: the modulated wave is continuous but the frequency is discontinuous at the significant instants.*

Frequency shift pulsing (FSP). A signaling technique which uses a frequency shift between 1070Hz and 1270 Hz. This type of signaling is used with narrow band systems, such as teletypewriter switching networks.

FSK. Frequency-shift keying, q.v.

FSP. Frequency-shift pulsing, q.v.

FTS. Federal Telecommunications System. A leased telephone network shared by U.S. federal government agencies.

Full duplex (FDX). Refers to a communications system or equipment capable of transmission simultaneously in two directions.

Functional test. Test carried out under normal working conditions to verify that a circuit or a particular part of the equipment functions correctly.

(G/T). Gain-to-noise temperature ratio of a receiving system where gain in dB is measured relative to an isotopic radiator and temperature in dB relative to 1° Kelvin.

Gain. A general term used to denote an increase in signal power in transmission from one point to another, and usually expressed in dB.

Gate. A basic logic circuit.

Gentrex. General telegraph service (public). Automatic routing and transmission of telegrams between telegraph offices.

Giro. A banking service provided by the Post Office (in several European countries, including the United Kingdom).

Go signal (start dial). A supervisory on-hook signal received by the calling end of a trunk after a stop signal has occurred. The go signal indicates that the called end is now ready to receive additional digits.

Grounded circuit. (1) A circuit in which energy is carried one way over a metallic path and returned through the earth. (2) A circuit connected to earth at one or more points.

Ground return circuit. A circuit which has a conductor (or a number of conductors in parallel) between two points and which is completed through the earth at these two points. Also called earth return circuit.

Ground-return ringing (divided ringing). A party line system where ringers in a telephone circuit may be connected between one or both sides of the line and ground, with a capacitor placed in series with each ringer.

Group. 12 voice channels frequency-division multiplexed together into a single frequency band. Sometimes called *Channel Group* in North America.

Group-busy tone (all trunks busy tone). A low audible tone on the sleeve of trunk jacks at cord switchboards indicating that all trunks are busy. An absence of this tone informs the operator that there is at least one idle trunk in a group.

Group delay or envelope delay. If a complex signal with a narrow bandwidth is sent down a transmission path, at the receiving end the envelope of the signal will appear to have suffered a delay, called the group or envelope delay.

Group link. The whole of the means of transmission using a frequency band of specified width (48 kHz) connecting two group distribution frames (or equivalents). It extends from the point where the group is assembled to the point where it is broken down. This expression is usually applied to the combination of "go" and "return" channels.

Half-duplex (HD or HDX) circuit. (1) CCITT definition: A circuit designed for duplex operation, but which, on account of the nature of the terminal equipments, can be alternately only. (2) Definition in common usage (the normal meaning in computer literature): A circuit designed for transmission in either direction but not both directions simultaneously.

Hamming code. A code using redundant bits to detect errors in data (transmission errors).

Hamming distance. The Hamming distance between two binary words (of the same length) is the number of corresponding bit positions in which the two words have different bit values. Also known as signal distance.

Handshaking.   Exchange of predetermined signals for purposes of control when a connection is established between two modems or other devices.

Hard copy.   A permanent, tangible record such as information printed on paper or recorded on film.

Harmonic distortion.   The presence of harmonic frequencies in a received signal (due to nonlinear characteristics of a transmission line). If a sine wave is sent these higher frequency components have the effect of flattening the peaks of the sine wave.

Harmonic ringing.   The technique of selectively signaling individual ringers on a party line using frequencies which are harmonics of fundamental frequencies. The fundamental frequencies used are 16 2/3Hz and 25 Hz; the harmonics are 33 1/3 Hz, 50 Hz, and 66 2/3 Hz.

HD or HDX.   Half duplex. (*See* Half-duplex circuit.)

Head-end.   Signal originating point in a CATV system.

Headset.   An operator's telephone set which consists of a telephone transmitter, a receiver, and cord and plug, associated, arranged to be worn so as to leave the operator's hands free.

Hertz (Hz).   A unit of frequency equal to one cycle per second. Formally CPS, cycles per second.

Heterodyne detector.   A nonlinear device used to mix two frequencies together so that their sum and different frequencies are generated at some desirable point in the spectrum.

Heterodyne radio.   A microwave receiver-transmitter in which the received signal is amplified at an intermediate frequency and retransmitted without being demodulated to baseband. Used for long-haul transmission.

Heuristic routing.   A routing method in which delay data produced by normal data-carrying packets coming in on different links from a given source node is used to guide the out-going packets as to the best link for getting to that node.

HF.   High frequency (3 to 30 Hz).

High D/low D.   An AT&T tariff structure for leased voice lines in which a lower charge per mile is made for lines on certain designated high-density routes, to that on all other (low-density) routes.

High density bipolar (HDB).   A modified bipolar code which avoids the long absence of pulses and thus eases clock recovery. There are many versions. Bipolar violations are used to signal strings of zeros.

High-low signaling.   Signaling in which a high resistance shunt indicates an on-hook condition, and a low resistance shunt an off-hook condition.

High tone.   A -17 dBm, 480Hz information signal that may, for example, be used as a partial dial tone.

High-usage trunk.   Direct trunks provided, where traffic volume warrants, to bypass a part of the DDD switching network.

Holding time.   The length of time a communication channel is in use for each transmission. Includes both message time and operating time, q.v.

Hollerith code.   An alphanumeric code used in card readers and sorters.

Home loop. An operation involving only those input and output units associated with the local terminal.

Hookswitch. A switch that is located within the supporting structure on which a telephone handset rests when it is not in use. When the handset is lifted, the switch closes the telephone circuit or loop.

Hot line. A line serving two telephone sets exclusively, on which one set will ring immediately when the receiver of the other set is lifted. Some common carriers offer a hot-line tariff.

Hot potato routing. Packet routing which sends a packet out from a node as soon as possible, even though this may mean a poor choice of outgoing link.

Hot standby. Alternate equipment stands by ready to take over an operation quickly if the equipment on which it is being performed fails.

Howler tone. A 480-Hz tone which progresses in successive levels from 0 to 120 dBa, and is used to alert a subscriber when his handset is off-hook.

Hum. Spurious electrical interference picked up from the conventional 60 Hz (or 50 Hz) alternating current power supply.

Hybrid circuit. A circuit having four sets of terminals arranged in two pairs designed so that there is high loss between the two sets of terminals of a pair when the terminals of the other pair are suitably terminated. Hybrids are commonly used to couple four-wire circuits to two-wire circuits.

Hz. Hertz—frequency of a waveform in cycles per second.

IC. Integrated circuit.

IF. Intermediate frequency.

IFRB. International Frequency Registration Board. Part of the International Telecommunications Union, ITU.

IMP. Interface message processor, a packet switching computer used in an experimental data communication network funded by ARPA.

In-band signaling. The transmission of signaling information via tones at some frequency or frequencies that lie within a carrier channel normally used for voice transmission.

Induction coil. An apparatus for obtaining intermittent high voltage consisting of a primary coil through which the direct current flows, an interrupter, and a secondary coil of a larger number of turns in which the high voltage is induced.

Infocom. An intracompany message switching service offered by Western Union.

Information bit. One of those bits which are used to specify the characters of a given code group (opposed to framing bits).

In-house. (*See* In-plant system.)

In-plant system. A system whose parts, including remote terminals, are all situated in one building or localized area. The term is also used for communication systems spanning several buildings and sometimes covering a large distance, but in which no common carrier facilities are used.

Intelligent terminal. A terminal that is programmable and can process its messages, for example to check validity, or to compress the data.

Intelsat. The International Telecommunications Satellite Consortium, formed in 1964 with the purpose of creating a worldwide communications satellite system. Membership in Intelsat now (early 1972) numbers 81 countries.

Intelsat I, II, III, IV, IVA and V. Names of communications satellites planned or launched by Intelsat.

Interface. A boundary between two pieces of equipment across which all the signals which pass are carefully defined. The definition includes the connector signal levels, impedance, timing, sequence of operation and the meaning of signals. The term has been extended to include the idea of a software interface, in some cases of high complexity. The essence of an interface is its accurate definition.

Interface–CCITT. The world recommendation for interface requirements between data processing terminal equipment and data communication equipment. The CCITT recommendation resembles very closely the American EIA Standard RS-232-B or C. This standard is considered mandatory in Europe and on the other continents.

Interface–MIL STD 188B. The standard method of interface established by the Department of Defense. It is presently mandatory for use by the departments and agencies of the Department of Defense for the installation of all new equipment. This standard provides the interface requirements for connection between data communication security devices, data processing equipment, or other special military terminal devices.

Interface computer. Part of a network which mediates between a network subscriber and the high-level or trunk network. It can be regarded as containing a local area switch and terminal processors.

Interface EIA standard RS-232 B or C. A standardized method adopted by the Electronic Industries Association (U.S.) to insure uniformity of interface between data communication equipment and data processing terminal equipment. Has been generally accepted by most manufacturers of data transmission and business equipment.

Intermodulation. The modulation of the components of a complex wave by each other in a nonlinear system, whereby waves are produced which have frequencies among others, equal to the sums and differences of those of the components of the original complex wave.

International telecommunication union (ITU). The telecommunications agency of the United Nations, established to provide standardized communications procedures and practices including frequency allocation and radio regulations on a world-wide basis.

Interoffice trunk. A direct trunk between local central offices (Class 5 offices), or between Class 2, 3, or 4 offices. (Also called intertoll trunk. See office classification.)

Interrupt. A jump out of one program into another due to an external event. A mechanism is usually provided to store the information needed for a return to the interrupted program. In addition to external events in the I-O system, interrupts are allowed from clocks and timers and for various malfunctions.

Interswitchboard line. (U.K.) A telephone line directly connecting either two private branch exchanges, or two private exchanges. Also tie line (U.K.) and trunk line (U.S.).

Inter-symbol interference. An (analog) waveform which carries binary data is generally transmitted as a number of separate signal elements. When received, the signal elements may be influenced by neighbouring elements. This is called inter-symbol interference.

Intertoll trunk. (*See* Interoffice trunk).

Interval timer. A timer can be set by a program to produce an interrupt after a specified interval.

Intraoffice trunk. The trunk connection within the same central office.

I-O channel. An equipment forming part of the input-output system of a computer. Under the control of I-O commands the "channel" transfers blocks of data between the main store and peripherals.

Isarithmic control. The control of flow in a packet switching network in such a way that the number of packets in transit is held constant.

ITU. International Telecommunication Union, q.v.

ITV. Instructional television. ITV is usually distributed over specially allocated broadcast television channels or by closed circuit television (CCTV).

Jack. A device used generally for terminating the permanent wiring of a circuit, access to which is obtained by the insertion of a plug.

Jumbo group. (U.S.) 6 U.S. master groups frequency-division multiplexed together in the Bell System. A jumbo group can carry 3600 telephone calls.

Junction. (U.K.) A line connecting two exchanges in the same multi-exchange area. Also called trunk.

Junctor. In crossbar systems, a junctor is a circuit extending between frames of a switching unit and terminating in a switching device on each frame.

Keyboard perforator. A perforator, q.v., provided with a bank of keys, the manual depression of any one of which will cause the code of the corresponding character or function to be punched in a tape.*

Keyboard send/receive (KSR). A combination teletypewriter transmitter and receiver with transmission capability from keyboard only.

Key pulsing. A manual method of sending numerical and other signals by the operation of nonlocking pushkeys. Similar to customer's push-button telephone service. Also called key sending.

Key pulsing signal. The signal which indicates a circuit is ready for pulsing in multifrequency and direct current keypulsing.

Key sending. Key pulsing, q.v.

Key telephone. When more than one telephone line per set is required, push-button or key telephone systems offer flexibility and a wide variety of uses, i.e., pickup of several central-office lines, foreign exchanges lines, PBX stations lines, private lines, and intercommunicating lines. Features of the system include pickup and holding, intercommunication, visual and audible signals, cutoff, exclusion and signalling.

Klystron. Thermionic tube used at microwave frequencies to deliver power outputs ranging from milliwatts to hundreds of kilowatts.

KSR. Keyboard (manual) send-receive teletypewriter machine, q.v.

Label. A set of symbols used to identify or describe an item, record, message, or file. Occasionally it may be the same as the address in storage.

LAMA. Local automatic message accounting, q.v.

Laser. Light amplification by stimulated emission of radiation. A device which transmits an extremely narrow and coherent beam of electromagnetic energy in the visible light spectrum. Coherent means that the separate waves are in phase with one another rather than jumbled as in normal light.

Latency. The time taken for a storage to reach the read/write heads on a rotating recording surface. For general timing purposes, average latency is used; this is the time taken by one half-revolution of the surface.

L-band. 390 to 1550 MHz (in the U.S.).

L-carrier. In the Bell System, a family of coaxial cable transmission systems.

Leak and loop tests. Two tests designed to simulate the worst-case loop pulsing conditions in the exchange network. The tests are made at the central office switching equipment to assure that it will accurately respond to any pulsing signals generated by telephones connected to that exchange network.

Leased circuit. A telecommunication circuit leased by a user or group of users for exclusive use between certain locations. No switching is employed, so the circuit is ready for immediate use. It may be point-to-point or multidrop. It is sometimes referred to as a *private* line, q.v.

LED. Light emitting diode. A small solid-state device which emits light when a current is applied.

Letters shift. A physical shift in a teletypewriter which enables the printing of alphabetic characters. Also, the name of the character which causes this shift. (*Compare with* Figures shift.) (1).

LF. Low Frequency (below 3 MHz).

LFU. Least-frequency-used. A replacement algorithm in which when new data has to replace existing data in an area of storage, the least-frequently-used items are replaced.

Library. (1) The room in which volumes (tapes and diskpacks) are stored. (2) An organized collection of programs, source statements, or object modules, maintained on a direct-access device accessible by the operating system.

Line broadcasting. The transmission, by a network of conductors, of a radiobroadcast program so as to permit any number of subscribers to receive it if they so wish. Also called wire broadcasting.

Line finder. A switch which finds a calling line among a group of lines and connects it to another device. Also called call finder.

Line hit. An electrical interference causing the introduction of spurious signals on a circuit.

Line speed. The maximum data rate that can be reliably transmitted over a line.

Line switching. Switching in which a circuit path is set up between the incoming and outgoing lines. Contrasts with message switching, and packet switching, q.v., in which no such physical path is established. Also called *circuit switching.*

Link. (1) A physical circuit between two points. (2) A conceptual (or logical) circuit between two users of a packet switched (or other) network permitting them to communicate (although different physical paths may be used).

List.   An ordered set of data items. A chain.

Loading.   Adding inductance (load coils) to a transmission line to minimize amplitude distortion.

Local automatic message accounting (LAMA).   A combination of automatic message accounting equipment and automatic number identification equipment in the same office. In such a system, a subscriber-dialed toll call can be automatically processed without operator assistance.

Local battery system.   A system in which the energy for speaking is drawn from a battery located at the subscriber's station.

Local call.   Any call for a destination within the local service area of the calling station.

Local central office.   (U.S.) A common carrier switching office in which subscriber's lines terminate. Also called local exchange, end office, Class 5 office.

Local exchange.   (*See* Local central office.)

Local loop.   That part of a communication circuit between the subscriber's equipment and the equipment in the local central office, q.v.

Local service area.   The area within which two subscribers can be connected on payment of a local fee.

Local trunk.   (U.S.) Trunks between Class 5 offices (local central offices) in the US. (*See* Office classification.)

Locap.   Low-capacity, low-loss paired cable used in the Bell System to transmit PCM carrier at 6.3 Mbps.

Logical.   An adjective describing the form of data organization, hardware, or system that is perceived by an application program, programmer, or user; it may be different to the real (physical) form.

Long distance.   Any telephone call, subject to charge, for a destination outside of the local service area of the calling station. Also called toll call and trunk call.

Longitudinal redundancy check (LRC).   A system of error control based on the formation of a block check following preset rules. The check formation rule is applied in the same manner to each character. In a simple case, the LRC is created by forming a parity check on each bit position of all the characters in the block (e.g., the first bit of the LRC character creates odd parity among the one-bit positions of the characters in the block).

Loop.   A local circuit between a central office and a subscriber telephone station. Also called subscriber loop and local line.

Loop checking, message feedback, information feedback.   A method of checking the accuracy of transmission of data in which the received data are returned to the sending end for comparison with the original data, which are stored there for this purpose.*

Loop circuit.   A circuit composed of continuous metallic conductors and not using an earth return. Also called metallic circuit. Generally refers to the circuit connecting the subscriber's set with the local switching equipment.

Loop disconnect pulsing.   (U.K.) A method of signaling which makes use of the break pulses in a loop circuit. Also called loop dialing.

Loop pulsing.   Signaling achieved by the repeated opening and closing of the loop at the originating end of the circuit. Rotary telephone dials are loop pulsing devices.

Loop signaling systems.   Any of three types of signaling which transmit signaling information over the metallic loop formed by the trunk conductors and the terminating equipment bridges. Transmission of the loop signals may be accomplished by (1) opening and closing the dc path around the loop, (2) reversing the voltage polarity, or (3) varying the value of the equipment resistance.

Loss (transmission).   The decrease in energy of signal power in transmission along a circuit due to the resistance or impedance of the circuit or equipment.

Low-high signaling.   A variation of high-low signaling. The on-hook condition is indicated by a low resistance shunt and an off-hook condition shows high resistance.

Low-tone.   A 480-Hz plus 620-Hz tone at 24 dBm. The 140-Hz difference $(620 - 480 = 140)$ frequency gives the tone its low-pitched sound. A low tone is used for line busy, reorder, and no-circuit tone signals that are reached by a subscriber.

LRC.   Longitudinal redundancy check, q.v.

LSI.   Large scale integration. The deposition of many electronic circuits, including transistors, on a small silicon chip by a single manufacturing process. A microminiaturized and reliable form of electronic circuit. It is used, for example, in pocket calculators.

LTRS.   Letters shift, q.v.

Magneto telephone set.   A telephone set in which the signaling current is provided by a hand magneto-generator.

Mailgram.   A "record" (message) communications service jointly provided by Western Union and the U. S. Postal Service. Messages received at Western Union offices or originated at telex stations are sent over the WU network to a post office near the addressee for the next day postal delivery.

Main distribution frame (MDF).   A distribution frame to which are connected on one side the line exterior to the exchange, and on the other side the internal cabling of the exchange.

Main station.   A subscriber's instrument (e.g. telephone set or terminal) connected to a local loop, which is used for originating calls and on which incoming calls from the exchange are answered. (As opposed to extension station.)

Mark.   Presence of signal. In telegraph communications a mark represents the closed condition or current flowing. A mark impulse is equivalent to a binary 1.

Marker.   A wired-logic control circuit that, among other functions, tests, selects, and establishes paths through a switching stage(s) in response to external signals.

Mark-hold.   The normal no-traffic line condition whereby a steady mark is transmitted. This may be a customer-selectable option. (*Compare with* Space-hold.)

Mark-to-space transition.   The transition, or switching, from a marking impulse to a spacing impulse.

Maser.   Microwave amplification by stimulated emission of radiation — The general class of microwave amplifiers based on molecular interaction electromagnetic radiation. The non-electronic nature of the Maser principle results in very low noise.

Mastergroup. A standard frequency-division multiplexed grouping of voice channels; 300 or 900 voice channels under CCITT standards; 600 voice channels under U. S. standard, i.e., 10 supergroups.

Master station. A unit having control of all other terminals on a multipoint circuit for purposes of polling and/or selection.

MDF. Main distributing frame where subscriber and trunk circuits appear in the central office.

Mean time to failure. The average length of time for which the system, or a component of the system, works without fault.

Mean time to repair. When the system, or a component of the system, develops a fault, this is the average time taken to correct the fault.

Measured rate. A message rate structure modified so that the rental includes payment for a specified number of calls within a defined area, plus a charge for additional calls.

Message. A sequence of characters used to convey information or data. In data communication, messages are usually in an agreed format with a 'heading', which controls the destiny of the message and 'text' which consists of the data being carried.

Message format. Rules for the placement of such portions of a message as message heading, address, text, end-of-message indication, and error-detecting bits.

Message numbering. The identification of each message within a communication system by the assignment of a sequential number.

Message switching. The technique of receiving a message, storing it until the proper outgoing line is available, and then retransmitting. No direct connection between the incoming and outgoing lines is set up as in line switching, q.v.

MF signaling. (*See* Multifrequency pulsing.)

Microwave (wavelength). Electromagnetic waves in the radio frequency spectrum above 890 MHz (the frequencies between 1 GHz and 30 GHz).

Millimeter (wavelength). Extremely high frequency (30 to 300 GHz). (*See* Frequency bands.)

Modem. Contraction of modulator-demodulator. A device which modulates and demodulates signals transmitted over communication facilities. The modulator is included for transmission and the demodulator for reception. A modem is used to permit digital signals to be sent over analog lines. Also called data set.

Modulation. The process of varying some characteristic of the carrier wave in accordance with the instantaneous value or samples of the intelligence to be transmitted.

Modulation, amplitude (AM). Form of modulation in which the amplitude of the carrier is varied in accordance with the instantaneous value of the modulating signal.

Modulation, differential. A type of modulation in which the choice of the significant condition for any signal element is dependent upon the choice for the previous signal element. Differential phase shift keying (DPSK) modulation is an example.

Modulation, frequency (FM). A form of modulation in which the instantaneous frequency of a sine wave carrier is caused to depart from the carrier frequency by an amount proportional to the instantaneous value of the modulating signal.

Modulation, phase (PM). Form of modulation in which the angle relative to the unmodulated carrier angle is varied in accordance with the instantaneous value of the amplitude of the modulating signal.

Modulation, pulse amplitude (PAM). The form of modulation in which the amplitude of the pulse carrier is varied in accordance with successive samples of the modulating signal.

Modulation, pulse code (PCM). The form of modulation in which the modulating signal is sampled and the sample quantized and coded so that each element of the information consists of one or more binary bits.

Modulation frequency (or modulating frequency). The frequency of the modulating wave.

Modulation with a fixed reference. A type of modulation in which the choice of the significant condition for any signal element is based on a fixed reference.*

Modulator. A device which converts a signal (voice or other) into a form that can be transmitted.

Monitoring key. A key permitting an operator to monitor or to listen on a circuit without sensibly affecting the transmission quality of that circuit.

Morse code. A two-condition telegraph code in which characters are represented by groups of dots and dashes, these groups being separated by spaces. Named after its inventor.

MOS. Metal oxide semiconductor. A type of LSI, q.v.

MSI. Medium-scale integration. Solid state circuitry with fewer components per chip than LSI, q.v.

MSK. Minimum shift keyed modulation.

Multi-access. The ability for several users to communicate with a computer at the same time, each working independently on his own job.

Multichannel. Use of a common channel in order to make two or more channels, either by splitting of the frequency band transmitted by the common channel into narrower bands, each of which is used to constitute a distinct channel (frequency-division multiplex), or by allotting this common channel in turn to constitute different intermittent channels (time-division multiplex).

Multidrop line. Line or circuit interconnecting several stations. Also called multipoint line.

Multifrequency pulsing. A method of transmitting address information and other signals for controlling the telephone network. The identity of each of the ten possible digits (0 to 9) plus the required supervisory functions is determined by a combination of two out of six possible frequencies. Referred to as MF signaling.

Multiple. A system of wiring so arranged that a circuit, a line, or a group of lines are accessible at a number of points, to any one of which connection can be made. Also called multipoint.

Multiplex, multichannel. Use of a common channel in order to make two or more channels, either by splitting of the frequency band transmitted by the common channel into narrower bands, each of which is used to constitute a distinct channel (frequency-division multiplex), or by allotting this common channel in turn, to constitute different intermittent channels (time-division multiplex).*

Multiplexer. A device which enables more than one signal to be sent simultaneously over one physical circuit.

Multiplex hierarchy. 12 channels = 1 group; 5 groups (60 channels) = 1 supergroup; 10 supergroups (600 channels) = 1 mastergroup (U. S. standard); 5 supergroups (300 channels) = 1 mastergroup (CCITT standard); 6 U. S. mastergroups = 1 jumbo group.

Multiplexing. The division of a transmission facility into two or more channels either by splitting the frequency band transmitted by the channel into narrower bands, each of which is used to constitute a distinct channel (frequency-division multiplex), or by allotting this common channel to several different information channels, one at a time (time-division multiplexing).*

Multipoint. A system of wiring so arranged that a circuit, a line, or a group of lines are accessible at a number of points, to any one of which connection can be made. Also called multiple.

Multipoint circuit. A circuit connecting 3 or more locations (known as a multidrop circuit in North America).

Multiprocessing. Strictly, this term refers to the simultaneous application of more than one processor in a multi-CPU computer system to the execution of a single 'user job', which is only possible if the job can be effectively defined in terms of a number of independently executable components. The term is more often used to denote multiprogramming operation of multi-CPU computer systems.

Multiprogramming. A method of operation of a computer system whereby a number of independent jobs are processed together. Rather than allow each job to run to completion in turn, the computer switches between them so as to improve the utilization of the system hardware components.

Multithreading. Concurrent processing of more than one message (or similar service-request) by an application program.

NAK. Negative acknowledge, q.v.

Negative acknowledge (NAK). In the method of error control which relies on repeating any message received with (detectable) errors, the return signal which reports an error is NAK, in CCITT alphabet No. 5 (the opposite to ACK, or *acknowledge*).

Network. (1) A series of points connected by communications channels. (2) The switched telephone network is the network of telephone lines normally used for dialed telephone calls. (3) A private network is a network of communications channels confined to the use of one customer.

Network terminating unit (NTU). The part of the network equipment which connects directly to the data terminal equipment. The NTU operates between the local transmission lines and the subscriber's interface.

Neutral transmission. Method of transmitting teletypewriter signals, whereby a mark is represented by current on the line and a space is represented by the absence of current. By extension to tone signaling, neutral transmission is a method of signaling employing two signaling states, one of the states representing both a space condition and also the absence of any signaling. Also called unipolar. (*Compare with* Polar transmission.)

No circuit (NC) signal (fast busy signal). A low tone (140 Hz) which is interrupted at 120 impulses per minute, and which indicates that there is no circuit available.

Node. In a topological description of a network a node is a point of junction of the links. The word has also come to mean a switching centre in the context of data networks, particularly in the context of packet switching.

Noise. Unwanted electrical signals, introduced by circuit components or natural disturbances, which tend to degrade the performance of a communications channel. Any unwanted background signal on a circuit, caused by system design limitations or by improper alignment, or by interference from an outside source, divided into classes.

Noise figure. Usually expressed in dB as the ratio of the actual output noise power to the noise power that would be obtained from an ideal noiseless network having the same gain characteristic.

Noise temperature. The noise temperature of a device is the temperature of a thermal noise source producing the same output noise power in the same bandwidth as the device under consideration.

Office alarm. An alerting signal indicating an abnormal condition in a central office.

Office classification. Numbers assigned to offices according to their function in the U. S. DDD network. The following class numbers are used:

Class 1: Regional center (RC)
Class 2: Sectional center (SC)
Class 3: Primary center (PC)
Class 4: Toll center (TC) if operators present, otherwise Toll point (TP)
Class 5: End office (EO)

Any one center handles the traffic from one to two or more centers lower in the hierarchy.

Office code. The first three digits of a seven-digit telephone number.

Off hook. Activated (in regard to a telephone set). By extension, a data set automatically answering on a public switched system is said to go "off hook." The off-hook condition indicates a "busy" condition to incoming calls.

Off-line. Pertaining to equipment or devices not under direct control of the central processing unit. May also be used to describe terminal equipment which is not connected to a transmission line. In telegraph usage, paper tapes frequently are punched "off line" and then transmitted using a paper tape transmitter.

On hook. Deactivated (in regard to a telephone set). A telephone not in use is "on hook." (1).

ONI. Operator number identification, q.v.

On-line. Connected to a computer so that data can pass to or from the computer without human intervention. Directly in the line loop. In telegraph usage, transmitting directly onto the line rather than, for example, perforating a tape for later transmission.

On-line computer system. An on-line system may be defined as one in which the input data enter the computer directly from their point of origin and/or output data are transmitted directly to where they are used. The intermediate stages such as punching data into cards or paper tape, writing magnetic tape, or off-line printing, are largely avoided.

Open-circuit signaling. Direct current signaling accomplished by opening and closing the telephone circuit. There is an absence of current flow when the circuit is in the idle condition.

Open wire. A conductor separately supported above the surface of the ground—i.e., supported on insulators.

Open-wire line. A line whose conductors are principally in the form of open wires suspended between telephone poles.

Operating time. The time required for dialing the call, waiting for the connection to be established and coordinating the forthcoming transaction with the personnel or equipment at the receiving end.

Operator number identification (ONI). At a local dial central office, that equipment which allows the operator to come in long enough to acquire the calling number so that it may be keyed into CAMA equipment.

Operator's telephone set. An operator's telephone set consists of a telephone transmitter, a receiver, and cord and plug, associated, arranged to be worn so as to leave the operator's hands free. Also called headset.

OTP. Office of Telecommunications Policy in the Executive Office of the U.S. President. This agency has no statutory powers of regulation but develops and recommends public policy in the area of telecommunications.

Out-of-band signaling. A method of signaling which uses a frequency that is within the passband of the transmission facility, but outside of a carrier channel normally used for voice transmission.

Overflow. Traffic, in excess of the capacity of the number of circuits on a particular route, which is offered to another (alternative) route.

PABX. Private automatic branch exchange—a private branch exchange that provides access to and from the public telephone network without operator intervention. (*See* PBX.)

Packet. A group of binary digits including data and call control signals which is switched as a composite whole. The data, call control signals and possibly error control information are arranged in a specified format.*

Packet switching. The transmission of data by means of addressed packets whereby a transmission channel is occupied for the duration of transmission of the packet only. The channel is then available for use by packets being transferred between different data terminal equipment. *Note:* The data may be formatted into packet or divided and then formatted into a number of packets for transmission and multiplexing purposes.*

Packet switching network. A network designed to carry data in the form of packets. The packet and its format is internal to that network. The external interfaces may handle data in different formats, and conversion is done by an interface computer, q.v.

Pair-selected ternary (PST). A pseudo-ternary code in which pairs of binary digits are coded together in such a way that the resultant signal has no long strings of zeros.

PAM. Pulse-amplitude modulation, q.v.

Panel. An early form of central office switching equipment. (*See* Panel switching system.)

Panel call indicator (PCI) pulsing. A dc pulsing system where information digits are each transmitted as a series of four marginal, polarized pulses, which are detected at the receiving end and registered on relays or switches at the terminating office. In manual offices, the pulses are displayed on a lamp in front of the operator.

Panel switching system. An automatic switching system which is generally characterized by the following features: (1) the contacts of the multiple banks over which selection occurs are mounted vertically in flat rectangular panels: (2) the brushes of the selecting mechanism are moved by a motor which is common to a number of these selecting mechanisms: (3) the switching pulses are received and stored by controlling mechanisms which govern the subsequent operations necessary in establishing a connection.

Parallel transmission. Simultaneous transmission of the bits making up a character or byte, either over separate channels or on different carrier frequencies on one channel. (1) The simultaneous transmission of a certain number of signal elements constituting the same telegraph or data signal. For example, use of a code according to which each signal is characterized by a combination of 3 out of 12 frequencies simultaneously transmitted over the channel.*

Parity check. Addition of noninformation bits to data, making the number of ones in a grouping of bits either always even or always odd. This permits detection of bit groupings that contain single errors. It may be applied to characters, blocks, or any convenient bit grouping.

Parity check, horizontal. A parity check applied to the group of certain bits from every character in a block. (*See also* Longitudinal redundancy check.)

Parity check, vertical. A parity check applied to the group which is all bits in one character. Also called vertical redundancy check. (1)

Partial dial tone. A high tone that notifies a calling party that he has not completed dialing within a specified period of time, or that not enough digits have been dialed.

Party line. A subscriber's line upon which several subscribers' stations are connected with, possibly, selective calling. Also called shared line.

PAX. Private automatic exchange. (*See* Exchange, private automatic.)

PBX. Private branch exchange. A telephone exchange on the user's premises with access to the public network. (In some countries it refers to a manually operated exchange; in modern American terminology it refers to a manual or automatic exchange.)

PCM. (*See* Pulse-code modulation.)

PDM. (*See* Pulse-duration modulation.)

Pel—Picture element. One of the many bright or dark spots that comprises a television, facsimile, or other transmitted picture. Also PIXEL.

Percent break. The percentage of the total time of a pulse interval during which dial contacts or relays remain open.

Perforator. An instrument for the manual preparation of a perforated tape, in which telegraph signals are represented by holes punched in accordance with a predetermined code. Paper tape is prepared off line with this. (*Compare with* REPERFORATOR.) (2)*

Peripheral device or equipment. A peripheral device of a computer is an input or output device or store (drums, discs, etc.) handled like external devices.

Peripheral interface. A standard interface used between a computer and its peripherals so that new peripherals may be added or old ones changed without special hardware adaption.

Phantom telegraph circuit. Telegraph circuit superimposed on two physical circuits reserved for telephony.*

Phase distortion. (*See* Distortion, delay vs. frequency.)

Phase equalizer, delay equalizer. A delay equalizer is a corrective network which is designed to make the phase delay or envelope delay of a circuit or system substantially constant over a desired frequency range.*

Phase-inversion modulation. A method of phase modulation in which the two significant conditions differ in phase by 180°.*

Phase jitter. A type of unwanted random distortion which results in the intermittent shortening or lengthening of the signals.

Phase modulation. One of three ways of modifying a sine wave signal to make it "carry" information. The sine wave or "carrier," has its phase changed in accordance with the information to be transmitted. For digital transmission, 2, 4 or sometimes 8 different phases are used. Bits are signalled by changes of phase, to avoid phase ambiguity.

Phototelegraphy. A system of facsimile having special regard to tone reproduction, in which the reception involves photographic processes. Also called picture-transmission.

Pictel. The name used by General Telephone and Electronics to describe its experimental video telephone.

Picturephone. AT&T's trademark for a video telephone service that permits the user to see as well as talk with the person at the distant end.

Pilot tone. Test frequency of controlled amplitude transmitted over carrier system for monitoring and control purposes.

PM. Phase modulation, q.v.

PMBX. Private manual branch exchange. (*See* PBX.)

PMS. Public message (telegram) service.

Point-to-point. A connection permanently established between two specific stations.

Polar keying. Form of telegraph signal in which circuit current flows one direction for marking, the other for spacing.

Polar relay. A permanent-magnet core relay which is designed to operate only when current flows in a specified direction.

Polar signal. A signal whose information is transmitted by means of directional currents.

Polar transmission. A method for transmitting teletypewriter signals, whereby the marking signal is represented by direct current flowing in one direction and the spacing signal is represented by an equal current flowing in the opposite direction. By extension to tone signaling, polar transmission is a method of transmission employing three distinct states, two to represent a mark and a space and one to represent the absence of a signal. Also called bipolar. (*Compare with* Neutral transmission.)

Polling. This is a means of controlling communication lines. When many stations are connected to the same circuit, polling from the center is used to ensure an orderly flow of data to the central location. The communication control device will send signals to a terminal saying, "Terminal A. Have you anything to send?" if not, "Terminal B. Have you anything to send?" and so on. Polling is an alternative to contention. It makes sure that no terminal is kept waiting for a long time.

Polling list. The polling signal will usually be sent under program control. The program will have a list for each channel which tells the sequence in which the terminals are to be polled.

Position. Part of a switchboard normally controlled by an operator.

POTS. Plain old telephone service. An acronym used by the telephone industry for conventional telephone service.

PPM. Pulse position modulation, q.v.

Primary center. A control center connecting toll centers; a class 3 office. (*See* Office classification) It can also serve as a toll center for its local end offices.

Primary group. A group of basic signals which are combined by multiplexing. It is the lowest level of the multiplexing hierarchy. The term is also used for the signal obtained by multiplexing these basic signals, or for the transmission channel which carries it. In FDM it usually comprises 12 voice channels. PCM speech channels are usually combined in 'primary groups' of 24 or 30.

Printing telegraph. Any method of telegraph operation in which the received signals are automatically translated as printed characters.

Private automatic branch exchange (PABX). (*See* Exchange, private automatic branch.)

Private automatic exchange (PAX). (*See* Exchange, private automatic.)

Private branch exchange (PBX). A telephone exchange serving an individual organization and having connections to a public telephone exchange.*

Private exchange (PX). An exchange serving a particular organization and having no means of connection with a public exchange.

Private line. Denotes the channel and channel equipment furnished to a customer as a unit for his exclusive use, without interexchange switching arrangements. More correctly called a leased line, q.v.

Processing, batch. A method of computer operation in which a number of similar input items are accumulated and sorted for processing.

Processing, in line. The processing of transactions as they occur, with no preliminary editing or sorting of them before they enter the system.

Propagation delay.   The time necessary for a signal to travel from one point on a circuit to another.

Protector.   Interface between inside and outside plant providing protection against hazardous voltages or currents.

Protocol.   A strict procedure required to initiate and maintain communication. Protocols may exist at many levels in one network such as link-by-link, end-to-end and subscriber-to-switch.

PSK.   Phase shift keyed modulation.

PTT.   Postal, telegraph, and telephone organization, usually a governmental department which acts as its nation's common carrier.

Public.   Provided by a common carrier for use by many customers.

Public switched network.   Any switching system that provides switching transmission facilities to many customers.

Public telephone station.   A station available for the use of the public, generally on payment of a fee which may be deposited in a coin box or paid to an attendant. Also called public call office and pay station.

Pulsating current.   Current which varies in amplitude but does not change polarity.

Pulse.   A brief change of current or voltage produced in a circuit to operate a switch or relay or which can be detected by a logic circuit. Also called signal pulse and signaling pulse.

Pulse-amplitude modulation (PAM).   Amplitude modulation of a pulse carrier.

Pulse-code modulation (PCM).   Representation of a speech signal (or other analog signal) by sampling at a regular rate and converting each sample to a binary number. 8000 samples per second is standard for telephone speech.

Pulse correction.   The restoration in a dial pulse repeater of signaling pulses which have been distorted during transmission.

Pulse-duration modulation (PDM).   Pulse-width modulation; pulse-length modulation. A form of pulse modulation in which the durations of pulses are varied.*

Pulse link repeaters.   A telephone repeater which connects one E & M signaling circuit, q.v., directly to another E & M signaling circuit.

Pulse modulation.   Transmission of information by modulation of a pulsed, or intermittent, carrier. Pulse width, count, position, phase, and/or amplitude may be the varied characteristic.

Pulse-position modulation (PPM).   A form of pulse modulation in which the positions in time of pulses are varied, without modifying their duration. (2)

Pulse regenerator.   A device for correcting the speed and signal shape of impulses.

Pulse stuffing.   (*See* Bit stuffing.)

Pulsing.   The transmission of address information to a switching office by means of digital pulses. Pulsing methods include multifrequency, rotary dial, and revertive, q.v.

Pulsing limits.   The maximum amount of pulsing distortion that a central office can tolerate in the dial pulses generated by a customer's telephone, before the switching equipment begins to make errors.

Pulsing signals.   Signals which are transmitted in the forward direction and carry the selective information to route the call in the desired direction.

Pushbutton dialing.   The use of keys or pushbuttons instead of a rotary dial to generate a sequence of digits to establish a circuit connection. The signal form is usually multiple tones. Also called tone dialing, Touch-call, Touchtone.

PX.   (*See* Private exchange.)

Quantizing noise.   When analog signals are encoded into digital form and reconverted into analog form, a noise association with the least bit resolution results.

Queue.   A collection of items which can be thought of as arranged in sequence, the two ends being the head and tail. New items are added to the tail. Items are removed from the head or tail.

Rack.   A framework or structure on which apparatus is mounted, usually by means of shelves, or mounting-plates. Also referred to as a bay.

Radiocommunication.   Any telecommunication by means of radio waves.

Radio wave.   Electromagnetic waves of frequencies between 10kHz and 3,000,000 MHz, propagated without guide in free space.

RAM.   Random access memory.

Random access.   The ability to directly establish a connection to any element in an array. In the computer field this term is applied to memory devices and is used especially to distinguish many forms of direct access storage from the linear access inherent in magnetic tape. In the telecommunications field, random access refers to the ability of any subscriber to reach any other subscriber by means of the telephone company switching equipment. The telephone industry prefers to call this latter property "switched access."

Random noise.   Noise due to the aggregate of the large number of elementary disturbances with random occurrence in time.

Rapid-connect circuit switching.   Fast-connect circuit switching, q.v.

RAX.   Rural automatic exchange.

Real time.   (1) Pertaining to actual time during which a physical process transpires. (2) Pertaining to the performance of a computation during the actual time that the related physical process transpires in order that results of the computation can be used in guiding the physical process. (3) Pertaining to an application in which response to input is fast enough to effect subsequent input, as when conducting the dialogues that take place at terminals on interactive systems. (4) Pertaining to transmission which occurs sufficiently fast that it is used in essentially the same manner as if it were instantaneous.

Rearrangeable switch.   A circuit switch which accommodates extra connections by rearrangement of some of the connections it already carries, so that they follow different paths through the switch. By such rearrangement a switch may become nonblocking.

Reasonableness checks.   Tests made on information reaching a real-time system or being transmitted from it to ensure that the data in question lie within a given range. It is one of the means of protecting a system from data transmission errors.

Receiving perforator (reperforator).   A telegraph instrument in which the received signals cause the code of the corresponding characters or functions to be punched in a tape.

Record completing trunk.   A trunk line outgoing from a local office to a toll office, used for recording the call and for completing the toll connection. Also called CLR trunk.

Recording trunk.   (U.S.) A trunk extending from a local central office or private branch exchange (PBX) to a toll office, which is used only for communication with toll operators and not for completing toll connections.

Recovery from fall-back.   When the system has switched to a fall-back mode of operation and the cause of the fall-back has been removed, the system must be restored to its former condition. This is referred to as recovery from fall-back. The recovery process may involve updating information in the files to produce two duplicate copies of the file.

Redundancy.   The portion of the total information contained in a message which can be eliminated without loss of essential information.

Redundancy check.   An automatic or programmed check based on the systematic insertion of components or characters used especially for checking purposes. (1).

Redundant code.   A code using more signal elements than necessary to represent the intrinsic information. For example, five-unit code using all the characters of International Telegraph Alphabet No. 2 is not redundant; five-unit code using only the figures in International Telegraph Alphabet No. 2 is redundant; seven-unit code using only signals made of four "space" and three "mark" elements is redundant.*

Reed switch.   Special type of relay consisting of fine moving reed-like contacts controlled by an electromagnet where the reeds themselves are part of the magnetic as well as the electrical circuit being controlled.

Reference pilot.   A reference pilot is a different wave from those which transmit the telecommunication signals (telegraphy, telephony). It is used in frequency-multiplexed carrier systems to facilitate the maintenance and adjustment of the carrier transmission system, for example, automatic level regulation, synchronization of oscillators, etc.

Regenerative repeater.   (*See* Repeater, regenerative.)

Regenerator.   Equipment which takes a digital signal that has been distorted by transmission and produces from it a new signal in which the shape, timing, and amplitude of pulses has been restored.

Regional center.   A control center (class 1 office) connecting sectional centers of the telephone system together. (*See* Office classification.) Every pair of regional centers in the United States has a direct circuit group running from one center to the other.

Register.   The apparatus, in an automatic system, which receives the dialed impulses and controls the subsequent switching operations. (1) A device capable of storing digital information. (2) The first unit in the assembly of common control equipment in an automatic central office. The register receives address information in the form of dial pulses or dual tone multifrequency (DTMF) signals, and stores it for possible conversion or translation. A register frequency operates in conjunction with a sender.

Relay.   A device, operated electrically, and causing by its operation abrupt changes in an electrical circuit (i.e., breaking the circuit, changing the circuit connection, or varying the circuit characteristics).

Remodulating baseband radio. A microwave receiver-transmitter in which the received signal is demodulated and then used to remodulate the transmitter. These radios are used for long-haul transmission.

Repeater. (1) A device whereby currents received over one circuit are automatically repeated in another circuit or circuits, generally in an amplified and/or reshaped form. (2) A device used to restore signals, which have been distorted because of attenuation, to their original shape and transmission level.

Repeater, regenerative. (1) A repeater utilized in telegraph applications. Its function is to retime and retransmit the received signal impulses restored to their original strength. These repeaters are speed- and code-sensitive and are intended for use with standard telegraph speeds and codes. (Also called regen.)
(2) A repeater used in PCM or digital circuits. It detects, retimes, and reconstructs the bits transmitted.

Repeater, telegraph. A device which receives telegraph signals and automatically retransmits corresponding signals.*

Repeater section. The section of line between any two adjacent repeater stations.

Reperforator (receiving perforator). A telegraph instrument in which the received signals cause the code of the corresponding characters or functions to be punched in a tape.

Reperforator/transmitter (RT). A teletypewriter unit consisting of a reperforator and a tape transmitter, each independent of the other. It is used as a relaying device and is especially suitable for transforming the incoming speed to a different outgoing speed, and for temporary queuing.

Residual error rate, undetected error rate. The ratio of the number of bits, unit elements, characters or blocks incorrectly received but undetected or uncorrected by the error-control equipment, to the total number of bits, unit elements, characters or blocks sent.*

Response time. This is the time the system takes to react to a given input. If a message is keyed into a terminal by an operator and the reply from the computer, when it comes, is typed at the same terminal, response time may be defined as the time interval between the operator pressing the last key and the terminal typing the first letter of the reply. For different types of terminals response time may be defined similarly. It is the interval between an event and the system's response to the event. Response time thus defined includes: (1) transmission time to the computer; (2) processing time at the computer, including access time to obtain any file records deeded to answer the inquiry; and (3) transmission time back to the terminal.

Restriction of message (toll diversion). A telephone arrangement where outgoing calls from a private automatic branch exchange (PABX) must either be routed through an operator or are limited to specified trunk groups.

Return loss. The return loss at the junction of a transmission line and a terminating impedance is the ratio, expressed in dB, of the reflected wave to the incident wave.

More broadly, the return loss is a measure of the dissimilarity between two impedances, being equal to the number of decibels which corresponds to the scalar value of the reciprocal of the coefficient, and hence being expressed by the formula:

$$20 \log_{10} \frac{Z_1 + Z_2}{Z_1 - Z_2} \text{ decibels}$$

where Z_1 and Z_2 are the two impedances.

Reverse-battery signaling.   A type of loop signaling in which battery and ground are reversed on the tip and ring of the loop to give an "off-hook" signal, when the called party answers.

Revertive pulsing.   Pulsing over a trunk in reverse direction. That is, from the terminating office instead of from the originating office. When the incoming office trunk is seized, it sends open or ground pulses to the office that originated the call. The originating office counts the pulses and opens the trunk when the correct number of pulses has been received.

RF.   Radio frequency.

RFI.   Radio frequency interference.

Ring.   The ring-shaped contact of a plug usually positioned between, but insulated from, the tip and sleeve. The audible alerting signal on a telephone line.

Ringback tone.   An interrupted low tone indicating that the called telephone is ringing. Now called audible ringing tone.

Ringdown.   A type of signal that uses either a 135-Hz or 1000-Hz signal, interrupted 20 times per second. The type of signaling employed in manual operation, as differentiated from dial signaling. Ringdown signaling utilizes a continuous or pulsing ac signal transmitted over the line. The term ringdown originated in magneto telephone operation, where cranking the magneto of a subscriber would "ring" its bell and cause a marker to fall "down" at the central switchboard. In ringdown signaling, a key is operated in a cord circuit to ring on a trunk. On intertoll trunks, ringers are used to transmit and receive the signals. While ringdown trunks are unsuitable for intertoll dialing, connection of dial trunks to ringdown trunks can be provided with operator intertoll dialing.

Ringing key.   A key whose operation causes the sending of a ringing current.

Ringing signal.   Any ac or dc signal transmitted over a line or trunk for the purpose of alerting a party at the distant end of an incoming call. The signal may operate a visual or aural device.

Ring trip.   The circuitry required to disable the ringing signal when the called telephone is answered (placed in the off-hook condition).

RO.   Receive only.

ROM.   Read only memory.

Rotary switching system.   An automatic telephone switching system which is generally characterized by the following features: (1) the selecting mechanisms are rotary switches; (2) the switching pulses are received and stored by controlling mechanisms which govern the subsequent operations necessary in establishing a telephone connection.

Routing. The assignment of the communications path by which a message or telephone call will reach its destination.

Routing, alternate. Assignment of a secondary communications path to a destination when the primary path is unavailable.

Routing indicator. An address, or group of characters, in the heading of a message defining the final circuit or terminal to which the message has to be delivered. Also called routing code.

Routing table. A table associated with a network node which states for each message (or packet) destination the preferred outgoing link that the message should use.

RT. Reperforator/transmitter, q.v.

SATT. Strowger automatic toll ticketing, q.v.

Saturation testing. Program testing with a large bulk of messages intended to bring to light those errors which will only occur very infrequently and which may be triggered by rare coincidences such as two different messages arriving at the same time.

S-BAND. 1.55 to 3.90 GHz (in the U. S.).

Scrambler. Coding device applied to a digital channel which produces an apparently random bit sequence. A corresponding device is used to decode the channel, i.e. the coding is reversible. By this means, harmful repetitive patterns in the data sent over a transmission line are avoided. They could still occur, but with low probability.

Sectional center. A control center connecting primary centers; a Class 2 office. (*See* Office classification.)

Seek. A mechanical movement involved in locating a record in a random-access file. This may, for example, be the movement of an arm and head mechanism that is necessary before a read instruction can be given to read data in a certain location on the file.

Seizing signal. In a semiautomatic or automatic working, a signal transmitted at the commencement of a call to initialize circuit operation at the incoming end of the circuit.

Selection. The process of indicating the number of the terminal being called. In the telephone case this would be termed 'dialing'. The term 'selection' is now preferred because the number may not come from a dial, but from push buttons, a keyboard or a computer, for example. 'Selection' has other meanings as in the selection of a peripheral on a channel.

Selective calling. The ability of the transmitting station to specify which of several stations on the same line is to receive a message.

Selective ringing. A system designed with the capability of ringing only the desired subscriber's telephone on a multiparty line. Ringers tuned to one of five possible frequencies are used to achieve this effect.

Self-checking numbers. Numbers which contain redundant information so that an error in them, caused, for example, by noise on a transmission line, may be detected.

Semaphore. A mechanism for the synchronization of a set of cooperating processes. It is used to prevent two or more processes from entering mutually critical equipment at the same time.

Semiautomatic system. A system in which the calling subscriber's order is given to an operator who completes the call through automatic switches.

Semi-selective ringing. A four-party line ringing arrangement in which each party hears his own ring plus one other (usually a one and two ring code).

Sender. A unit which receives address information from a register or routing information from a translater, and then outpulses the proper routing digits to a trunk or to local equipment. Sender and register functions are often combined in a single unit.

Separate-channel signaling. A type of CCSS, q.v. A carrier system signaling arrangement where the signaling for several channels is multiplexed on a single voice channel.

Serial transmission. (1) Used to identify a system wherein the bits of a character occur serially in time. Implies only a single transmission channel. (Also called serial-by-bit.) (2) Transmission at successive intervals of signal elements constituting the same telegraph or data signal. For example, transmission of signal elements by a standard teleprinter, in accordance with International Telegraph Alphabet No. 2; telegraph transmission by a time-divided channel.*

Service terminal. The equipment needed to terminate the channel and connect to the station apparatus or customer terminal. In some cases it may include channelizing equipment or broadband modems where these are an integral part of system operation.

Shared line. (U.K.) A party line; in Britain, has two parties only.

SHF. Super high frequency (3 GHz to 30 GHz).

Sideband. The frequency band on either the upper or lower side of the carrier frequency within which fall the frequencies produced by the process of modulation.*

Sidetone. The reproduction in a telephone receiver of sounds picked up by the associated microphone. The microphone may pick up either the voice of the speaker or the room noise.

Signal. Aggregate of waves propagated along a transmission channel and intended to act on a receiving unit.

Signaling. The process by which a caller or equipment on the transmitting end of a line informs a particular party or equipment at the receiving end that a message is to be communicated. Signaling is also that supervisory information that lets the caller know that the called party is ready to talk, that his line is busy, or that he has hung up. Signaling is used for setting up a trunk circuit of many sections, and ensuring that the path is complete while a transmission or conversation continues.

Signaling, in-band. Signaling which utilizes frequencies within the intelligence band of a channel, usually within the voice channel.

Signaling, out-of-band. Signaling which utilizes frequencies outside the intelligence band (usually the voice band). Also used to indicate the use of a portion of a channel bandwidth provided by the medium such as a carrier channel, but denied to the speech or intelligence path by filters. It results in a reduction of the effective available bandwidth.

Signal-to-noise ratio. The ratio, expressed in dB, of the usable signal to the noise signal present.

Simplex circuit. (1) CCITT definition: A circuit permitting the transmission of signals in either direction, but not in both simultaneously. (2) Definition in common usage (the normal meaning in computer literature): A circuit permitting transmission in one specific direction only.

Simplex signaling (SX). Signaling over a trunk circuit. Signaling information is sent over the circuit when a signal-transmitting relay at the calling end is energized. A signal-receiving relay at the called end provides the signaling information to the voice channel. Both relays are connected to the midpoints of repeating coils or retardation coils at each end of the circuit.

Single-current transmission (inverse); Neutral direct-current system. A form of telegraph transmission effected by means of unidirectional currents.*

Single-frequency signaling. A method of signaling in which a single frequency tone, 2600 Hz for example, is placed on the voice path. The tone is on during the idle condition, pulsed during dialing, and off when the circuit is being used. This condition is known as tone-on-when-idle.

Single threading. A program which completes the processing of one message before starting another message is called 'single threading'. (*See also* Multithreading.)

SITA. Societe Internationale de Telecommunications Aeronautiques, which operates a message and data network for a large group of airline companies.

Sleeve. (1) The third contacting part on a telephone plug-preceded in the location by the tip and ring. (2) The sleeve wire is the third control wire of each telephone in an automatic switching office.

Soft copy. A temporary record, such as information presented by a visual display unit but not printed.

Software. A term that originally referred to computer programs but has increasingly been used to refer to anything that cannot be considered hardware.

Space. (1) The opposite signal condition to a "mark." A space impulse is equivalent to a binary 0. A mark is equivalent to a binary 1. An impulse which, in a neutral circuit, causes the loop to open or causes absence of a signal, while in a polar circuit it causes the loop current to flow in a direction opposite to that for a mark impulse. (2) In some codes, a character which causes a printer to leave a character width with no printed symbol.

Space-division multiplexing. Multiplexing in which separate signals travel over separate physical paths such as wire pairs within a cable.

Space-division switching. Method for switching circuits in which each connection through the switch takes a physically separate path. (*Contrast with* Time-division switching.)

Space-hold. The normal no-traffic line condition whereby a steady space is transmitted. (*Compare with* Mark-hold.) (1).

Space-to-mark transition. The transition, or switching, from a spacing impulse to a marking impulse. (1).

Spacing bias. (*See* Bias distortion.)

Spade. Single-channel-per-carrier, PCM multiple access demand-assignment equipment. A FDMA, q.v., scheme used on Comsat satellites, which permits geographically dispersed users to share a transponder.

Speaking key.   A key permitting an operator to speak on a circuit. Also called talking key.

Specialized common carrier.   (U.S.) A common carrier offering specialized type of service or serving a specialized market. A term refers to a new breed of common carriers which grew up in the 1970s giving a new form of competition to the traditional telephone and telegraph companies. Most of them operated microwave transmission facilities.

Spectrum.   (1) A continuous range of frequencies, usually wide in extent within which waves have some specific common characteristic. (2) A graphical representation of the distribution of the amplitude (and sometimes phase) of the components of a wave as a function of frequency. A spectrum may be continuous or, on the contrary, contain only points corresponding to certain discrete values.*

Spectrum roll-off.   Applied to the frequency response of a transmission line, or a filter, it is the attenuation characteristic at the edge of the band.

Speech-simulated signal.   A signal made up of those components of a voice signal which will cause the false operation of tone-operated supervisory equipment.

SSB.   Single sideband modulation.

SSB-SC/AM.   Single sideband, suppressed carrier amplitude modulation.

Start-dialing (start pulsing) signal.   The on-hook condition indicating that the receiving end is ready to receive pulsing information.

Start element.   The first element of a character in certain serial transmissions, used to permit synchronization. In Baudot teletypewriter operation, it is one space bit.

Start-stop (signaling).   Signaling in which each group of code elements corresponding to an alphabetical signal is preceded by a start signal which serves to prepare the receiving mechanism for the reception and registration of a character, and is followed by a stop signal which serves to bring the receiving mechanism to rest in preparation for the reception of the next character. (*Contrast with* Synchronous system.) Start-stop transmission is also referred to as asynchronous transmission, q.v.

Start-stop transmission.   Transmission using start-stop signaling, q.v.

Station.   One of the input or output points of a communications system — e.g., the telephone set in the telephone system or the point where the business machine interfaces the channel on a leased private line.

Station equipment.   Telephone company instruments and associated equipment furnished to subscriber.

Status information.   Information about the logical state of a piece of equipment. Examples are: (a) A peripheral device reporting its status to the computer. (b) A network terminating unit reporting status to a network switch.

Status information is one kind of control signal.

Status maps.   Tables which give the status of various programs, devices, input-output operations, or the status of the communication lines.

STD.   Subscriber trunk dialing, q.v.; called direct distance dialing, DDD, in North America.

Step-by-step switch (SXS). A switch that moves in synchronism with a pulse device such as a rotary telephone dial. Each digit dialed causes the movement of successive selector switches to carry the connection forward until the desired line is reached. Also called stepper switch. (*Compare with* Line switching and Cross-bar switch.)

Step-by-step system. A type of line-switching system which uses step-by-step switches.

Stop bit. (*See* Stop element.)

Stop element. The last element of a character in asynchronous serial transmission, used to ensure recognition of the next start element. In Baudot teletypewriter operation it is 1.42 mark bits. (*See also* Start-stop transmission.)

Store and forward. Applied to communication systems in which messages are received at intermediate routing points and recorded (stored). They are then retransmitted to a further routing point or to the ultimate recipient.

Strowger. A type of step-by-step switching system, named after its inventor. (*See* step-by-step switch.)

Strowger automatic toll ticketing. A system which, when a customer dials a toll call, automatically makes a record of the calling number, the called number, the time of day, and the duration of the call.

Stunt box. A device to (1) control the nonprinting functions of a teletypewriter terminal, such as a carriage return and line feed; and (2) a device to recognize line control characters (e.g. DCC, TSC, etc.).

Subcontrol station. The repeater station through which an international telephone circuit passes, and which is responsible within its own country for the state of the circuit under the direction of the control station.

Subscriber line concentrator. A remote switching system providing the first stage of switching near a group of subscribers.

Subscriber's extension station. A telephone station associated with a main station through connection to the same subscriber's line.

Subscriber's line. The telephone line connecting the exchange to the subscriber's station.

Subscriber's loop. (*See* Local loop or Subscriber's line.)

Subscriber trunk (toll) dialing (STD). The direct calling by the calling subscriber of the distant called subscriber, the switching being effected automatically. Also called direct distance dialing, DDD.

Subset. A subscriber set of equipment, such as a telephone. A modulation and demodulation device data set.

Subvoice-grade channel. A channel of bandwidth narrower than that of voice-grade channels. Such channels are usually subchannels of a voice-grade line.

Supergroup. The assembly of five 12-channel groups occupying adjacent bands in the spectrum, for the purpose of simultaneous modulation or demodulation, i.e. 60 voice channels.*

Supergroup link. The means of transmission using a frequency band of specified width (240kHz) connecting two supergroup distribution frames which multiplex and demultiplex 60 voice channels (or equivalent). This expression is usually applied to both-way channels.

Superimposed circuit. An additional circuit obtained from one or more wires provided for other circuits and arranged so that all the circuits may be used simultaneously without mutual interference.

Superimposed ringing. A type of semi-selective ringing which uses a combination of ac and dc.

Super master group. 900 voice channels in the CCITT definition. (*See* Group.)

Supervisory control. Characters or signals which automatically actuate equipment or indicators at a remote terminal.

Supervisory lamp. A lamp illuminated or darkened by the operations or switchings during a call and indicating to an operator the state of the call.

Supervisory programs. Those computer programs designed to coordinate service and augment the machine components of the system, and coordinate and service application programs. They handle work scheduling, input-output operations, error actions, and other functions.

Supervisory signal. (1) A signal which indicates whether a circuit is in use. A signal which gives an indication of status or change of status in a telephone system. For example, a signal used to attract the attention of an operator. (2) A signal used to indicate the various operating states of circuit combinations.

Supervisory system. The complete set of supervisory programs used on a given system.

Support programs. The ultimate operational system consists of supervisory programs and application programs. However, a third set of programs are needed to install the system, including diagnostics, testing aids, data generator programs, terminal simulators, etc. These are referred to as support programs.

Suppressed carrier transmission. That method of communication in which the carrier frequency is suppressed either partially or to the maximum degree possible. One or both of the sidebands may be transmitted.

Switchboard. Equipment on which switching operations are performed by operators.

Switch hook. A switch on a telephone set, associated with the structure supporting the receiver or handset. It is operated by the removal or replacement of the receiver or handset on the support. (*See also* Off-hook and On-hook.)

Switching center (office). A location which terminates multiple circuits and is capable of interconnecting circuits of transferring traffic between circuits; may be automatic, semiautomatic, or manual.

Switching plan. Office classification, q.v.

Switchover. When a failure occurs in the equipment a switch may occur to an alternative component. This may be, for example, an alternative file unit, an alternative communication line or an alternative computer. The switchover process may be automatic under program control or it may be manual.

SXS. Step-by-step switch, q.v.

Synchromonic ringing. A type of party line selective ringing which uses ringing frequencies of 16 Hz, 30 Hz, 42 Hz, 54 Hz, and 66 Hz.

Synchronous. Having a constant time interval between successive bits, characters, or events.

Synchronous network. A network in which all the communication links are synchronized to a common clock.

Synchronous system. A system in which the sending and receiving instruments are operating continuously at substantially the same frequency and are maintained, by means of correction, if necessary, in a desired phase relationship. (*Contrast with* Start-stop.)*

Synchronous transmission. A transmission process such that between any two significant instants there is always an integral number of unit intervals. (*Contrast with* Asynchronous transmission or Start-stop transmission.)

Talking battery. The dc voltage supplied by the central office to the subscriber's loop to operate the carbon transmitter in the handset.

Talking path. In a telephone circuit, the transmission path consisting of the tip and ring conductors.

Tandem office. An office that is used to interconnect the local end offices over tandem trunks in a densely settled exchange area where it is uneconomical for a telephone company to provide direct interconnection between all end offices. The tandem office completes all calls between the end offices but is not directly connected to subscribers.

Tandem office, tandem central office. A central office used primarily as a switching point for traffic between other central offices.*

Tandem trunk. A trunk between Class 5 offices (end offices—*see* Office classification) which can connect a trunk to another Class 5 office. This permits calls to be routed from A to B or A through B to C on the same trunk.

Tariff. The published rate for a specific unit of equipment, facility, or type of service provided by a communications common carrier. Also the vehicle by which the regulating agencies approve or disapprove such facilities or services. Thus the tariff becomes a contract between customer and common carrier.

TASI (Time assignment speech interpolation). Because audible speech on the average voice circuit is only present approximately 45% of the time, the efficiency of the expensive overseas channels can be improved if some of the remaining 55% of the time could be utilized. TASI switching equipment connects a party to an idle circuit while speech is taking place, but disconnects the party when speech stops, so that a different party can use the same circuit. During periods of heavy traffic, TASI improves line efficiency from 45% to 75–80% (*See also* Digital speech interpolation.)

TAT. Transatlantic cable (followed by identifying digit; as TAT-6).

T-carrier (Bell System). A hierarchy of digital systems designed to carry speech and other signals in digital form, designated T1, T2, and T4.

The T1 carrier has 24 PCM voice channels. The signal is sampled 8000 times per second into 7 bits + 1 signaling/supervision bit. Framing digit added after 192 pulses. Line rate = [(24) (7 + 1) + 1] 8000 = 1,544,000 bits/second. The T2 carrier has 96 PCM voice channels, or equivalent with a 6.312 megabit line rate. The T4 carrier transmits 274 million bps.

TCCS. Western Union service whereby telex subscribers can send messages to TWX subscribers.

TD. Transmitter-distributor, q.v.

TDM. Time division multiplex, q.v.

TDMA. Time division multiple access, q.v.

Technical control center. In telecommunications, a facility for supervision and control of the transmission network. It is principally used as the office of personnel who monitor and test the quality of lines, taking corrective action as necessary to maintain service.

Telecommunication. Any process that permits the passage from a sender to one or more receivers of information of any nature delivered in any usable form (printed copy, fixed or moving pictures, visible or audible signals, etc.) by means of any electromagnetic system (electrical transmission by wire, radio, optical transmission, guided waves, etc.). Includes telegraphy, telephony, video-telephony, data transmission, etc.

Telegram. (1) Documentary matter, whether in written, printed, or pictorial form, entrusted to the general telegraph service with a view to its transmission by telegraphy and delivery to the addressee. (This term also covers radio telegrams.) (2) The document itself and any reproduction made in the course of transmission, if necessary, or made for delivery to the addressee.

Telegraph alphabet. A table of correspondence between written characters together with some of the functions (e.g., spacing, line feed, inversion, etc.) and the telegraph signals which represent them.

Telegraph channel. (U.S.) The transmission media and intervening apparatus involved in the transmission of telegraph signals between two terminal sets, or, more generally, between two intermediate telegraph installations.

Telegraph distortion. Distortion which alters the duration of signal elements.

Telegraph repeater. A device which receives telegraph signals and automatically retransmits corresponding signals.

Telegraphy. That branch of telecommunications concerned with processes providing reproduction, at a distance, of documentary matter such as written, printed, or pictorial matter, or the reproduction at a distance of any kind of information in such form.

Telephone circuit (line). An electrical connection permitting the establishment of a telephone communication in both directions between two telephone exchanges.

Telephone exchange. A switching center for interconnecting the lines which terminate therein. Also called central office.

Telephone frequency. Any frequency within that part of the audiofrequency range essential for the transmission of speech of commercial quality, i.e., 300 to 3000 hz. (*See also* Voice frequency.)

Telephone receiver. An electro-acoustic transducer to convert electrical energy into sound energy and designed to be applied to the ear.

Telephone repeater. A term used to denote, in certain instances, a combination of one or more amplifiers, together with their associated equipment, for use in a telephone circuit. Also called telephone amplifier. (*See* Repeater.)

Telephone set. An assembly of apparatus including a telephone transmitter, a telephone receiver, and usually a switchhook, dial, and the immediately associated wiring and components.

Telephony. A system of telecommunications set up primarily for the transmission of speech.

Teleprinter. A start-stop apparatus comprising a keyboard transmitter, together with a printing receiver. Also called a teletypewriter.

Teleprinter exchange service. A service provided by communication common carriers to interconnect teleprinters. Similar to regular telephone service: customers dial calls from station to station, but communicate using teleprinter equipment rather than telephone. Examples are TWX (U.S.) and telex (worldwide).

Teleprocessing. A form of information handling in which a data-processing system utilizes telecommunications facilities.

Teletype. Trademark of Teletype Corporation, usually referring to a series of different types of teleprinter equipment such as tape punches, reperforators, page printers, etc., utilized for communications systems.

Teletypewriter exchange service (TWX). A public switched teletypewriter service in which suitably arranged teletypewriter stations are provided with lines to a central office for access to other such stations throughout the U.S.A. and Canada. Both Baudot- and ASCII-coded machines are used. Business machines may also be used, with certain restrictions.

Telex service. A dial-up telegraph service enabling its subscribers to communicate directly and temporarily among themselves by means of start-stop apparatus and of circuits of the public telegraph network. The service operates world wide. Baudot equipment is used. Computers can be connected to the Telex network.

Telpak. A service offered by U. S. telephone companies for the leasing of wide band channels between two or more points. Telpak channels can be leased in groups of 60 or 240 voice grade channels.

Teltex. Western Union service whereby telex subscribers can originate telegrams.

Terminal. (1) A point at which information can enter or leave a communication network. (2) Any device capable of sending and/or receiving information over a communication channel. The means by which data are entered into a computer system and by which the decisions of the system are communicated to the environment it affects. A wide variety of terminal devices have been built, including teleprinters, special keyboards, light displays, cathode ray tubes, thermocouples, pressure gauges and other instrumentation, radar units, telephones, etc.

Terminal processor. A small computer connecting terminals to a telecommunications network, or providing other functions to aid the operation of the terminals.

Ternary. Having three possible values. There are ternary number representations using, for example, the digits 0, 1, 2 and there are ternary signals which nominally take 3 possible values, for example +1 volt, 0, −1 volt.

Test board. A switchboard equipped with testing apparatus, so arranged that connections can be made from it to telephone lines or central office equipment for testing purposes. Also called Technical control center and trunk test rack.

Text. That part of the message which contains the substantive information to be conveyed. Sometimes called "body" of the message.

Throughput. The total useful information processed or communicated during a specified time period.

Through supervision. Total supervision of a toll call by the originating operator through any intermediate switching points to the called telephone.

Tie line. A private-line communications channel of the type provided by communications common carriers for linking two or more points together. Called interswitchboard line and tie trunk.

Tie line network. A network of tie-lines serving a corporation or other organization; usually a CCSA network, q.v.

Time-derived channel. Any of the channels obtained from multiplexing a channel by time division.

Time-division multiple access (TDMA). Communicating devices at different geographical locations share a multi-point or broadcast channel by means of a technique which allocates different time slots to different users (*Compare with* Frequency-division multiple access.)

Time-division multiplex (TDM). A means of obtaining a number of channels over a single path by time-dividing the path into a number of time slots and assigning each channel its own intermittently repeated time slot. At the receiving end, each time-separated channel is reassembled. The system is ideally suited for the transmission of digital data, and is now used for digitized speech and other signals. The time-slot allocation may be repeated regularly (fixed cycle) or may be made according to demand (dynamic).

Time-division signaling. Signaling over a time division multiplex system in which all voice channels share a common signaling channel, with time division providing the separation between signaling channels.

Time-division switching. Switching method for a TDM channel requiring the shifting of data from one slot to another in the TDM frame. The slot in question may carry a bit or byte (or, in principle, any other unit of data).

Time-out. In a communication procedure, one party may have to take action if it gets no response from the other within a specified time. This occurrence (exceeding the allowed time) is called a time-out.

Time sharing. A method of operation in which a computer facility is shared by several users for different purposes at (apparently) the same time. Although the computer actually services each user in sequence, the high speed of the computer makes it appear that the users are all handled simultaneously.

Tip. The contacting part at the end of a telephone plug or the top spring of a jack. The conductors associated with these contacts. The other contact is called a ring.

TIP (Terminal interface processor). A small computer connecting terminals to a packet-switching network. A term used on the ARPA network.

Toll board. A switchboard used primarily for establishing connections over toll lines. Also called trunk switchboard.

Toll center. Basic toll switching entity; a central office where channels and toll message circuits terminate. While this is usually one particular central office in a city, larger cities may have several offices where toll message circuits terminate. A class 4 office. Also called "toll office" and "toll point." (*See* Office classification.)

Toll circuit. A long-distance circuit connecting two exchanges in different localities. Also called trunk circuit.

Toll connecting trunk. A trunk used to connect a Class 5 office (local central office) to the direct distance dialing network.

Toll office. (U.S.) (*See* Toll center.)

Toll switching trunk. A line connecting a trunk exchange to a local exchange and permitting a trunk operator to call a subscriber to establish a trunk call. (British: trunk junction.)

Tone dialing. (*See* Pushbutton dialing.)

Tone signaling. The transmission of supervisory, address, and alerting signals over a telephone circuit by means of tones.

Torn tape switching. A manual form of message switching in which incoming messages perforate paper tape. Operators tear off this punched and printed tape, and transfer it to paper tape readers on the requisite outgoing circuits.

Touch-call. GT&E term for pushbutton dialing, q.v.

Touchtone. AT&T term for pushbutton dialing, q.v.

Trace packet. A special kind of packet in a packet-switching network which functions as a normal packet but causes a report of each stage of its progress to be sent to the network control center.

Traffic matrix. A matrix of which the (i, j) element contains the amount of traffic originated at node i and destined for node j. The unit of measurement could be calls, or packets per second, for example, depending on the kind of network.

Transceiver. A terminal that can transmit and receive traffic.

Transducer. A device for converting signals from one form to another, such as microphone or a receiver.

Transit network. The highest level of a switched network. The transit network is well provided with links between its switching center so that it rarely needs intermediate switching.

Translator. A device that converts information from one system of representation into equivalent information in another system of representation. In telephone equipment, it is the device that converts dialed digits into call-routing information.

Transmitter-distributor (TD). The device in a teletypewriter terminal which makes and breaks the line in timed sequence. Modern usage of the term refers to a paper tape transmitter.

Transparency. If a signal passes through a network or facility unchanged, that network or facility is said to be transparent to it.

Transponder. A receiver-transmitter combination that retransmits the received signal greatly amplified at a different frequency. Communication satellites contain several transponders.

Transreceiver.   A terminal that can transmit and receive traffic.

Transversal filter.   In this filter the input is passed through a delay network and delayed versions of the signal through suitable attenuators, are added to generate the output. The attenuators must be able to invert the signal.

Trunk.   Cables that contain numerous shared telephone circuits that are used to interconnect telephone switching centers.

Trunk call.   (U.K.) Term for toll call; i.e., any telephone call, subject to charge, for a destination outside of the local service area of the calling station.

Trunk circuit (British), Toll circuit (American).   A circuit connecting two exchanges in different localities. Note: In Great Britain, a trunk circuit is approximately 15 miles long or more. A circuit connecting two exchanges less than 15 miles apart is called a junction circuit.

Trunk exchange (British), Toll office (American).   An exchange with the function of controlling the switching of trunk (toll) traffic.

Trunk group.   Those trunks between two points both of which are switching centers and/or individual message distribution points, and which employ the same multiplex terminal equipment.

Trunk junction (British), Toll switching trunk (American).   A line connecting a trunk exchange to a local exchange and permitting a trunk operator to call a subscriber to establish a trunk call.

Trunk switchboard.   (U.K.) A switchboard used primarily for establishing connections over toll lines. Also called toll board.

Trunk test rack.   (U.K.) A switchboard equipped with testing apparatus, so arranged that connections can be made from it to telephone lines or central office equipment for testing purposes. (*See* Test board and Technical control center.)

TSPS.   Traffic service position system; a stored-program computer with telephone operator consoles permitted calls needing operator intervention to be handled as efficiently as possible.

TTY.   Teletypewriter equipment.

Turnaround time.   The actual time required to reverse the direction of transmission from send to receive or vice versa on a half-duplex circuit.

Two-wire circuit.   A circuit formed of two conductors insulated from each other, providing a "go" and "return" channel in the same frequency. Metallic and not grounded.

TWT.   Traveling wave tube. A device used to amplify signals of high frequency, e.g. microwave signals.

TWX.   Teletypewriter exchange service, in which a customer rents one teletypewriter and is connected to a switching exchange which permits him to interconnect to other such stations. (Formerly Bell System, now a Western Union Telegraph Company service.)

UHF.   Ultra high frequency (300 MHz to 3 GHz).

Unattended operations.   The automatic features of a station's operation permit the transmission and reception of messages on an unattended basis.

Uniform-spectrum random noise. Noise distributed over the spectrum in such a way that the power per unit bandwidth is constant. Also called "White" noise.

VACC. Value-added common carrier, q.v.

Value-added common carrier. A corporation which sells services of a value-added network. Such a network is built using the communications offerings of traditional common carriers, connected to computers which permit new types of telecommunication tariffs to be offered. The network may be a packet-switching or message-switching network. Services offered include transmission of data charged for by the packet, and transmission of facsimile documents.

VAN. Value-added network. (*See* Value-added common carrier.)

Vertical parity (redundancy) check. (*See* Parity check, vertical.)

VFT. Voice frequency telegraph.

VHF. Very high frequency (30 MHz to 300 MHz).

Via net loss. This term describes the net losses of trunks in the long distance switched telephone network of North America. The trunk is said to be in a via condition when it is an intermediate trunk in a longer switched connection.

Video signal. A signal comprised of frequencies normally required to transmit pictorial information (1 to 6 MHz).

Video telephone set. A telephone service that permits the user to see as well as talk with the person on the distant end. (*See* Picturephone, Vistaphone, and Pictel.)

Videovoice. The name of the slow scan video telephone manufactured and marketed by RCA Global Communications Corp.

Virtual. An adjective implying that something is different in physical reality to the way it is made to appear to programs or users. The following definitions contain examples.

Virtual circuit. A proposed CCITT definition for a data transmission service. The user presents a data message for delivery, with a header of a specified format. The system delivers the message as though a circuit existed to the specified destination. One of many different routes and techniques could be used to deliver the message, but the user need not know which is employed. It appears to him as though a virtual circuit exists.

Virtual storage. A computer user may employ a computer as though it has a much larger memory than its real memory. The difference is made up by software rapidly moving pages in and out, to and from a backing store. The apparent memory which the user can employ is called virtual memory.

Virtual terminal. A terminal which is defined as a standard on a network which can handle diverse terminals. Signals to and from each non-standard terminal are converted to equivalent standard-terminal signals by an interface computer. The network protocols then operate as though all terminals were the standard "virtual" terminals.

Vistaphone (TM). The registered trademark of the video telephone manufactured by Stromberg-Carlson Corp.

VNL. Via net loss, q.v.

VOGAD. Voice-operated gain-adjusting device. A device somewhat similar to a compandor and used on some radio systems; a voice-operated device which removes fluctuation from input speech and sends it out at a constant level. No restoring device is needed at the receiving end.

Voice frequency, telephone frequency. Any frequency within that part of the audio-frequency range essential for the transmission of speech of commercial quality, i.e., 300–3000 hz.*

Voice-frequency carrier telegraphy. That form of carrier telegraphy in which the carrier currents have frequencies such that the modulated currents may be transmitted over a voice-frequency telephone channel.

Voice-frequency multichannel telegraphy. Telegraphy using two or more carrier currents the frequencies of which are within the voice-frequency range. Voice-frequency telegraph systems permit the transmission of up to 24 channels over a single circuit by use of frequency-division multiplexing.

Voice-grade. A telecommunications link with a bandwidth (about 3kHz) appropriate to an audio telephone line.

Voice grade channel. A channel suitable for transmission of speech, digital or analog data, or facsimile, generally with a frequency range of about 300 to 3000 Hz.

Voice-operated device. A device used on a telephone circuit to permit the presence of telephone currents to effect a desired control. Such a device is used in most echo suppressors.

Volume. An electrical quantity which is related to speech power and is measured by means of a specified instrument (e.g. a volume indicator) at a stated point in a telephone circuit.

VRC. Vertical redundancy check. (*See also* Parity check.)

VSB. Vestigial sideband modulation.

VTAM. Virtual telecommunications access method, IBM teleprocessing input/output software using virtual techniques, q.v.

Watchdog timer. This is a timer which is set by the program. It interrupts the program after a given period of time, e.g., one second. This will prevent the system from going into an endless loop due to a program error, or becoming idle because of an equipment fault. The watchdog timer may cause a computer interrupt if such a fault is detected.

WATS. Wide area telephone service. A service provided by telephone companies in the United States which permits a customer by use of an access line to make calls to telephones in a specific zone in a dial basis for a flat monthly charge or to receive "collect" calls at a specified number at a flat monthly charge. Monthly charges are based on the size of the area in which the calls are placed, not on the number or length of calls. Under the WATS arrangement, the U.S. is divided into several zones to be called on a full-time or measured-time basis. (1).

Wet circuit. A circuit which carries direct current.

Wideband channel. A channel wider in bandwidth than a voice-grade channel.

Wink operation. In wink operation, trunk equipment sends on-hook signals toward each end during the idle condition. When a collect signal is received, the called office requests a register or sender. However, the on-hook signal to the calling office remains until the register or sender is connected at the called office. At that time, the idle on-hook signal changes to off-hook. The register or sender retains the off-hook signal for not less than 140 milliseconds, then returns an on-hook signal to the calling end, indicating that it is ready to receive pulses from the calling office.

Wink signal. A short interruption of current to a switchboard busy lamp to indicate that the circuit is busy. On key telephone sets, the wink signal indicates that a line is being held. It is also an indication of change of state between an on-hook and off-hook condition.

Wire broadcasting. The transmission, by a network of conductors, of a radiobroadcast program so as to permit any number of subscribers to receive it if they so wish. Also called line broadcasting.

Wired city; wired nation. The concept of a city or nation in which every home, institution, and commercial establishment, is connected to a CATV or other broad bandwidth telecommunications systems.

Wire pair. Two conductors, isolated from each other, and associated to form a communication channel.

Word. (1) In telegraphy, six operations or characters (five characters plus one space): "Group" is also used in place of "word." (2) In computing, a sequence of bits or characters treated as a unit and capable of being stored in one computer location.

WPM. Words per minute. A common measure of speed in telegraph systems.

XB. Crossbar switch.

X-band. (U.S.) 6.2 to 10.9 GHz.

INDEX

662